The Sword in Anglo-Saxon England

from the 5th to 7th century

Edited by

Paul Mortimer

and Matt Bunker

Published 2019 by
Anglo-Saxon Books
Hereward, Black Bank Business Centre
Little Downham, Ely, Cambs. CB6 2UA

www.asbooks.co.uk

PL

ISBN 9781898281764

Dedications

Paul Mortimer dedicates his parts of the book to his daughter Lauren.

Matt Bunker dedicates his parts of the book to the four women in his life.
His mother Karen for giving him the love of history,
his partner Anna for being his rock,
his daughter Maddy for still being his light and joy and
his grand-daughter Poppy who has yet to hear all the stories.

Contents

Left: Replica based on the sword from Snape, grave 47. Blade and hilt by Paul Binns. Matt Bunker made the scabbard. Commissioned by Tim Edwards.

Middle: Replica of the sword from Mound 1 at Sutton Hoo. Blade by Vince Evans, hilt and scabbard jewels by Dave Roper, scabbard by Matt Bunker and braid by Dariusz Wiewiorka. Commissioned by Paul Mortimer.

Right: Replica of the sword from Saltwood Grave 3944. constructed by Vince Evans. Commissioned by Tom Dowling.

Cover Picture Photo by Matt Bunker

Paul Mortimer with replica helmet and sword from Sutton Hoo burial mound 1

Cover Picture. Photo by Lee Widegren Lundin

The Contributors

Matt Bunker has been interested in early medieval history and archaeology since his childhood which largely consisted of being taken around castles and museums. This led to a degree in History, a hobby in re-enactment and a side-line in reconstructing sheaths, scabbards and other artefacts which are based as closely as possible on archaeological examples.

Having once illustrated a children's book, he assumed that helping to write this one wouldn't be much more difficult. He was wrong.

He lives in Somerset with his partner, Anna, too many books and a room full of weapons.

Matt Easton is an experienced researcher and instructor of historical European martial arts. He is the founder of Schola Gladiatoria, a founder of the Historical European Martial Arts Coalition, a qualified instructor of the British Federation for Historical Swordplay and a founding member of the London Historical Fencing Alliance. He has taught and competed all over Europe and has won and ranked highly in a number of international martial arts tournaments, using a variety of historical weapons. Matt is a Whitehall civil servant, as well as running a popular YouTube channel on historical warfare, and also collects and sells antique weapons. He is a medieval history and archaeology graduate of the Institute of Archaeology, University College London.

Vince Evans was born and raised in Hawaii in an artist family. His mother was a portrait artist and his father an ornamental blacksmith and metal sculptor. H grew up around his father's shop and became familiar with metalworking at an early age. He began doing restoration work on antique swords in 1980 and has been a full-time bladesmith since 1981.

"Schools and classes that teach many of these Old World arts did not exist so I have learned through experimentation and examining original pieces in museums and private collections. Handling original swords and seeing the craftsmanship that went into each piece inspires me to keep learning and experimenting; to understand the craftsman; and in some way, touch the past."

Jesus Hernadez was born in Zamora, Spain and grew up in Castilla surrounded by the mysticism of castles. During his early years, he walked the streets of Salamanca, a town of medieval enchantment, where he became a doctor in medicine. He moved his practice to the United States after meeting Christine, his wife. He now lives in Roanoke, Virginia furthering his knowledge of the traditional crafts of Japanese bladesmithing and metalworking.

"For a long time I have been fascinated with the ability to create both useful and beautiful objects by hand. The craft of a blade is an expression of that ability. Every aspect of this art is capable of bringing a sense of satisfaction to the maker and to the owner of such creations. The craft of bladesmithing has provided me with a path of self-reflection.

"I am mostly self-taught. My journey has been one of admiration for the simple beauty of lines and character of the Japanese blades which led to a pursuit of the same esthetics through the recreation of the traditional metalworking techniques. From making steel out of iron ore and charcoal, to forging the hot metal into a blade, to the always mysterious process of hardening the steel and finally polishing the metal to reveal its inner beauty, its soul."

Paul Mortimer is a retired secondary school teacher. He has now fully recovered and it able to pursue his interest in history and archaeology, focussing largely on the magnificent ship-burial in Mound 1 at Sutton Hoo. Paul has commissioned museum-quality replicas of many of the finds – the helmet, sword, shield, the Stone and other celebrated items - which he uses to illustrate his talks. Paul's first book, *Woden's Warriors*, was published in 2011 and a second (with Stephen Pollington) *Remaking the Sutton Hoo Stone*, followed in 2013. He has appeared on several television programmes including *Britain's Secret Treasures* and *Saxon Hoard: a Golden Discovery*.

Stephen Pollington has been involved in writing and speaking about the Anglo-Saxons for more than two decades. He has appeared in numerous television programmes – most recently with Michael Wood in *The Story of England* and *The Great British Story* – and has written more than a dozen books on aspects of the subject as varied as medicine (*Leechcraft*), runes (*Rudiments of Runelore*), language (*First Steps in Old English*, *Wordcraft*), warriorhood (*The English Warrior*), art and material culture (*Wayland's Work*) pre-Christian religion (*Elder Gods*) and feasting/social structure (*Meadhall*). He is a researcher and translator, a part-time consultant to a London auction house and a member of the Company of Art Scholars, Dealers and Collectors. (www.stevepollington.com)

Stuart Davies developed an interest in Anglo-Saxon history at a young age and studied classical and medieval history at higher level education. His interests now range from Classical Greece to the Wars of the Roses and he works across the country delivering historical education programmes to primary age children.

Lee Sauder began blacksmithing when he was 12 years old, and has supported himself as a full-time smith for all of his adult life. In recent decades, his primary interest has been in reviving and utilizing ancient iron smelting techniques.

Acknowledgements

It was Steve Pollington who suggested writing a book on swords from England.

Although only a few of us actually wrote the chapters, this book is a collaborative work and would not have been possible without the active help from the following, who have all contributed in different ways.

Irene Barbina
Patrick Barta
Paul Binns
Paul Browne
Sue Brunning
Owen Bush
Jenni Butterworth
Anni Byard
Tobias Capwell
Joan Cooper
Tom Dowling
Tim Edwards
Grace Evans
Chris Fern
Peter Fischer
Leszek Gardela
Helen Geake
Brian Gilmour
Dana Goodburn
Adrian Green
Pieta Greaves
Imogen Gunn
Alexander Haigh
Matthew Harffy
Brett Hammond
Dieter Huggins
Peter Johnsson
Thomas Kamphuis
Dirk Kennis
Lindsay Kerr
Aksel Klausen
Alison Klevnäs
Janet Lang
Scott Lankton
Kevin Leahy
Antony Lee
Ulrich Lehmann
Wayne Letting
Lee Widegren Lundin
Stefan Mäder
Mick Maxen
Johann Matzke
Reto Marti
JT Pälikkö
Caryl Pike
Andy Pilkington
Mike Poole
Su Poole
Pernille Richards
Dennis Riley
Dave Roper
Mark Routledge
Wendy Scott
Kim Siddorn
Craig Slater
David Stříbrný
Andy Ternay
Bert Tessens
Æd Thompson
Leo Todeschini
Angus Wainright
Roland Williamson
Roger Wikell
John Wills
Robin Wood
Arian Xilliox
Torun Zachrisson
Gabriele Zorzi

Martine C. van Haperen
Alice Howard (Ashmolean)
Guy and Carolyn Lown
Lars Lüppes (Hakun Risti)
Maidstone Museum & Bentlif Art Gallery
The Ashmolean
The British Library
The British Museum
The Museum of Anthropology and Archaeology, Cambridge
The National Trust
The Portable Antiquities Scheme
The Royal Armouries, Leeds
The Sutton Hoo Visitors' Centre
The Sutton Hoo Society
The Trustees of the British Museum
The Wallace Collection
TimeLine Auctions
Wulfheodenas
Ken McDevitt and Travis A. Markel of Classical Numismatic Group, Inc.
Tillerman Beads
Bushfire Forge
Ganderwick Creations

Abbreviations

CH: Crossley-Holland
F: French
L: Latin
OE: Old English
OF: Old Frisian
OHG: Old High German
OI: Old Icelandic
ON: Old Norse
OS: Old Swedish
PAS: Portable Antiquities Scheme

Glossary

*adapted from Gilmour
^ adapted from Wiktionary

Álfar(ON) / ælfe(OE) / elves: a type of supernatural being found in Germanic folklore often possessing magical powers.

Alloy: a metal made by combining two or more metallic elements, especially to give greater strength or resistance to corrosion.

Angles, Anglii: a Germanic people who formed part of the influx of peoples into the British Isles from the fifth century onwards. England is named after them.

Anglo-Saxon: used to describe the inhabitants of much of England during the 5th to 7th centuries; most, if not all of whom would have spoken a Germanic language. The term is unlikely to have been used by the peoples themselves at that time.

Annealing: a general term applied to several softening operations affecting crystalline structures within a particular steel. The temperatures applied are higher than those used for tempering and the effects more pronounced. Microscopy of a sample of the metal will reveal which operation has taken place. The steel is heated in a hearth or furnace and held for a while at a temperature just high enough to alter its structure, the length of time depending on thickness, followed by very slow cooling, usually in the furnace. Various annealing procedures are likely to have been experimented with by early ironsmiths. See also under **tempering**.*

Atheling, æðeling (OE): a prince-- a man from a noble family that had a claim on kingship

Auto-tempering (self-tempering): see 'slack quenching'.

Austenite: the face-centred cubic lattice (gamma) form of iron which can contain up to 2% carbon in solid solution.*

Baldric: a suspension method for a scabbard, the baldric is a long strap made of leather or textile.

Banding: in this book, this refers to lengths of metal laid side-by-side along the length of the blade that give the appearance of 'stripes'.

Bainite: a range of microconstituents found in ancient steels, the formation of which is dependent both on the composition of the steel and the size and shape of the object, as well as how rapidly it has been cooled. It can be observed in alloy steels or carbon steels that have been rapidly cooled from the austenite range, but not rapidly enough to form marten site. In some instances bainite is found in early steels subjected to intermediate continuous cooling condition (such as slack quenching) although the reasons for this are not clear. One form -upper bainite -has a feathery appearance which is also laminated and so resembles pearlite, whereas another form -lower bainite -has a dark, acicular appearance similar to lightly tempered martensite, although bainite is less hard. Structures such as bainite suggest that early ironsmiths had the specific intention of producing a steel of intermediate hardness and brittleness.*

Bloomery iron: the product of the single-stage method of directly smelting or reducing iron from its ores using charcoal as the fuel and reducing agent.*

Bloomery steel: the direct production of steel from iron ore by making the conditions in a bloomery smelting furnace more reducing than is necessary to produce plain or low-carbon iron. A higher charcoal to ore ratio is required and control over the air blast is more critical, factors which make the production of bloomery steel a more difficult and expensive operation than that of plain iron. In practice, the difficulty of maintaining even reducing conditions in early bloomery furnaces probably resulted in the production of blooms with varying carbon content.*

Bluing: Turning steel a blue-black colour by heating it over a bed of charcoal and then quenching when the required colour has been reached. Steel can also be blued, browned, blacked or greyed by chemical processes. Bluing was widely used in India to provide contrast to the silver inlay or overlay used to decorate steel objects.*

Bracteate: a pressed medallion, usually of thin gold, sometimes deposited as a sacrifice in the late 5th and early 6th centuries.

Carbide: a compound composed of carbon and a less electronegative element.

Carburisation: the process by which carbon atoms are made to diffuse in from the surface of a piece of iron. Under reducing conditions carbon monoxide gas is produced which on decomposition forms nascent carbon atoms which are absorbed at the surface of iron.*

Cast iron: iron-carbon alloys containing approximately 2-5% carbon are classed as cast irons.*

Celts: Those tribes and peoples who spoke what is now called a 'Celtic' language. Mentioned by Caesar and Tacitus, the impression given by those authors is that most of the Celts dwelt west of the Rhine, whereas most Germans lived east of it; otherwise, speakers of a language recognised as coming from the group of Celtic languages.

Cementite: a white, very hard, brittle compound of iron and 6.7% carbon ($Fe3C$).*

Cristobalite: The mineral cristobalite is a high-temperature polymorph of silica, meaning that it has the same chemical formula as quartz, $SiO2$, but a distinct crystal structure. Both quartz and cristobalite are polymorphs with all the members of the quartz group, which also include coesite, tridymite and stishovite. Cristobalite occurs as white octahedra or spherulites in acidic volcanic rocks and in converted diatomaceous deposits in the Monterey Formation of the US state of California and similar areas. Cristobalite is stable only above 1470 °C, but can crystallize and persist metastably at lower temperatures.

Chape: a fitting sometimes found at the bottom of a scabbard.

Charcoal clamp: a method of making charcoal by heating in oxygen deprived atmosphere.

Cold working: the strengthening of a metal by plastic deformation, often by hammering. This strengthening occurs because of dislocation movements and dislocation generation within the crystal structure of the material.

The Continent: in this book refers to mainland Europe.

Crucible: a vessel in which substances are heated to high temperatures. Before the development of modern steel-making the crucible was a vessel made of refractory clay.*

Crucible steel: steel made in by heating ore to a liquid in the presence of carbon in an enclosed container.

Damascus steel: steel made using *wootz*, a crucible steel.

Damascening: A misleading, relatively modern term, which has been used to refer, amongst other things, to a variety of decorative processes linked to the production of iron and steel objects, in particular those made of crucible steel.* It can also be used to describe the technique of inlaying one metal of one colour into another through carving lines into the host metal and forcing the other into the grooves that have been formed. This technique can also be called *tauschieren*.

Danes: Originally a people of southern Scandinavia, whose power grew in the Late Roman Iron Age to include Jutland and its adjacent islands. Modern Denmark is named after them.

Decarburisation: loss of carbon from the surface of a piece of steel or cast iron by oxidation, either accidental or deliberate.*

dryhten Drihten (OE): a warlord.

Ductile: in the case of metal, capable of being bent or deformed without losing strength. Metal capable of being drawn as a wire.

Duguð (OE): a doughty warrior, experienced, proven and to be relied upon.

Ealdorman (OE): a noble with rank, later earl or jarl

Ensis: (L) Sword. (proto-Indo European roots)

Eutectic: describing the thermodynamic equilibrium conditions where a liquid coexists with two solid phases. ^

Eutectoid: For steel this refers to the laminated structure pearlite, and is also used to describe a steel containing approximately 0.8% carbon, where the entire micro-structure consists of pearlite.*

Fabrica (L): (plural *fabricae*) a Roman manufacturing centre.

Ferrite: The body-centred cubic lattice (alpha) form of iron (q.v.), stable at lower temperatures.*

Filigree: decorative technique using fine twisted wire, often of precious metal.

Flipe: an airgap left around the base of a charcoal clamp.

Folctogan (OE): a commander, a general.

Forge welding: the joining of separate bits of metal by heating and hammering.

Foundation moulding: A decorative technique in leatherwork in which the leather is formed over a patrix..

Framea (L): spear.

Francia: the area of Europe controlled by the Franks.

Franks: Germanic people of the Rhineland, usually divided into the Salian (coastal) and Ripuarian (river-dwelling) branches, eventually became dominant in parts of France (named after them) and parts of Germany. Founded the Merovingian dynasty.

frið (OE): 'peace' state of hospitality in which violence is not permitted.

Fuller: a longitudinal depression, (can be more than one) along the centre of a blade.

Fylgja (ON): *fylgja*, also *fylgjukona* 'fetches. The souls of people but separate from their bodies. The fylgjur in Old Norse literature are only seen in dreams or else by people capable of seeing them with supernatural powers.[1]

Geat: a Germanic speaking people from southern Scandinavia.

Germanic: used to describe peoples who spoke or speak a Germanic language or to describe artefacts or other aspects ascribed to them.

Geoguð (OE): a youth, a young soldier with little experience.

Gladius (L): usually applied to the short stabbing sword employed by Roman legionaries.

Guldgubber: Danish name for small, figurative, pressed, gold foil plates used as sacrifices and deposited in the post holes of houses and other dry land sacrificial places (Gullgubber in Norwegian, Guldgubbar in Swedish)

Grip: The part of the sword hilt assembly that is held in the hand while it is in use.

Gromps: lumps of iron and slag found apart from the bloom in a furnace.

Halstatt: more properly Halstatt Culture, was the predominant Western and Central European culture of Early Iron Age Europe from the 8th to 6th centuries BC, developing out of the Urnfield culture of the 12th century BC (Late Bronze Age) and followed in much of its area by the La Tène culture. It is commonly associated with Proto-Celtic and Celtic populations in the Western Hallstatt zone and with (pre-)Illyrians in the eastern Hallstatt zone. ^

Hammer welding: uniting two or more pieces of metal using heat and the hammer.

Hardness: increases resistance to wear, resists penetration but may increase susceptibility to sudden shock; can increase brittleness.

Heregeate(OE): the return of military equipment to a lord after the retainer's death, can also be a payment to a lord after the death of a follower.

Hilt: the top of a sword including the grip.

Indirect method:

Jutes (OE *Eote*): one of the Germanic peoples established in Britain, traditionally confined to Kent and the region of the Solent and New Forest.

Kenning: a circumlocution used instead of an ordinary noun in Old English, Old Norse and later Icelandic poetry.

La Tène Period: period beginning about 450 B.C. to the Roman conquest. The art style that came from this time remained influential for centuries.

Ledeburite: in iron and steel metallurgy, ledeburite is a mixture of 4.3% carbon in iron and is a eutectic mixture of austenite and cementite. ^

Limes: The limits of the Roman Empire – the border area. In Germany, it follows the river Rhine for its first section (the Lower Germanic Limes) before becoming a 550km long land border between the Rhine and the Danube (the Rhaetian Limes).

---

[1] Simek,1984. 96.

Männerbund: a military and religious community bound by a common goal and allegiance.

Mantle (for smelting):

Martensite; a supersaturated solution of carbon in gamma iron, produced by very rapid cooling such as is obtained by quenching a steel in water. The carbon present does not have time to separate as pearlite and the distorted lattice structure of martensite results. Its structure means that it is very hard and non-ductile: the higher the carbon content, the greater the distortion to the lattice structure and the greater the hardness of the martensite. The hardness increases most markedly for steels with a carbon content of up to about 0.7% but the effect is rather less thereafter.*

Merovingian: Frankish dynasty who ruled from the 5th to the 8th century.

Mild steel: a form of low-carbon iron made as a liquid, only possible with the introduction of the Bessemer converter and the Siemens open-hearth process, both circa 1860. It is characteristically more homogeneous than low-carbon iron produced by earlier methods (e.g. bloomery, finery) and contains fewer non-metallic inclusions but is usually recognisable by the presence of impurities such as manganese sulphides.*

Motty peg: a large peg around which wood is stacked prior to igniting a charcoal clamp. The motty-peg is withdrawn just before burn begins.

Niello: compound of silver, copper and lead used decoratively in a technique whereby some areas are deliberately and permanently blackened.

Normalising: normalising is carried out to relieve the stresses set up during forging iron or steel and is particularly needed if the piece is going to require more hot forging.

Odin, Óðínn: equivalent of Woden (OE) and Wotan (OHG), god of magic, wisdom, battle, hanged men, a shape shifter.

Other, the; the world of the supernatural, gods, spirits, ancestors and the realm of the dead.

Pattern-welding: method of making blades with patterns using different metals and hammer or forge welding.

Pearlite; a very regular, laminated structure, consisting of alternate plates of cementite and ferrite giving a characteristic mother-of-pearl appearance. Its formation is greatly affected by the rate of cooling. It is most commonly or easily visible in early steel objects cooled in air.*

Phase: A general term referring to the metallurgical structure of a metal. In most earlier iron/steel objects more than one phase (i.e. structure) is usually found to be present.*

plano-convex: convex shape – lenses are plano-convex..

Pommel: the top of a sword which sits on the upper guard.. The tang of the sword is sometimes peened over the pommel.

Pressbleches (G): pressed metal foil plate used for decoration on a helmet, and other military equipment. Also used to manufacture the majority of guldgubber

Quenching: the process of rapidly cooling a re-heated metal object by plunging it in water or another liquid, such as oil. Steels can be hardened to varying extents depending on their carbon content and the rate or speed at which the quench occurs, which varies with the thermal capacity of the quenching liquid as well as the mass of shape of the object being quenched. In an air-cooled steel the cooling rate is sufficiently slow to allow the structure of the steel to transform to one at least partly made up of pearlite. When this critical cooling rate is exceeded, the transformation to pearlite no longer has time to take place and, depending on the rate of cooling, one of several less stable structures is the result. The more efficient the quench the less stable and correspondingly harder and more brittle is the resultant structure. When the rate of cooling is rapid enough a fully quenched structure of martensite is the result.*

Reduction – the process of reducing an ore and decompose the unwanted elements, leaving the metal behind. Carried out in a furnace through smelting.

Roasting: the heating of ore for a considerable time in order to drive out moisture, change its chemistry and get it ready for crushing.

Salin Style. A system of classification for Germanic animal art in the 5th to 10th centuries devised by Bernhard Salin (1861-1931), a Swedish archaeologist, cultural historian and museum curator. He identified three distinct phases of development, two of which are relevant to the period encompassed by this volume: -

Style I. Appears in Northwest Europe and is first used in the chip carving techniques applied to bronze, silver and or gold decoration on belt buckles, brooches, etc during the 5th century. It is characterized by animals and humans whose bodies are divided into sections, and typically appear at the fringes of designs whose main emphasis is on abstract patterns. Features disjointed animal and human forms.

Style II. Most commentators feel that after the middle of the 6th century the use of Style I, began to decline and tended to be replaced with Style II. Style II usually includes whole beasts whose bodies are intertwined in a form of complex and asymmetric interlace with no attempt at realism. Although the animals often resemble serpents, other animals are actually represented, such as boars, wolves and horses, and body parts such as limbs, hips heads and teeth can be discerned.

Sammel: clay-rich undersoil used to coat the charcoal clamp and thus control the temperature inside and isolate the contents from the outside atmosphere.

Scabbard: the protective container for a sword blade.

Seax, sax (OE): A knife,varying in size and blade shape, used by the northern peoples.

Scop/scops (OE): poet or story teller.

Slack quenching: quenching carried out using a cooling liquid, such as oil, with a lower thermal capacity than a fast-cooling liquid like water. Depending on the carbon content of a steel quenched in this way, an intermediate transformation product such as bainite will be formed, rather than the fully quenched, much harder and more highly stressed transformation product, martensite. More gentle quenching avoid excessive distortion or even breakage, although an insufficiently well quenched steel might be too soft and perhaps too easily blunted or bent.*

Slag: the glassy waste by-product of smelting. A small proportion of this inevitably remains trapped within iron or steel produced as a bloom however well this has been consolidated during manufacture and is a particular characteristic of directly smelted iron and steel.*

Smelting: to obtain iron from its ore, the iron oxide in the ore must be reduced to metal. The lowest temperature at which this can be achieved is about 800° C, the melting-point of pure iron being about 1535° C.*

Spatha (L): often refers to the longer sword (as opposed to the gladius) used by Roman cavalrymen. Has also been used to describe the swords used by the Germanic peoples and others at least up to the 7th century.

Steel: carbon steel is usually referred to as an alloy of iron and carbon containing up to about 2% carbon. For convenience, in the present work steel is defined as that alloy of iron and carbon which can be hardened by quenching, which means a practical lower limit for carbon content of approximately 0.3%. An iron-carbon alloy with less carbon than this would normally be referred to as a low-carbon or mild steel, but here is referred to as low-carbon iron.*

Svear: the Swedes (Swedish: *svear*; Old Norse: *svíar / suar* Old English: *Sweonas*) were a Germanic speaking tribe who would eventually give their name to Sweden. During the 6th to 7th centuries they appear to have had control of that part of Sweden around Lake Mälaren.

Sword-bead: a form of decoration attached to some swords during the 5th to 7th centuries. Beads could be made of different materials including glass, amber, horn, etc.

Sword-boss: a form of sword decoration which may have sometimes had a practical function. The boss was usually made from metal, often gold and garnet cloisonné, and frequently supported by some pale coloured material.

Sword-pyramids: a form of sword decoration, most frequently encountered as a true pyramid shape, although there are other variations. Most grave finds are made from a metal but pyramids made from organic materials and stone have been found.

Tang: iron extension of a sword that protrudes above the blade, enters the hilt and facilitates the fixing of the grip by peening.

Tauschierung(G): inlaying metal wives into a different base metal to produce a decorative effect.

Tempering: Process for relieving the highly stressed condition of a fully hardened (i.e. martensitic) steel that is otherwise very hard but also brittle and liable to cracking. The objectives of tempering are to both, a) relieve internal stresses caused by quenching and, b) to toughen the steel.*

Thegn (OE): "servant, attendant, retainer", "one who serves", often used to describe either an aristocratic retainer of a king or nobleman in Anglo-Saxon England, or, as a class term, the majority of the aristocracy below the ranks of ealdormen and high-reeves. ^

Thrall: a slave, a bound person, one who was not free.

Toughness: the ability to resist sudden shock loads.

Tuyère (F): a pipe through which air is blown into a furnace while smelting ore.

Valkyrie, Valkyrjar (ON): Handmaidens of the god Óðinn.

Vickers Pyramid Hardness Scale (VPH): The **Vickers hardness test** was developed in 1921 by Robert L. Smith and George E. Sandland at Vickers Ltd. It uses a diamond pressed into the material to gauge hardness.

Vikings: a term that has many meanings depending on who is using it – probably, originally meant nothing more than 'pirate' but refers to a whole group of peoples from Scandinavia from 793 A.D. to 11th century. However, it was not an ethnic term and should not be used as such.

Viking Period /Viking Age: roughly 793 A.D. to about the 11th century. The dates can vary depending on the point of view of the relevant expert.

Welding: before the late nineteenth century (introduction of liquid phase) welding inevitably referred to the hot, solid phase process for compressing and fusing together two pieces of iron or steel. This process is often referred to as hammer-welding or forge-welding.*

Woden: see Odin

Work-hardening: some metals, including iron, can be hammered whilst cool and this will result in them becoming harder.

Witan or *witena gemot* (OE): a council of important men and possibly the king's advisory council in later Anglo-Saxon England. They were responsible for electing the new king after the death of the old.

Wootz: a corruption of ukku (pronounced wukku), a word meaning steel in Telangu-Kannada, a Dravidian, non Indo-European language of central-southern India. Possibly the earliest known reference to wootz is in a letter sent in 1794 from Bombay by Dr Helenius Scott to Robert Pearson of the Royal Society of London.*

Wrought iron: plain iron produced indirectly (from its ore) by the decarburisation (firstly by fining and later.*by puddling) of cast iron.

Wüstite: Iron oxide (FeO).*

Foreword

Brian Gilmour

Before we consider the significance of long swords of the Anglo-Saxon period we perhaps need to think about when, where, how and why swords came in to existence and their early significance. We know little about the origin of swords and how they were first used but we do know that they were, and continue to be, one of the most enigmatic of artefacts. But why should we consider them as enigmatic at all? This book considers the early development of swords and follows that through to the Anglo-Saxon period, which is regarded as the peak of the metal-workers art.

Where do swords come from and what is the first evidence of their existence? Arguably the forerunners of what we call swords can be traced back some 6,000 years to the eastern Mediterranean or Black Sea region in the late Neolithic or early Bronze Age. They came into existence not as metal objects but as very long flint flakes, which must mark one of the peak achievements of the flint knapper's art. Good, well dated, examples of these have been found accompanying burials such as the Chalcolithic (late Neolithic/early Copper Age) cemetery at Varna near the Black Sea coast of Bulgaria and in some early dynastic tombs in Egypt. By sometime in the 4th millennium BC copper alloy long swords started to appear. A very long example (more than 1.2m) is possibly made from arsenical copper and like much copper alloy work of this period it is from the Caucasus region. It is currently on display in the Georgian National Museum in Tbilisi.

But how are we to interpret the very long early flint blades which are the probable forerunners of copper alloy and iron long swords? How and why the flint blades were made probably has a bearing on how we interpret the very long successor blades made in metal. This is so whether we are talking about the early Bronze Age or the transitional late Bronze Age to early Iron Age (from which there are various examples of long swords) or beyond up to the time of the long swords of the Anglo-Saxon period, which are the subject of this book.

The longest flint blades exceed 30cm in length and would have needed a correspondingly large flint core of sufficient quality to make such a flake possible. It would have also required exceptional skill to make them. Why were these extremely long flint blades made? One clue is that they are invariably found (except in the case of stray finds) in very high status graves, which suggests they were artefacts of high status. Given the hard and brittle nature of these long flint blades it is unlikely they were used as weapons in combat, which suggests they had a mainly ceremonial role. The rarity of long flints of the quality required plus the skill needed to make the blades suggests they were a status symbol, hence the longer they were the more prized they may have been.

The heyday of these very long flint blades, the possible progenitors of copper alloy and then iron swords, seems to have been 5,000 to 6,000 years ago. It was towards the end of this period that what we now think of as long swords first appear and become more common across Europe, both as longer examples of broad-bladed swords and very narrow rapier-like blades. The new metal technology allowed the longest swords to be up to about 120cm in length, and occasionally more, which was about four times the length of their flint predecessors.

One interesting idea that has been floated in recent years (by the prehistorian Dr Francis Prior) is that there may have been great ceremonial importance placed on the casting of swords in the Bronze Age – hence the 'sword in the stone' myth possibly being a distant echo of the ceremonial importance of the act of casting a sword in a stone mould. If this is so, it is likely that the ceremonial or cultural significance did not stop there.

Many Iron Age long swords approached or exceeded one metre in length and some argue that this is completely unnecessary, or even counter-productive, for general practical combat use and these could be seen perhaps as symbols of authority rather more than as practical fighting weapons, particularly if their ownership was restricted to those with the highest status in the contemporary societies. There seems to be little or no evidence that very long swords played any part in the standard part of the military equipment of the Roman Army – the earliest fully professional fighting force – while they were certainly widespread in the regions north of the Roman Empire, from where the idea and actual forerunners of Anglo-Saxon long swords appear to have come.

Roughly contemporary long swords of late Iron Age type or date (contemporary with but culturally separate Roman society) are widespread in northern Europe from as far afield as Anglesey to eastern Poland. Occasionally these are found in Roman contexts but in all cases other associated evidence points towards the origin of these swords as being from outside the Roman Empire, even where inscriptions point towards some degree of cultural interaction as in the case of some swords, for instance, from the very large peat bog/pool votive deposit at Nydam in Jutland where some inscriptions would appear to show Romanised versions of local names. In these cases it seems clear that these names simply reflect a local fashion for associating Roman style names – perhaps as a reference to the makers – with some of these sword blades. Many of the long swords found in the votive peat-bog/pool deposits in Denmark have also been found to have been made using an earlier form of 'pattern-welding' which is clearly ancestral to that used on the majority of Anglo-Saxon long swords. This technique also may go back to the earlier stages of the Iron Age.

In England a few forerunners of the typical Anglo-Saxon long sword have been found in Roman contexts and these examples have mostly been made using similar pattern-welded techniques and styles to those found in the contemporary votive peat bog/pool deposits in what is now Denmark, but at the time was territory occupied by the Angles and associated neighbouring peoples. Similar swords have also been found at other sites further to the east. Two of these type of swords can be dated to the late 2nd century AD and – along with other pieces of equipment which identify their owners as probable officers of German origin but serving in the Roman army – were found dumped in a shallow irregular grave, along with their murdered owners on a site on the outskirts of the then Roman town of Canterbury. The fragmentary remains of three (or possibly four) further long swords were found as a part of a probable votive deposit again of artefacts of Germanic origin but this time buried under a newly constructed rampart at the Roman supply fort (now) of South Shields (south of the estuary of the River Tyne) in *c*205 before the visit to the area of the Emperor Septimus Severus in 208. In all these cases the swords seem unlikely to have formed part of the standard equipment of the Roman army but were brought in and worn by their owners who came from areas near or beyond the Roman frontier.

Swords such as these may represent a very specialized and specifically Germanic/north European tradition of sword manufacture and ownership that came from Iron Age and earlier traditions and led ultimately to the particularly insular tradition of Anglo-Saxon long swords that form the subject of this book. Fortunately for us the earlier long swords of the Anglo-Saxon period become common from about the middle part of the 5th century and many have been found, almost exclusively in cemeteries dating to the next 200 years. Following the adoption of Christianity in the mid 7th century long swords again become rare for about another 200 years. A re-emergence of earlier pagan votive practice of depositing weapons, especially swords – and again this is likely to be significant – probably due to Scandinavian/Viking influence from about the middle of the 9th century resulted in many more long swords being deliberately placed in rivers, pools and other watery places, possibly seen as liminal places or a gateway to the 'next world'. Again this practice seems to have been at its height over the following two centuries right up to the end of the Anglo-Saxon period.

Given the clear cultural significance or importance of long swords in Anglo-Saxon society it is surprising that no comprehensive or detailed study of these artefacts has appeared since Hilda Ellis-Davidson's book of much the same name came out in 1962 (although a slightly updated edition of this book came out in 1994). Thus these enigmatic artefacts have long deserved a large and detailed study such as this although it is to be hoped and expected that the subject matter presented and interpretation discussed herein will be debated and will lead to further studies in the future.

1. Introduction

Opening thoughts

> *There is no single object among the possessions of a modern man which may be compared to it. It demands, therefore, a determined effort of study and imagination if we are to realize the significance of the sword, so powerful a weapon in men's hands and so potent a symbol in men's minds for hundreds of years. But such an effort is indeed worthwhile, for much of the life and vigour of our early literature must be lost forever when the memory of the sword has faded.*
>
> *Hilda Ellis Davidson.*[1]

With these words, with which we can only agree, Hilda Ellis Davidson ended her seminal book on the sword in Anglo-Saxon England. It was published in 1962, more than fifty years ago at the time of writing this preface and is still very much in-print despite the fact that much of the research that went into the book was carried out during the 1950s or earlier. Since then there have been more discoveries, new books and papers, new methods of gaining information and new ways of thinking about objects and their contexts and a great deal of new research.[2]

Davidson's work remains important, not least because of the insights that she provided, many of which remain of value, but also because it set a standard for others to try and emulate. However, in light of all the developments which have taken place since that publication, it is time to try to bring knowledge of the sword in Anglo Saxon England up to date. We also hope that this volume will help to promote a similar level of curiosity in the way that facts are examined and assessed and in how conclusions are reached as Davidson's work did. Where uncertainties remain, we will attempt to make this clear in the text.

Davidson's book was mostly the work of a single individual, although she acknowledges many people who assisted her with the project. The present volume is the result of a number of people working together in order to provide as comprehensive an analysis of as many aspects of the sword from the 5th to 7th centuries in England as possible. Some of them have contributed in writing, but many have suggested ideas, provided insights and shown how the technologies of the period developed and could be utilised, while others have generously allowed their research to be used by the authors of the chapters. Another way in which this book differs to Davidson's is that we will not attempt to cover the whole history of the Anglo-Saxons and their swords in detail, but will call a halt at the end of the 7th century.[3] We will consider swords well before the coming of the Anglo-Saxons to these shores in order to provide a context for later developments and, as Davidson did, we will step outside the geographical confines of the British Isles where ever it is necessary to demonstrate relationships and artistic continuity where we believe they exist and where information is available.

At times, we will have to include discussions on subjects other than weapons if we are to show how and why swords were important as potent social, cultural and religious symbols to the Anglo-Saxons and their

---

[1] Davidson, 1962. 216.

[2] A corrected and slightly updated edition was published in 1994.

[3] That will not be the case for the catalogue of patterns, as this will include information until the end of the 11th century. The reasons for this change in policy will be explained in that chapter.

neighbours, why they made such auspicious gifts, why they often figured in the literature of the period and why they continued to do so for many centuries.

It would be wonderful to think that we could provide the definitive history of the Anglo-Saxon sword to the end of the 7th century but that is not possible; much more work still needs to be done as there are many gaps in our knowledge. No doubt new discoveries will be found which will challenge some of the ideas presented within the pages of this book and yet more ways of considering the evidence will emerge in the years to come. However, we hope that we have made a contribution to the continuing discussion of the sword and its history and that we have come some way to satisfying Davidson's suggestion that the sword, "...demands, therefore, a determined effort of study and imagination if we are to realize the significance of the sword,..."

Structure of the book.

The main thrust of this volume is to consider as many dimensions of the sword as possible, from the 5th century until about the end of the 7th century. The reason for the artificial end to our inquiry is because that is roughly when goods stop being deposited in graves, at least in England, probably because of the conversion from the older ideas and customs to new ways of thinking and in part to the seeming general acceptance of Christianity. Intriguingly, most of the finds of swords after the 7th century are associated with rivers and watery places, perhaps because fighting often took place in such locations, or because weapons and other items were deliberately deposited in the water, perhaps as a sacrifice to bring victory or as a mark of thanks for victory being granted. Or perhaps casual loss is the explanation, although it is difficult to explain why people after the 7th century became more careless with their processions.[1]

The book is divided into a number of parts, each with their own chapters. The *Introduction* begins our exploration with a survey of the works of those writers who have written about the swords of our chosen period. It will also briefly look at the structure of society of the 5th to 7th centuries and contains a note about the Staffordshire Hoard.

The next section is titled *Production* and contains chapters that will explain how a sword is made, together with its ingredient, the role of the smith and the importance of pattern-welding.

This is followed by *Analysis*, which considers the data that has been collected from a survey of cemeteries from as many 5th to 7th century cemeteries in England as we could locate reports for. Much of the information is supported by the appendices.

Typology includes consideration of parts of the sword, such as hilts, scabbards and sword decorations and provides an insight into the development of the sword over time. It will look, too, at the making of scabbards and how swords were worn by their owners.

Language and Runes surveys some of the ways that swords are referred to in the literature and the use of runes on a very few swords. There is a brief consideration of the other marks on sword blades.

Within *Case Studies* there are a small set of reports on some re-made swords, beginning with the Mound 1 Sutton Hoo sword and each has some notes on what the processes and the careful examination of the originals has shown us.

[1] Theuws and Alkemade do not believe that many of these finds are accidental or due to a tendency to fight by rivers, (Theuws and Alkemade, 2000. 426.) nor does Brian Gilmour (personal comment). This is an idea that we will be returning to later in this book.

The Sword in Practice attempts to examine the way that swords were used in combat during the 5th to 7th centuries. The section has chapters on *Wear and tear* and on the trauma that swords could inflict on their victims.

Finally, the *Conclusion*, which only contains one chapter, will attempt to bring much of the evidence together and explain the significance of the sword to the societies of the 5th to 7th centuries, partly in the light of new research carried out by others and a little by the contributors to this book.

One further matter before we proceed is to mention that we have quite deliberately stolen the title from Davidson's book. This is partly to show that we want to build upon her work, but also because it was a very good title; it is about swords in England, not the English or Anglo-Saxon sword; as far as we can tell, there was no such thing during the 5th to 7th centuries. There were no such people as 'Anglo-Saxons'[1] either, at least for most of that time, and the term, when it is used in this book, is only used as a term of convenience.

Notes:

1. Each chapter has the author's name clearly stated and the book can be read as a series of independent essays, if the reader so desires. We may not have agreed with each other on every aspect of the sword.

2. Paul Mortimer has used two separate databases on blade details to glean information for this book. One is based largely on the work of Dr. Brian Gilmour but that has been supplemented with some additional material and consists of a sample of two-hundred swords and their blade patterns. The second database is the result of surveying two hundred and thirty-four cemeteries and burial places in England dated to the 5th to 7th centuries. From this survey Paul found one hundred and sixty-nine swords. Some of the same swords occur in both databases, but because of different methods of labelling the swords and the cemeteries by the scholars involved, it has proven difficult to reconcile the two sets of swords with each other. Throughout, Paul will try to make it clear from which data set the information has been gleaned. In the pattern data base, most entries are accompanied by the museum acquisition reference, whereas in those that come from the cemetery survey, the cemetery and grave number have usually been given. This is not entirely satisfactory, but Paul found that he could not do it any other way.

3. We have used other databases, datasets and sources to assist in the project, most notably the corpus compiled by Doctor Esther Cameron for her study of scabbards and sheaths[2] and that of the Portable Antiquities Scheme (PAS) which is an invaluable resource and has greatly increased our awareness of the distribution and number of hilt fittings, particularly pommels.

Paul Mortimer and Matt Bunker

England, October 2018

[1] 5th to 7th century England was made up of different and competing kingdoms, chiefdoms, etc. Their ethnic or tribal origins would have differed from place to place and much more complex than the word 'Anglo-Saxon' implies.

[2] Cameron 2000

2. The Sword in Anglo-Saxon England: a survey of the literature

Paul Mortimer

There has been a surprising variety of literature devoted to the subject of the Anglo-Saxon sword and its relatives. What follows is a chronological survey of many of the more important contributions.

Elis Behmer 1939

Elis Behmer's study of Germanic swords from the age of the migrations (*Das Zweischneidige Schwert der Germanischen Völkerwanderungszeit*)[1] was the first serious study of the subject and brought together an enormous amount of material for the first time. The study considers swords from all the Scandinavia countries, Germany, France, England, Belgium, Switzerland, Italy, Russia, Poland and Hungary, from the 2nd century A.D. until the end of the 7th century A.D.

Behmer begins by looking at the Roman forerunners of the Germanic sword and considers the Sarmatian and Alannic influences on its early development.

She establishes a typology and explains how she separates the sword into its components, blade, pommel, grip, scabbard, etc. The typology runs from type I to type IX and these are collected together into three groups; group A types 1 – IV, group B types V – VII and group C types VIII and IX. The rest of the text is devoted to explaining when and where each type occurred, how it developed and the characteristics of the elements that make up the sword in each type.

A set of find charts follow the text, one for each type I to IX, and the charts refer directly to the photographs of sword that follow the charts, so that the reader can see several good examples of each type of sword and understand where they were found.

Behmer's book is a pioneering, systematic and illustrated study that set standards for all subsequent commentators to emulate. It is significant that many scholars still use her typologies in their work.

Hilda Ellis Davidson 1962

Until now, there has only ever been one book length treatment of the subject of the Anglo-Saxon sword and that is Hilda Ellis Davidson's, *The Sword in Anglo-Saxon England: Its Archaeology and Literature*. Strangely the book's subtitle does not appear on the cover, but only on the title page, although it aptly describes what Davidson set out to do; that is to explain what insights we can gain from a study of the literature associated with the sword and from looking at finds of actual swords, their associated materials and their contexts. As has been seen above, Davidson was not the first scholar to consider the Germanic swords of northern Europe. Elis Behmer, a Swedish scholar, published her book in 1939 which looked at swords from the 2nd century to the end of the 7th from all over Europe. An earlier book by the Norwegian scholar, Jan Petersen, *De Norske Vikingesverd* (1919), had considered swords of the Viking period. Both Behmer and Petersen established typologies of the swords that they wrote about and studied within their respective time periods. Davidson relied on both authors for her classification of the swords that she writes about in her own book and she was the first scholar to put the study of the Germanic sword into an Anglo-Saxon context. Many subsequent works have relied to a lesser or greater extent on the foundations that she laid down in her pioneering work.

---

[1] 'The two-edged sword of the Germanic Migration Period'.

Davidson explains that the book grew out of a lecture that she gave on pattern-welding in the early 1950s and that the volume was a response to the growing interest in that particular subject. She provides, for the time, a comprehensive consideration of her chosen subject and initially sketches out for us the general circumstances of sword finds and describes the types of literary sources available. She then proceeds to go into a little more detail regarding the circumstances of sword finds within graves and reports on the many finds of swords in bogs, mainly in Denmark, and in rivers. She outlines some of the stylistic developments of the Anglo-Saxon sword and uses continental and Scandinavian comparisons throughout her book where appropriate. She explains that it is difficult to see the Anglo-Saxon sword as an isolated phenomenon and that it can only be understood in connection with discoveries elsewhere.[1] She also attempts to see trends in the way that swords are dealt with during the burial rite.

Davidson considers the attachments that are associated with some swords and returns several times to discuss ring-hilts and their possible purposes as well as various forms of bead and a short consideration of sword pyramids. She looks at the significance of the inscriptions, sometimes runic, that are carved into hilts and scabbard throat bands and includes a very interesting and thoughtful analysis of their use. She briefly discusses, too, the materials that scabbards are made from.

She is concerned to track down where the swords were made and returns to this theme several times during the book. Early on she says, "Now that widespread interest has been aroused in pattern-welded swords, we shall probably gather sufficient evidence in time to give us a definitive answer to the question of where the finest blades were produced,..."[2] More than fifty years later we are only a little wiser. One area that she identifies as having a real possibility of being associated with sword making is the Rhine region and another is Namur, in present day Belgium.[3] Davidson said of pattern-welded swords, "...as yet there is no evidence that they were ever produced in England or Scandinavia, though there seems no convincing reason why they should not have been." [4]

In her survey of the literature she includes Latin writers such as Cassiodorus, Sidonius and Tacitus as well as Old English poetry, riddles and wills. She goes on to explore the Norse sources including Skaldic and Eddic poetry and the Sagas. Davidson, as a folklorist, points out that, while some of these sources are quite late, they were written when the sword was still an important weapon and whatever was being said about swords would have been familiar to those to whom the stories were being related to and therefore the sword details within the tales would still have to have been current. She explains, too that some of the material within this literature preserved older traditions and memories.[5] In addition, Davidson uses the writings of Arab chroniclers and scholars to add useful insights into the making of swords and their uses. She moves on to discuss legendary swords and their attributes and what they can tell us about real swords from the period. She looks at how swords were used by considering duelling from the later period, stories from the legends and some accounts of battles, although she never relates how swords must have felt in the hand or discusses fighting styles that would have been adopted by the swordsman.

One of Davidson's achievements was to make the subject of pattern-welding available to a wider public and it is probably through reading her work that many people have come to some understanding of what the processes involved and how skilled were the smiths of the time. She was aware of, and was able to articulate to some extent, the high level of skill involved in making patterned swords and explain why they were so valued

---

[1] Davidson, 1962 (1994). 2.

[2] Davidson, 1962 (1994). 38. There is probably as much discussion regarding the location where the swords were made today as there was in the 1950s and 1960s.

[3] Davidson, 1962 (1994). 40 and 91.

[4] Davidson, 1962 (1994). 34.

[5] Davidson, 1962 (1994). 3.Davidson agrees that the amount of older traditions included is very much a matter for debate.

by their owners: "It is no longer difficult to understand the popularity of the pattern-welded blade in North-Western Europe for so many centuries."[1] However, as she points out, pattern-welded swords were not always of the best quality. She does not really explore other methods for making Anglo-Saxon or Germanic swords, but does describe the damascening process and explains how that differs from pattern-welding. Davidson does not go into much detail regarding techniques for constructing the hilts of swords.

Another of her achievements was to discuss the period literature with reference to swords and many of her insights and ideas have been of great value to subsequent workers within the field.

However, since the 60s the story has moved on, there have been many new archaeological discoveries and some of the assumptions made within Davidson's book are very much open to challenge. One such is her repeated assertion that the Anglo-Saxons always fought on foot, ".....since the Anglo-Saxons never developed the art of fighting on horseback,...". [2] Work by many scholars since then has shown that the Anglo-Saxons were just as capable of fighting on horseback as other contemporary peoples in northern Europe and this has implications for the employment of the sword.[3] Davidson believed that the Anglo-Saxon sword of the early period was mainly a slashing weapon. When she is discussing later developments of the sword she says, "But by about A.D. 900 there was a change in the type of blade, and swords of this time are found to be better balanced, with the centre of gravity no longer halfway down the blade but nearer the hilt, so that they must have been more efficient weapons both for thrusting and cutting."[4] It is certainly true that the later swords were different in some ways and possibly had properties that the earlier swords did not but this should not imply that the earlier swords were not versatile instruments that were able to be used flexibly by their wielders. Some aspects of sword design did change, but Davidson was wrong about their 'balance'; no serious attempt was made to balance the blade with a pommel cap until well into the later medieval period, if then. Lower guards do not become significantly wider until quite late and this could be to offer the hand a little more protection, but even then, those swords were unsuitable for fencing. The changes that occurred to the sword are most likely due to changes in fighting styles and techniques that came into play as warfare took on a different character in the later Middle Ages.[5] As for thrusting with the early sword, Davidson herself provides some examples from (admittedly later) literature when Sigmund pierces the dragon,[6] and reports that *Offa of Angeln* kills his opponent with a 'piercing blow'[7].

Davidson felt that the frequency of swords in graves reflects the number of swords available to the population and that only certain men were allowed to equip themselves with swords, "The general assumption is that a man buried with a sword is likely to have been one of rank and wealth, while the spear and shield without a sword mark the less distinguished follower."[8] As Davidson says, most archaeologists and historians accepted this proposal, almost without question until fairly recently. However, there are problems with this idea. The first is the belief that cemeteries reflect what goes on in real life; this is unlikely and recent studies of finds by metal detectors have shown that the frequency and types of items found outside graves are very different to those deposited deliberately into the grave, which is perhaps not

---

[1] Davidson, 1962 (1994). 30.

[2] Davidson, 1962 (1994). 10. See also:189.

[3] See for example, Cathers, 2002. Halsall, 2003. 181 to 182. Glover, 1996. 177.

[4] Davidson, 1962 (1994). 39. Davidson was mistaken and we will discuss this in the chapter, *The Sword as a Weapon.*

[5] The actual handling and use of swords of the 5th to 7th centuries will be discussed later in the chapter, *The Sword as a Weapon.*

[6] Davidson, 1962 (1994). 198. The episode is included in *Beowulf.*875 to 892 (CH).

[7] Davidson 1962 (1994). 200. The passage comes from Saxo Grammaticus Book 4 Davidson and Fisher, 1980. (92 to 93 for introductory comments) 106 to 109. Offa is called Uffi in Saxo.

[8] Davidson, 1962 (1994). 10.

surprising.[1] Perhaps an Anglo-Saxon funeral was more like a modern day, western wedding, where although the general format is the same, departures from the basic script are designed to make the event as memorable as possible in the minds of those taking part? This would mean that no funeral was really 'typical' and what was deposited would depend on a number of factors, such as the wealth of the individual and his/her family as well as local customs, beliefs and adherences. There is a lot of evidence that funerals were planned quite carefully before the event could take place. There could be many reasons why a sword was included or not. One suggestion is that those buried with swords will have one because they are of a particular age and that older men who are now 'landed' became farmers and fought less often and so had passed the age when they had been 'swordsmen'.[2] The scarcity of swords within graves might be accounted for by the fact that swords did not always belong to those who wielded them, but to the lord or the family and would outlive the deceased. It is possible too that, generally, only worn-out or broken swords tended to be buried.[3] Whatever the reality, sufficient doubt exists to make Davidson's (and others) assertion unsafe and the recent discovery of more than eighty, mostly high status, sword pommels in the Staffordshire Hoard has shown that swords would have been more common than many people have previously thought. Another factor is that men buried seemingly without grave goods might have been well equipped with gorgeous textiles and other organic materials that have since disappeared; we cannot say with certainty that an absence of grave goods necessarily denotes someone of lesser wealth or rank.

Davidson goes some way to discussing the effectiveness of swords in combat but does not really consider evidence of wounds from Anglo-Saxon graves, relying instead on quotes from the stories and a brief consideration of skeletal remains from the Battle of Visby in Sweden in 1361.[4] This is a little disappointing as Anglo-Saxon evidence was there and could have been utilised.

Despite these fairly minor criticisms, Davidson's book is remarkable for its scope and for the insights that she provides and I hope to take some of the discussions that she began in her work a little further in the present volume.

Vera Evison 1967

In 1967 Evison published a short paper entitled *The Dover Ring-sword and Other Sword-rings and Beads*.[5] Her paper was the first summary in English of all known sword rings found within England; at the time that she was writing there were eighteen plus the ring knob from Sutton Hoo Mound 1. The inclusion of a catalogue of all known ring-swords from Europe[6] and sword-bead finds from England makes the paper even more valuable. The catalogue also includes a list of scabbard throat bands with perforations or lugs and a full discussion on the runes of the Gilton sword pommel.

Evison does ponder on the purpose of the rings and takes into consideration the thoughts of other scholars, but after deciding that they have no practical purpose she makes several suggestions connected to their possible symbolism. She felt that the interlinked rings probably indicated a bond, perhaps to another warrior or to the lord. Another suggestion is that the rings might show the sword was given by a leader to his thegn and that the ring-sword was a reward for service or perhaps it was a badge of office or rank and that only certain levels of followers were able to have one. She suggests, too, that it could just symbolize the mutual

---

[1] Geake, 2011.

[2] Personal comment by S. Pollington. We will be considering the reasons behind the inclusion of swords in graves, in the chapter, *The Meaning of the Sword.*

[3] Mortimer, 2011. 110.

[4] Davidson, 1962 (1994). 196 to 197.

[5] In Archaelogia 101, 1967.

[6] Evison, 1967. The catalogue lists 12 from mainland Europe (France 4, Germany 5 and Italy 3) and 40 from Scandinavia (Norway 5, Denmark 5, Sweden 16 and Finland 14). This includes finds of isolated ring knobs.

loyalty of lord and man.[1] She thinks that as there was no ring-sword at Sutton Hoo that they were, perhaps, not appropriate for kings.[2] She notes that some swords had had their rings removed before being committed to the earth and that sometimes rings are located without swords.

Evison extends the typology created by Montelius[3] and attempts a chronology, both of the rings and of the sword pommels that are found on swords from the 5th to 7th centuries. She links this information to an interesting discussion on the chronology of scabbard mounts. Evison discusses the numbers of ring-sword finds from the continent and Scandinavia[4] and tries to discover where the custom began and although not entirely successful she does state in her conclusion that the sword from Petersfinger 21 is the earliest known based on stylistic evidence and that the custom probably started with the Franks.[5]

Evison's short exploration of the custom of wearing sword-beads is interesting and she links that to the lugs and holes that are occasionally found on the rear of some scabbard throat bands.[6] She links the practice, again, to introduction by the Franks.[7]

We will be returning to discuss many of the aspects of rings, beads and scabbards raised by Evison, elsewhere in this book.

Vera Evison 1976

Evison wrote a further paper on ring-swords, beads, etc, with an article published in Archaeologia (volume CV, 1976) for the Society of Antiquaries of London adding some other items that she was unaware of previously. Throughout, she compares continental and Scandinavian finds to their English equivalents and everything is subject to the, then, new dating of the Sutton Hoo, Mound 1 ship burial.

She begins with a discussion of some ring-sword pommels from France and continues with the rings from the Sutton Hoo shield and notes the reasons why they are thought to belong to the shield;[8] she recalls that a loose ring ensemble from Helgō, in Sweden is in some ways similar, as does the one from Niederstotzingen, grave 9, which was not actually found next to the sword in the grave.

She comments on several other ring-swords, but also notes a pommel which closely resembles several from Kent[9] but was found in Langeneslingen, Westphalia, Germany.[10]

Evison carries out a re-examination of the reports on the Coombe sword, and reports, "There is no reasonable basis for the assumption that all the objects came from a single burial, or that the second sword no longer extant was a figment of the excavator's imagination." She also disputes the claim regarding the cremation and says that it is more likely that inhumations and cremations were excavated by the labourers together. She is critical of some of Davidson's and Webster's[11] assumptions – concludes that "…it would be prudent the Coombe sword as unassociated."[12]

---

[1] Evison, 1967. 63. We will discuss ring-swords further later in this volume.
[2] Evison, 1967. 63.
[3] Evison, 1967. 66 to 67.
[4] Evison, 1967. 78 to 79.
[5] Evison. 1967. 80 to 81. See also 67.
[6] Evison. 1967. 65.
[7] Evison. 81. Based on Werner's work.
[8] Evison, 1976. 304.
[9] It's shape is very similar and on one side it has the three ring-and-dot motif.
[10] Evison, 1976. 305 to 306.
[11] Davidson and Webster, 1967.
[12] Evison, 1976. 307.

She notes interesting design parallels between the Style II interlace design on the Crundale Down sword and that on one of the swords from Valsgärde 8 in Sweden.

Evison also points out that some of the white 'beads' – cylinder shaped – used to seat cloisonné bosses on (as Niederstotzingen grave 9) can also be found as beads on women's necklaces and gives examples.[1]

Evison adds some new beads to those she mentioned in her previous paper, including a square one with central circular perforation of bone from Brighthampton grave 15. She says, that, "...the sword bead custom already current on the European mainland could have been brought to this country by a Germanic swordsman as early as the beginning of the fifth century."[2]

Wilfried Menghin 1983

Wilfried Menghin's book *Das Schwert im Frühen Mittelalter*[3] began its development as Menghin's doctoral dissertation and was originally called, "*Zweischneidige Langschwerter aus Germanischen Grabern des 5. Bis 7. Jahrhunderts n. Chr*".[4] The work was completed in 1971 and after some revisions and additions it was published in 1983 as a book. Although since 1983 much has happened and there have been new discoveries and possibly some changes in thinking, it is still the most comprehensive survey of the Germanic sword that has been written and has to be consulted by anyone seriously looking at the subject.

The book includes a brief consideration of the history of the long sword (spatha) until the end of the Roman period. Menghin then divides his chosen period, 5th to 7th centuries, into 6 time periods (*zeitgruppe*) A to F and provides corollary evidence to link the sword finds to. Next he moves on to discuss types of sword grip concentrating at first on pommel types and he identifies seven main types. Menghin then takes a look at handgrips and hand guards before going on to make a very detailed consideration of various aspects of scabbards. He considers the taping of scabbards, throat decoration, edge-strengthening, chapes, suspension methods as well as other decorations. Menghin carefully provides classification, often of his own, for all of the above aspects of swords and their furniture[5] and illustrates the book throughout with excellent line drawings, photographs, pictures of grave assemblages, chronological charts and distribution maps.

Menghin records sword-beads and sword-pyramids finds but does not really attempt to discuss their origins or purposes, nor does he really try to explain the ring-sword phenomenon.

What he does magnificently is to include pictures, diagrams and information on all the notable Germanic swords in northern Europe and Italy known at the time of writing, including those from Anglo-Saxon England. Details of each grave and associated finds are also provided. His classifications of sword parts and scabbard furniture must surely replace the much older and incomplete ones by Behmer and will be used in this book. Menghin's charts and illustrations are very clear and provide an excellent starting point for any study of sword of this period.

R. F. Tylecote and Brian J. Gilmour 1986

Tylecote and Gilmour published their book, *The Metallography of Early Ferrous Edge Tools and Edged Weapons* in 1986, and it considers a selection of edged tools and weapons found within the British Isles. About a third of the book is given over to the study of swords, some 39 in all, 33 of which are Anglo-Saxon. At least 25 of those studied are relevant to the main theme of the current volume – swords from the 5th to 7th century.

---

[1] Evison, 1976. 312 to 313. The examples of similar beads on women's necklaces provided are Risely (Derbyshire), Horton Kirby (Kent), Riby (Lincolnshire) and Staxton (Yorkshire).

[2] Evison, 1976. 315.

[3] 'The Sword in the Early Middle Ages'.

[4] 'Double-edged long swords from Germanic Graves of the 5th to 7 Century AD.'

[5] Most of which we will adhere to in this book.

Tylecote and Gilmour provide a very detailed metallographic examination of all the selected swords, providing full descriptions and drawings of the construction of the examples chosen, taking full advantage of radiography. They determine the metallic composition of their samples and the Vickers hardness from the edge and the core of each blade. They do not show any of the patterns from the blades in full length, which is unfortunate, but are able to give some idea of the variety of construction methods used to manufacture the weapons and an insight into the variety of patterns used by the smiths of the time and provide some indication of the skills needed to make them.

Two of the swords that they look at are non-patterned, according to their definition, although those swords are 'piled' or laminated. They would appear to be the only authors in this survey of literature to provide information regarding 'non-patterned' swords, which is very useful and possibly shows that all swords of the period were very complex, even those without a 'pattern'.[1]

Although they are only really concerned with blades, as far as swords are concerned, this is a pioneering work.

Janet Lang and Barry Ager 1989

Janet Lang's and Barry Ager's important paper appears (along with several other interesting essays) in the 1989 book, *Weapons and Warfare in Anglo-Saxon England,* edited by Sonia Chadwick Hawkes. Their study subjected 142[2] sword and seax blades from the whole of the Anglo-Saxon period, mostly belonging to the British Museum, to radiography in order to see how many were pattern-welded to establish what patterns were involved and to attempt to determine whether the blades were of insular or continental construction.

The essay includes a history of pattern-welding scholarship as well as a brief survey of the development of the process up to and including the Anglo-Saxon period. The process of radiographing the swords is described in some detail as is the machine and the film that was used. The weapons included in the study are all listed with find details and whether they were patterned or not.

The results of the radiography are included in tabular form and the frequencies of pattern-welding according to date are included in a number of different charts. Interestingly, the frequency of patterned swords rises from around 44% during the 5th century to 100% by the 7thand then falls off to 45% during the 9th to 10th centuries.[3] Ager and Lang were aware that the three types of patterns described by Neumann in 1927[4] were not sufficient to describe all the patterns that they were seeing on the swords that they looked at and so they established a new system of their own. Their classification makes the number up to seven different patterns and will be discussed elsewhere in this book.[5] The frequency and type of patterns on continental swords are compared to those found in England and this subject is discussed and some conclusions reached as to which came from France and German and those that may have been made by Anglo-Saxon smiths.[6]

The authors were interested in the purpose of pattern-welding. They carefully consider the suggestions of other writers such as Tylecote, Salin, France-Lanord, Gilmour and others and come to the conclusion that the process was 'largely decorational' although much skill was required in fabricating this type of sword.[7]

---

[1] We will discuss this further in the chapters called, *Pattern-welding* and *Sword Blade Patterns.* Gilmour includes some non-patterned swords in some of his later work.

[2] Lang and Ager, 1989. 85. Some blades were too fragile and only 139 swords are listed as being used in the concluding table (116 to 199).

[3] Lang and Ager, 1989. 107. There are no swords from the 8th century included in this study.

[4] Lang and Ager, 1989. 87. This subject and its history is described in the chapter on Pattern-Welding in this book.

[5] Lang and Ager, 1989. 87 to 88.

[6] Lang and Ager, 1989. 110 to 111.

[7] Lang and Ager, 1989. 110.

They also spend time considering the inlays and inscriptions that are sometimes found on later blades of the 9th and 10th centuries but they, because of their late date, are outside the scope of this book.

Peter Bone 1989

Peter Bone's chapter, *The Development of the Anglo-Saxon Swords from the fifth to the Eleventh Century*, is also included in Hawkes's, *Weapons and Warfare in Anglo-Saxon England*, and provides a reasonable, if very brief, summary of the history of the early English sword.

Bone begins by stating that the weapon was a development of the Romano-Celtic and Sarmatian cavalry swords used from around 100-400 A.D. and that these were frequently of diamond cross-section and could be used for both cut and thrust.[1] He proceeds to explain that the blades change very little in shape throughout the whole of the Anglo-Saxon period, so that it is the hilt that tends to provide dating and ethnic origin evidence. Bone says that Anglo-Saxon swords could be in service for a long time and could become 'precious heirlooms', he does remark that during the long life of a sword it will probably require re-hilting and that even when new hilts and swords were not necessarily both made by the same man. He believes that there is evidence of a 'thriving export trade in both blades and finished swords from the Frankish Rhineland' and 'growing evidence' for sword-blade production in England, especially Kent during the 6th and 7th centuries, based on the work of Lang and Ager.[2]

Bone then provides a diagram which illustrates the sword terminology that he uses and moves on to discuss the way that the hilt changes during the 5th to 11th centuries. He includes ring-swords but does not attempt to discern their purpose. He also mentions pattern-welding. Bone briefly describes scabbard construction and methods for wearing the sword and suggests that they were mostly worn using a baldric.

Included within this short chapter is a paragraph on the *Sword in Society*, where he concentrates on the high value of some swords that have decorated hilts and accoutrements. He says that the poetry tells us that the sword is the weapon of the hero but does not give any examples and thinks that their relative scarcity in pagan burials probably does illustrate their relative scarcity in life. He concludes this particular paragraph by stating that as the sword cannot be used for hunting it "…would have been a potent symbol of the aristocratic warrior class, marking its bearer as both wealthy and a warrior."[3]

Bone's final paragraph looks at how the sword performed in use and he is really the first of our commentators who may have actually handled realistic recreations of such a sword. He remarks that the slashing blow can be devastating to a human and that there is evidence of its efficacy from cemeteries. He feels that it is primarily a cutting weapon although thrusts are possible but could be tiring. He does say that using a sword in any way is 'tiring' and while that is the case, it is true of any hand held weapon of the early Middle Ages.

---

[1] Bone, 1989. 63.

[2] Bone is referring to Lang and Ager, 1989.

[3] Bone, 1989. 69.

Brian Gilmour 1990

Brian Gilmour's unpublished, doctoral thesis, *Developments in Iron Smithing and Decorative Welding Techniques Found in Anglo-Saxon Swords and Related Edged Weapons,*[1] takes much of the information included within Tylecote and Gilmour, 1986 and extends it. The work closely studies and examines the same 39 swords and uses many of the same illustrations and diagrams to show how they were made, although there is considerably more data included in this work. As with the previous work, Gilmour reports on 7 Anglo-Saxon seaxes, which he refers to, often, as scramasaxes, 4 spearheads and one Malayan *kris*. Although he does mention other parts of swords he focuses most of his attention on the blades of the weapons that he studies.

The thesis is in two volumes. The first contains a brief history of the development of swords from the Bronze Age up to and including the Anglo-Saxon period, the manufacture of iron swords, including the materials they were constructed from, a description of the methods used to examine the blades, including their metallography, the scientific analysis of the selected weapons and Gilmour's catalogue of a total of 319 weapons that he had examined (which includes some spearheads and seaxes). Perhaps one of the most important aspects of Gilmour's work in the catalogue was the careful recording of the blade patterns that he had seen. Volume 2 includes the records of data, charts, tables and 3D construction diagrams of many of the blades.

Within the thesis, Gilmour discusses many important aspects of the sword including the presence or absence of fullers on blades and the contexts in which swords are usually found. He also discusses where swords might have been made and believes that there are good reasons to think that many of the swords found in England were made there. He describes fully the processes involved in pattern-welding and explains how that differed from the eastern style of blades made from *wootz*. He writes about sword references in Old English literature and Icelandic sagas and looks at what Arab and Middle-Eastern commentators had to say about swords, their manufacture and the materials that made them.

The work contains a very full description of the chemistry of iron and its alloys and how they relate to the making of the sword and of how different heat treatments would have been used to improve the blades, both the theory and examples from surviving weapons. He explains how such treatments leave traces, many of them not visible to the naked eye, within the metallurgy of the blades and how they can be detected using modern techniques. Gilmour explains what smelting techniques were available to peoples of that time, in Northern Europe, along with the types of ores that would have been available. He shows, too, that it is possible that there was a sophisticated trade in types of iron alloys from a fairly early stage.

Gilmour's work has provided a solid scientific basis upon which we can build our knowledge of the sword in Anglo-Saxon England.

Ewart Oakeshott 1964, 1974 and 1991

Ewart Oakeshott wrote two different books about the medieval sword, the first, *The Sword in the Age of Chivalry* is devoted to weapons from 1100 to 1500 A.D. but he does briefly consider the forerunners of the later sword in the first chapter, although not in any depth. In his introduction he tells us, "The "Knightly" sword is derived, via the swords of the Viking and Migration periods, from the long iron swords of the pre-historic Celts."[2] He explains that as Behmer, Petersen and Davidson have comprehensively dealt with the earlier swords in their work, he will only discuss this class of sword where it is directly relevant to the development of the later sword.[3]

---

[1] PhD Thesis, University College London, Institute of Archaeology.

[2] Oakeshott, 1964. 12. He tells us, too, (15) that Behmer and Petersen hardly considered blade types – they concentrated on hilt types and scabbard decoration. He could have included Davidson in that comment as well. See also, Oakeshott, 1991. 1

[3] Oakeshott, 1964. 13.

Oakeshott discusses what he refers to as 'Migration period' swords and reduces Behmer's hilt typology of nine types and sub-types to four,[1] and proceeds to state that although the hilts varied a great deal, the blades were much less diverse and remained fairly standardised, apart from a broad, parallel sided type with 'spatulate' point and a, comparatively slender type with a more acutely pointed tip. He believed that the, "...Norse peoples themselves made a distinction, for they had four words for a sword; two of them, *svaerd* and *maekir*, seem to be applied to these two kinds of blade; svaerd, the more usual term, seems to apply to the broad slashing blade, while maekir denotes the slender pointed type."[2] He mentions, too, that fullers could be singular, multiple or absent and that the swords did vary in size and weight, as did the men who used them. He moves on to look in great detail at his real area of interest, the 'Knightly' sword.

In *Records of the Medieval Sword,* Oakeshott tells us more or less the same things about the earlier swords, although he does mention pattern-welding, which was apparently so "...rich and rare that they were considered to be royal treasures..."[3] and doesn't discuss the *svaerd / maekir* words. He does add seven photographs of swords from bogs but says nothing else about swords from the 5th to 7th centuries, or even swords from the Viking period.

However, Oakeshott did provide much fuller discussion of 'Migration period' swords in another of his books, *The Dark Age Warrior*, which was written, primarily for a younger audience and looks at many aspects of the warrior and his weapons during the 5th to late 7th centuries. He begins his exploration of swords of the period by looking at how the hilt would be constructed and then discusses pattern-welding at some length, providing a retelling of the Weland/Velent myth along the way. He does say that some patterns were given names, *Blodida*, or 'Blood-eddy' which is a pattern that looks like waves in water, *Ann*, the herringbone pattern and *Idvarp*, which is, apparently, the name for the pattern that has long parallel stripes 'down the length of the fuller' and means 'gut weaver'.[4] Unfortunately, he provides no reference for these imaginative interpretations.

Oakeshott, again, claims that blades were fairly uniform and were made in only a few centres of production and that it is only the hilt forms that can tell us the swords date and their regional use.[5] He uses his simplified Behmer typology to show development of the hilt types. In this book he repeats his claim that the names *Svaerd* and *Maekir* differentiate two types of blade[6] He mentions ring-swords and 'peace-bands', and the possibility that ring-swords were the swords of kings, then largely dismisses that idea and comes to the conclusion that the rings were a reward for valour or were a badge of rank.[7] He moves on to give a fairly full account of scabbards, their construction and decoration, including sword-beads, 'life stones' and sword bosses, all well illustrated with his own drawings.

Oakeshott tells the story of a number of legendary swords and heroes, mostly, but not all, from the later Norse sagas. Finally, he considers marks on swords, together with the ideas of naming a sword and maker's marks.

In many ways, this account of the 'Dark Age' sword is a comprehensive one and provides an easily readable and short introduction to the subject.

---

[1] Oakeshott, 1964. 25.

[2] Oakeshott, 1964. 26. We will discuss this idea further, elsewhere in this book and also look at the vocabulary of the sword.

[3] Oakeshott, 1991. 1.

[4] Oakeshott, 1964. 56.

[5] Oakeshott, 1964. 57 to 58.

[6] Oakeshott, 1964. 63.

[7] Oakeshott, 1964. 65.

Anne Nørgård Jørgensen, 1999

Anne Nørgård Jørgensen's book, *Waffen und Gräber: Typologische und Chronologische Studien zu Skandinavischen Waffengräbern 520/30 bis 900 n.Chr. (Weapons and Graves; a typological and chronological study of Scandinavian Weapon graves from 520/30 until 900 A.D.)* is an extremely useful consideration of martial material found within graves in Norway, Denmark and Sweden. Unfortunately Finland is not included. Although the book does not discuss much Anglo-Saxon material it does cross reference the Scandinavian finds to those from Merovingian areas.

Jørgensen provides many charts showing the chronology and find types from all the regions included and provides welcome illustrations of many of the finds themselves. The book is about all weapon types, not just swords, but it does relate them to categories of seax, spear, axe, shields and belt and baldric fittings, over time and it does this regionally and sometimes even on a more local level, for instance many of the Norwegian finds are compared closely to those from the island of Gotland situated in the Baltic Sea. She refers to the work of previous scholars and also to recent correspondence analyses of the weapon finds and uses the charts within the book to show the correlations. Jørgensen demonstrates where the scholars have differences over their chronologies within her charts and provides a catalogue of all the Scandinavian finds that she refers to, most of which, if not all, are included as line drawings. She discusses, too, the military systems and societal structures that would have prevailed during the period under study and how that will have altered over time.

Jørgensen's work provides important chronological and typological references that can be used to help illustrate sword development in Anglo-Saxon England and contrast it with developments in Scandinavia during the same period.

Richard Underwood 1999

Richard Underwood's book, *Anglo-Saxon Weapons and Warfare*, is not just the result of reading a great deal of literature and examining museum specimens, but also of practical experiments and the handling of reconstructed weapons. This makes Underwood's writings particularly valuable.

Although his well-illustrated sword chapter is relatively short, he covers quite a lot of ground, from pattern-welding techniques, simple hilt and grip classifications, runes on swords, ring-swords, scabbard construction and scabbard decoration, the method of attaching the sword to the belt and the handling and use of swords. Underwood has included two interesting sequences of photographs which show how the weapon may have been used in combat.

Esther Cameron 2000

Esther Cameron's *Sheaths and Scabbards in England AD400 – 1100* tells us a much about surviving sword scabbards, their construction, decoration and quite a bit about the swords themselves. Her work in this book is limited to finds from England. Cameron explains in detail the cellular structures of leather and wood and how they degrade over time in soil and explains what we can learn about the originals from the remains that are found in graves and elsewhere.[1] She tells us, too, how contact with the metals that the swords were constructed from affects the way that the scabbard materials change during their time in the ground,[2] and how the metals themselves are altered. She includes a thorough discussion on the types of wood, leather and vegetable and animal fibres used to construct scabbards.[3]

---

[1] Cameron, 2000. 8 to 11.

[2] Cameron, 2000. 11 to 13.

[3] Cameron, 2000. 17 to 21.

Cameron devotes a significant amount of her study to the scabbards of the 5th to 7th centuries and covers scabbard bindings, carving, foundation moulding, strap slides, chapes, edge reinforcers, throat bands as well as decorative additions, such as beads and raised discs. She brings in evidence from earlier traditions where relevant and from outside England when there is a relationship to the Anglo-Saxon culture.

She concludes her study with a discussion of changing trends and with an illustrated corpus of known scabbards and sheaths. Her contribution to the knowledge of sword development and its accoutrements is important and indispensable.

Paul Hill and Logan Thompson 2003

Paul Hill's and Logan Thompson's short paper, *The Swords of the Saxon Cemetery at Mitcham,*[1] is a study of the seven surviving swords from the cemetery at Mitcham, Surrey. The authors set out the circumstances of the cemetery's discovery and exploration and show that it is likely that many swords and other objects were lost to science.

They look at each sword in detail and use some comparisons with other contemporary finds to further both their and our understanding. They make some comments about how the weapons would have been used and believe that they were 'point heavy and rather clumsy'[2] and agree with Härke's suggestion that warrior types were of larger physiques.[3] They also believe that most swords at this time probably had fullers [4]and call for a 'nationwide project' to x-ray surviving swords of the period.[5] However, there is no guarantee that such a method would reveal fullers but it would, in many cases, show if there was patterning on the blade.

The importance of the paper is in making the details of the swords studied within the paper available.

Lena Thålin Bergman and Birgit Arrhenius 2005

Lena Thålin Bergman began working on *Excavations at Helgö XV. Weapon Investigations: Helgö and the Swedish Hinterland* more than 30 years before it was finally published (it was originally intended for publication during the 1970s) and it has apparently suffered from being worked on for this length of time.[6] Birgit Arrhenius edited Thålin Bergman's writings and notes and has turned them into a coherent whole. Nevertheless this is an important work and helps link the making and using of swords from Scandinavia to Anglo-Saxon England and the Merovingian Continent.

The work concentrates on weapons from various parts of Sweden including the island of Helgö. The author reports mainly on swords and spearheads but does include some other weapons, too. The periods covered range from the Roman Iron Age to the late Viking period but most of the weapons included come from the later end of the spectrum.

A variety of techniques were used to study the weapons from x-rays, spectroanalysis and metallography and the work is well illustrated with photographs, diagrams and charts. Some intriguing samples mentioned within the text are not easy to locate and are not catalogued properly, which is a pity. Nevertheless, much of the work is carefully explained and clear.

The author refers to an impressive databaseof Swedish swords that have been examined as well as those that she has considered herself.

---

[1] Published in the Surrey Archaeological Collections, 90. 2003.

[2] Hill and Thompson, 2003. 158.

[3] Härke, 1989.

[4] Hill and Thompson, 2003. 158.

[5] Hill and Thompson, 2003. 159.

[6] See, Editor's foreword, in Thålin Bergman and Arrhenius, 2005. 7. Bergman's treatise was never published and, as far as I know, did not have a finalised title. She did publish a paper, Svensk fïrhistoriskt järn – ett forskningsprojekt, in Jernkontorets Annaler 151, Stockholm, under the name Lena Thålin.

The book discusses the chemistry of iron and its alloys and points out that smiths were very capable of discerning different types of metals for different purposes and makes some valuable comments on the development of pattern-welding.

While the book does not directly refer to the swords in Anglo-Saxon England often, it does underline the high quality of, 'a high percentage of the weapons found in Sweden,[1] and shows that similar styles of work were being used and probably being made there.

Marcin Biborski and Jørgen Ilkjær 2006

The two volume work by Biborski and Ilkjær on the swords from the Illerup bogs constitutes an extremely important contribution to our knowledge of the pre-Anglo-Saxon, continental sword and its relationship to the earlier Roman weapon. *Illerup Adal: Die Schwerter: Textband / Katalog, Tafeln Und Fundlisten: volumes. 11 & 12,* is part of a multi-volume exploration of the finds from the bog.[2]

The first task volume 11 deals with is to describes all the different concentrations of finds of swords and other associated artefacts. The work proceeds to looks at all aspects of the recovered weapons and supplies full explanations of the characteristics of blade types, hilt furniture, scabbard fittings and decoration and suspension methods. All weapons are assigned to particular categories and fully discussed within that specification. Volume 11 concludes with a discussion of the Illerup swords put into the context of other bog finds from modern Denmark and Northern Germany.

Volume 12 contains the catalogue and tables of all the finds. Both books are well illustrated, but volume 12 systematically provides images of all the sword finds and their constituent parts; photographs and superb line drawings are both provided.

One particularly valuable contribution the work makes is the way that it shows and discusses the large variety of patterns the blades were made from. Many of the swords were deposited separately from their scabbards and the bog at Illerup, for the most part, provided ideal conditions for the preservation of the blades.

Janet Lang 2007

Janet Lang's unpublished, doctoral thesis, *The Rise and Fall of Pattern Welding: an Investigation into the Construction of Pre-medieval Sword Blades*, is an investigation into the practice of making pattern-welded and other sword blades, from the earliest days of the sword-making. She sets herself a number of tasks and says that, "The aim of the present study is to trace the development of the construction of iron swords up to AD 1000, especially with regard to the number of components and the technical changes in construction, together with some consideration of the social changes which could have influenced them."[3]

Lang investigates the way that iron and iron alloy swords were constructed and uses a number of methods during her investigation. She selected a number of swords from the British Museum collection including those from the then recent dig at the Buckland Anglo-Saxon cemetery near Dover, but also some from the Iron Age together with some Roman swords. Carefully selected samples were taken from the swords and these were subjected to a number of techniques, including x-ray dispersive energy spectroscopy (EDXA), X-ray fluorescence spectrometry (XRF), micro-hardness testing and some mechanical tests. The samples were also carefully examined using high power microscopes to discover their metallographic characteristics and the processes that had been used during the blades fabrication.

Lang had made a series of test bars that, as far as possible, mimicked the structures that she found within the blades and she then subjected these to a series of mechanical tests to see how they behaved. She admits that

---

[1] Thålin Bergman and Arrhenius, 2005. 82.
[2] At the time of writing, 14 volumes have been published.
[3] Lang, 2007. 12.

there are a great many limitations to this process, partly because of the difficulty of obtaining ferrous alloys of the same types as in the ancient swords, partly because of the limitations of the modern smith (it seems that she made most of the samples herself), the type of hearth that she used and because the mechanical tests did not directly relate to the way that the swords would have been used in life.[1] Nevertheless, she believes that some useful insights were gained into the mechanical advantages of some forms of pattern welding over wrought iron.

Lang discusses the history of pattern-welding and describes the attention that it has received from scholars and spends some time showing how the processes involved in manufacturing 'damascus steel' differ from pattern-welding. She explores the social aspects of the sword, including how violence was a part of life and how warrior culture was a fixed part of society. She considers smith myths from various parts of the World and how the smith worked and the metal trade was organised in ancient Europe.

A clear and comprehensive look at the way that iron was made and its metallurgy is included, together with the processes and techniques used in making swords during the extensive time period that she covers.

Despite its title, this thesis is about a great deal more than pattern-welded swords and is one of the only works listed here that gives non-patterned swords of the 2nd to 7th centuries a fairly detailed consideration. One other intriguing section deals with the actual handling of a sword based on a scientific, not a personal analysis, taking into account several factors, including blade shape, length and width, pommel weight, size of the grip and other factors that she considers relevant. While this cannot tell us exactly how a particular sword would perform, it is a brave attempt to tell us something about sword use and is the only such account in all the works listed here. If her formula had been married to the actual handling of accurately produced replicas by an experienced swordsman, the results might well have been more revealing.

Overall this is a valuable and well-illustrated contribution to the study of the sword, although she strangely neglects the 5th to 7th century swords in some of her discussions, preferring to write about aspects of La Tène, Roman and Viking ones.[2]

Christian Miks 2007

Christian Mik's two volume study of Roman swords, *Studien zur Römischen Schwertbewaffnung in der Kaiserzeit*[3], is an extremely comprehensive and thorough study of Roman swords, mainly from the beginnings of Imperial Rome (50 B.C.) until the start of the 6th century. Miks does include a survey of the earlier swords of the Roman Republic the beginning of his book, but this area is not the main focus of his work. His efforts are particularly valuable for us in our pursuit of the Anglo-Saxon sword, in that the book covers the development of the early Germanic two-edged weapon right up to the early Anglo-Saxon period.

Miks has catalogued and illustrated all known 'Roman' swords, put them into developmental sequences, considered the two main types of sword – gladius and spatha and produced well-illustrated typologies for the blades, hilts, scabbards and their associated furniture and fittings. He provides an analysis of the techniques that Roman smiths used to construct their blades and the types of iron that they utilised and gives cross-sectional diagrams to indicate construction.

In his typologies and classifications Miks uses charts to show where the swords were found and their dating and their numbers in his catalogue, so that it easy to cross reference the finds and follow his commentary. This scheme is also used when he looks at other aspects of the swords, such as blade inlays, scabbard and hilt

---

[1] Lang, 2007. 175.

[2] Lang, 2007. For details see, 119, 120 and 121. Lang goes into considerable detail concerning blade shape and dimensions and only rarely refers to early Anglo-Saxon swords in the discussion.

[3] 'Studies on the Roman sword weaponry under the Empire'.

fittings. Each part of the sword has its own full length chapter in which it is discussed along with its associations, such belts, baldrics and suspension methods.

His typology for the Danish bog swords is different to that provided by Biborski and Ilkjær and is a little simpler. He, like many of our other authors, considers all such swords to be 'Roman'; this is possibly not the case, at least, they might not all have been made under Roman control.

Miks's two volumes are a wonderful resource and an example of great dedication and scholarship.

Brian Gilmour 2007 - 2010

Brian Gilmour was responsible for at least two seminal papers concerning Anglo-Saxon swords during the years 2007 to 2010. The first, *Swords, Seaxes and Saxons: Pattern-Welding and edged Weapon Technology from Late Roman Britain to Anglo-Saxon England*,[1] discusses why some weapons have survived and shows that 'Germanic' swords were certainly being deposited before the 5th century,[2] presumably by 'Germans' employed within the Roman armed forces, although he doesn't show that they were ever included within burial rites.

Gilmour goes on to consider iron alloys, welding and composite construction and discusses the fact that forge-welded blades were being made in Iron Age Europe and suggests ways in which the benefits of pattern-welding could have been recognised and the techniques developed. He takes a detailed look at the construction of some pattern-welded 'Roman' swords that have survived, before moving on to discuss Anglo-Saxon swords of the 5th to 7th centuries. He establishes a continuity between the earlier bog swords, and the sword in Anglo-Saxon England and its contemporaries on the Continent and proceeds to think about the way that some patterns may have been achieved, and comes to the same conclusions that Lang and Ager came to regarding the supposed Continental habit of grinding fullers as opposed to the Anglo-Saxons using fuller blocks.[3]

Gilmour writes about the technique in later centuries, too, and in other areas, such as the Middle East as well as presenting the contemporary written evidence for consideration. He does provide some excellent diagrams of specific Anglo-Saxon swords which indicate the types of ferrous metals that they were constructed from and fully explains the methods used to arrive at his conclusions.

One of his most interesting insights comes towards the end of this paper when he makes connections to the visual effects of pattern-welding and its similarities to eddies in water.

In 2010 Gilmour published his essay, *Ethnic Identity and the Origins, Purpose and Occurrence of Pattern-Welded Swords in Sixth Century Kent: The Case of the Saltwood Cemetery*.[4] In this work Gilmour analyses the eleven swords found in the Saltwood cemetery in close detail, and attempts to come to some conclusions concerning where and when the swords were made. He grades them by the quality of their workmanship into three categories and discusses the varying types of ferrous materials within their structures. The swords are quite remarkable for the variety of different patterns that they contain.

He does determine that the patterns were no mere by-product of the manufacturing process but were meant to be seen and shown off. He suggests, too, that by the 5th century there were specialised centres producing particular alloys of iron for the purpose of fabricating swords.

Gilmour, again, speculates that there could have been a pagan connection with the patterns and water and that is a possible reason for the technique becoming less popular during the Christian period.

---

[1] In Henig and Smith, 2007.
[2] Gilmour, 2007. 92.
[3] Lang and Ager, 1989. 112 and Gilmour, 2007. 101.
[4] In Henig and Ramsay, 2010.

Svante Fischer 2008

Svante Fischer's book, *"Les Seigneurs des Anneaux" (The Lords of the Rings)* concerns itself, mainly, with ring-swords and swords with runic inscriptions and catalogues thirty-seven such examples from graves in Kent, Oxfordshire and the Merovingian areas of Northern France and Germany. It also considers the phenomenon of balances found within weapon burials.

Fischer uses Evison's categorisation of rings and provides a full discussion on this subject. Wherever possible, he provides photographs and line drawings of each of the graves concerned with all their contents displayed and commented on when relevant.

The book brings together a great deal of comparative information.

Mike Loades 2010

Mike Loades book, *Swords and Swordsmen* is a personal commentary on types of swords and their development from Ancient Egypt to the American Civil War. He discusses the uses and the contexts in which the swords were used, bringing in the personalities of historical figures wherever he can and he has often had replicas of various types of weapon made.

He devotes a chapter of his book to *The Sword of King Raedwald, King of the East Angles* and uses the chapter to discuss Anglo-Saxon swords from the 600 years that they controlled what became England. He had a fairly typical early Anglo-Saxon sword made by the smith, Hector Cole, although Loades does not reveal what find his replica is based upon.

He considers the Mound 1 sword in some detail and attempts to put its constituent parts, such as the decorated pommel and scabbard, into a historical context. Some of the things that he mentions concerning the construction of the Sutton Hoo sword and its elements are not included in the report by Bruce-Mitford. He does, to some extent, describe the process of pattern-welding, something that he experienced for himself while filming Hector Cole making his blade.

During his discourse, Loades produces some interesting insights regarding the handling and use of the Anglo-Saxon sword in combat and brings in evidence from the sagas and other literary accounts to reinforce what he says. Loades has hands-on experience with many types of weapon and he uses that practical experience to interpret how swords were used, both on their own and in combination with other weapons both in single combat and in battles.

Simon James 2011

Simon James's book, *Rome and the Sword*, is quite a remarkable history of not just the Roman sword, but of Rome, its Empire and its relations with its neighbours. He concentrates much of his attention on the Roman soldier and his equipment, while at the same time relaying the events that took place and how the weaponry responded to them and the changing conditions in which they were employed.

James points out early on in his work that there was never such an organisation as the 'Roman Army', rather that there were many independent units paid for by powerful men and that the Roman Empire was largely run on private enterprise together with a hefty helping of violence used against both Rome's enemies and, quite often, other Romans.

He touches on, and often fully discusses, many topics within the book, including manufacturing techniques, mass-production, pattern-welding, armour, methods of fighting, the military reforms, Rome's enemies and the way that their equipment affected the development of Roman arms. He explains the evidential basis for all of his claims and is able to show how the Germanic sword developed from those of the Romans and thereby demonstrates that the Roman weapon did not die out. He asks, "...but at what point did such weapons cease to be Roman swords and become Roman-style Frankish or Anglo-Saxon weapons, in a process of convergence, continuity and oblivion?" He then says, "A fine pattern-welded blade was buried early in the seventh century with an Anglo-Saxon king at Sutton Hoo in eastern England. Its blazing gold hilt reflects the glittering weapons and obsession with sword-play of the epic poem *Beowulf*, set in heroic, pagan Scandinavia; the blood-red garnets set into the pommel attest tastes which spread to the Germanic world from the steppe. Yet the Roman association may not have been entirely forgotten, for with the sword was buried a helmet clearly imitating late-Roman precursors."[1]

James, like Miks, has assumed that nearly all of the swords found in the Danish bogs were Roman.

Janet Lang 2012

Janet Lang's careful and exhaustive metallographic study of five swords[2] from the most recent excavations at the Dover Buckland cemetery is an extremely important addition to the knowledge base of the metallurgy of early Anglo-Saxon swords. As she points out, the study has great limitations, only small sections of metal were allowed to be cut from the often severely corroded bade remains, but much was achieved by Lang.

She describes the analytical tools employed and the processes that the samples went through. Each sword is dealt with and discussed in turn and its constituent metals analysed together with the blade structure. Lang provides a summary for each blade and includes comments concerning the perceived effectiveness of each weapon.

The individual analysis is followed with a very full treatment of the swords, the chemical composition of the blades, their inclusions, the methods of producing iron, the processes of making each type of metal and the methods of construction employed. The discussion also contains descriptions of possible heat treatments and where there is evidence of this, together with other processes such as etching, grinding and polishing.

In the same volume, Barry Ager's report on the swords finds and comparisons with other cemeteries, and the work carried out on the scabbards by Fleur Sherman, Hayley Bullock, Eric Nordgren and Heidi Lesseur, together with Lang's work, provide a model of sword analysis and reporting that it is to be hoped will be followed by all investigators in the future.

Alan Williams 2012

The Sword and the Crucible; a History of the Metallurgy of European Swords up to the 16th Century by Alan Williams investigates methods of making swords from iron and steel and their metallurgy from the beginning until well after the medieval period. He explains the various methods of smelting iron and steel not just in Europe, but also elsewhere in the world and how some of those ideas might have travelled. However, his study confines itself to the metallurgy of sword from Western Europe only and does not include Muslim Spain.

He uses a number of methods, including radiography and microscopy to analyse the metallurgy of the sword blades and he provides a very thorough description of the ferrous particles that can occur in the metals and believes that the smiths often had a quite sophisticated knowledge of the chemistry. He discusses various qualities of metals, too, such as hardness and toughness and how these can be achieved through different heat treatments and shows where there is evidence for it and where it was lacking. His description of the processes involved in pattern-welding and the ways in which it could have developed is fairly thorough and he considers some examples of the swords from the Nydam bog deposit and then moves on to look at a few swords from the Migration Period.

Although, to some extent, his main area of interest appears to be the mid to late Viking Period swords his careful consideration of the metallurgy of earlier swords adds to what we know of blades of our period.

---

[1] James, 2011. 276. See Mortimer, 2011. 24, for arguments disputing the supposed Roman origin of the Sutton Hoo helmet.

[2] In Parfitt and Anderson, 2012.

Susan Brunning 2013

Susan Brunning completed her doctoral thesis, *The 'Living' Sword in Early Medieval Northern Europe: An Interdisciplinary Study*, in 2013 and within it she set out to explore one main theme; that swords could be perceived as 'living' artefacts with their own biographies during the Early Medieval period, and her work seeks to establish what insights that concept can provide. [1] The thesis considers the physical and visual characteristics of swords, as well as their life histories and personalities.[2] Brunning examines the nature of the connection between warriors and their swords, paying particular attention to the idea that the identities of sword and owner could become intertwined, transforming swords into an 'extension of self'.[3]

Another task that Brunning sets herself is to compare and contrast the Anglo-Saxon material with that from Scandinavia and she demonstrates that she is well aware of the problems and possible pitfalls involved in such an approach and accounts for them in her reasoning.[4] She considers, too, how the uses of swords and their portrayal in iconography and within the literature may have changed and altered over time and in the two different regions.[5]

As far as actual swords are concerned, Brunning restricts herself to looking closely only at swords that have survived with some, or all of their metal fittings, including hilt and scabbard furniture.[6] These were examined to search for clues to see how they were used and curated.

Brunning provides a very valuable survey of the portrayal of swords in the iconography throughout the whole Anglo-Saxon and Viking periods in both England and Scandinavia; occasionally she does also reference relevant examples from other areas such as France and Germany. Her survey of the literature, which includes wills and laws besides poetry and other forms of writing, is equally thorough and useful. Both studies, along with what she was able to glean from the archaeological survivals enable her to suggest some insights.[7] One such is the extra value that appears to have been attached to old and ancient swords,[8] another is that men wore their swords frequently and certainly not just during times of war.[9] Her thorough study of wear and the curation of swords provide opportunities for taking her ideas further.

Brunning believes that the evidence shows that not all warriors were sword users but that among their number there existed different grades of sword carriers; she speculates that ring-swords are one example of swords that were only carried by a few, higher grade warriors.[10]

Brunning suggests that, at least as far as the poetry is concerned, Scandinavian poets provided a more detailed description of swords than is included within the surviving corpus of Old English poetry and that this might have been due to the possibility that the Old English examples were copied down by monks who lacked hands-on experience.[11] She finds, too, that Old English poetry is more linear than its Old Norse skaldic equivalent; in Old English, swords are portrayed in peace and war during the telling of the story, whereas in the Old Norse they tend to be referred to during episodes of action. She also found differences in the way that sword kennings are composed. Her conclusion is that swords are perceived more as 'battle tools' than as a

---

[1] Brunning, 2013. 14.
[2] Brunning, 2013. 186 to 187, 23, 233.
[3] Brunning, 2013, 261.
[4] Brunning, 2013. 50.
[5] Brunning, 2013. 51.
[6] Brunning, 2013. 60, 101,
[7] Brunning, 2013. 171 to 198.
[8] Brunning, 2013. 117 to 118, 137, 189, 235, 236,237, 238, 255 to 256, 260..
[9] Brunning, 2013. 133.
[10] Brunning, 2013. 245.
[11] Brunning, 2013. 217 to 218.

personal attribute in the Old Norse versions, where battle, heroism and glory are more closely assimilated, while the Old English poems tend to be more tragic and emphasise suffering.[1] Brunning is aware that this difference in emphasis could just be due to the limited range of sources that has survived.

As there are more surviving swords from Scandinavia than from England, she does speculate that Scandinavians possessing swords were probably more numerous than sword-bearing warriors in Anglo-Saxon England,[2] and that the weapon might have signified just a 'warrior' in Scandinavia, but an 'elite warrior' in England. Unfortunately, she provides no figures and dates here for comparison, and as burial with goods ceases during the 7th century in England but continues for several centuries more in Scandinavia it is perhaps as straightforward a comparison as it at first seems.

Another important finding is that swords, "… transmitted messages about status and identity…".Even when just being worn or carried, they were never 'passive'.[3] Well-known swords would have been immediately recognisable,[4] with hilts that could have actually been the 'face' of the sword, while the blade and scabbard formed the body.[5]

Brunning returns to the connection between swords and women several times during her writing,[6] and explains that where such associations are made, the women should be considered, 'irregular or even extraordinary' in literature, iconography and in the archaeology. Miniature sword amulets are found within women's graves, but there appear to be no examples of women buried with an unadapted sword, so far, in Anglo-Saxon England or Scandinavia, unless the woman was buried with a male.[7] Swords re-made into weaving battens are sometimes found in women's graves.

This thesis represents important progress in the way that we think about these ancient artefacts and will affect all future studies of swords of the period.

Ulrich Lehmann 2016

Ulrich Lehmann's work has added to our knowledge of the sword of the early middle ages considerably. Since 2009 Lehmann has been making papers of his work available while he worked towards gaining his doctorate. His book, Wurmbunte Klingen. Studien zu Konstruktion, Herstellung und Wetigkeit der frühmittelalterlichen Spatha in Westfalen, is an extension of that work and of his thesis.

Lemann's extremely well illustrated volume looks, in considerable detail, at a number of swords that were recovered from graves in Westphalia (Westfalen). The graves are dated from the early 6th century to the beginning of the 9th, although, throughout the study, Lehmann consistently divides the swords into an older group (6th and 7th centuries) and a younger one (8th and 9th centuries).

The components of the sword, the blade, scabbard, hilts and decorative elements are all extensively explored and this includes examinations of the manufacturing techniques and the materials involved in their fabrication, including findings on the various iron-alloys used within a blade and how heat treatments have been involved, or not, with the weapons. Lehmann, considers too, evidence of use and signs of repair during the lives of the weapons.

Ulrich Lehmann carefully explains his methodology and examines the use of x-radiography in looking at blades and considers its limitations. Where his work is of fundamental significance is in his use of computer

[1] Brunning, 2013. 219 and 258.
[2] Brunning, 2013, 258.
[3] Brunning, 2103. 226 and 246.
[4] Brunning, 2013. 231.
[5] Brunning, 2013. 233.
[6] Brunning, 2013. 35, 152, 153, 210, 219, 247
[7] Brunning, 2103. 248 to 249.

tomography. He discusses and uses two different techniques -- 3-D Röntgen computer tomography and 3-D neutron computer tomography to gain new insights into the manufacture and construction of patterned blades. Using these methods, in which many x-ray photographs are taken whilst the object is revolving, Lehmann has been able to build up layers of information concerning each of the blades in his study. When all the images are combined, in effect, a movie taking the observer through the various elements of the blade results and the construction can be seen without destroying any part of the blade.

Although Lehmann's book is limited in geographical scope it has advanced our knowledge of the construction techniques used by smiths to make sword blades throughout the 6th to 9th centuries and its lessons are applicable to all areas where such swords were made and used during those periods.

Other works with relevance to our subject

Three other books that have a bearing on the Anglo-Sword during the 5th to 7th centuries and its development are I. M. Stead's *British Iron Age Swords and Scabbards* (2006), Ian Peirce's *Swords of the Viking Age* (2004) and *Viking Swords; Swords and Social Aspects of Weaponry in Viking Age Societies* (2014) by Fedir Androshchuck.

Stead's book has a great deal of information about the cultural and technical development of the sword in Britain, up to the coming of the Romans. He surveys 274 British Iron Age swords and scabbards, and divides them into northern (43 swords) and southern types (171 swords) while the remainder are of a mixed tradition, with some Roman influences (42).

Pierce's book looks at the way that the sword developed from the 8th century until 11th. It looks at the development of the blade and parts of the grip and catalogues some 70 swords. All are beautifully illustrated. One, very well preserved blade, however, is dated to the 5th or 6th century.[1] Unfortunately, no real provenance is provided or rationale for the dating is provided.

Androshchuck's work is mainly focussed on looking at hundreds of swords from the Viking age found in Sweden and puts them into their wider European context. He considers most components of the swords including, blades, tangs, grips, scabbards, belts and baldrics and their decoration. A consideration of art styles is included as is the place of the sword in society, which is helpful for our study, too. He does briefly visit pre-Viking weapons in his discussion of the development of swords within the region. The book is well illustrated with photographs, line drawings, tables and charts.

[1] Pierce, 2002. 26.

3. The background: warfare and society, kings and the aristocracy and a bit about beliefs

Paul Mortimer

"Love, loyalty, honesty beyond the most niggling technicalities, were only for one's kindred, chieftain, and closest friends. The rest of mankind were foemen or prey. And often anger or treachery broke what bonds there had been." Poul Anderson.[1]

Warfare and the military world

Having experienced or been taught about the horrors committed during the 20th century and the heavy loss of civilian life that came from them, it is rare to find anyone today who is prepared to glorify war for its own sake.[2] Such a view is understandable but it makes it difficult to fully appreciate the way in which warfare, and what goes with it, was viewed and understood by people in the past. It is cetain that their beliefs and attitudes to fighting and dying were different to ours, even if the motivations and emotions of young warriors are much the same now as they were then. In order to better understand their attidudes it is useful to attempt to gain an insight to their way of thinking because without that we will not be able to even begin to appreciate what the symbolism of the sword meant to the peoples of the past.

Judging by the literature, law codes, histories, tales and poetry of the period, and the later Norse sagas, it is easy to detect an attitude to everything to do with warfare and combat that is fundamentally at odds with modern concepts, at least in the higher levels of society and probably also in other strata. This may in part be because war was simply unavoidable for many; especially for those living in strategically important areas and disputed lands.

Warfare has always been an expensive business and there have always been those who profit from it, and in the distant past they would have tended to be those who won.[3] The prospect of loot and plunder was a major part of going to war and it was also a way for young men, and perhaps those not so young, to get on in the world. War thus provided considerable opportunities for social mobility.

The populations of most of Europe, let alone the northern parts, would have been regularly affected by wars and rumours of wars and could be subjected to changes in leadership from time to time which may have involved conflict.[4] Davidson says, "...warfare permeated the social structure of north-western Europe, influencing art, symbolism and literature; it played an important part in the pre-Christian religion and left its mark on popular beliefs and customs.",[5] certainly the symbols of war(swords, spears and shields) would have

---

[1] Poul Anderson, 1973. Hrolf Kraki's Saga. xix

[2] There are, of course, exceptions and it is accepted that a few people may not share the modern consensus and that there are whole communities in some parts of the world that view things very differently.

[3] In modern times, it is not that unusual to find examples of arms dealers selling to both sides in a conflict.

[4] Steuer, 2006. After surveying the behaviour and development of war bands from, Early and Middle La Tène period to the time of the Normans, explains that he sees set stages in the way that successful warbands evolve and eventually, through 'ethno-genesis', become peoples.

[5] Davidson, 1989. 11.

been seen in every village. Hedenstierna-Jonson remarks, "Societies were in many ways rooted in warfare and the warrior mentality or warrior ideology saturated ideals, morals and actions, leaving no part of society untouched by the concept of war."[1]

Northern European society during the early medieval period was highly organised and a large part of that organisation was focussed around its military. James describes the characteristics of a militarised society and says that it is, " … a society in which there is no clear distinction between soldier and civilian, nor between military officer and government official; where the head of state is also commander-in-chief of the army; where all adult free men have the right to carry weapons; where a certain group or class of people (normally the aristocracy) is expected, by reason of birth, to participate in the army; where the education of the young thus often involves a military element; where the symbolism of warfare and weaponry is prominent in official and private life, and the warlike and heroic virtues are glorified; and where warfare is a predominant government expenditure and / or a major source of economic profit."[2] It can certainly be argued that much of what was happening in our period, as well as before and after, fits much, perhaps all, of what James described and that society in most or all of Europe was 'militarised'. From what we know of events at the time it probably needed to be.

The frequency of warfare

Records are incomplete and do not cover every region and those with any reliability begin quite late, however, between 600 and 850 A.D. fourteen wars were recorded where the kingdoms of Wessex and Mercia fought each other. During the same period the Mercians fought eleven wars with the Welsh and went to war with other opponents on eighteen further occasions.[3] That would appear to be forty-three wars in all, roughly one major conflict every five years nine months. In other parts of England there would have been other struggles between and within competing kingdoms. Halsall tells us that during the 7th century in England, there were twelve battles in which a king or major royal figure was killed and remarks that there were ten other battles recorded and at least eleven other instances of warfare.[4] In addition to this skirmishes, raids, ambushes and harrying operations would have been used to train young warriors, keep them occupied and sharpening their skills. Such activity would have been used to put or keep an enemy or subordinate kingdom in its place.

On the Continent the Frankish records from 581 to 590 A.D. show that there were thirty-seven occasions when the army was called out. From a later period we know that Charles Martel (717 – 741 A.D.) and his grandson, Charlemagne, went on campaign almost every year of their adult lives.[5]

Returning to the British Isles, it is with some justification that Abels says, "War was endemic to the kingdoms of seventh and eighth century Britain. An Anglo-Saxon ruler of this period was above all else a warlord, a *Drihten*, (*sic*) as the Old English sources put it. His primary duty was to protect his people against the depredations of their neighbours and to lead them on expeditions of plunder and conquest."[6] So it would seem that war was a feature of just about everyone's lives in one way or another. It is no wonder that the trappings of warfare, including, and probably especially the sword, had significance and meaning beyond their practical functions.

---

[1] Hedenstierna-Jonson, 2006.p 24. The author is specifically writing about Viking age Scandinavia, but I believe that the evidence shows us that the mores of 5th to 7th century Europe were similar.

[2] James,1997. 19.

[3] Halsall, 2003. 2.

[4] Halsall, 2003. 3.

[5] Halsall, 2003. 2. Bachrach, Merovingian Military Organisation. 1972. 99 to 106

[6] Abels, 1998. 11. Abels is referring to Anglo-Saxon England, but the same would have applied to many parts of the European mainland and to Scandinavia.

Kingship, nobility and aristocracy

Kings

Society was fairly stratified, in that there were different levels and hierarchies, and we will now spend a little time exploring what that meant during the 5th to 7th centuries. There is only space in this book to write about it in fairly general terms and I am certain that there must have been many variations in custom and practice over time and in different regions. However, some principles appear to have operated in most areas.

Although much of the relevant literature was written before or after our period there are some contemporary accounts from the Continent. Taken together the sources provide clues which help us build a reasonable picture of how societies where organised and enable us to gain some understanding of the status of kings, their authority and its limits. Likewise, information about the associated aristocracy is important when looking at how swords circulated and their significance within the 'economy of honour'.[1] Later Medieval society was, perhaps, much more strictly layered and had the full support and authority of a powerful international and supernatural body to enforce it; the Catholic Church. In our period the conversion was beginning to happen in England, and the Church was attempting to spread its influence, but it had not really changed the nature of kingship, at least not in most areas of what would become England.

There would have been groups of people who did not quite fit into the hierarchy discussed below, such as travelling merchants, entertainers and other itinerants. We will not be discussing them here but it is necessary to account, in at least a small way, for those outside the main 'system' and to acknowledge that some of them would have been responsible for spreading news from one region to another; it is possible that a few regular travellers acted as paid messengers or even spies. Other 'outsiders' may have included some smiths. We have devoted a whole chapter to the way that they may have been viewed.[2] There were also slaves, or the unfree, although it is not entirely clear what that status would have meant at different times and in different places.

The office of king was important but it was not, generally, a hereditary one passed from father to son. In most areas a king had to have the right blood line and be noble, but was elected by the important men in his kingdom from a pool of possible successors. There could be a number of eligible candidates, besides the immediate family of the dead king, and, it seems that a council of important men (the *Witan*, in the case of later Anglo-Saxon England) would choose who they thought was the best man for the job. Obviously, this competition could lead to resentment and even strife if a failed candidate could gain enough backers. Almost any ambitious and thrusting young man (perhaps even quite old chaps, such as a brother or uncle of a previous king) would be able to conjure up a convincingly good ancestry if he put his mind to it. In much of northern Europe it seems to have been of great benefit to be descended from Woden,[3] but another god may have done at a pinch.

Some of the processes of becoming a king in the northern lands seem to have been established fairly early on, Tacitus writing of the Batavian king, Julius Civilis tells us what was involved in his ascension. Civilis was a Batavian of noble blood who went through a set of initiation tasks when he became a military commander. He took an oath and participated in ritual dancing, human sacrifice was involved in the rites and he seems to have been elected by the people. Once king, he grew his hair long and dyed it red while the Batavians, who had traditionally been allies of the Romans, went to war with them. The revolt of 69 A.D. against the Romans was quite successful, but eventually the Batavians and their allies settled and, it would appear, Civilis

---

[1] This concept will be discussed in the chapter, *The Meaning of the sword*.

[2] The Mysterious Smith?

[3] Pollington, forthcoming. 238. Pollington discusses the reasons why Woden may have been so important as a divine ancestor.

was allowed to live out the rest of his life unmolested by the Romans.[1] I am not suggesting that all subsequent kings until the 7th centuries went through the exactly same processes, but the idea of 'noble blood', initiation, ritual, election and being a war leader (*dryhten* OE) has many echoes in the later literature. Incidentally, the Batavians, or at least some of their descendants, may well have become part of the later migrations to Britain.

One of the king's most important jobs was to act as the intercessor between this world and the 'Other' and to ensure that relationships remained positive between the two (or more) worlds.[2] This was done by enacting the correct rituals at the right times of year and by appeasing the gods, by speaking the 'words of the ancestors' and by good, kingly behaviour. Young kings, especially, would have to prove their worth in the military world by leading their men into battle, showing courage and by winning most of their battles. A good king collected wealth, through warfare, tributes, taxes and tolls, but did not hoard it all; it was his job to see that it circulated, and the giving of gifts,[3] including swords, was a part of that function.[4] Generous kings were lauded, mean ones were not, and may as a consequence not have lasted long.

The king was also responsible for the protection of his people and his treasure and for ensuring communications and the harvests; if he failed he would not remain in post for very long, particularly if his lack of success came early in his career before he had had time to build a significant supporting infrastructure.[5] The story of the legendary Svear, king Domaldi, is instructive. Although he may never have actually existed, the story illustrates that a king should take his responsibilities and duties very seriously. During Domaldi's reign there was a bad harvest one year, so they sacrificed animals to the gods for an improved one the following year. Unfortunately, the harvest was poor again, so that year they killed men to appease the powers of the Otherworld. Again the harvest failed, so they sacrificed Domaldi.[6] Incidentally, there are stories of other kings in the literature who were sacrificed for failure and for other reasons when they had upset enough people.

Part of protecting the people included ensuring 'peace' (*frith* OE) for his people. This peace did not mean that his troops did not go to war, it just meant that the fighting was taking place elsewhere, and that no other force was able to, or would dare to raid the homeland. A king who could not protect his people was not much use and there would be others who would believe that they could do a better job and might well garner sufficient support for a coup. An unsuccessful king would be removed one way or another.

One important aspect of kingship to note is that the powers of the king were not absolute, he always had to negotiate with others and was subject to the laws like everyone else, but the more he was perceived as a 'good' king, the better and more secure his position would be. Perhaps one of the attractions of converting to the new religion was the possibility of making the king's position more secure, perhaps by making it more 'divine' and seemingly ordained by God?

In England and in many other areas of the north, during the 5th to 7th centuries there were numerous little kingdoms competing with each other and some inevitably grew larger at the expense of their neighbours, perhaps absorbing smaller kingdoms where a ruler could not protect his people and territory. As states grew larger, so more power and wealth became concentrated in fewer hands. However, the king would still have had to keep his people happy by allowing much of that wealth to trickle down to them; keeping the more

---

[1] Wallace-Hadrill, 1971. 6 and 7. Wallace-Hadrill is referring to Tacitus's *Histories*.

[2] Theuws and Alkemade, 2000. 447. Bazelmans, 1999. 9, 19, 36 and 128.

[3] Härke, 2000. 379.

[4] For extended discussions on the importance of gift giving and its reciprocal nature see Bazelmans, 1999 and Nicolay, 2014. 7 to 12. For a full discussion on 'speaking the words of the ancestors' see Sundkvist, 2002 and for a briefer version see Mortimer, 2011.

[5] Bazelmans, 1999. 128.134.

[6] Sundqvist, 2002. 241 to 242. Perhaps present-day politicians should be treated in a similar way?

powerful of his followers rewarded adequately would have been wise and they had their own supporters to look after in turn. Being a king must have been quite difficult at times and if he got the balance between accumulating wealth and re-circulating it wrong, he could be in trouble. Kings may well have occupied a special position deserving of respect, honour and privilege but they were always exposed, and in the end were just a tool that the people used to obtain a good relationship with the Otherworld so as to get reliable harvests and to provide for them protection and 'peace'.

Being a noble; becoming part of the aristocracy

In the rank below the king there were many people who, as we have mentioned above, needed to be considered, looked after, or perhaps in extreme cases, got rid of. They were the members of the aristocracy; that is the nobles who had some political power. They served the king and acted as administrators and military commanders of his kingdom. The so-called 'high status' graves that are well furnished are usually thought of as representing members of the 'aristocracy', that is, those who were 'noble'.[1] Some commentators seem to think that someone buried in a sword-grave must be from the aristocracy and it is possible they are correct, although it is a difficult notion to test. However, I must make it clear that I do not believe that all sword users in **life** were nobles, even if the ones buried with a sword were.[2]

As for the nobility, what did it take to be a noble? Could a non-noble become one? Was the system similar to that in modern day England, where the proper aristocracy is pretty much fixed and difficult for a commoner to penetrate?[3] Were there degrees of 'nobility'?

Rives tells us that in Rome at the time of Tacitus (1st century A.D.) that the concept of ***nobilitas***, formed from the adjective, ***nobilis***, meant well known or had renown but it could also have something to do with political power and distinguished ancestry. In Rome the word was typically used of someone who could count among his ancestors the holder of a high public office.[4] It may have been a similar situation for the early Germans as Tacitus tell us, "Distinction of birth or great deeds done by their fathers confer on mere lads the rank of chief;"[5] Tacitus also tells us that the more young men someone could attract into their personal following, the more important that he would appear to be. Tacitus stresses that valour was an important quality in a leader, and that he must not be less valorous than his men.[6]

These same ideas occur again in other literatures, and quite often in the much later Icelandic sagas. It would seem that again an ambitious, intelligent young man could become, by his own efforts and a bit of luck, a noble; that through warfare he could make his name and fortune. In *Beowulf* we have the example of Weohstan who received a sword from the Swedish (Svear) king, Onela, for slaying his seditious nephew (*Beowulf* 2610-2625). According to Hill this act raises the simple and unrelated retainer Weohstan to almost familial and kinship status.[7] As mentioned above, just having a famous ancestor may elevate a young man to be considered 'noble'; however, to retain that status he would presumably need to meet the expectations that

---

[1] 'High status' graves are those which are well-furnished. Archaeology is limited to revealing what is left in the grave. Much of the organic materials that were deposited will often have completely vanished, leaving little or no trace of their existence. It may well be that what seems to be a plain grave, actually had gorgeous textiles, cunningly worked artefacts of bone or wood and they have just dissolved. Härke, too, points this out. Härke, 2014. 44.

[2] This idea will be discussed elsewhere within this book.

[3] I am specifically writing about the aristocracy in modern Britain. The situation may well differ in other lands. Being ennobled by gaining a title of entering the House of Lords, does not really make someone a member of the traditional aristocracy, but, ironically, may well be closer to the original concept of becoming 'noble'.

[4] Rives, 1999. 146. Wickham, (2009. 29) explains that it was not necessary to have a particularly long ancestry.

[5] Rives, 1999. 82.

[6] Rives, 1999. 83.

[7] Brunning 1999. 28. Brunning is explaining Hill's belief here – for more see Hill, J. M., 2000. *The Anglo-Saxon Warrior Ethic: Reconstructing Lordship in Early English Literature*. Gainsville.

went with the position. We can see this with the concept of 'worth' (*weorð* OE) which is embedded in much of the literature, especially *Beowulf* and is closely connected to the idea of being 'noble'. A youth from a 'noble' family inherited a certain amount of *weorð*, as Beowulf does when at age seven he leaves his childhood behind and goes to the court of king Hreðel and receives treasures from the king.[1] At this stage Beowulf had been publically rated as having *weorð* and potential, but eventually, once he had received his man's weapons, it was up to him to prove himself and maintain or improve his status.[2] So it must have been with any ambitious young man.

The early medieval concept of nobility does not seem to be the fixed and impenetrable barrier that it may have become during the later Middle Ages, perhaps social mobility was possible and occurred often and almost anyone with a bit of determination and luck could rise in rank. We may be seeing an example of this in Hrolf Kraki's Saga when Hött, a farmer's son who was generally held in low regard and persecuted, was made to drink a monster's blood and then became a great hero called Hjalti.[3]

The warriors

> *...a world of thin skins, short fuses and physical violence.* *D R Lesnick*[4]

The king does not protect his people by himself; he required a strong and effective band of warriors to do that. If he was himself a famous warrior, then young warriors, especially those 'on the make' would be attracted to his service.

The warrior tradition in Europe has a very long history, as it did in other regions of the world and in other traditions.[5] In northern Europe it was highly developed and complex long before our period and although there would have been variations in different places and times, there are many features that were probably fairly universal. Concepts of 'honour', loyalty and proper behaviour would have dominated the way that a warrior acted, being noticed and remembered would have been important to such a man. Shared hardships, eating together, caring for each other, learning each other's strengths and weaknesses would bring them closer, increasing the bonds of 'brotherhood'. They would have been inspired by heroic tales of great achievements and even greater deaths. Many would want to emulate such feats and some would see anything but death in battle as shameful. Warriors would hope that the *scops* (OE) would sing of their deeds so that they would be remembered by their folk. While still alive, gaining and displaying tokens of their lord's, and others, high regard would have been a key feature of their lives while in his service. As for the afterlife, we do not have a clear and unambiguous picture of what they were expecting in the pre-Christian belief system. We gain some glimpses from the later Scandinavian literature, concerning how the Norse may have viewed such things and there are hints in the art and some earlier literature that in some ways the Anglo-Saxons may have had similar ideas, but that is all.[6] That the pre-Christian Anglo-Saxons had notions of what happens to

---

[1] *Beowulf*, 2425 to 2431.

[2] Nicolay, 2014. 8 to 9.

[3] Byock, 1998. 23.

[4] Lesnick, D R. 1991. 72. *Insults and threats in Medieval Todi*. Quoted in Halsall, 1998. 4. The quote is not directly referring to warriors of this period or region, however, it seems apposite.

[5] Shepherd, 1997. 1.

[6] There is work on this, see Pollington, 2011, Elder Gods. Carver, Sanmark and Semple, 2010, Signals of Belief. Dubois, 1999, Nordic Religion in the Viking Age. Price, 2011. The Viking Way; Religion and War in Late Iron Scandinavia. Even in the Viking period the ideas about the afterlife is not at all comprehensive, there are many questions still be answered.

someone after they had died is apparent from their burial customs and their art. Of course, for those who had become Christians, dying in the service of their lord would probably ease their passage into heaven.

Not all fighters and warriors were on an equal footing with each other that is apparent from the poetry and other literature;[1] Beowulf, although legendary, for instance was quite clearly special in many ways. For example, he was from an aristocratic background whose blood line qualified him for kingship; he was also stronger and mightier than any of his companions. The poet goes to great lengths to impress upon us how honourable his behaviour always was. His companions may have also been from an aristocratic background, but some might not have started their lives with that status. There were different types of warrior/fighter, men who came from varied backgrounds represented by Beowulf and his retinue. They were expected to take part in fighting, and many, like Beowulf, would have travelled to gain experience, wealth and reputation, or die in the attempt. Those who were better at it, or luckier, would, of course, be more honoured, have more prestige and survive longer, but it is unlikely that all fighters were full-time professionals.

A kingdom would need to be able to call on more than just the king's war band when things became really serious, we have already seen that Anglo-Saxon England was a place where kingdoms/chiefdoms were vying with each and trying to gain ascendancy. A large force would have been required as that competition intensified. Where would a king raise sufficient troops?

Davidson separates fighters into categories, beginning with the elite, soldiers who fought for a while in their youth and then returned to their estates but who may return when the leader needed them.[2] These would have been the aristocratic warriors and would probably form the backbone of the army and most likely most of its officer corps. She also lists mercenaries and other professional warriors who fought on a contractual basis for a period of time. These may not all be aristocrats and Beowulf's troop may well be viewed as this type of warrior body whilst in the land of the Danes. There would have been systems of raising a levy from the villages to pay for men and their equipment. These would have formed the main body of troops, much as they did in later Anglo-Saxon England. They would have received some training throughout their upbringing but for most of their lives they would have been mostly farm-workers and small holders. Those among them with special talents and who preferred it may have been able to permanently swap the plough for the sword. Some men would usually be left on the land to keep the crops and animals safe, except perhaps when things became really extreme. It is likely that all able-bodied men received some training in military pursuits, and many young men, whatever their social origins and throughout history, have tended to display a liking for some form of fighting.[3]

The Scandinavian scholar, Jørgensen, when writing about Iron Age Scandinavia, recognises three categorises of military men: warriors, whose vocation was war; soldiers, whose job or profession was war; and conscripts, who fought when ordered to,[4] which seems to fit well with what Davidson suggests.

The training of 'warriors'

The training of combatants was and is still, absolutely critical, it contributes to morale, strengthens the bonds between men, teaches co-ordination, familiarises everyone with what is expected of them and their weapons and increases trust in the abilities of comrades and commanders. It is brought about by drill, discipline and much practice, until everyone knows how to instinctively react in a given situation. An untrained man would have been a liability to himself and his comrades. It seems that for some, at least, training began at a very

[1] Härke, 1997b. 95.

[2] Davidson, 1989. 11.

[3] See also the discussion in *The sword as a weapon.*

[4] Jørgensen, 2001. 9.

young age and there are some interesting bits of information in the Anglo-Saxon literature.[1] The examples that we are provided with are likely to have mainly applied to youths from established, aristocratic families, but it is likely that most youths from free families were expected to become familiar with many aspects of the military in their early lives.[2] We are told by Bede that in the 7th century, St Cuthbert joined the boys' company in his eighth year, and took part in running, jumping, wrestling and gymnastics.[3] In the poem, Beowulf was sent away from home at age seven to be with king Hrethel.[4] This seems to reflect a long standing practice from within Indo-European societies. Kris Kershaw, who published a study on the *männerbund* and its links to the god, Woden (Wotan), suggested that generally males were divided, by age, into four groups:

i. That little boys belonged to the world of the women; ii. That at a certain age they were taken to be trained in military matters as scouts and raiders, and in the ways of their ancestors; iii. Once they reached about twenty years old, they would re-join society, could marry and hold property; such men were the disciplined troops of the tribe; iv. Those that reached middle-age would become the elders of the tribe and the tribe would be guided by their experience and accumulated wisdom.[5]

Kershaw provides evidence from a number of European cultures to support her work, including examples from Tacitus's *Germania*.[6] To add weight to her suggestions she enlists images from the 5th to 7th centuries which include helmet plates (pressbleches), such as those found at Sutton Hoo and from Vendel and Valsgärde in Sweden,[7] and scabbard mounts and other items from the Continent and England. She shows, too that there would appear to be strong links between the warrior bands and the god, Woden.[8]

We are told that at age fourteen, St Wilfrid was given his arms and a horse, although he eventually made the decision to enter a monastery instead of the fighting profession.[9] Again this accords with other evidence and we are informed that both St Wilfrid and St Guthlac, when they reached 'warrior age', gained their own bands of armed followers.[10] This seems to echo what Tacitus said earlier about young nobles gathering a band of followers.

Our word 'youth' comes from the Old English, '*geoguð*', which meant a group of young warriors. A young man who joined the geoguð would be one who has received his arms but is still learning from his training, experiences and through watching how others perform. Such young men were probably kept busy and away from home as much as possible; such youths, full of testosterone and ambition, have always been likely to get into mischief, but led by a seasoned warrior they could perform useful service, receive more training and get 'blooded' at the same time. Examples of such service could be border patrols, scouting expeditions, tribute collecting, raiding and skirmishes with the enemy. While still quite young, they were probably kept away from the main body of troops during battles and may have used bows and arrows and other missiles to harass the

[1] Vita Sancti Wilfrithi – the life of Saint Wilfrid and Bede's History of the English Church and People, and Bede's Life of St Cuthbert, especially are useful, as is the Anglo-Saxon Chronicle.

[2] Matzke, 2011. 7, 10, 37, 38, 42, 43 and 97. Matzke was writing about the later medieval period, but I feel that much of what he says can be extended back in time. See my comments in the chapter on *The sword as a weapon,* where the discussion on training young warriors is taken a little further.

[3] Davidson, 1989. 20.

[4] Curiously, there is still a tendency in a certain strata of British society to send their young boys away to school at around age seven, surely just a coincidence?

[5] Kershaw, 2000. 24 to 25.

[6] Kershaw, 2000. 62 and 82.

[7] Kershaw, 2000.

[8] Kershaw, 2000. 25 and 212. Mortimer, 2011. 205, briefly looks at the etymology of the name, *Woden,* as does Pollington, forthcoming.

[9] Davidson, 1989. 20. See also Halsall, 2003, 58.

[10] Halsall, 2003. 57.

enemy whilst keeping their distance from the main action; this would familiarise them with the sounds, smells and sights of battle, keep them alive and let them know what was expected of them as they got older. It is very unlikely, that any of these would be allowed to marry or hold land in their own right until they had progressed through the ranks. Some may have been sent away to serve other lords for a time, or to be fostered, or as a hostage. As the youngsters matured and became experienced they would become '*duguð*', doughty men; reliable, proven warriors some of whom would be chosen to lead others. Possibly around their late teens or early twenties some of them would marry, acquire property and settle down, it is probable that their service from then on would be part-time as they would have other duties to perform, including on their own land. It is possible, too, that a few would not want to stop soldiering because they liked it and would become full-time professional soldiers, as Tacitus tells us was happening during the 1st century with the Germanic tribes on the Continent.[1]

A few members of the *duguð* would become *folctogan* (OE), senior officers in the king's forces and take the role of commanders, perhaps the ancient equivalent of generals; these would have been important men outside the military sphere as well, and would be part of the body advising and assisting the king in administering his realm.

It is important to understand that folk from one kingdom would have regarded those from outside not just as potential enemies but possibly as not properly human, as Shepherd says, "Since resources on the Outside by definition belong to no human, the claims of the possessors of such resources have no validity. Raiding cattle, pillaging settlements, pirating ships--these are all honorable (*sic*) and right actions for the warrior as long as they are directed toward the Outside."[2] Her thoughts echo the extract written by the novelist, Poul Anderson, with which we began this chapter. Julius Caesar, in *The Gallic Wars*,[3] writing of an earlier time, explained how 'Outsiders' were seen when he wrote that "German youths were not punished for robberies committed outside the boundaries of their own kingdom, since they claim that these are carried out for the purpose of disciplining their youth and preventing sloth." Shepherd explains more fully: "He (the warrior) protects the boundaries of Home from the unknowns and terrors of the Outside. With this protection, a sense of Territory is achieved. In so doing, the warrior continually reaffirms group cohesion. The group being defined, the Other—the people on the Outside—become defined as well. The people of the Other are not part of the true social order. Therefore, ethical and moral rules of behaviour do not apply to them, and they may with impunity be treated as non-human. To outlaw an undesirable person from the social order is to expel him into the midst of this not-human Other."[4]

The size of armies

We will briefly consider the sizes of armies as for many years there has been a tendency to suggest that the armies of the Anglo-Saxons must have been quite small and possibly only consisting of a few tens of men. Such a size, to me, would seem to make a raiding party and not an army.

Despite the fact that Bede tell us that that Penda of Mercia took thirty 'battle hardened legions' (*legio*) led by 'famous commanders' to attack Oswy of Northumbria 655 AD,[5] the idea that the Anglo-Saxons were only capable of raising small armies has persisted. This is partly because many have preferred to regard the Laws of

---

[1] Rives, 1999. 90.

[2] Shepherd, 1997. 1.

[3] Hammond, 2008.

[4] Shepherd, 1997. 1. Please note that when we refer to the Other elsewhere in this book, we are not using it in the sense that Shepherd does here, which correlates the Other with outsiders; elsewhere in the book, the Other refers to the world of deities, spirits and the supernatural.

[5] Sherley-Price, 1955. 183. Petersen, 2013, discusses at length the history of scholarship on the size of armies and their capabilities in Europe. See for example pages 1 to 5. Unfortunately, Petersen does not deal with Anglo-Saxon England.

Ine (688 to 726 A.D.), clause 13.1 as some sort of definitive guide. The law says that, "We call up to seven men thieves; from seven to thirty five is a band; above that is an army."[1] This would indicate that an army could be a minimum of thirty-six men! Not a terribly significant force and certainly not one that would be capable of striking fear into a decent king's war band.

Thirty legions, on the other hand, sounds formidable, and Bede certainly thought so, even though Penda's forces were 'miraculously' defeated at the Battle of Winwaed. We do not know what constituted a 'legion' in the 7th century and while accepting that it probably was nothing like the size of a Roman legion from a previous time, it is likely that each such unit contained hundreds of men. A king who could only muster around thirty-six men, I suspect, would not reign for an extended period. The tale related in the next paragraph may illustrate this. Unfortunately, it has also been used to reinforce the idea that Anglo-Saxon armies were tiny.

The episode is contained within the Anglo-Saxon Chronicle entry for 757 A.D.[2] It is the story of Cynewulf and Cyneheard, where Cyneheard the Atheling, with eighty-four men, carried out an action in the kingdom of Wessex. Cynewulf was the king of Wessex and Cyneheard was the brother of Sigeberht, the previous king,who had been killed by Cynewulf. Cyneheard was told that the king was with a small troop of men at the house of his mistress in Merton,[3] and so he rode there with his men, surrounded the house and confronted the king. The king fought bravely but was killed after he had wounded Cyneheard. The king's men now alerted and awake, were offered money, land and their lives by Cyneheard to go away and leave their dead king; they refused the offer and preferred to die fighting the king's assassin. Cyneheard's force was sufficiently large to defeat them and they all died except for one British hostage, who was badly wounded.

As soon as the rest of the king's men, presumably still at his residence, heard what had happened, they lost no time in assembling a force that would be able to deal with Cyneheard's, which may have been somewhat smaller than it had been before dealing with the king's troop, and travelled quickly to the place where Cyneheard was. They found the 'stronghold' locked against them. The dead king's men were led by the king's ealdorman, Osric and his thegn, Wigfrith. The Atheling, Cyneheard, offered them money and land if they would give him the throne; he also told them that many of the men in his force had relatives in the king's forces outside. The king's men refused the offer and said that no relative was dearer to them than their king Cynewulf and that they must avenge him. However, they offered freedom to those of their relatives inside the stronghold who wanted to leave with their lives. Again the offer to leave was refused and so they set to. Cyneheard and all his men died in that place, except for one, who though wounded was allowed to live; he was the ealdorman's godson.

This tale illustrates that the warrior ideal was still very much alive in the 8th century, what it doesn't do is show the size of an army. Halsall explains that this was a *putsch*, an attempted surprise take over with a small force. There was no attempt to overwhelm the kingdom, eighty-four men would not have been sufficient to achieve that , and the dead king's supporters had no trouble in quickly raising a sufficient force to deal with the rebel force.[4]

As for Ine's definition of an army of thirty-six men, Halsall again demonstrates that this is not an indication of the size of a real army. Ine's predecessor (Cadwalla)[5] had violently seized the throne of Wessex in 685 A.D. and Ine had succeeded a few years later and was conscious that the armed retinues of powerful followers were

---

[1] Halsall, 2003. 59 to 60.

[2] Swanton, 1996. 46, 48 and 50 (Winchester manuscript A), 47and 49 (Peterborough E)

[3] Swanton explains that we do not know which of the many Mertons it was. Swanton, 1996. 47. fn 12.

[4] Halsall,2003. 123.

[5] Sherley-Price, 1995. 232.

a danger to his position; he was all too aware of recent history and must have had concerns that a relatively small force could attempt a *putsch* and cause considerable damage, even if unsuccessful. His laws are an attempt to establish royal prerogatives and the control of armed force.[1] The clause giving numbers of men is only a part of the law and there are other clauses providing the penalties for those who may disobey the law; those involved in raids committed by 'armies' are much more severely punished than those doing something similar in 'bands'. As Halsall explains, "In effect, any magnate with a following of three dozen or more men who commits a breach of the peace will be treated as leading an attacking army and severely dealt with."[2] Again, this example should not be used to calculate the size of an army in Anglo-Saxon England.

How many men in an army?

There are indications from elsewhere that powerful men could raise forces of considerable size. For instance some Frankish armies were claimed to be of 10,000 men and, considering the wealth at the disposal of some of the Merovingian kings, there could be some truth in the claims.[3] However, Halsall looks at the high numbers provided in the texts from the early medieval period critically and comes to the conclusion that any numbers given are unreliable.[4] He thinks that ancient authors used particular numbers because they had biblical significance, or just rounded them up to the nearest 10,000 or so, and some reporters give very different figures to another commentator when writing about the same event. After considering other forms of evidence, Halsall comes to the conclusion that, in the more economically developed parts of the Continent, armies of 5,000 to 6,000 were feasible at times, but that in post-Roman Britain, Germany and Scandinavia it is unlikely that armies often, if ever, exceeded 2,000 men.[5] Halsall's relatively modest suggestions regarding the numbers in Continental armies are supposedly based on his idea that rural settlements in Gaul were small.[6] Petersen takes issue with Halsall's interpretation of army numbers in several parts of his book and says, "What is often forgotten in such arguments, however, is that a very large number of these villages belonged to much larger estate complexes, whose cultivators paid dues and/or performed services for their lord, depending on the nature of the estate organization."[7] Petersen believes that there is good evidence that Wamba's[8] army that put down a rebellion in Septimania during the late 7th century must have numbered about 20,000 men,[9] and suggest that, from the early 6th century onwards, "The scale and complexity of Merovingian warfare is often severely under-estimated."[10] He goes on to explain that, "...the Frankish army that was sent to aid the in 590 under *twenty duces* could hardly have numbered less than 20,000 men."[11] Petersen also records that wealthy individuals could muster relatively large bodies of men, one such example is Theudis who was able to recruit a private army of 2,000 professional soldiers and a force of bodyguards from the estates of his Hispano-Roman wife during the early 6th century, and this appears not to be an isolated case.[12]

---

[1] Halsall, 2003. 59.

[2] Halsall, 2003. 59.

[3] Bacrach, 1972. Petersen certainly thinks so. Petersen, 2013.

[4] Halsall, 2003, 121 to 123.

[5] Halsall, 2003. 132.

[6] Petersen's reference here is puzzling, as he claims that Halsall (2003, 126) believes that some settlements in Gaul after 400 A.D. may have numbered no more than fifty souls. Earlier (Halsall, 2013. 123) he had suggested that an 8th century settlement in England may have numbered less than eighty-four people.

[7] Petersen, 2013. 219. Petersen is referring to Gaul.

[8] King of the Visigoths 672 to 680 A.D.

[9] Petersen, 2013. 175 to 176.

[10] Petersen, 2013. 192.

[11] Petersen, 2013. 213 to 214. Petersen provides a great deal of evidence in his book to support his claims about the capacities and capabilities of Continental armies.

[12] Petersen, 2013. 166 to 168. Petersen refutes Halsall's (and others') suggestion that these were slaves in a footnote.

In any case, it is not just the numbers of men that was the limiting factor, it is logistics; the problems of transporting the force and keeping it fed and healthy that were (and still are) the main difficulty in maintaining a large force for any length of time.[1]

There is other evidence from Denmark that would appear to provide added credibility at least to Halsall's figures about Scandinavia and possibly England; it is from the bog deposits.[2] Albrethsen looked at the deposits in one section of one set of bogs; that found at Illerup A near Aarhus in Denmark. The bog has only been partially excavated, with some 40% of the deposit explored so far and the equipment of five senior officers has been found there. Alongside the gear of three hundred fighters there are also weapons and paraphernalia for forty officers of middle rank. Albrethsen suggests that if and when the rest of the bog is dug out (60%), there is likely to be a total of twelve senior offices, one hundred middle rankers and seven hundred and fifty men represented. The bog deposit is thought to represent the arms of a defeated enemy, and Albrethsen feels that 50% of that army probably escaped, so the true size of the losers' force would have contained twenty-four seniors, two hundred officers of middle rank and 1,500 men. It is possible to argue with Albrethsen's surmise and his figures, but Illerup A certainly does represent a sizeable force and Albrethsen's calculations accord with the suggestions made by Halsall.[3]

In conclusion, I would suggest that the early Anglo-Saxon kings and chieftains could muster considerable numbers of troops when required and, although at times that may have been difficult for some monarchs, it was an absolute necessity; if they were unable to do so, a king in the competitive situation of the 5th to 7th centuries would have found himself in a very precarious situation indeed.[4] Of course, I am not suggesting that armies were always made up of large numbers, and actions with a few score or low hundreds was probably the norm. Wise generals would avoid expensive pitched battles if they could, preferring raiding, ravaging, skirmishes and other nuisance tactics where smaller numbers could be employed and limited objectives achieved.[5]

Beliefs

Throughout the above discussion the pre-Christian ways of looking at the world have hovered in the background and now we need to consider them briefly as, during our examination of many aspects of the sword, whether considering ornament, symbolism, value, burial practices or the part that swords played in the community, those ideas about the world will be colouring our discussion. There is a difficulty though, and that is we do not know enough about what many of those ideas were, or exactly how the pre-Christians viewed much of their everyday lives. That they had customs, practices and beliefs in the supernatural is evident, we just do not know enough of the details. For instance, we have almost no actual evidence about what they thought would follow death; many of the ideas on that have been extrapolated from much later, mostly Norse material. It is likely that such ideas may well have altered over time and space and even from family to family and that there never was a single set of accepted ideas adhered to by all. Whatever their world-view was, it permeated every aspect of their lives and, occasionally, we are able to catch a glimpse into their world through their practices, their artwork and both contemporary and later literature. It is apparent that what they believed was very different to the dogma taught by the highly centralised religion, Christianity, which followed and, to a large extent, supplanted it. Some of the older ideas, however, remained and were absorbed by the Church, or operated outside of it and were more-or-less sanctioned or at least ignored by the Church, although some practices were forbidden, especially the worship of certain gods and sacrifices to

[1] Mortimer, 2011. 191.

[2] Albrethsen, 1997. 216.

[3] Steuer, (2006. 229) agrees on the numbers constituting armies at Illerup.

[4] For a fuller discussion and more evidence of army size see Mortimer, 2011.(189.) The Size of Armies.

[5] Tyerman, 2007. 22. Tyerman's insight is a useful one, even though he is writing of armies centuries later than our period.

them.[1] In some ways Christianity had to adapt[2] as it was changed by its contacts with Northern ideas; remnants of the older ways survived and are, perhaps, still with us.[3]

Russell tells us that there are fundamental difference between Christianity and what preceded it, and possibly the main difference is the way that Christianity rejects the world and its temptations as at least distracting, but often evil. Indeed, it requires its adherents to be 'saved' from the world,[4] whereas the older systems of ideas, the 'world-accepting' ones (which Russell calls 'folk-religions') meant that the world was just what it was (and is), that it was accepted, lived with, explained through myth and accommodated. It was the human who had to bend and adapt to meet the Other, not the other way around.

Another illustration of the way that the two outlooks, the Old Way and Christianity, differed can be illustrated by their attitude to animals. Of course participants in both systems ate and used animals in many ways, but according to the Christian Bible, God made man in his own image and gave him dominion over the animals (Genesis 1:26). In other words, man was the master over all things. By contrast, animals in the northern world were not always in a subordinate position. For instance gods could appear as animals, and Óðinn, in the myths had a habit of turning himself into a snake, an eagle or a four footed animal.[5] Óðinn was not alone, other spirits and even humans could take animal form,[6] while wolves, bears, boars, and birds of prey were regarded as particularly powerful animals connected to the elite.[7] Those animals, together with serpents and horses regularly appear in the art work of the 5th to 7th centuries.[8] Animals could inform the observer about the future by their movements, as Hedeager reports: "In the cosmic world of the Old Norse system of belief, animals were an integral part because they possessed knowledge of nature and conveyed it to humans and gods."[9] All the signs are that the northern peoples during the 5th to 7th centuries saw animals similarly.

A further contrast could be the way that the two sets of adherents viewed the countryside around them; Christians would only regard those parts of the landscape associated directly with worship or saints as 'sacred', whereas to Old-believers the whole could be numinous including the air, the wind, trees, stones, etc.[10]

Price explains that the northern gods did not require 'adorative' worship: "The religion of the Æsir and Vanir demanded only recognition that they existed as an integral and immutable part of human nature and society, and of the natural world, and that as such they possessed an inherent rightness –perhaps even a kind of beauty. If one wished to avoid disaster, it was necessary to come to terms with the gods, and the terms would be theirs, not those of their followers."[11]

---

[1] See Russell, 1994. 123. Some forms of sacrifice were just altered and were turned into celebratory feasts for Christians. Russell, 1994. 186.

[2] Russell, 1994. 7, 23 to 24, 38. Christianity accepted the 'heroic ideal' and used it to bring the 'Germans; into the fold. Ruseell, 1994. 40, 207, 212 to 213.

[3] Russell, 1994, 33, 36, 39, 44, 203, 212 to 213. It would appear that many of the 'converted' were never educated in what that actually meant and just absorbed a new deity into their old ways.

[4] Russell, 1994. 4, 49, 52 to 55 (for definition of 'world-rejecting' and accepting), 102 to 103. All the Abrahamic faiths are 'world-rejecting' as is Buddhism, they all preach that a better world is elsewhere and can be ours, provided certain conditions are fulfilled.

[5] Hedeager, 2011. 81 to 82.

[6] See Hedeager's discussion of *hamr*, *hugr* and *fylgjur*, Hedeager, 2011. 82 to 83.

[7] Hedeager, 2011. 83 to 84.

[8] Hedeager, 2011. 85 to 95.

[9] Hedeager, 2011. 98.

[10] Wickham, 2009. 55. Wickham is specifically contrasting classical paganism with Christianity, but I believe that it applies just as much to northern beliefs.

[11] Price, 2002. 55. Of course, it has to be acknowledged that the earlier views in England may have been different to the later Norse one.

One more contrast between Christianity and what went before it is that Christianity assumed (and still does) that all had 'sinned', whereas that concept was alien to a non-Christian at that time, ideas concerning wrong doing were tied up with notions of honour, loyalty and behaving correctly.[1]

We know that they had gods,[2] some of them have left their names as the names of certain days of the week, and we know that the gods were not all-powerful, or all-seeing; we are given hints that there may have been other beings that were stronger and wiser than the gods. It is possible that most of the gods may have had little relevance to many of the people much of the time; there were other levels of the Otherworld that were, perhaps, closer to the everyday, mundane world. The beings that assisted with childbirth, or not, those in the fields that guarded the harvests, local spirits, guardians of the family's 'luck', some nameless, others not, some forgotten, some still remembered and the ancestors may all have played much bigger roles in the daily life of someone on a farm or settlement than the major gods. There were also spirits that haunted the Outside and the Dark, some were just fickle, while others, like Grendel from *Beowulf*, were deliberately malevolent and wanted to cause harm, others, such as the *ælfe* (OE),[3] were indifferent to people, or could be either malicious or beneficial depending on how they were treated. Superstition and ritual would have been part of everyone's lives every day and all would have been linked to the natural world and its phenomena.

To call the older ideas about the Otherworld and its relationship to the everyday a 'religion' is a moot point and may not be accurate. Neil Price, writing about beliefs during the Viking Period, says that "'Religion' to us conjures up something orthodox, a creed, with more-or-less rigid rules of behaviour that usually embody concepts of obedience and worship. These tenets are often set out in holy books, with holy men and women to interpret them, with all that implies in terms of social differentiation and power relationships......In Scandinavia before the coming of Christianity, however, no-one would have understood this concept. For the late Iron Age it is instead more appropriate to speak of a 'belief system', a way of looking at and explaining the world. What we would now isolate as religion was then simply another dimension of daily life, inextricably bound up with every other aspect of existence."[4] His words are certainly applicable to ideas about beliefs in England and elsewhere during much of the 5th to 7th centuries.

As mentioned, very briefly above, the Old Ways may have been slower to fade than is often assumed and both ideologies may have mixed and intermingled well into the period of conversion and beyond and Christianity did little to alter the warrior ideal, so attitudes to weapons and war gear would have remained the same throughout the 'conversion' and into the much later period and beyond.

---

[1] Russell, 1994, 120 and 204.

[2] Like the word 'religion', the word 'god' in our modern world has connotations that may not have been applicable to the older belief system.

[3] The elves.

[4] Price, 2002. 26.

4. The Staffordshire Hoard

Paul Mortimer and Stuart Davies

Anglo-Saxon society was so violent that a central fact of its politics, its way of life, even, was fighting and making war.

Eric John.[1]

The Staffordshire Hoard was discovered on 5th July 2009 in a field by Terry Herbert while using a metal detector on farmland near the village of Hammerwich in Staffordshire.[2] Fortunately, Mr Herbert reported the find to the local Finds Liaison Officer of the Portable Antiquities Scheme (PAS) and a full investigation of the site was launched. The Hoard contains more than 4,000 separate pieces, including over five kilograms of gold and almost one and a half kilograms of silver.[3] Originally, it seems the treasure had been quite a way below the surface, but erosion and the ploughing of 2008 had spread the artefacts over an area of about one hundred and fifty-five square metres.[4] All the evidence suggests that the Hoard was deliberately buried in that location. Incidentally, most of the items that are damaged were probably buried that way as very little harm was caused by the plough.[5]

The vast majority of the items appear to be connected to weaponry and warfare; many of the pieces are from swords, helmets[6] and shields. The latest writing suggests that the treasure was buried between 650 and 670 A.D.,[7] but of course many of the items will have been of considerable age when buried and it is not impossible that future research could affect that date range.

Unfortunately, there are not really any other hoards from England dating from the 5th to 7th centuries to compare it to, but its composition is strange in that it is almost completely (perhaps totally) made up of items torn from war gear.[8] Some pieces appear to have come from Christian crosses and other Christian regalia, but these, too, could have been used as symbols for war or as supernatural protection for a Christian war band. There appear to be no items related to females in the Hoard.[9]

The Hoard has already revealed a considerable amount about warfare and warriors in Anglo-Saxon England and our understanding is still developing; it has already altered some of our ideas about society during that time. However, one of the most important facts about the Hoard is that it includes eighty-five gold or silver

---

[1] John, Eric.1966. 132. English Feudalism and the Structure of Anglo-Saxon Society. Quoted in Halsall, 1998. 2.

[2] Leahy and Bland, 2009.6.

[3] Symons, 2014. 9.

[4] A further excavation was carried out during 2012 and more objects were retrieved. http://www.staffordshirehoard.org.uk/about/new-discovery-november-2012

[5] Symons, 2014. 63. Cool, 2015. 104.

[6] Symons, 2014. 13. Fern and Speake, 2014. 6. Cool, 2015. 87.

[7] Symons, 2014. 13.

[8] Leahy and Bland, 2009. 9. Leahy and Bland do mention the Crondall hoard from Hampshire consisting of 99 gold coins thought to have been deposited in 645 A.D. see Leahy and Bland, 2009. 12.

[9] Leahy and Bland, 2009. 11.

sword pommels and pommel.[1] There are also a number of seax hilt fittings but they are not included in that count.[2] Many of the pommel caps are decorated with gold and garnet cloisonné or gold or silver-gilt filigree and had been parts of very valuable weapons.[3] The pommels and caps are not the only sword related items within the Hoard, other fine items include sword collars, quite a few finely decorated with garnets and filigree, lower and upper guard plates, many elaborately embellished, decorated sword-pyramids and probable sword-bosses. Some of the other objects could come from scabbards.

The Hoard has helped to demonstrate that swords might not have been such rare objects as some commentators would have us believe, especially as many of the sword parts come from high status weapons; if there were so many high quality items circulating, perhaps there were even more of a more modest character?[4] Whatever, the sword evidence contained within the Hoard is of great importance and it needs to be absorbed into the way we regard swords in Anglo-Saxon England during the early period. It could help if we knew why the Hoard was buried.

There has been much speculation concerning why the Hoard came to be deposited where it was found, ranging from being lost loot from a battle, a hidden and perhaps forgotten chest intended to purchase someone's ransom, part of a royal treasure hoard, plunder left by raiders, or buried to protect it from raiders, an offering to the gods,[5] a trophy hoard, or a king's hoard intended to reward young warriors.[6] Leahy and Bland point out that that there have been at least nine other finds of isolated sword pommels and parts of swords between 1999 and 2009, possibly stripped from blades and abandoned and wonder whether these represent mini-hoards,[7] even personal sacrifices to a deity.

Leahy and Bland noticed that the Hoard's location could have been significant, in that it was situated on high ground,[8] but that may not be the only important fact about where the Hoard was discovered. We will return to that idea a little later. Several commentators have wondered if the Hoard could be somehow linked to Penda[9] (632 to 655 A.D.), Wulfhere (658 to 675 A.D.) or Æthelred (675 to 704 A.D.), all kings of Mercia but, so far, no direct connection has been found.[10]

Stuart Davis, the co-author of this essay, noticed an interesting comment in an old gazetteer that could have some relevance to the Hoard. The entry was in *An Historical, Topographical, and Descriptive view of the county of the Northumberland* by E. Mackenzie and it refers to a place called St. Oswald, which lay north of the 'military road' and about four and a half miles north of Hexham and more than two miles west of Portgate. The entry tells us that, "A defaced Roman altar stands in the church yard. In a field near it, skulls

---

[1] Cool, 2015. 87 to 88. Cool says that, "In addition, three silver fragments are tentatively identified as rivet housings, and there are also two gold panels (K5 and K136; that may be from the sides of pommels. If included, this would take the total to ninety"

[2] Cool, 2015. 94. There are fittings from at least 5 seaxes. Pieta Greaves, the lead conservator of the Hoard has said that the Hoard may represent nearly two hundred swords. (Personal comment).

[3] In fact 63 are of gold and 22 of silver. Cool, 2015. 88. She adds a caveat that, the number of pommels does not equal the amount of swords represented as some might have had metal fittings but an organic hilt. Cool, 2015. 89.

[4] This issue is discussed more fully in the chapter called *Pattern-welding*.

[5] Symons, 2014. 13.

[6] Leahy and Bland, 2009. 11 to 13.

[7] Leahy and Bland, 2009. 12. Losing a sword pommel by accident is unlikely. Sword-pyramids and other loosely attached items can become disconnected and be lost without the owner noticing. This is further discussed in the chapter, *The Riddle of the Pyramids* where it is pointed out that around 60 loose pommel finds are listed by the PAS and also in Chapter 11, *Adorning the Sword*. See also *Appendix 13 Sword Pommel Finds Recorded by the PAS*.

[8] Leahy and Bland, 2009. 12.

[9] Symons, 2014. 14 to 15.

[10] Leahy and Bland, 2009. 14.

of men and hilts of swords are frequently ploughed up."[1] There is no more information and there are no clues in the text as to what period the hilts or the skulls came from, but it is relatively close to Hallington (Halyton, Haledon) or Holy Hill which was, according to Mackenzie, said to be the site of the battle of Hefenfelth or Heavenfield, "...the vale where Oswald conquered."[2] The Battle of Heavenfield was fought in 633 or 634 A.D. between a Northumbrian army under the Christian king Oswald of Bernicia and a Welsh army under Cadwallon ap Cadfan of Gwynedd. The Northumbrians defeated the Welsh decisively and it is possible that the 'hilts and skulls' belonged to the latter. It is significant that they are found quite close to the churchyard and that the information is very specific, 'hilts and skulls', nothing else. The dead have been stripped of their identity and their power! We shall see later in the chapter on *The Meaning of the Sword*, that swords in graves were targeted for reasons other than to re-use a weapon, almost certainly to do with the depriving the dead of their power.

Of course we must treat this entry with some caution, the finds might be nothing to do with Oswald, or even the Anglo-Saxons, it is possible that the remains could be much earlier, or later, perhaps the result of some unrecorded event; but they may relate to Heavenfield, or some other incident involving Angles and so it is worth spending a little time speculating. Incidentally, there are numerous examples from history demonstrating that Christian kings could be no less ruthless than their non-Christian counter parts, so the fact that Oswald was a Christian king does not mean that he would have stayed his hand, besides there would have been many in Oswald's army who still remembered the Old Ways. Indeed Christianity selected and adopted some of the Old ideas, even going to the extent of taking over old sacred sites and using them as churches;[3] it is not impossible that Oswald allowed his non-Christian followers to select items from the defeated and bury them, it might be that it was not done for non-Christian reasons but just to dishonour the dead or to act as a battle field marker; it could even be that some of his followers performed a non-Christian rite without Oswald knowledge.

There is of course ample earlier evidence from Denmark of weapon sacrifices in the bogs and many of these will be referred to later in this book. We also know that there was a tendency to move from wet-land sacrifice to sacrifice on dry-land from about the early 5th century onwards,[4] and both the entry by Mackenzie and the Hoard could be part of this later tradition. It could be argued that the Danish bog deposits are just of weapons, but recent discoveries of large amounts of human remains believed to come from more than 1,000 men at Alken Enge, could counter that because they bear the marks of a violent death but are mostly without weapons.[5]

Again a note of caution, skulls tend to survive longest in the ground, with the skull being the single most massive part of the skeleton and would, therefore take the longest to decay, and if the hilt fittings were of gilded copper alloy, silver or even gold they would not tend to deteriorate as quickly as an iron sword blade would, and might survive to be ploughed up by a team of horses after most of the iron had gone.[6] It could be that other unrecognisable bits of rust or broken bones were found but not remarked upon.

---

[1] Mackenzie, 1825. 300 to 301.

[2] Mackenzie, 1825. 301.

[3] Russell, 1994. 185. Pope Gregory to Bishop Augustine concerning Augustine's mission to England, "...The idols are to be destroyed, but the temples themselves are to be aspersed with holy water, altars set up in them, and relics deposited there..."

[4] Sundqvist, 2002. 72 to 73. Summarised in Mortimer, 2011. 211.

[5] http://videnskab.dk/kultur-samfund/nyfunden-massegrav-vidner-om-barbarisk-krigsskik The find of skeletal material is dated to an earlier period than most of the weapon finds. However, there are numerous finds of single 'bog body' sacrifices from many places in northern Europe and they come from an extended period of time.

[6] Matt Bunker, personal comment.

Some 8 years after the Battle of Heavenfield, Oswald and his Northumbrians met Penda and his Mercian army at the Battle of Maserfield (Hatfield) in Shropshire and was defeated and killed. Penda was not a Christian and it is possible that the Hoard is comprised of artefacts taken from the defeated at that battle, brought back to home territory deposited as a 'thank you' to the gods of the Mercians, or was lost en-route, etc.

Interestingly, and again a link to Penda, at Cuckney, in Nottinghamshire there is tradition of the Battle of Hatfield Chase taking place there. The battle was fought during October 623 or 633 A.D.[1] and the Northumbrians were led by Edwin and the Gwynedd-Mercian alliance was led by Cadwallon ap Cadfan and Penda. It was a decisive victory for Gwynedd and the Mercians; Edwin was killed and his army defeated, leading to the temporary collapse of Northumbria. In 1951, two hundred male skeletons were dug up beneath the foundation layers of the church,[2] and the burials were reckoned to be pre-1100.

Returning to the Hoard, the site where the Hoard was found could help us to decide why it was left there.

We have already noted that the site selected was on high ground, in fact it is the <u>only</u> high ground that immediately overlooks the old Roman road called Watling Street that runs east to west at this point, through Mercia towards Wales. To make certain that this was the case, Stuart Davies inspected the area by driving along Watling Street from its junction with Icknield Street, another Roman road that at this intersection travels mostly north/south through the area. There are also a number of ancient track ways that pass nearby and if you add to that the fact that the area is close to an important Mercian stronghold, the site begins to look very significant indeed and was never just an isolated place. There are five Iron Age forts within a three mile radius, one of which is Cannock Chase; this is, possibly, the densest clustering of important Iron Age remains in Mercia, all of which would have been visible in the 7th century. We know that the Anglo-Saxons had regard for such monuments which would have provided the area around where the Hoard was buried with extra meaning and perhaps inspired reverence. Exactly what it would have meant to a 7th Mercian can only be surmised, but it would have been a special place and a liminal one; that is outside of the 'normal' and between worlds of man and the Other.

In an age when warfare was endemic, society was organised for war, as we have seen,[3] and the heroic ideal was important an appeal or sacrifice to the gods for victory in war is perfectly possible.

---

[1] Sherley-Price, 1955. An alternative site for the battle is in south Yorkshire.

[2] St Mary's church, Cuckney.

[3] See chapter entitled *Background*.

5. Iron and steel

Paul Mortimer

Ancient bloomery sites are outstanding examples of a skilled production based upon tacit knowledge acquired by our forefathers through practice.

Arne Espelund.[1]

Production Sites and their nature in some Germanic areas

In Western Europe it appears to have been the people of the Hallstatt culture (800 to 500 B.C.[2]) who first invested in iron working. It is possible that the skills and technology for producing their weapons and tools were imported from elsewhere but they may have originated locally. These people were the first in Europe to occasionally include iron swords as grave goods.

Iron technologies developed through the succeeding centuries and were used extensively by many peoples, including the Romans, and not just for weapons but for domestic and many other uses. However, there is ample evidence that iron smelting and forging was happening throughout most parts of Europe long before the Romans built their empire in the west.[3] For the purposes of our study we will first look at examples of areas inhabited by Germanic-speaking peoples during the 2nd to 7th centuries who were producing iron. The types of iron produced in these areas may have gone on to become the sword making materials used by the Early English, their ancestors and neighbours. There are, of course, other areas within the countries mentioned that produced iron; all I am doing is providing a sample to illustrate that iron making took place at an early stage. Later, we will consider how iron technologies were incorporated into sword manufacture.

Denmark

It would appear that from around 500 B.C.,[4] farmers in what became Denmark had been exploiting their iron resources as a sideline to their agricultural pursuits. This activity seems to have been carried out during late summer on a small scale and was to continue for centuries;[5] it was part of the normal yearly routine where there was sufficient access to good quality bog iron and enough wood. The main objective of the farmer smelter was to produce a low carbon billet of iron either for local use or to be traded elsewhere.[6] According to Lyngstrøm the situation altered, somewhat, during the 6th century A.D. when forge techniques became more complex and imported metals are used in conjunction with native metals to provide, presumably, a better product.[7] However, even then the production of iron seems to have been small scale and was not allowed to interfere with the main business of food production.[8]

---

[1] Espelund, 2003. 195.

[2] Buchwald, 2005. 113. Buchwald calls them 'Celts' as do many others so, for convenience, we shall too. Gilmour, 2006, 48, suggests that iron working began some 4,000 years ago in Asia Minor (Turkey).

[3] Dinnetz, 2003.

[4] See also Buchwald, 2005. 181 to 194, for a more comprehensive survey of ancient iron working in Denmark. See also La Salvia, 1998. 24 to 25.

[5] Lyngstrøm, 2003. 21.

[6] Lyngstrøm, 2003. 22.

[7] Lyngstrøm, 2003. 23. Lyngstrøm comments on the progress being made in the ability to locate the provenance of iron and steel from its slag inclusions. 22.

[8] See also La Salvia, 2007 (25), for more information on iron working in Denmark.

Sweden

An example of an area producing iron in Sweden is the Red Earth area (Röda Jorden) of south central Sweden (Västmanland) where the smelting of iron seems to have started not later than 700-600 B.C.[1] and continued until about 200 A.D. The remarkable thing about the smelting here is that it took place miles away from any settlement areas. Supplies of ore and of wood were and still are abundant in this area. The type of furnace used was of the shaft furnace type.[2] (figures 5.3 and 5.4). Most of the examined remains have been of low carbon steel.[3] From 0-600 A.D. production in Sweden expanded and new areas became involved; Buchwald lists Väster Gotland, Småland,[4] Scania, Halland, Gästrikland and Jämtland.[5]

Germany

The iron working area of Wolkenberg, approximately 120 kms south-east of Berlin, would seem to have been a fairly large scale operation with, so far, more than 1,000 direct-process furnaces having been excavated in a fairly small area and the prospect of many more being found as the project progresses.[6] The smelting was performed exclusively in, "shaft furnaces with sub-surface combustion chambers.", and dates to the Late Roman Imperial period.[7] The iron in this area is found in a depression, not far from the village, and is comprised of bog iron in the form of sand, clay and iron oxide which arrives there through the precipitation of ferrous water through fine sand.[8] Wood and clay for furnace mantles are easy to find in this region. Most of the activity seems to have taken place during the 4th century A.D. and must have been quite sophisticated, as the smelters were able to partly remove phosphorous from their product.[9] Stenvik tells us that according to an 18th century Norwegian source, craftsmen, without resource to modern chemical analysis, could identify at least eight different qualities of bog iron ore. They did this by considering the colour, grain size, but above all, the taste of the ore. It would be interesting to determine whether these skills were available wherever bog iron was used and to see just how far back into the past those skills can be traced.[10]

Frank Nikulka and his team have surveyed the literature on iron working in North-western Germany from 2nd/1st century B.C. to 1400 A.D., around 286 sites were considered. Some of the sites exploited bog iron while others were in hilly areas or dry land and had to use other sources of ore.[11]

---

[1] Grandin and Hjärthner-Holdar, 2003. 33.

[2] Grandin and Hjärthner-Holdar, 2003. 33

[3] Grandin and Hjärthner-Holdar, 2003. 35. Buchwald, 2005. 197 to 199. Buchwald surveys other, ancient, iron producing areas within Sweden, Buchwald, 2005. 241 to 248.

[4] See also La Salvia 1998. 24.

[5] Buchwald, 2005. 241.

[6] Spazier, 2003. 38.

[7] Spazier, 2003. 38.

[8] Spazier, 2003. 38.

[9] Spazier, 2003. 41.

[10] Stenvik, 2003. 79. Buchwald, 2005. 143. Provides descriptions of the different tastes and what they indicate, from written, 18th century accounts. Presumably, Buchwald, who names his sources, is quoting from the same sources as Stenvik, who doesn't. Buchwald, (2005. 146 to 157) describes, in detail, the colours and grain sizes that the iron master would be looking for in bog ores and lake ores, he also provides an analysis of the chemistry involved.

[11] See also La Salvia, 2007 (25 to 26) for more on iron working between the Elbe and Weser rivers where fifty-nine sites have been identified, dating from the 1st century A.D. and even more between the Elbe and Oder rivers. La Salvia aslo summarises iron production in the Netherlands and Poland (26 to 27).

Northern Italy[1]

Excavations at Ponte di Val Gabbia, Brescia, northern Italy has provided some very important finds. In a sunken hut that had been used as a forge, a chisel was found made from tempered steel obtained from the decarburisation of pig iron (cast iron).[2]

Tizzoni and Tizzoni report on the debate surrounding the introduction to Europe of the indirect method for iron reduction which involves cast iron and its subsequent decarburisation to form a high carbon steel. It had been thought that this method dated to the 13th century but Tizzoni and Tizzoni have been able to show that it was happening in Lombard Italy during the 5th and 6th centuries.[3] This discovery will have implications for other areas of Europe too, we will have to see what further research reveals.

England

Iron working had been carried out in the British Isles for centuries before the Anglo-Saxon period began during the 5th century,[4] but we will confine ourselves, largely, to evidence from the 5th to 7th centuries. Unfortunately so far, few ironworking sites from the earlier Anglo-Saxon period have been identified. Slag blocks, from the 5th and 6th centuries have been found at Little Totham in Essex and at Mucking, also in Essex.[5] Mucking, apparently, represents small-scale production of materials almost certainly for local needs. Mucking is both a settlement and cemetery and dates to the very early Anglo-Saxon period.[6] Another settlement and cemetery site at Bloodmoor Hill in Suffolk has yielded considerable evidence of metal working, particularly with iron. Smelting slags, remains of *tuyères* and hammer scale have been identified at several places within the settlement.[7] Cowgill remarks, "..it is evident that both iron smithing and non-ferrous metal casting were occurring at Bloodmoor Hill at a scale not previously recorded on a site of this date in England."[8] Romsey, Hampshire (probably 6th and 7th centuries) is situated near significant iron deposits, and provided material that showed that smelting technologies were used there. Hamwic, Hampshire, was an urban site and its beginnings are roughly contemporary with the other sites mentioned and high carbon steel may have been produced there.[9] Lang and Ager refer to smithing and smelting operations taking place at Aylsham and Witton, both in Norfolk, Lyminge, Kent and at Shakenoak in Oxfordshire. [10]

A seventh century smithy has been excavated at Wharram Percy, East Yorkshire, from which a considerable amount of smithing debris was recovered.[11] Larger scale activity might have taken place near the 5th to 7th

---

[1] La Salvia (2007. 29 to 33) provides evidence of sophisticated iron production in the areas where the possible forerunners of the Lombards, and other groups, are thought to have originated, dating from the 1st century A.D. La Salvia 1998 (23 to 27) provides a brief but useful survey of iron production in Northern Europe.

[2] Tizzoni and Tizzon, 2003. 49.

[3] Tizzoni and Tizzon, 2003. 49. Also noted in Lang, 2007. 80 and Lang, 2012, 259. Lang, 2007, explains that this could be achieved either by the co-fusion process (heating white cast iron with low carbon bloomery iron) or by burning the carbon off with an air blast. She believes that co-fusion to be the most likely method used.

[4] See Cleere, 1981, for numerous examples.

[5] McDonnell. !989. 374. Hinton, 1998. 3 and 4. Hinton tells us that most excavations of larger settlement sites have produced evidence of smithing.

[6] For a full discussion on the date of Mucking see Hamerow, 1993. 93 to 97, and Hirst and Clark, 2009. 727 to 729.

[7] Cowgill, 2009. 372 to 374. One pile of slag weighed, 32.8kgs while another 26.6 kgs. Cowgill expresses regret at the speed at which the excavators had to work.

[8] Cowgill, 2009. 380.

[9] McDonnell, 1993. 83. McDonnell, 1989. 375. dates the workings to 7th century onwards, although on pages 378 and 379 McDonnell lists Hamwih (*sic*) as 6th to 9th centuries. Lang, 2102, 259 Quoting Mack, *et al*, 2000. notes that there is a possibility that liquid steel was produced from cast iron in this area. See also Lang, 2007. 80.

[10] Lang and Ager, 1989. 112. They also mention iron working at East Grinstead, but this may be very late in the Anglo-Saxon period.

[11] McDonnell, 1993. 83. McDonnell. 1989. 375.

settlement and cemetery at Cleatham, North Lincolnshire. A series of sites, known locally as the 'cinder hills' represent slag remains of ironworking that took place over a considerable period of time; the iron could have been bog iron extracted from the nearby heathlands at Manton Warren.[1]

An important 8th to 9th century site used for smelting iron in considerable quantities and over an extended period was discovered in Ramsbury, Wiltshire.[2] Apparently, its discovery, "…came as a total surprise…"[3] and although it is outside our time period, it does demonstrate that the exploitation of iron was happening in the west of England in a very organised way during the slightly later period.

In addition to the above, Hamerow provides a very interesting summary of settlement during the 5th to 10th centuries in North-Western Europe and, in her work, discusses iron production and working, together with trade and exchange networks in the region during that period.[4]

The fact that ancient iron has not been regarded as at all a 'glamorous' subject to study, and that it and its associated slag, etc, tend to look unprepossessing, may help explain why few scholars have engaged with it to study it and that many such remains may have been overlooked and unreported, not just in England but elsewhere too. Thus the frequency of reported finds may not provide an accurate picture of the amount of iron working sites that existed nor of the level of technical expertise achieved by smiths during the 5th to 7th centuries; this situation could possibly change in the future as new discoveries are made and old ones re-evaluated.

Methods of production

The study of early iron and steel has undoubtedly been hampered by a modern dual perception of this metal as being first a utilitarian, mass-produced commodity, which secondly, is transformed in the ground into rusty lumps which do not tend to stimulate our interest, or bring out quite the same sense of awe as do well preserved gold, silver or even copper alloy objects from earlier times. This has led to an assumption that iron technology, unless proved otherwise, was carried out at a rudimentary level until relatively recently.[5]

Although iron is a very common element and constitutes around 5%[6] of the weight of the earth's surface it is almost never found in its pure state and is comparatively difficult to purify and obtain by the methods available to smelters and smiths in ancient times.[7] As it is usually found combined with other elements it must also have been quite hard to recognise at first, although many of the iron bearing ores are red-brown, this is not always the case. Some other metals, such as copper, tin, silver, lead and gold are fairly consistent, visually, in their ore forms and all of these have relatively low melting temperatures and can, usually, be extracted from ores simply by heating.[8] All of these metals can be cold worked to some extent, although, except for gold, they will become work-hardened and require *annealing*[9] from time to time. By contrast, iron ores exist in a myriad different forms and the ability to recognise them as iron bearing, and then extract the iron must have developed over a long time, probably, at first, by accident and then eventually via much experimentation.

---

[1] Leahy, 2007. 7.

[2] Haslam, Biek and Tylecote, 1980.

[3] Haslam, Biek and Tylecote, 1980. 3.

[4] Hamerow, 2002. 172 to 176.

[5] Gilmour and Hoyland, 2006. 48.

[6] Buchwald, 2005. 63. Cleere, (1981. 22.) says' "Iron is the second most common metal on the earth's surface, but it does not always occur in a form that is readily reducible so as to produce pure metal." Sim and Kaminski, (2012. 1) give a figure of 5.6%. Aluminium is the most abundant metal in the Earth's crust.

[7] Buchwald, 2005. 69. In any case, Buchwald points out that "..pure iron is an academic abstraction." Cleere, (1981. 46.) supports Buchwald.

[8] Buchwald, 2005. Chapter 2. Williams, 2012. Chapter 1.

[9] Heating to soften the metal.

According to Williams, metallurgy, and hence the exploitation of all metals, originally comes from the practice of making and firing pots.[1]

Iron is a peculiar element and is very different to metals like copper, lead, tin, silver and gold; it has radically different characteristics and melts at a much higher temperature than any of those metals, around 1538°C. Iron is allotropic -- it exists in more than one form based on its temperature; at temperatures up to 910°C it is ferritic iron;[2] hard, unyielding and magnetic. Above this temperature it is austenitic iron; soft, malleable and non-magnetic, this transition allows iron to be heavily hot worked by the smith and it can then be drawn into wire or shaped just like softer metals such as silver and gold.[3] One important factor concerning the two forms of iron at different temperature is that when it is austenite, iron will much more readily absorb carbon;[4] ferrite hardly absorbs carbon at all and its lattice can only hold up to 0.4% carbon as a maximum.

Steel is an alloy of iron and carbon, although other elements might be present. Steels usually range from 0.4% to 1.7% carbon .[5] Iron with carbon content below 0.4% is not really steel because it behaves more like iron and cannot be hardened by heat treatment. The word 'steel' is just a name given to a particular set of alloys of iron and carbon. Cast irons have carbon contents in the range 2.5 to 4.5% carbon. As the carbon content increases the melting point of the metal decreases and cast iron, for example, will melt at 1130°C. Most cast iron is very brittle and is not terribly useful for weapon or tool production, unless some of the carbon can be extracted. Steel is harder and can be tougher than iron, hence its desirability. Steel undergoes a *eutectoid* reaction[6] on cooling; this change in phase allows steel to be hardened through heat treatment. As it cools the austenite with high carbon content breaks down into ferrite and *cementite* (iron carbide) which together form a compound called *pearlite* because it resembles mother-of-pearl when viewed through a microscope.[7] Below 0.8% carbon steels are *hypoeutectoid* and above this point they are *hypereutectoid* steels.[8] Iron is a single phase material and does not undergo a eutectoid reaction so cannot be hardened by quenching but only by work hardening when cool. Steel, in contrast, can be further hardened and toughened by suitable heat treatment, which we will consider in due course.

In Europe, It is possible that reliable methods of producing larger quantities of steel were not available to smelters until 15th century, because the *bloomery* process seems to have been used exclusively there;[9] in other continents, where *crucible* steel was manufactured, the situation could have been different.[10] However, smelts conducted in a shaft furnace are quite capable of reaching temperatures where steel and even iron can be melted and exceeding them,[11] so perhaps further thought and research is required, especially as new

---

[1] Williams, 2012. 3.

[2] Sim and Kaminski, (2012. 25) say the transition occurs around 912°C. La Salvia, 2007. 8.

[3] Gilmour,1990. 38. Ferrite or alpha iron (also known as α-Ferrite or α-Fe) has a body-centred cubic structure, whereas gamma iron (austenite or γ-Fe) is face-centred and is only stable at temperatures above 910°C.

[4] Gilmour, 1990. 38.

[5] Buchwald, 2005. 66.

[6] The eutectoid reaction describes the phase transformation of one solid into two different solids. For more detail see http://www-g.eng.cam.ac.uk/mmg/teaching/typd/addenda/eutectoidreaction1.html

[7] Gilmour, 1990. 39.

[8] Gilmour, 1990. 40. La Salvia, 2007. 8. Williams. 2012. 19 and 188. According to Williams the temperature at which steel melts decreases as carbon is absorbed until it reaches about 2% and then will go no lower. At 2% steel melts at 1150°C. La Salvia, 1998. 37.

[9] Williams, 2012. 202. Tizzoni and Tizzoni have cast some doubt on this idea, at least as far as Lombard Italy is concerned. Salvia, 2007 (31) reports on virtually slag free iron tools found in Březno in Bohemia and dating to around 500 A.D.

[10] See Gilmour, 2006, for information on the Middle Eastern and southern Asian crucible steel, for example.

[11] Personal comment Owen Bush. Owen has carried out many smelts (over 30 at the time of writing) and kept accurate records of the way that the smelts were conducted and tested the results. He also belongs to a network of smiths, who between them have built up an enormous bank of experience and discuss, regularly, what they do. Sim and Kaminski (2012. 15) agree, they discuss

discoveries, like those in northern Italy are made? Sim and Kaminski have pointed to the virtually slag free examples of iron and steel used in Roman armour and wonder if this is evidence of an ability to acquire and maybe manufacture steel using the indirect method by adding carbon to liquid metal in an oxygen reduced atmosphere (crucible steel).[1] This would seem to provide a basis for what could have been happening in Northern Italy some centuries later.

Iron working appears to have first started in Anatolia,[2] at least in a fairly organised way and over time, at least three different methods evolved to produce steel. The method used in Europe (and in Japan)[3] to obtain iron and then steel was the *bloomery* process. Buchwald lists four steps in this process the first of which is *roasting* where the temperature of the ore would be raised to around 400-500°C for several hours. The intention was to drive out moisture and get the ore ready for crushing into smaller pieces. The smelter would also have to select the best bits of ore to use during this process. The ore would need crushing to a coarse powder before being smelted. The next stage is *production* where the pre-heated bowl, or shaft furnace (figures 5.2, 5.3 and 5.4) was charged with alternating batches of iron ore and charcoal. In some areas iron rich sand could be used instead of ore. Eventually a *bloom* of iron or steel, plus some slag will form through chemical reactions; the iron never melts;[4] the bloom forms via a chemical reaction. The third stage of the process involves the removal of the bloom and the slag and the final stage is the forging of the metal into a manufactured item.[5] The bloom could be left within the furnace and heated for longer, perhaps even during the formation of other blooms, so that parts of it absorbs more carbon but this will not be a uniform process and part of the skill of the smelters and smith is in recognising which parts of the bloom are useful for which purposes. The bloom would have to be broken up, perhaps by dropping while still hot into cold water, to get at the higher quality pieces which could then be forged together into larger packets of steel.[6] Occasionally, lumps of high carbon steel and slag, called 'gromps', by some, might form elsewhere in the furnace; these would have been highly prized, although they would need work in order to get rid of most of the slag. Gilmour adds other stages of preparation that would be carried out prior to roasting and that is washing and weathering. Washing involves rinsing the ore to remove all the soft soil and clay, while weathering require leaving the ore in the open, possibly for quite a while so that the weather can remove the 'earthy parts'.[7] Owen Bush reports that when using ores with a high sulphur content, which can make the results brittle, it might be possible to wash much of the sulphur out by leaving it in a running stream of water for days or weeks.[8]

The second method of producing steel was the *crucible* process. In this method, small pieces of iron were separated from the bloom and, together with carbon, were heated in a closed crucible, made of clay, until the

---

experiments by several researchers, including Pleiner, Tylecote and Rehder and say that a simple shaft furnace can reach temperatures of 1600°C. La Salvia, 1998 (34) states that, "...it was impossible to reach iron's melting point (1,530 C) with the types of furnaces then available." It would appear that La Salvia was incorrect.

[1] Sim and Kaminski, 2012. 12, 20 to 23. They point out that repeated heating and hammering will only get rid of slag inclusions until they are at about 5%. After that the inclusions are surrounded by metal and cannot escape. There are many examples of Roman work that have a much lower level of inclusions. Sim and Kaminski, 2012. 16 and 18.

[2] Buchwald, 2005. 86. Buchwald provides a very brief, but insightful, history of iron production from page 63 to 113. Others have different ideas, see Cleere, (1981. 58.) for a list of commentators. Cleere (1981. 67 to 68) is more cautious about an ultimate origin, but does discuss possible routes that iron-working may have taken to reach the British Isles.

[3] Williams, 2012. 24.

[4] Although there is no doubt that the bloom forms via a chemical reaction, the furnace, as we have seen, can melt iron to a liquid state.

[5] Buchwald 2005. 90 to 94.

[6] Williams, 2012. 16.

[7] Gilmour, 1990. 51.

[8] Personal comment, Owen Bush. Sim and Kaminski (2012. 5) report that such a process as Bush describes can turn sulphide ores, such as marcasite into limonite, eventually.

components had melted and formed steel ingots. Controlled cooling and forging will then produce the famous *wootz* cakes or ingots which would have a potential pattern within its grain structure.[1] This wootz came to be labelled *damascus* steel because of the patterns, which has led to some confusion when discussing pattern-welding. It must be said that not all steel produced in this way was wootz.[2] It is possible that some of this crucible steel could have found its way to Europe and may be the steel that was used in the original *Ulfberht* swords (+VLFBERH+T).[3] According to Williams the crucible method was used first in Persia and Central Asia and only later in India.[4]

The third method of obtaining steel was simply to leave the bloom in the furnace until it melted and produced cast iron, which was the usual starting point for iron and steel working in China.

As all these methods seem to depend upon the bloomery process to a lesser or greater degree it is difficult to see why European smelters and smiths, perhaps working in isolation would not have experimented, at least on a small scale with method number three and perhaps even the second one. Certainly, something different was happening in Italy by the time of the Lombards and perhaps much earlier under the Romans. And then there are the Ulfberht swords themselves which show that at least some European smiths knew how to work such metal wherever it was produced but these swords occur well after the period that this book is devoted to.[5]

Furnaces

Producing iron and steel was labour intensive and needed abundant resources.[6] These resources included a suitable site, wood fuel for charcoal, clay and other materials to build the furnaces, iron ore, a water supply, a great deal of preparation and many man hours of work to extract the iron and steel. We will consider some of these requirements below, but first we will look at the types of furnace used.

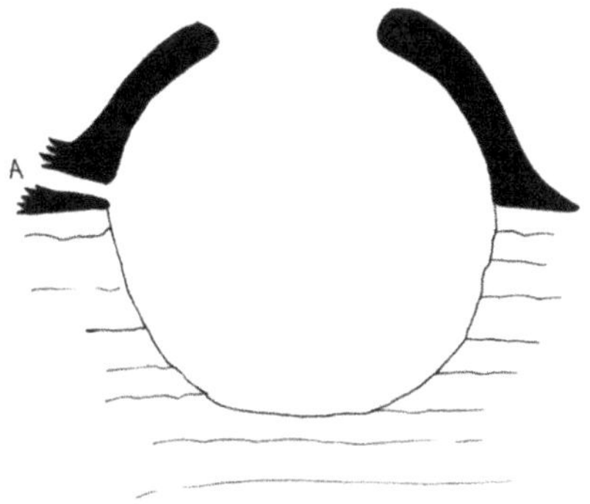

5.2. An attempt to reconstruct a simple bowl furnace. The black areas would be made from clay and would keep heat in but still allow the furnace to be regularly charged. 'A' would allow air to be inserted via a tuyère. (Author's drawing). However, such a furnace would be inefficient and there was not really any need for them as metal workers had been using shaft furnaces to smelt copper long before the Iron Age. (Brian Gilmour, personal comment). In addition the bowl in the ground could have had a shaft above it and not a dome.

[1] Williams, 2012. 24. Presumably the bloomery process is used to obtain iron in the first place.

[2] See Williams, 2005 chapters 3, 6, 7 and 8 for further discussion of this topic, including some ancient recipes. According to Gilmour, the term 'wootz' just means steel. This is borne out by wikipeadia, which says that it is probably a mis-transcription of the Dravidian word, '*urukke*', the word for steel: http://en.wikipedia.org/wiki/Wootz_steel This subject will be discussed further in the chapter on Pattern-welding.

[3] See Williams, 2012. 116 to 183 for a full discussion of what appear to be original Ulfberht swords and their copies.

[4] Williams, 2012. 31. Prior to the 6th century A.D. the bloomery process, only, was used in India. Sim and Kaminski (2012. 5, 58, 61 and 64) report that the Romans valued such trade with the East.

[5] See Williams, 2012.

[6] See Cleere, 1981. 77 to 78 for some examples and arithmetic.

There is not a great deal of information regarding the types of furnaces used to smelt iron by the 5th to 7th century Anglo-Saxons. During the later period at Ramsbury bowl[1] furnaces (figure 5.2) were still being used effectively.[2] It is likely that shaft furnaces were also being used as the Romans in Europe and in Britain had been using both bowl and shaft furnaces and both types would have had tapping arrangements to remove slag.[3] Bowl furnaces are basically a dome shape of clay and other materials, sometimes, but perhaps not often, built over a small depression or pit in the ground. These, as we have seen, were still used as late as the Middle Saxon period (650 to 850) in England and elsewhere but, perhaps, were less than predictable in their results and relatively inefficient in their use of resources.[4] It is possible, too, that this type of furnace was not capable of making very much iron on each firing and might not have produced consistent results. It would seem that the addition of a shaft made it much easier to use resources more efficiently and gain a better level of control over the production process. Certainly, tall, slag-tapping shaft furnaces had been used by the Romans in Britain as had low-shaft dome furnaces, probably by the Britons.[5] Stenvik tells us of Northern Europe that, "The shaft furnaces thus represent an old tradition extending back to the Bronze Age."[6] He reports that remains of ancient shaft furnaces have been found in Denmark, Sweden, Norway,[7] Poland, Germany and the Netherlands.[8] One of the sites in Germany contained a complete shaft in situ which has been very informative. There would seem to be few obstacles to accepting that the inhabitants of England, during the 5th to 7th centuries, would have been aware of the technology and were putting it to use.

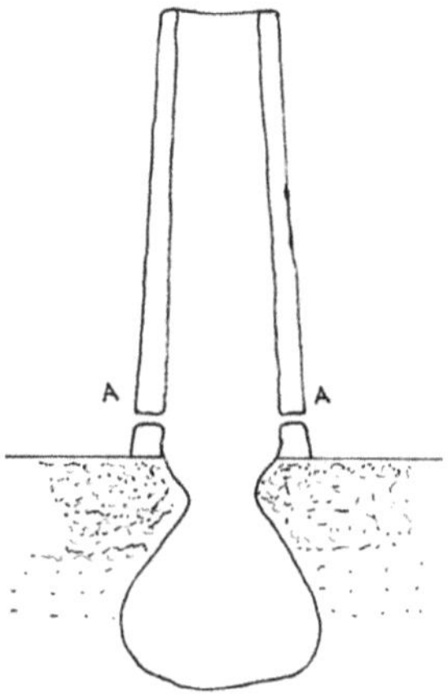

5.3. A slag tap furnace. The bowl dug out of the earth below the shaft would be filled with 'straw' before firing. During the working of the furnace, liquid slag would drip into the void and collect in the bowl. The bloom, which would fill close to the tuyères (A), would not be able to pass into the bowl because of the narrow neck. For fuller details of the working of this type of furnace see, Mikkelsen, 2003. 43 to 48. (Author's drawing, based on Voss 1993).

[1] It is difficult to see how a 'bowl' furnace can be distinguished from a shaft furnace archaeologically; once the 'dome' is removed from the top, if it was ever present, what remains in the ground could have had a shaft over it. Brian Gilmour, personal comment.

[2] Haslam, Biek and Tylecote, 1980.

[3] Cleere, 1981. 58 to 59. Cleere thinks that furnaces without a slag tap are unlikely to have been used during the Roman period and I think that it is just as unlikely that most furnaces during the 5th to 7th centuries would have been without either. Sim and Kaminski (2012. 16 to 17) report that bowl, shaft and pot (domed) furnaces were used by the Romans.

[4] Ottaway, 2012. 2. Again, the 8th to 9th century furnaces at Ramsbury appear to have been bowl furnaces, see

[5] Lang, 2007. 40. Cleere, 1981. 68.

[6] Stenvik, 2003. 78.

[7] Stenvik, 2003. 77 to 78. From the Early Iron Age.This type of slag-pit furnace seems to have operated with only natural draft and was reusable.

[8] Stenvik goes further and says that so far this type of furnace has been found on sixty site representing more than 20,000 furnaces. Stenvik, 2003. 79.

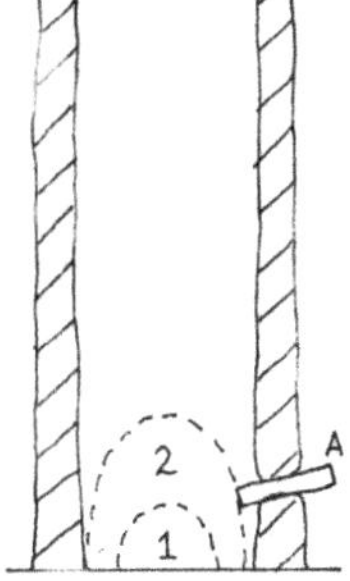

5.4. A simple shaft furnace of the type used by Owen Bush. 'A' is where the tuyère would be inserted. There are two pre-pared weak areas, '1' would be opened when the slag is ready to be evacuated and '2' when the bloom is taken out. This type of furnace can be re-used with minimal repairs. (Author's drawing).

Slag tap furnaces consisted of a shaft, some form of ventilation, presumably some form of forced air, probably via bellows, in order to provide sufficient oxygen for the high temperatures required for the process of making iron to be achieved (figure 5.3).If the shaft was tall enough, it could be that bellows would not be needed as sufficient air could be drawn through the chimney to obtain the higher temperatures.[1] Slag tap furnaces have a hole near the base which would be opened up, at the appropriate time to release the liquid slag to flow out and much of the more solid pieces to be raked clear. Such a furnace has the potential to be used many times, although it would need to be maintained through regular repairs.

According to Stenvik, the shaft on both types of furnace would be made from clay, reinforced with straw, stones and bits of slag and the walls would be reasonably thick.[2] If using charcoal, the shaft does not need to be more than around one-and-a-half metres high and about one metre in diameter at the base. The dimensions might have to increase if using pine wood as fuel. [3] The shaft will narrow as it rises in order to form a chimney and encourage the intake of air from the base. Around the base would be a number of openings called, *tuyères*, (a modern term) which would allow the intake of air, probably by using a bellows, although normally aspirated intakes were certainly used in historical times.[4]

Certainly, the Lombard furnaces mentioned above, were slag tap furnaces, as there are indications that they had a tap at their fronts and were used many times.[5] Furnaces from Heek, North Rhine Westphalia examined by Nikulka were found to have had induced air and tuyères by the Roman Iron Age.[6] By the Migration period (late 4th to 6th century) production of iron intensified in the north Rhine area with sites of up to 50 furnaces.[7] Further east, at Wolkenberg, an area never dominated by the Romans, relatively large scale exploitation of the local resources was happening. At Wolkenberg, they appear to have used slag pit furnaces of a special type; the slag pit was dug into the ground and was large in capacity so, presumably, could be used several times until

[1] Personal comment, D. Riley. African furnaces in country areas in the modern day have eight foot chimneys and do not require bellows, but they are probably working in higher ambient temperatures than were available in most of early medieval Europe.

[2] See comments below about a smelt in practice and the thickness of the walls.

[3] Stenvik, 2003. 78. Cleere reports shaft furnaces of 3metres going up to possibly six metres in height, but from a much later period than ours, see Cleere, 1981. 57.

[4] Some of the lumps of slag kept in the National Museum of Denmark from around our period are huge, and must have come from much larger furnaces than the one described here.

[5] Tizzoni and Tizzoni, 2003. 51.

[6] Nikulka, 2003.63 to 63.

[7] Nikulka, 2003. 64.

full.[1] The shaft would have to be removed after each bloom was produced and there is good evidence for the regulated entry of air.[2] It would appear that charcoal was produced nearby as there are a large number of remains of kilns for the creation of the fuel on site.[3]

Within Denmark variations of both basic types of furnace were used up to the seventh century.[4] Major discoveries of iron producing sites from the Roman Iron Age have been located in parts of Norway, at Trøndelag near Trondheim in the north, at Dokkfløyvann in Oppland and at Eg, near Kristianssand in the south.[5] The Norwegian furnaces are mostly of the slag tap type and some of the accumulations of slag are impressive and show that the industry was carried out on a large scale.[6]

As previously mentioned evidence for iron production during the early Anglo-Saxon period, is scant, but not non-existent, perhaps due to the fact that such sites have not been discovered or recognised. Where there is evidence, such as the 'Cinder Hills' in Lincolnshire it would appear that it was carried out on a sizeable scale and over a long period of time. Certainly, the early Anglo-Saxons needed iron and steel, just as any other European people did at that time and were as familiar with the techniques required to make the material as any other post Roman group.

A smelt in practice

On Monday 21st July 2014, I was fortunate to witness a smelt directed by Owen Bush in Kent. Owen was using siderite from Kent that he had collected himself, under the auspices of the Wealden Iron Research Group. To the siderite Owen added an equal amount of magnetite as an experiment. The siderite had been roasted for a considerable time and then crushed by hand. Before the roasting, the siderite was yellowish on the surface and light grey inside and quite hard; the roasting altered the ore considerably, the colour changed and it became much easier to crush being reduced, relatively easily, to a dark orange-brown gravel.

The furnace was built from clay that had been put through a pugmill and had had grog added.[7] The furnace was about one point four metres tall, the aperture at the top was around fifty centimetres wide, with walls that were approximately six cm thick, but much thicker at the bottom. If the walls are too thick, the furnace would be so well insulated, that the heat inside would start to burn the walls; it is, therefore, important that some of the heat is able to permeate through and escape. The furnace tapered slightly as it rose from its base. The base of the furnace was lined with ash to which was added a layer of charcoal dust.

The furnace was preheated for about an hour before being charged with ore, once heated to the correct temperature, charges of charcoal together with an equal weight of ore were added until the furnace was full. Air was fed into the furnace near the base using a *tuyère* that was powered by a compressor, instead of hand driven bellows, otherwise the whole procedure was much as it was in ancient times. About every thirteen minutes the charge would drop about six cm and then equal weights of ore and charcoal were added each time until the furnace was full again.

The furnace had to be tended constantly and any escapes of hot gases near the base were treated straight away with brushed-on wet clay to seal them.

---

[1] One contained 194 kgs of slag. Spazier, 2003. 40. Another contemporary site nearby, Metzdorf, had three hundred furnaces of a similar sort.

[2] Spazier, 2003. 38. There appear to have been very similar 'shaft removal' and large slag pit furnaces in Norway, see Stenvik, 2003. 78. These were normally aspirated and seem to have used pine wood as fuel.

[3] Spazier, 2003. 40.

[4] Stenvik, 2003. 78 to 79.

[5] Buchwald, 2005. 230.

[6] Buchwald 2005. 231. The estimated weight of slag for the five furnaces at Storbekken, for example is twenty-four to forty-eight tons.

[7] 'Grog' is the crushed remains of a previous furnace.

This process continued until about twelve kilos of ore and twelve kilos of charcoal had been put into the furnace. At this stage a small aperture was made in a pre-prepared area at the bottom of the furnace to let the slag out. Initially this was copious and free-flowing, just like liquid lava, but gradually slowed as the slag became thicker and more like a stiff porridge. For the rest of the smelt, Owen would occasionally rake out more slag. The slag would always be shovelled clear straight away, so that the area immediately outside the aperture was kept free of obstructions. After the initial slag removal, the contents of the furnace dropped considerably and it was, again filled to the brim with equal amounts of ore and charcoal. Now that the bottom was opened, the drops in the contents of the furnace became more frequent and it had to be refilled more frequently, around every eight minutes.

Once the full charge of thirty-eight kilos of ore and thirty-eight kilos of charcoal had been put into the furnace it was allowed to burn down. A small amount of extra ore was placed on top of the burn as a cap to help exclude oxygen.

As the smelt neared its conclusion, Owen began carefully to enlarge the hole at the base of the furnace in order to remove the bloom with tongs. The bloom had formed through a chemical reaction just below where the air from the tuyère entered the furnace. Eventually the furnace released the bloom, and some of the agglomerations clinging to the main section of the bloom were immediately removed by Owen's students using sledge hammers. Owen then cut the bloom into two main sections, one of which was further split into two smaller segments; one of these, the smaller, was hot forged into a flat disc and the other into a roughly square section bar for testing.

The two sections were cooled and then spark tested and found to be partly wrought iron and partly low carbon iron. The bloom yielded more than ten kilos of iron, although there were many smaller pieces (gromps) that later provided another five kilograms of bloomery iron. The carbon content of the bloom could have been increased if the proportions of the charcoal and ore had been altered; a larger proportion of charcoal would have led to greater absorption of carbon.[1]

Owen, who has had considerable experience of smelting and belongs to a community of others who carry out similar smelts, and discuss and record their experiences, says that various subtleties in furnace design and procedures can help to bring about different results, particularly when using different types of ores. For instance, a pure magnetite smelt by itself produces much less slag and the resulting metal is always steel, but the smelt would have to be conducted differently to the one that I witnessed in order to get good results. The ancient smiths would, perhaps, have been even more aware of the need to alter the way things were done in order to get the results that they required.

Essentially, the main difference between the smelt that I witnessed and one conducted in the Anglo-Saxon period is that the tuyère was powered by a machine pump and not by hand; hand pumping would require at least two men taking turns on the bellows, probably more, as getting sufficient air into the furnace has to be continuous and is hard work and perhaps at least one more tuyère.[2] As it is, the furnace had to be attended to for a period of about five to six hours without a break. It was hot, and quite hard work, particularly wrestling the extremely hot bloom from the furnace with tongs.

Charcoal production

Almost everything to do with making iron is hard work, including building the furnace and making the charcoal, which we will briefly discuss here. It is possible to use other fuels, but charcoal was in regular use by

---

[1] Personal comment, Mick Maxen. Sim and Kaminski (2012. 16) Confirm this and say, "The carbon content of the iron will be dependent on the ratio of iron ore to charcoal when the material was smelted."

[2] Although Sim and Kaminski (2012. 15) report that, "Rehder (1986) argues that a simple charcoal furnace thirty cm in diameter with a single tuyère easily powered by one man can reach a temperature of 1600^{0}C."

the Roman period and has far better characteristics than any other fuel,[1] if consistent results were to be achieved. According to Cleere, its main advantages are high carbon content, no sulphur, high calorific value, it is readily available and easy to produce.[2] I suspect that 'easy' is a relative term as the people who made it had to work fairly hard for a sustained period.

Incidentally, Cleere reports that charcoal suitable for smelting does not have to be made from hardwood as conifers are quite suitable.[3]

Charcoal has to be produced in an oxygen reduced atmosphere in a 'clamp', (figure 5.1) which is basically a wood pile covered over with soil, turf and other materials to keep most of the air out. First a sufficient area of land has to be cleared, the Regia Anglorum re-enactment group[4] when they re-created a charcoal burn, cleared a circle about nine feet across of turf, topsoil and stones. They had about three to four tons of wood to include in the burn.

Once cleared a central stake called a *'motty-peg'* was staked into where the middle of the pile would be and the wood was built around it in such a way that it was self-supporting and would not collapse when the motty-peg was removed.

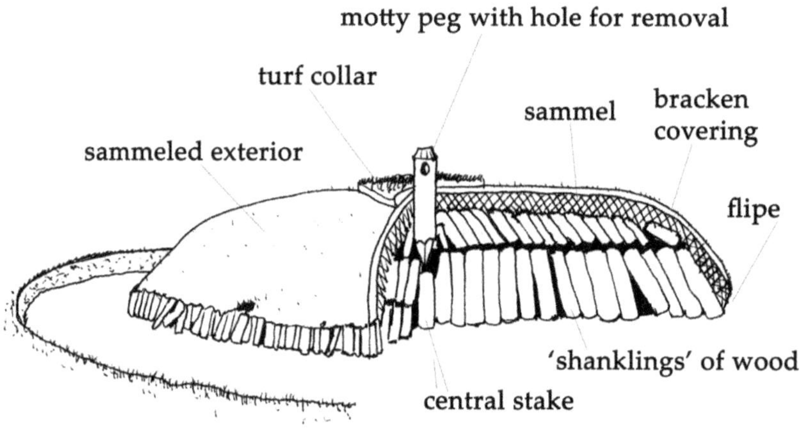

5.1. A charcoal clamp. (Author's drawing after Ben and Colin Levick used with the kind permission of Regia Anglorum)

A considerable amount of digging by Regia members produced piles of clay-rich undersoil, which is called *'sammel'*. After the wood pile had first been covered with layers of bracken, the sammel was used to

---

[1] Until the modern period, of course.

[2] Cleere, 1981. 49.

[3] Cleere, 1981. 57. Apparently early Slavic iron production used exclusively conifers, but study of the Romano-British site at Bardown, Sussex, showed that all the species growing locally were used, including, oak, ash, beech, hornbeam, birch, hazel, hawthorn, and elder.

[4] For the full account see: http://www.regia.org/research/life/charcoal.htm Many thanks to Regia for allowing me to use this information, and to Dennis Riley for additional details.

completely cover the wood pile and exclude the air. The sammel cover stopped at about six inches (fifteen point four two cm) above the ground, to leave an air gap, called a '*flipe*'. The flipe was left uncovered all the way around circumference of the clamp.

The protruding motty-peg had a collar of turf around it to finish it off. The motty-peg was removed and blazing coals were put into the resulting hole to begin the burn. Once fully alight, the hole was packed with charcoal and capped off with turf to seal the hole and the clamp from the outside air.

The clamp has to be continuously attended while the burn is taking place; if a fire breaks out it will have to be put out with water, covered with more sammel or turf and smothered. Meanwhile, the wood inside is being roasted and the moisture driven out.

After about twenty-four hours the burn needed to be halted which meant stripping the covering off the top half of the clamp, dousing it with copious amounts of water and then being recovered with sammel. The bottom half would receive the same treatment next.

Eventually the clamp has to be allowed to cool down as even when the burn has been put out, the hot charcoal, if exposed to air, could catch fire. The cooling can take quite a while and involves a similar process to putting out the burn that is removing the sammel, adding water and then covering over again with sammel.

The charcoal production took the Regia team five days and was reasonably successful, apparently a full size clamp would have been five times larger and could take even longer, perhaps as much as two weeks or more to complete. Of course the ancient charcoal burners were specialists and probably would have been brought up in the trade and would have been able to streamline their methods a little.

Iron ores

According to Cleere, Britain is remarkably rich in iron ore, although he does qualify that statement a little when he reports that, "...the British ores are for the most part not rich ones; they cannot compare with the magnetites of Lapland or the Carinthian ores. They range from the Cumberland and South Wales hematites containing 50-80% iron to the lean Jurassic ores of Lincolnshire and Northamptonshire, with no more than 20-30% iron."[1] Cleere provides a reasonably comprehensive list[2] of where the ores can be located and almost every area of Britain is fairly close to sources of iron.[3]

Cleere explains that there are four main types of iron ore found in Britain, and these are listed here with his comments:[4]

Magnetite is the richest source of iron. It is a black magnetic mineral containing very high proportions of ferroso-ferric oxide (Fe_3O_4). It is usually found in contact with or included in igneous or metamorphic rocks.[5]

Hematite is principally ferric oxide (Fe_2O_3); when it is pure the iron content is 60-70%. In colour it is mainly reddish-brown, and its colour gives rise to its name (from the Greek word for blood). The iron content of hematite deposits varies widely. The pure form is comparatively rare in Britain; the best deposits are those in Cumberland, the Furness area of Lancashire, and parts of South Wales, all of which average about 50% Fe. Pure hematite (e.g. the kidney ores of Cumberland) is somewhat difficult to reduce with the temperature and reducing conditions that could be achieved in the bloomery smelting furnaces of the type used in antiquity.[6]

---

[1] Cleere, 1981. 22.
[2] He excludes very small deposits.
[3] Cleere, 1981. 22 to 35.
[4] Cleere, 1981. 22.
[5] Cleere, 1981. 22.
[6] Cleere, 1981. 22. As discussed above, this may not be as big a problem that Cleere suggests.

Hydrated ores cover a number of minerals containing large proportions of hydrated ferric oxide, the principal ones being limonite ($2FE_2O_3.3H_2O$) and goethite ($Fe_2O_3.H_2O$). These ores usually occur as bedded masses; the so-called 'bog ores' are of this type. They are relatively easy to reduce in bloomery furnaces.[1]

Carbonate ores cover those important deposits containing iron carbonate ($FeCO_3$); they are variously known as chalybite, siderite, and spathic ore. The iron content of the pure ore is high, but pure ore is very rare indeed; the carboniferous clayband and blackband ores of Britain generally contain about 30% Fe and that from the Wadhurst Clay in the Wealden series around 40%.[2]

Cleere does explain that deposits of iron ore in Britain rarely occur in neat homogenous deposits and that particular finds of iron might be composed of mixtures of different types of ore.[3]

As to how the ore was collected, it is not impossible that it was mined, that is pits were dug into the ground to follow an orebody, after all, as Cleere points out Neolithic peoples in Britain had been quite capable of going fairly deeply into the ground in their quest for suitable stone. Sim and Kaminski agree.[4] However, he also tells us that, "It should also not be overlooked that iron ores are very common in Britain. Their abundance was such that, when extraction became too difficult, it would have been simpler to move to another site where it was easier to mine."[5]

Smithies and tools

But the ancient blacksmith learned his business by trial and error and very rapidly made good progress. His major problem may have been to separate the solid iron from the furnace bloom and slags, and consolidate the iron into a handy bar. When he had mastered this step, subsequent shaping and welding were relatively easy. Vagn Fabritius Buchwald.[6]

A smith's job was to take the product from the furnace and turn it into usable objects, tools, utensil and weapons, including swords. It could be that some smiths specialised and only did certain kinds of work; it is unlikely that the average village blacksmith would have been capable of making the best pattern-welded sword, for example. Certainly, blacksmiths, and other craftsmen and women, were grouped together in *fabricae* by the Romans in order to serve the legions. Known sites are, Wiesbaden, Mainz, Trier, Lorch, Nierderbieber, Köln and Neuss, in Germany and Benwell, Houseteads and Corbridge, in England.[7] It is possible that some of these centres stayed in operation for a considerable time after the Romans had officially left those areas.[8]

The tools of a smith haven't change very much in 3,000 years of working, apart from maybe the anvil, which probably started as a boulder then became a large piece of shaped iron fixed in a heavy wooden base.[9] Today, they are made of cast steel. On this much of the smith's work would be carried out. Ancient anvils were

---

[1] Cleere, 1981. 22.

[2] Cleere, 1981. 23.

[3] Cleere, 1981. 22 to 23.

[4] Sim and Kaminski, 2012. 9 to 10.

[5] Cleere, 1981. 51.

[6] Buchwald, 2005. 70.

[7] Buchwald, 2005. 284.

[8] Petersen, 2013. Petersen provides a great deal of evidence to show that much of the infrastructure, including administration in much of Western Europe was maintained and continued.

[9] A sarsen stone seems to have been used as an anvil at Ramsbury in Wiltshire. See Haslam, Biek and Tylecote, 1980. 17.

probably rarely as big as an anvil is today and only needed to be around ten cm in either direction in order for the forging to take place.[1]

The smith has two main operations to perform on the metal. The first of these is shaping, whether hot forging with a hammer, bending, cutting or filing with other tools while hot or cold, the aim is to achieve the shape of the object that he is making. The second operation is to weld different pieces of metal together either because he needed a larger piece of metal or because he wanted a different type of metal in different parts of the object, for instance a steel edge on a blade.[2] We will return to the subject of hammer welding below.

Another skill that the smith had to develop would be to distinguish between the different types of iron and steel that he was presented with. For instance, it is important that he be able to tell the difference between unalloyed iron, that is iron with little carbon in it, and carbon rich iron, with around 4% carbon. They both look similar and even taste the same, according to experiments by Riley.[3] However, the smith can, by experience, tell the difference when forging; the carbon free iron will feel soft when hammering while the carbon rich material will be more resistant.[4] Buchwald suggests the smith would also be able to work out the approximate levels of carbon and phosphorous within the material by its colour and tendency to give off sparks[5] when forging and by filing when cool, although, of course, he would not know what those elements were or even their names, but he would know what he needed for particular purposes. He would gain an estimate of its toughness by looking at the grain size on a fractured surface.

The smithy would be deliberately dimly lit in order for the smith to be able to see the colours of the material that he was using when heating as this would inform him how and when to perform different actions, such as hammering, annealing or hardening. Understanding the surprising variety of colours came through lessons passed down and experience.[6] Lang reports that a rural blacksmith would take a strip of steel to white heat at one end noting the variation in colours along its length and then quenching it. He would then break bits off along its length and examine the grain size; the desired temperature (colour) would be where the grain size was the finest.[7]

Heat treatment

According to Williams, swordsmiths of the past may have been confused by the fact that steel, despite being harder than iron, has to be forged at a lower temperature because of its lower melting temperature.[8] It was certainly not the only awkward fact about hardening steel that smiths had to contend with. Until recent centuries there was no way they could gain a complete understanding of the physics and chemistry of the metals they were working with. To start with they relied on the knowledge passed down to them by those who trained them and to this they added experience gained from their own trial and error. Craft skills had value as a means of earning an income so smiths would have been reluctant to share their discoveries with

---

[1] Personal comment, Dennis Riley.

[2] Buchwald, 2005. 274 to 275.

[3] Riley, 2011. 20 to 26. Riley actually considers three types of metal and attempts to use colour, filing, sparking, touch stone marking, tasting, etching and forging. Only forging and sparking with a modern grinding wheel allowed differentiation. Riley's experiments were all carried out using modern materials. Stenvik too talks about methods of determining types of iron that include taste. Stenvik, 2003. 79.

[4] Buchwald, 2005. 282.

[5] Spark testing is regularly used by modern smiths; the shape of the spark plume and amount of sparks indicate the type of alloy. At the smelt that I attended, Owen Bush uses spark testing to find out what alloys different parts of the smelt are composed of.

[6] Buchwald, 2005. Buchwald lists some of the colours and their corresponding temperatures, which, of course, the ancient smiths could not have known, on pages 132 and 133.

[7] Lang, 2007. 72. She is reporting on the work of Bealer, 1995.

[8] Williams, 2012. 25, (footnote 1) and 37.

'outsiders' or those they saw as competitors. Their trade might have been regarded as 'magical' or 'supernatural'[1] by the general populace and I suspect that some smiths were very happy to keep it that way; indeed, many of them, with their use of special liquids, additions of particular plant or animal materials harvested at auspicious times, might well have believed their craft to have divine origins. Ing-Marie Back Danielsson has even suggested that crushed human bone could have been added to the metal when heating.[2]

Successful methods of *hardening* and *tempering* steel developed through time and were reasonably understood by the end of the middle ages in Europe, but it is apparent from the study of blades from the European Iron Age to the 16th century that smiths were not always able, or willing, to harden their blades consistently – or perhaps didn't feel the need to. Often, it might not have mattered that much, after all even a bloomery or wrought iron blade will still cut flesh, and a brittle cast iron dagger would be fine as a stabbing weapon and if swords were rarely, if ever, used for 'fencing' ultimate hardness would not have been the main requirement.[3] However, ideally, a sword blade should be both hard and tough; that is, it should be able to hold an edge and not break whilst in use. It is relatively easy to make a blade that could be sharp or tough, much more difficult to design and make one that can be both.

Smiths learned by experience that repeated heating and forging will help remove, break up or re-distribute slag inclusions, which otherwise would affect the performance of a blade, but methods of *carburising* or *quenching* the metal to harden it were much more difficult to understand and the task seems to have been made harder because it was not possible to make large quantities of useful steel. Therefore most blades were of composite construction, using, perhaps an iron core with a steel edge and even then the steel might be made up from several smaller pieces forge-welded together.[4] Such a blade consisting of very different qualities of metal could be difficult to heat treat effectively when the makeup of the metal was not fully understood.

Steels of more than 0.3% Carbon can be hardened by quenching rapidly in a coolant such as water, oil or brine, higher carbon steels of around 0.6 to 0.8% carbon will harden in air in thin sections.[5] If carried out correctly the act of cooling will ensure that the crystals of austenite will form *martensite*.[6] Martensite can be very hard indeed, VPH (Vickers Pyramid Hardness Scale) 700 to over 1200.[7] Martensite crystals won't allow much movement against each other and so provide rigidity and hardness, the problem is that martensite can be brittle, so it is important that this is concentrated at the edge of the blade where sharpness is more critical, and those other parts of the blade are less hard in order for the bulk of the sword to be tough and protected from fracturing. A martensite edge could be so hard that it could do with tempering by gently heating followed by cooling in air which will decrease the hardness but increase the toughness of the blade

---

[1] Lang and Ager, 1989. 111. We will discuss this idea more fully in the chapter, *The Mysterious Smith?*

[2] Danielsson, 2007. 93, 94, 117, 126, 174, 242, 243, 246, 247, 248. And perhaps semen. We will return to this idea of 'bone coal' below. See also Gansum, 2004.

[3] See chapter, The Sword as a Weapon.

[4] Williams, 2012. 16. This may be a simplistic view if it is applied to the Roman and post-Roman periods

[5] Quenching steel is mentioned in Homer's Odyssey. The episode is quoted in Buchwald, 2005. 72. Also in Sim and Kaminski, 2012, 30. Davidson, 1994. 19., mentions that honey could be and was used in Finland. Pliny also shows that the Romans were fully aware of the possibilities of quenching in different mediums, see Sim and Kaminski, 2012. 30. They point out, too, that Philon of Byzantium (280 to 220 BC) revealed that there was considerable understanding of the processes of heat treatment and work hardening amongst contemporary Iberian smiths. Sim and Kaminski, 2012. 38.

[6] Gilmour, 1990. 43. LaSalvia, 2007. 9.

[7] The Fachhochschule in Flensburg states martensit in modern materials has a Vickers-hardness of 700-1200 (link: http://www.fh-flensburg.de/wt/metalle2.html). Williams, 2012. 21. Gives figures of 200 to 800 VPH21, but this does seem very soft.

considerably.[1] However, these techniques, as mentioned above depend upon knowledge of what is happening to the particles within the blade so hardening and tempering appears to have happened fairly haphazardly, although we shall see there is evidence that some smiths mastered the processes.

Full quenching was possibly relatively rare until later in the Middle Ages but one technique used was *slack quenching* where the blade was plunged more slowly into the coolant or removed from it quickly. Such a delayed or interrupted quench will mean that not all the steel becomes martensite but that the rest of the steel will form *pearlite* and perhaps, *bainite*. This mixture will be less hard but much tougher than pure martensite and, depending on how the slack quench was performed; the heat from the interior of the blade will travel out towards the edges and temper the martensite.[2]

Two other forms of heat treatment seem to have been common and still are. The purpose of 'annealing' is to render the metal into its softest possible state, probably after it has become work-hardened while working at a cool temperature. With ferrous metals this is achieved by heating the piece to a suitable temperature (depending on carbon content) and holding it at that temperature for a prolonged period followed by a quenching in cold water.[3] The item can then be worked cold again if desired. 'Normalising' is carried out to relieve the stresses set up during forging and is particularly needed if the piece is going to require more hot forging; the metal will be heated to red heat - 700°C to 900°C, depending on carbon content and thickness and then cooled in air. Sim and Kaminski report that, "This process produces maximum grain refinement, and consequently the steel is slightly harder and stronger than in the fully annealed condition."[4]

Carburisation

One way that could be used to increase the carbon content of iron was *carburisation*, where the desired parts of the blade would be heated adjacent to a source of carbon for a long period in order for the edge to absorb extra carbon.[5] This process is very slow[6] and will only form a thin layer of carbon steel that would not take many sharpenings, but it might have been possible to occasionally repeat the process, provided the hilt assembly could be stripped off and perhaps the core of the blade would be protected with a coat of clay while the edges received the heat-treatment. The iron has to be covered in a carbon rich material such as leather, charcoal, hoof and horn, all of which were readily available throughout our period and beyond.[7]

Early iron swords up to the Roman Age

The first iron swords were made in Europe by the 'Celts' around 7th century BC. Radomir Pleiner published reports on ninety-three 'Celtic' swords from all over Europe. Fifty-nine of these were examined in cross section and twenty-one were found to be made of iron,[8] most of which consisted of several different pieces of metal hammer welded together. Some of these swords had *piled* blades, a term that we will discuss more fully elsewhere in this volume. Thirty-eight of the blades had either one or both edges made of steel. Some of the blades consisted of a sandwich of iron and steel or a thin steel layer wrapped around an iron core, although

---

[1] Williams, 2012. 21. This would still be important in a blade made from a single piece of steel – something which does not seem to have been possible in Europe until the Ulfberht blades arrived in the 10th century. Buchwald, 2005, page 67 explains that the tempering for martensite requires gentle reheating to 100° to 600°. See also La Salvia, 2007. 9.

[2] Williams, 2012. 22. Lang, 2006. 86. Slack quenching seems to be reported in the treatise on swords by Arab commentator, Al-Kindī during the 9th century. See Gilmour, 2006. 58.

[3] Sim and Kaminski, 2012. 31.

[4] Sim and Kaminski, 2012. 31. Gilmour, 1990. 41.

[5] Williams, 2012. 16. La Salvia, 2007. 8. Carburisation can also be referred to as 'cementation'.

[6] Sim and Kaminski, (2012. 59) say that, "…iron heated to approx. 900 - 1500°C it will absorb carbon. The rate of absorbtion is approximately 1.0 mm per hour for first three to four hours and this rate gradually falls off."

[7] Sim and Kaminski, 2012. 59. The authors comment that such methods were still being used well into the 20th century.

[8] This includes swords with less than 0.3% carbon (mild steel) behaves more like iron and cannot be appreciably hardened by quenching. Williams, 2012. 49.

sometimes the smith had problems positioning the steel pieces in the best places.[1] Forty-six of the original blades examined by Pleiner had steel edges that could be hardened, and only eight[2] of them had in fact been hardened by quenching.[3] Twenty-three of the swords were re-examined by Buchwald who reports that his sample were mostly of wrought iron and had all been work hardened, cold, as part of the finishing process by the smith.[4]

Williams examined Roman swords and was generally less than impressed with their quality, finding from the specimens that they were no better than the 'Celtic' swords. Roman swords seem to be utilising the same techniques in manufacture that was used in the Celtic swords although there is a higher ratio of hardened edges in that around half of the swords that could be hardened had experienced some form of successful heat treatment. He suggests that, "It seems likely that the uncertainties entailed in quenching an unknown metal led to a preference for a reliable sword of moderate sharpness over an unreliable sword of greater sharpness."[5]

Williams points out that many Roman age Gallic swords were made of wrought iron but says that Polybius's comment that at the battle of Telemon in 225 BC the Gauls had to keep straightening their long swords by placing them on the ground and bending them back to shape is, probably, an exaggeration. He says that an iron sword may bend a few centimetres but will still be perfectly serviceable[6] and that Roman swords were not a lot better in quality. Buchwald goes further and says that the, "Celtic sword and Celtic warrior were feared by the Romans..." and uses a different quote from Polybius to make his point.[7] Buchwald categorically states that it is virtually impossible to bend a 'Celtic' sword in the way that Polybius describes (presumably at a drastic angle) and it is certainly not possible to bend it back without reheating and forging.[8]

The Arab writers

There is very little in the surviving literature concerning sword making from Europe, and nothing from the 5th to 7th centuries. However a number of writers from the Islamic world have left us their writings. The earliest paper that survives was written by an Arab philosopher, Yaᶜqub ibn Isḥāq al Kindī, writing in the 9th century for the Abbasid caliph Muᶜtaṣim (833 – 842 A.D.) provided two works on sword making techniques and the types of swords available in the Islamic world.

Kindī's highly detailed work demonstrates a number of important aspects of sword making and the knowledge of iron and its alloys; some of these could be relevant to sword making in Europe two hundred years, and more, earlier. Eventually as research progresses, we will know more.

Firstly, smiths knew a great deal about the way that various types of iron could be used and how they would behave in a blade and where to get the metals from. They knew how the best results could be achieved and were aware of the limitations of the materials that they were working with and that sometimes compromises were necessary. They had, too, an established vocabulary that they could use to refer to types of metals, processes, and blade characteristics. Swords were graded into many different types,[9] including 'Frankish'

---

[1] Reported in Williams, 2012. 49 to 50.

[2] Williams, 2012. 59.

[3] Williams, 2012. 50.

[4] Buchwald, 2005. 113 to 124.

[5] Williams, 2012. 58.

[6] Williams, 2012. 18. Buchwald also finds Polybius's statement difficult to reconcile with what he has learned of Celtic swords. Buchwald, 2005. 127. However, Vince Evans knows of an instance where a friend, using a modern swords, unwittingly caused it to bend some 6 cms out of true during a bout. Personal comment.

[7] Buchwald, 2005. 127. The quote is taken from Polybius, Book 2, 29.

[8] Lang does not agree with either Williams or Buchwald on this point, "The comments may or may not have been accurate, but the archaeological evidence of deliberately bent, sacrificial swords suggests that it was not difficult to bend the swords, despite Buchwald's remarks to the contrary (2005,127-8)." Lang, 2007. 167. However, the buried swords may well have been heated to soften them before being bent and then buried.

[9] Gilmour, 2006. 52 to 82 and figure 8.

swords,[1] from Europe that were pattern-welded and unlike swords made from crucible steel, unlikely to become brittle and break in very cold weather.[2] Incidentally, Islamic smiths also made composite blades, but most of their swords were made using crucible steel and are therefore 'damascus' blades.

Secondly, they were quite advanced in their use of heat treatments to condition their blades and while they knew about quenching and tempering, often seem to have used slack quenching as a preferred option, as did, it seems, many of their earlier European counterparts.[3]

Thirdly and perhaps most significantly, the trade network in alloys of iron and sword parts were extensive and included parts of Europe, Egypt, Asia Minor, India, Sri Lanka and beyond.[4] How extensive contacts were for European smiths in the 5th and 7th centuries still needs investigation.

Finally, Gilmour says that, "…Kindī has provided a clear, well thought out, wide ranging and detailed description of various aspects of sword technology of the day. Nothing is portrayed as mysterious or secret." Which would seem to be in opposition to ideas about the mysterious smith discussed elsewhere in this book, but Kindī was no ordinary man coming from quite a high level of his society and was learned, unlike many of his compatriots. We do not know how the ordinary person in this society would have viewed the smith and his craft. Even if the smith did not view his profession as being linked with the Other,[5] and I am sure that many practical working men did not, it does not mean that others were not in awe of them or their craft.

[1] Gilmour, 2006. 77 to 78.

[2] Gilmour and Hoyland, 2006. 168. Gilmour, 2006. 62. Swords made from crucible steel are more prone to breakage during Russian winters due to their high carbon content.

[3] Gilmour, 2006. 58.

[4] Gilmour, 2006. 53 to 72.

[5] The supernatural.

6. Pattern-welding

Paul Mortimer

Sword carrying warriors in Denmark seem to have been relatively well informed about customs and practises, and very much so about the development of weapons.[1]

Academic interest

Academic and archaeological interest in the process of pattern-welding perhaps began when Carl Engelhardt excavated some of the bogs in the area around Schleswig and published his findings in 1863.[2] He mentions, in his reports, the finding of *damasceret Jernsværd* [3]or 'damascened iron swords'. He was unaware that the swords he was finding were not made by the process of 'damascening' but by an entirely different process; that realisation would come some years later, but not to him. It would appear that Engelhardt had an incomplete understanding of how true Damascus swords were made as he assumed the patterns were probably created by chiselling the iron.

A colonel in the Danish army, O. Blom found, through destructive experiments, that some of the blades were made of 'excellent steel' while others were of 'soft iron'; he even made himself a little knife from one of the blades.[4] He published his findings in 1868. However, it seems from his writings from 1884 to 1903, that it was L. Beck, in Germany, who first realised that the northern swords were made from a different process to that of damascus steel blades from the east and that it was the practice of welding different pieces of metal together that had formed the blades of the bog swords. Beck made models using wax to try and understand how the welding and twisting worked and he was at least partially successful in explaining the processes involved.[5] However, Beck, too, used the word '*damast*' to describe the blades.[6]

In 1927 Neumann [7] published the first scientific study on pattern-welded swords based on a number of swords from the Nydam bog and Davidson reports that he had distinguished three main patterns which Neumann called, *streifendamast*, *winkeldamast* and *rosendamast*. Davidson uses the English terms 'straight', 'herringbone' and 'curving' as translations,[8] although the latter doesn't really mean 'rose like', which is what Neumann seems to have intended. However, Williams says that Neumann called the third category *dopplewinkeldamast* or 'double herringbone' and that the first category, *streifendamast*, is just a reference to a 'piled' structure.[9] Davidson, when writing her book on Anglo-Saxon swords is content to use the three

[1] Lyngstrøm, 2003. 24. Lyngstrøm writing about 6th century Danish warriors; she could just as easily have replaced the word, 'Denmark' with 'England'.

[2] Engelhardt, 1863. An English version was published in 1866.

[3] Engelhardt, 1863, see for example page 40.

[4] Williams 2012. 64.

[5] Williams, 2012. 65.

[6] Beck used the term 'Damaszenerklingen' (Damascene sword) Buchwald, 2005. 282.

[7] Lang and Ager 1989. 87.

[8] Davidson, 1994. 23.

[9] Williams, 2012. 66. However, Lang and Ager, 1989. 87., report Neumann using identical terms to Davidson. Both Williams and Lang and Ager are referring to the same work by Neumann (Davidson doesn't provide full details). It is likely that *dopplewinkeldamast* is a subcategory of *winkeldamast*. I have not been able to obtain a viewing of Neumann's work namely,

categories of patterns that she says Neumann used and we shall see that things have moved on a little since she wrote her book and that the number of pattern types has grown.

The blades from Nydam had been immersed in anaerobic, waterlogged conditions, which protected them from corrosion but, during their stay in the bog the acidic waters had slowly but surely continued the etching process so that by the time that they were recovered the patterns in the blades were easily visible to the excavators, much more so than with many other finds.[1]

Neumann also made attempts at a metallurgic assessment of the blades and we will return to that aspect below when we consider the effectiveness of pattern-welded blades.

It was Herbert Maryon, in 1947, who apparently coined the term, 'pattern-welding' when writing about a sword found near Ely, Lincolnshire. He was able to describe aspects of the process of making the blade fairly accurately but was unable to find a smith who could replicate the effects that he had seen on ancient blades.[2]

In 1948, Albert France-Lanord published a report on an Alemannic sword from a cemetery at Vieil-Aître that had had sections removed which were then etched and polished in a laboratory, he too, used coloured wax to try and unravel the secrets of the process.[3]

Others were working on the problem of revealing how pattern welded blades were made and in 1951, a Norwegian, Aslak Liestøl, sectioned a blade, and worked on it, took very good photographs and made clear diagrams based, this time, on plasticine models. In this way Liestøl was able to make his work more readily available to others.[4]

John Anstee, in England during 1955, was the first to be able to actually produce a modern version of a pattern-welded blade relatively accurately in England. He worked alone in a small forge and conducted a series of experiments until he was successful. He was able to demonstrate how most of the processes involved in obtaining the various patterns were utilised. He insisted on using a small forge and only the methods that he thought would have been available to a smith from the period.[5] Anstee later worked with another experimenter, Leo Bieck, and they were able to demonstrate that 'curving' patterns can be achieved by grinding the surface away in 1961, from that time most of the processes in fabricating a patterned blade were understood.

In 1989, an American smith, Scott Lankton, made a replica of the blade that was found in Mound 1 at Sutton Hoo and presented it to the Director of the British Museum, Sir David Wilson; since that time the recreated blade has been on display at the museum. Lankton, helpfully, wrote about the experiments and the work that ended with his recreation.[6]

That same year, Lang and Ager[7] published a paper about Anglo-Saxon swords and listed many other papers that had been written on the subject of pattern-welding and experiments that had been carried out between 1961 and 1982, before their own paper was published. Since the paper by Lang and Ager, 1989, and the Lankton replica was installed at the British Museum, interest in the subject has increased enormously and there have even been several programmes made for television that have attempted to explain and illustrate the

---

Neumann, B. 1927-8. *Romanische Damastahl*, Archiv für das Eisenhüttenwesen (Dusseldorf). We will discuss and try to define the term 'piled', at least as far as this book is concerned, in the chapter called *Patterns and Pattern Categorisation*.

[1] Gilmour, 2007. 96.

[2] Davidson, 1994. 24. Lang and Ager, 1989. 87.

[3] Davidson, 1994. 24.

[4] Davidson, 1994. 25

[5] Davidson, 1994. 25 and for a full account of Anstee's work see appendix A, page 217.

[6] Engstrom, Lankton and Lesher-Engstrom, 1990. Lankton, 1989.

[7] Lang and Ager 1989. 87.

process and shown smiths making such blades.[1] Interest from academics, staff from the heritage industry, smiths and reenactors, have all stimulated the growth of literature and there are even internet forums where the techniques are regularly discussed and new work shown. Many new pattern-welded blades are being recreated as some very talented smiths have become fascinated by the processes and are enthusiastically learning how to reproduce some of the finds; their thoughts, experiences and experiments provide us with valuable insights into how the smiths of northern Europe went about the creation of such weapons. It should all be documented, but that task is beyond the scope of the present volume. However, we will be considering some of the work of some smiths, including Vince Evans and Jesus Hernandez, two of the most able and experienced pattern-welding smiths in the World, and finding out from them, something about how they approach the planning, designing and making of a blade.

Earlier commentators assumed that the skills involved in pattern-welding and forge welding had died out within Europe by the 20th century due to the adoption of autogenous welding techniques and cheap electric welding transformers that were introduced into workshops all over the world.[2] This might not have been the case. There could have been smiths capable of using pattern-welding techniques throughout the middle ages and down to modern times; certainly shotgun barrels were being fabricated using pattern-welding in the 18th and 19th centuries[3] and military, ceremonial daggers and swords were crafted using similar methods in Germany during the 1930s and 1940s and probably elsewhere.[4] It is possible that had certain sections of the smithing world and archaeologists been able to talk to each other, a good deal of time and effort could have been saved in working out how the earlier smiths had accomplished what they had. However, Gilmour suggests that European smiths were no longer using pattern-welding techniques by about the 11th century, but that the method was used elsewhere in the World, such as Malaysia and the Middle East and that it was re-introduced, in a limited way, into Europe via the Ottoman Empire later, probably during the 18th century.[5] This is an area where it may be worth carrying out further work in order to see just how persistent pattern-welding in Europe, really was. Certainly hammer and forge welding continued in Europe without interruption until modern times and pattern-welding is just a matter of taking those skills a little bit further.

The attitudes of commentators towards the efficacy and purpose of pattern-welding swords has changed somewhat over the years. There tended to be an almost uncritical and reverential acceptance that the patterned blades were superior weapons because of the stages that the components of such a blade went through, e.g. France Lannord, but later writers like Tyelecote and Gilmour have suggested that such swords were little more than elaborate status symbols to be worn rather than actually used as a weapons and have pointed out that there were often shortcomings in the hardening and tempering stages of the swords' production.[6]

'Damascus Steel'

Before proceeding any further it is probably just as well to determine what we mean by the terms, 'damascus steel' or 'damascene'. It is unfortunate that the early writers used the term, but they did and it still causes confusion and is still used to describe pattern-welding in some circles.

First of all it is important to mention that the steel has nothing to do with the city of Damascus; it would appear that it was given this exotic name because such steel came from the east and Damascus was also in the

---

[1] See for example, 'Saxon Gold' made by National Geographic in 2011 and screened around the world.

[2] Buchwald, 2005. 274.

[3] Davidson, 1994. 35.

[4] Johnson, 1995. 315. Both the shotgun barrels and the dress swords and daggers were mentioned by Jones, together with the possibility that the techniques of pattern-welding were never 'lost'. 1997.

[5] Gilmour, 2007. 94 and 107.

[6] An idea that we will explore later.

'East'.[1] A damascus steel sword blade will be decorated and good smiths could manipulate the metal to produce a variety of patterns, but the patterns were potentially already in the cake of metal before it was forged because of the way that the metal and its inclusions are dispersed in relation to each other.[2] Other techniques of layering and cutting could add extra patterns or develop them further. According to Davidson, the Romans appreciated its qualities and imported it. They called it *Seric Iron*, because they thought that it came from China, in fact the steel that the Romans liked to import almost certainly came from India.[3]

The crucible steel used in the Damascus process can be made in a number of different ways, all involve enclosing the materials within a sealed container, often made from clay. Small pieces of cast iron (this has a high carbon content) can be mixed with pieces of bloomery, plain or low carbon iron and heated until molten within the crucible and then extracted and allowed to cool slowly. Another method uses bloomery or low carbon iron to which is added carbonaceous material, such as plant matter and then heated until molten within the crucible.[4] The resulting metal cakes or ingots were sometimes (at least in more recent times) called wootz.[5] The material may have undergone further processing, including reheating, and was difficult to forge, but a skilled smith who had learned how to work with it, could produce an excellent blade and one with patterns; the patterns would be revealed after careful etching had taken place.[6]

As far as I am aware, no sword that was used during the 5th to 7th centuries in northern Europe has ever been found that was made using the wootz process. Even if European smiths knew of it and could successfully forge it, it might not have been suitable for swords in northern Europe because of the nature of the winters and the tendency that crucible steel swords have to snap in such cold conditions.[7]

The early history of pattern-welding

The majority of swords found in England that date from the 5th to 7th centuries were made using the techniques of pattern-welding in order to construct the core of the blade.[8] It is a complex process and can produce a visually striking pattern; such blades were highly valued in the north of Europe, and not just Anglo-Saxon England, for several centuries, at least from the 2nd century AD and possibly until as late as the 10th. In this section we will consider how they were made, how the processes were possibly developed over time and attempt to explain why they were so treasured.

---

[1] Gilmour, 2007. 94. Gilmour, 2006. 54. It would appear that good swords had once been made in Damascus, but that by Al Kīndi's time (9th century A.D.) production there had ceased see also Gilmour and Hoyland, 2006. 172.

[2] Williams says that the pattern was caused by large crystals of cementite (Fe_3C). Williams, 2012. 36. Gilmour reports that the 'watering' patterns are caused by impurities such as vanadium or phosphorous "Their presence on solidification gives rise to a segregation effect within the metal that persists through subsequent heating/cooling and forging cycles and results in a watered effect on the surface of the blade after this has been polished and etched." Gilmour, 2006. 52. Wootz is characterized by a pattern of bands or sheets of micro carbides within a tempered martensite or pearlite matrix.

[3] Davidson, 1994. 20. Davidson explains how the name *Seric* arose. See also Sim and Kaminski, 2012. 58 and 64.

[4] Gilmour and Hoyland, 2006. 172. Some smelters apparently used pomegranate skin to provide the carbon. Sim and Kaminski (2012. 58) add that leather, charcoal, hoof and horn were used and easily available.

[5] See chapter on Iron and Steel for etymology of *wootz*. See also, La Salvia, 1998. 37.

[6] Davidson, 1994. 20. See also Lang and Ager, 1989. 86. Mäder believes that there is little evidence for etching and the patterns would have been revealed through a high level of polish. Mäder, 2001.

[7] Gilmour and Hoyland, 2006. 168.

[8] Gilmour posits that 90% of swords may have been pattern-welded during the Late Roman period in Northern Europe. Gilmour, 2007. 99. I suspect that based on the evidence, there is no reason to suggest that this proportion decreased until the mid-7th century at the earliest. See also La Salvia, 1998. 41.

The technique did not begin in England, or with the Anglo-Saxons or their forbears; the consensus is that it was the 'Celtic' peoples of central Europe who took the first steps.[1] Smiths certainly knew that some metals were better at carrying a cutting edge than others, because of their hardness while others were 'tougher' and could resist shock more easily; the trick was to get both characteristics into one blade and produce a sword that wouldn't break when used but would perform effectively as a weapon. It is highly likely that smiths had great expertise in identifying which results of the smelting process were suitable for which function and we should not underestimate their abilities in this area.[2] By the time pattern-welding is being used extensively, iron working had been taking place for more than a thousand years in Europe and much experience had been gained. We know that the peoples of the La Tène Cultures[3] had experimented with the manufacture of iron sword blades and were probably using fairly small pieces of suitable iron and steel to make their blades, which may have been all that was available to them after their bloomery process was performed.[4] It is possible that in Europe we have them to thank for developing the art of forge or hammer welding different pieces of metal together, whilst the iron was in its austenite phase, more specifically white hot, and hammering the metal on an anvil to join the parts, permanently, together. They will have also noted that folding and re-forging metal repeatedly would have produced a 'layered' effect that was aesthetically pleasing and improved the performance of the metal;[5] they might have seen, too, that the natural grain in iron could be quite attractive. Over time, it seems that plates of metal of differing qualities were systematically superimposed upon each other to form stacks.[6] Later several rods of iron or carburised iron[7] were being laid side by side and welded together to form a 'striped' or 'banded' blade so that the weld lines ran parallel to edge of the blade.[8] These methods would, produce a better blade if the process was executed properly and, it would seem that the aesthetic effect was pleasing to the eye.[9] According to James, the Romans appreciated the technique and copied it for themselves.[10]

It is not possible to pinpoint when smiths began to experiment with patterns and to produce the more elaborate designs, but as James points out smiths, possibly working for the Romans, around the 2nd century A.D were experimenting with the manufacturing and design boundaries in order to try and produce a better weapon. He suggests that it was then that twists were added to the bars by smiths in the Northern provinces, and full pattern-welding was born.[11] He feels that this development probably took place in areas like Gaul that were used to large troops of mobilised soldiers and had long traditions of iron working.[12]

---

[1] James, 2011. 184. 213.Williams, 2012. 62. Bone, 1989. 63. Lang and Ager, 1989. 86. Davidson, 1994. 34. Buchwald, 2005. 282. No attempt to rationalise what is meant by the term 'Celtic' will be attempted here as it is not really relevant to our study.

[2] Gilmour, 1990. 59 and 227. Lang, 2007. 34. See, also, chapter on *Iron and Steel* and below for further discussion.

[3] Often considered to have been 'Celtic'. Davidson, (1994, 17.) reports that the 'Celts' were making good, steely iron from around 500 BC from the excellent ores at Noricum. The comment regarding quality may be somewhat contradicted by recent work. See Williams, 2012. 16, 59 and Gilmour, 1990. 56.

[4] See Lang's comments. Lang,2012. 237.

[5] Gilmour, 1990. 63 and 64.

[6] Lang, 2007, 10. Lang calls this technique 'edge-to-edge layering'. I call it 'stacking'.

[7] Lang and Ager, 1989. 87.

[8] Lang and Ager, 1989. 87. Davidson, 1994. 34. Suggests that the 'Celts' were making 'piled' blades at least as early as the 8th century BC and probably earlier. McDonnell. (1989. 378), suggests that piling could, originally, have occurred unintentionally. Lang calls this technique 'surface-to-surface banding', Lang, 2007. 10.

[9] Williams says that the 'Celts' were producing iron swords of varied construction from the 7th century BC. Williams, 2012. 49. Gilmour, (2007. 94.) claims that pattern-welding was occurring on swords from around 500 B.C. Good examples of such swords can be seen in Stead, 2006. 46 to 47, 57. Gilmour, 2010. 60.

[10] James, 2011. 147.

[11] James, 2011. 184. Not everyone agrees that it was Roman smiths that were doing this; Brian Gilmour, personal comments.

[12] James, 2011.212. He also, quite reasonably, points out that we have little information concerning what was happening with smithing in the Eastern Empire at this time. There are no definite pattern-welded blades in the excavation from Dura Europos,

James cites two sword blades buried with two murdered soldiers in Canterbury, Kent England as, perhaps being, the earliest examples of fully pattern-welded weapons found in the British Isles[1] and says that by the 3rd century A.D. pattern-welding was 'commonplace' on Roman swords.[2] James remarks, "Pattern-welding imparted better strength and resilience than earlier methods to weapons made mostly from softer iron alloys with only limited proportions of harder to produce steel. It also offered additional aesthetic attractions which remained a significant element of sword manufacture thereafter."[3] As we have seen in the previous chapter, Williams was not impressed with the general quality of most Roman or Celtic swords that he had examined,[4] so perhaps James is correct, however, we shall consider strength and resilience when we look (below) at the effectiveness of these swords but there is no doubt about the aesthetics: a well-made and richly patterned blade can be very striking indeed. We will also ponder on just how 'commonplace' pattern-welded swords were throughout the late Empire.

What is certain is that by the 3rd century pattern-welding had captured the imagination of many northern smiths,[5] and/or their clients, and the smiths were trying to find out just what was possible with pattern-welding and were producing some extremely complex blades as evidenced by the bog finds in Southern Scandinavia and Northern Germany, most notably from Illerup, where there is a relative profusion of swords that have been thoroughly examined and analysed using modern methods.[6] Nydam was one of the earliest bogs to be excavated and yielded one hundred and six swords ninety-three of which were pattern-welded, some exquisitely.[7] It is possible that this explosion in production was under the control of the Romans, at least during the 2nd to 4th centuries. Production was probably taking place along the northern parts of the *Limes*.[8] Buchwald suggests that the *fabricae* were often part of legionary camps and that there were known ones at Wiesbaden, Mainz, Trier, Lorch, Niederbieber, Köln and Nuess.[9] During this period, the iron probably came from Siegerland and Sauerland; the former could, reportedly, produce iron that was relatively phosphorous free.[10] Of course, it is possible that these types of swords were being made elsewhere in the Empire and brought to the North, although that would have been a lot less convenient. Certainly we know that the Roman authorities tried to maintain bans on exporting swords to the *Germanii*.[11] It would appear from the evidence of Illerup, (if they are Roman swords) and elsewhere that the Roman border controls were not up to the task.

Syria, for example. James, 2010. Davidson, (1994. 34.) seems to agree. Lang and Ager mention a sword with twists from Llyn Cerrig, Anglesey, Wales, dating to the 2nd century A.D. Lang and Ager, 1989. 87. James, (2011. 149), gives details of arms production in an area of Gaul. Neither sets of authors provide us with any evidence that any blades produced were patterned.

[1] Gilmour, (2007. 92, 93 and 96), notes the two swords from Canterbury and mentions more, perhaps five, from South Shields,(2nd or early 3rd century) some of which are extant and at least two of which were pattern-welded. Gilmour regards them all, including the Canterbury swords as belonging to Germanic soldiers in Roman service. See also, Gilmour, 2010. 60. See Davidson, (1994, 32.) for more information on two of these swords. The two Canterbury blades are included in our pattern sample.

[2] James, 2011. 184.On page 33, James says that such swords were' widespread' by 200 A.D.

[3] James, 2011. 184.

[4] Williams, 2012. 16, 51, 59 and 60. The swords that Williams considered were not pattern-welded. See also Kmetič, Horvat, and Vodopivec. 2004. The 2nd/3rd century, Roman sword they examined was not a first class weapon, but they do mention the existence of some excellent quality Roman swords but do not provide details. McDonnell, (2007. 92.) also reports on the variability of Roman weapons.

[5] James, 2012. 33 and 184. Williams, 2012. 62. Davidson, 1994. 37. Lang and Ager, 1989. 87.

[6] Biborski and Ilkjær, 2006. Gilmour, 2007. 96. Gilmour, 2010. 60.

[7] Gilmour, 2007. 96. Buchwald, 2005, 285. Buchwald says, "…some 90 were pattern-welded."

[8] Buchwald,2005. 283.

[9] Buchwald, 2005. 284.

[10] Buchwald, 2005. 285.

[11] James, 2011. 210 and see footnote 100. Buchwald, 2005. 285.

It is probable that the swords were being made in the lands of the Germanic peoples outside the Empire, and by Germanic smiths. As we have seen iron smelting was happening on a large scale in the north.[1]

From the 3rd century onwards, pattern-welded swords are mostly found in the British Isles, Scandinavia, Germany, France, parts of Central Europe and Northern Italy. Most of these swords have never been comprehensively analysed and it had been thought that smiths had settled down to producing a comparatively small number of still quite complex patterns by the 5th century and that they continued with these and a few variants from then on.[2] However, judging from the evidence of those swords that have been x-rayed from England, Scandinavia and the Continent and some recent discoveries, such as those at Saltwood, Kent, it appears that ambitious artist-smiths were continuing to produce some of the more elaborate blades a good deal later.[3] The difficulty with blade patterns is that it isn't always possible to tell when a blade was actually made; a sword blade might be in a 6th century grave and could have 6th century grip furniture, but the blade itself may be an heirloom from an earlier period.

Making a blade

So far I have said that the process of making a Damascus steel blade was different to that of making a pattern-welded one but have not described the latter techniques in any detail. I will try to remedy that situation here and include a brief idea of the sorts of metals that a smith might include in his blade and provide some reasons for their choice.

The basic ideas of pattern-welding are to choose alloys that will perform well in a sword blade when combined and that will provide a pleasing or meaningful pattern after all the production work is complete. However, it is a far from simple task and all aspects of making the blade have to be designed in from the start or the finished item will be a poor one. Exactly how earlier smiths went about their design work is not known but probably, the first thing that the smith would have to do is to plan the design of the patterns and work out how each element will be incorporated into the geometry of the blade. (figure 6.1) A modern smith is likely to draw out the overall design at one to one scale on paper (an ancient smith would probably have drawn his on a wooden board or scratched it on a hard surface, such as stone) so that he can see what to do and where. If he is to include the herringbone patterns and the bars are to be twisted, this has to be thought through; the number of twists and how much to twist each rod can be quite critical to the design and to the limits of the materials that he is working with. Whether the design will be different on each side of the blade is another factor that will affect the way he goes about his task as will the number of layers to be incorporated into the blade. It is possible that the early medieval smith's designs were limited, sometimes, by the materials that were immediately available to him or perhaps some smiths would only work with patterns that they knew well. It could also be that clients demanded certain patterns because they had some significance or meaning that they wanted to convey. This is an idea that we will look at later.

---

[1] Brian Gilmour feels that this is certain. Personal comment. I do too; see the far from complete survey of iron-making in the chapter, *Iron and Steel.*

[2] Gilmour, 2007. 101 and 102. These would be the patterns referred to in Lang and Ager, 1989. 87 to 89.

[3] See also Lehmann, 2016, for more examples from Westfalen, Germany.

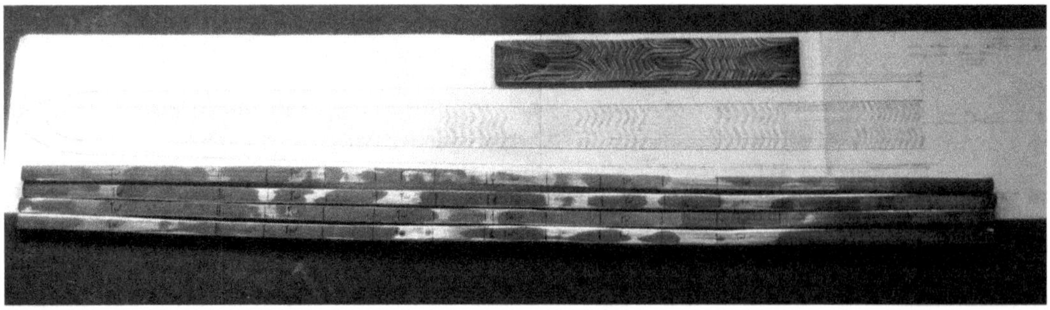

6.1. The pattern design for the Sutton Hoo Mound 1 blade. (Photo courtesy of Vince Evans)

Once the design was settled, the smith would then have to decide how he was going to make the patterns show up and select his materials accordingly, always bearing in mind that the finished weapon should be both tough and hard.[1] Ideally, he would save the hardest metals for the edge and as is the case with all known pattern-welded (and non-patterned) swords from our period, the edge would be a separate construction from the core or centre of the blade.

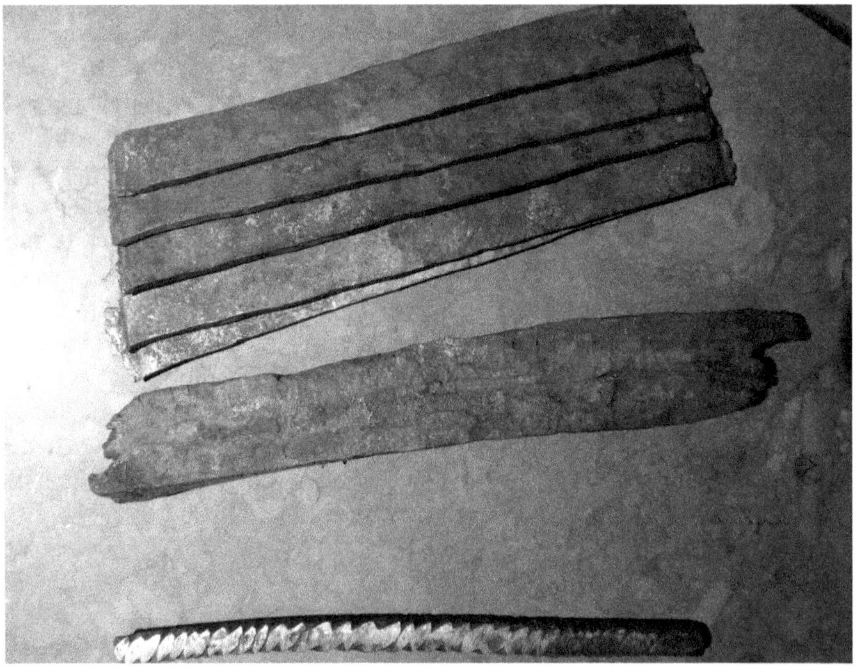

6.2. In this picture can be seen the components for a billet to make a rod for a sword, together with a forged billet, below the rods. These examples have been prepared by Jeroen Zuiderwick and are on display in the Museum of Leiden. At the bottom of the picture can be seen a sample rod that has been twisted. (Photograph Paul Mortimer)

[1] Lang, 2006. 86.

6.3. One of the billets is being extruded by forging until it reaches the required length and thickness. (Photo courtesy of Vince Evans)

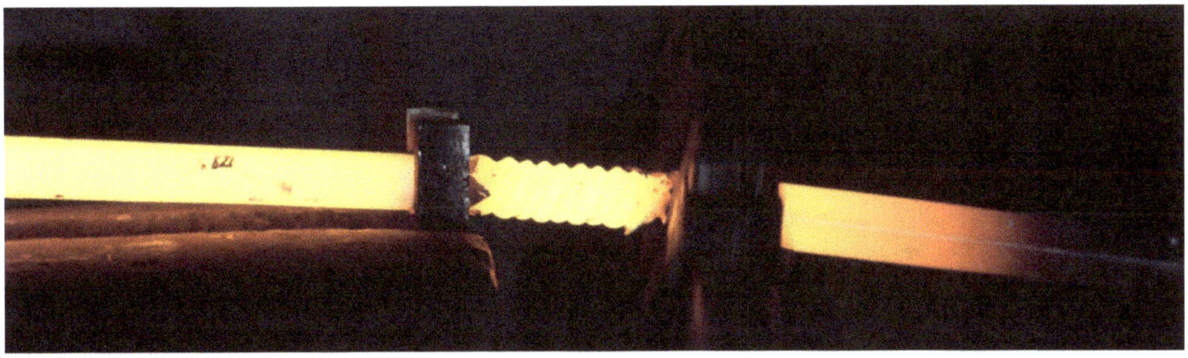

6.4. The planned twists are being put into a rod. (Photo courtesy of Vince Evans)

Smiths would work their metals into shape by repeated heating, forging and folding, which would help to homogenise the metal and break down remaining area of inclusions from the bloomery process. We do not know exactly what procedures ancient smiths followed to make the rods that would form the core of the blade, but most modern smiths take their selected metals and after shaping them into rectangular plates about the size of a man's hand, stack them on top of each other, until the required number of layers is achieved. (figure 6.2) Most sword rods seem to have odd numbers of layers from three to nine[1] and the stack would, usually, consist of alternate layers of low carbon iron plates and ones of phosphoric iron.[2] Paul Binns, the English swordsmith, has a reason for the odd number of layers; he always puts low carbon iron as the top and bottom layers as it has a much higher melting point and will resist the rigours of the fire and forge much better than most other types of iron-alloy and this process will always result in an odd number of layers. Although much of the bottom and top layer will be lost to oxidisation, using low carbon iron on the outside of the rod helps to limit the loss.[3] The stack would be returned to the fire and heated and then forged into a long rod until it reached the required length and thickness.[4] (figure 6.3) The completed rods would then be

---

[1] Lehmann, 2013. 2.

[2] Modern smiths usually use iron with relatively high nickel content rather than phosphoric iron.

[3] Paul Binns, personal comment.

[4] Most rods seem to have been around 8mm to 10mm in square section. It may be that some smiths would have used more than two different metals. It is possible that more than nine layers could have been used, but as far as I am aware none have been recorded, although some modern smiths go well beyond this number.

cut to a bit longer than the sword would end up being. The smith would have already decided how complex the centre of the blade would be and the number of rods that he would need for the centre of each side of the blade.[1] Next, he would heat the rods to red heat and then put the required amount of twists in the appropriate places on each rod in turn. (figure 6.4)

Whilst forming the rods and subsequent operations it is highly likely that the smith had assistants or apprentices aiding him; it would certainly have made some aspects of the work easier. Some master smiths probably had other smiths, as well as other workers, working with them in their workshops. Incidentally, there is evidence from the later middle ages that some smiths were female, but these seem to have been family members helping the master smith.[2]

The rods would then be clipped together and hammer-welded into one piece after being heated again. (figure 6.5) It would appear that, in some cases, at least, with those swords that had differing or off-set patterns on each side, that the smith could have split the core length ways along its thin edge and then the two sides would be placed back together after he had made sure that they were configured in the way that the smith desired.[3] An alternate method is to make two separate cores and grind them to the required thickness; in some types of pattern the grinding is essential to achieve the required pattern. Swords with a third layer of softer iron between the patterned sides would have it added at this stage and those swords that would consist of two entirely different patterns on each side would have to have a different half-core added. The whole would then be heated again and welded together to form the completed core.

6.5. In his case the four rods have been prepared and have been clipped together for hammer welding into a single piece. (Photo courtesy of Vince Evans)

The smith would also have to forge and fold the selected materials for the edge together several times and then to shape it so that it will wrap around the outside of the core. (Figure 6.6) The edge would then be tied to the centre of the blade, probably with thin wire and the whole blade returned to the heat until the welding temperature was reached and then the sword could be completed by forge-welding.[4] (figure 6.7) The cutting edge and tip typically consisted of a welded composite of plain iron and medium-to-high carbon steel (0.5 to 0.8% carbon).[5] As can be seen from the accompanying illustrations, the construction methods used to form the edge of a blade could be anything but simple, and varied a great deal.

[1] Almost all swords so far examined have at least two layers as part of the blade structure, many have three with a soft iron core between the two patterned sides.

[2] Hinton (ed), 2000. 111.

[3] This is what Vince Evans believed happened with the Sutton Hoo Mound 1 blade and is discussed more fully in that chapter and a brief explanation follows shortly below.

[4] Lang, 2012. 263.

[5] Gilmour, 2007. 107.

6.6. Forging the edge bar that will be welded to the edge of the completed centre or core. (Photo courtesy of Vince Evans)

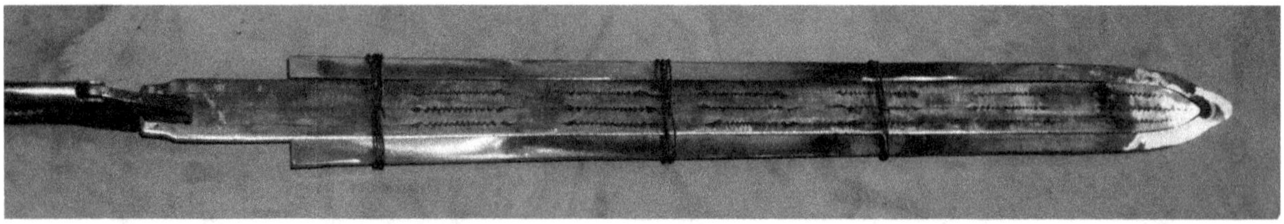

6.7. The centre and edge clipped together ready to go back and be heated and hammer-welded into a single and complete blade. (Photo courtesy of Vince Evans)

The temperature that the metal has to reach in order for hammer welding to be successful is critical: too hot and some of the elements will burn, too cool and the welds will not take, in either case the blade could be ruined. The smith would know by the colour and behaviour of the metal when it had reached the correct temperature; some smiths claim that the weld should take place when the metal is giving off sparks, sometimes called taking 'Hitz', however at that stage the metals is actually burning and it is likely to be too hot and, as Buchwald points out, unless great care is taken much metal will be lost.[1] Sim and Kaminski explain that the temperature range between 760°C and 1537°C, depending on carbon content, is when the metals become plastic and behave like a semi-liquid.[2] A measure of the smith's skill was (and still is) his ability to successfully weld metals with different melting and burning temperatures.[3] During welding the smith might use sand[4] as a flux to sprinkle on the soft metal; modern smiths tend to use borax.[5]The flux could be visible when the blade has been completed as very fine lines between the welds.[6] The melted flux coats the metal and helps prevent oxidisation;[7] too much oxidisation and the smith's work could be ruined.

Until recently, it had been thought that if a reverse or off-set pattern was required for the back of the blade, it needed double the amount of rods; if four rods were required for the front, then another four would be needed for the rear.[8] However, when Vince Evans carefully examined the x-rays of the sword from Mound 1 at Sutton Hoo, he realised that if that was the case, the pictures would show the patterns differently to the way that they are shown in the images and suggested that the original four bars made for the core may have been split, lengthwise and then one of the resultant pieces would have been turned end over end to provide the reverse pattern.[9](figure 6.8) The two sides could then be hammer welded back together. Vince Evans discussed his interpretation of the x-rays with Janet Lang at the British Museum and she was sufficiently impressed to mention the idea in her recent work.[10] Vince Evans has since made a number of blades using this technique and has shown that the welded core can be split when hot with, a chisel lengthwise quite effectively and accurately. This method means that there is less metal to deal with than if two sets of rods had been used. The method drastically

---

[1] Buchwald, 2005. 281. Lang, 2012. 265.

[2] Sim and Kaminski, 2012. 27.

[3] Lang, 2012. 265.

[4] La Salvia, 2007 (37) reports that, "Silica sand was used by blacksmiths in order to dissolve the oxidic scales which formed on the surface of the heated iron semi-products or objects." He is referring specifically to smiths in Lombard Italy and in Hungary.

[5] Lang, 2012. 265. Gilmour, 1990. 64.

[6] Gilmour, 2010. 65.

[7] Pelsmaeker, 2010. 13. Sim and Kaminski, 2012. 27.

[8] See for example, Engstrom, Lankton and Lesher-Engstrom, 1990.

[9] See *Sutton Hoo Style Pattern-welded Blades* by Vince Evans, and *The Sutton Hoo Mound 1 Sword the* Replica, both in this volume. Ulrich Lehmann, in his work using CT scanning of blades, also came to the conclusion that half-rods were being used on Merovingian blades, he is certain that in some cases the rods were being ground but says that Jaap Ypey had suggested splitting with a chisel (Personal comment Ulrich Lehmann, Lehmann, 2013. 8 and Lehmann 2016).

[10] Lang, 2012. 264 and 266. In theory it would be useful to x-ray Vince Evans's blade and see if it can match the images of the original, however, as the new blade is not corroded, the structures are unlikely to be revealed by conventional x-rays.

cuts the post forging work required, including the need for filing and while grinding to shape, it saves expensive files from being worn out, and perhaps more critically, saves wasting valuable metal.

6.8. With the Mound1 sword, we believe that the half-rod thickness was achieved through splitting. The two sides would then be offset against each other and welded back together. See text for further comments and chapter on *The Sutton Hoo Mound 1 Sword*. If ancient smiths used this technique, they would have used a chisel, not a band-saw as in the picture. Vince did experiment with a chisel and found it quite straightforward. (Photo courtesy of Vince Evans)

Some suggestion[1] had been made that the more complex blades, those with four centre rods or more tended to be two layered blades, while the simpler blades with three or fewer rods consisted of three layers, however the data does not bear this out as can be seen in the chapter, *Sword blasé patterns* and in the data in, *Appendix 7 the Patterns Database*. Blades with only one layer were constructed but are rare in the archaeology.[2]

When the smith had completed the construction of the blade he might decide to add a fuller that could be forged while the blade was hot using fuller blocks. Alternatively, fullers can also be ground in and this will affect the pattern in a different way from using blocks, usually by making the patterns appear curvy.[3] Lee Sauder has explored a third method of making a fuller by upsetting the sides of the core and then tacking the edge on via forge-welding and this is discussed by Lee as an addenda to this chapter.

As mentioned above, grinding can produce variations in the pattern depending on how much metal is ground away. Then work can be done to add more interest to the blade or to achieve a particular effect, which may include adding a fuller. The 'curvy' stage or half circles in the twists begin to appear when a twisted rod is ground down to about 48% or more of its original thickness. If the grinding is continued the diagonals of the twists will change direction. (figure 6.9) Further subtleties of the pattern can be added by having thinner or

---

[1] See for example Gilmour, 2010. 65..

[2] Personal comment, Ulrich Lehmann, who is aware of two examples and Vince Evans who knows of a Viking blade with one layer. There are five in our pattern-welded sample based largely on Brian Gilmour's work.

[3] See comments above regarding the work of L. Bieck. And see Lang, 2012. 266. Gilmour, 2007. 101. Lehmann, (2013. 9.) provides examples of both ground-in fullers and forged ones from the blades that he has studied.

thicker layers within the rods.[1] After its final heating the blade might be quenched, and go through other appropriate heat treatment. As previously discussed it is likely that many early medieval smiths relied on slack quenching[2] at this stage, although there are swords that appear not to have been subjected to any type of additional heat treatment.

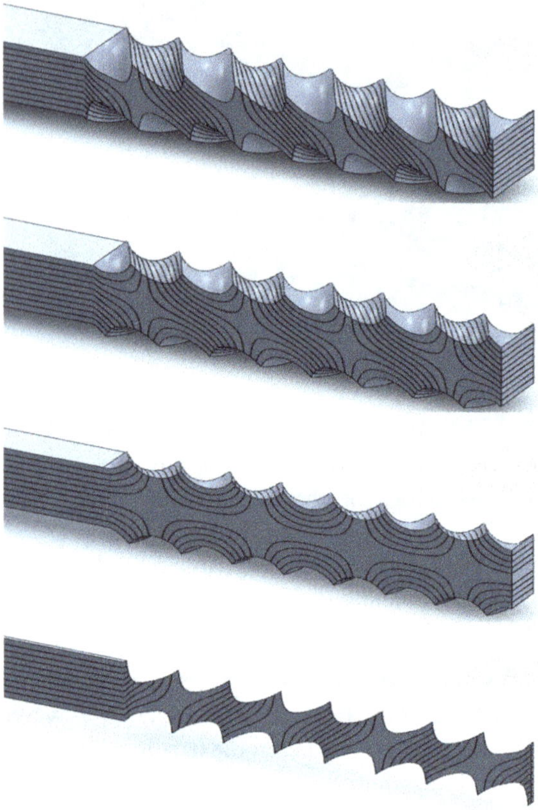

6.9. The effects of grinding a twisted rod to enhance or alter the pattern. The first illustrations shows the effect of a 5% grind, the second 25%, the third 50% -- note that the half circles have appeared, the fourth 95% -- note that the direction of the diagonals has reversed. (Diagram courtesy of Jesus Hernandez)

The sword would then require filing and grinding to final shape and profile, with polishing and sharpening following and then the blade would be etched with a relatively mild acid to bring out the pattern. The blade would need a hilt to make it a useable weapon and a scabbard to protect it from the environment. It is certainly possible that all these operations could be carried out in the same workshop and modern smiths, such as Vince Evans, Jesus Hernandez, Patrick Barta, Robert Môc, Peter Johnsson and Paul Binns have demonstrated that they are able to manufacture a complete sword and all its accoutrements. It is possible though, that the finished blade might have been sent to other specialists who would make the grip, pommel and scabbard.[3] Some swords were re-hilted during their lifespan, which almost certainly meant work by a different craftsman, and a metal blade would last longer than a wooden and leather scabbard, which was designed to take all the knocks and suffer the depredations of the weather while protecting the blade, so a sword was likely to require re-scabbarding occasionally during, what could be, a long operational career. A

[1] I thank Jesus Hernandez for this insight.

[2] Or 'auto-tempering as Lang calls it. Lang, 2007. 73 and 211. See also the very clear illustrations in Jones, 1997. See also the section on *Heat-treatment* in the preceding chapter, *Iron and steel.*

[3] Lang, 2012. 266.

sword owner might even have had more than one scabbard for his weapon at any one time, perhaps one for damper months another for the summer and one for storage? Scabbards would have provided more than just a convenient way of carrying a sword around, they could display allegiance, ethnic origin, wealth, achievement or even just the latest fashion. They are discussed elsewhere in this book.

The pattern-welding process could take many hours; Patrick Barta, who made a sword for me, reported that the making of the blade took him one hundred and fifty hours; that is about three week's work just making the blade and he is a very experienced smith. Scott Lankton wrote that the polishing of the blade that he made for the British Museum alone took seven days.[1] The process is also intensely resource hungry, while making just the eight rods for the core of his blade, Lankton used one US ton of coal while working and forming the metal.[2] Lang tells us that a smith (Crew) used 100kg of charcoal and put in about twenty-five man-hours of work to produce a kilogram of fully smithed bar of iron from bog ore.[3]

Appreciable amounts of valuable metal could be lost as hammer scale, as each time the iron and steel is heated to red heat or above, the surface reacts with the oxygen in the air and this scale will need to be brushed off before work continues, this would matter to a smith a great deal as iron and steel were, and are, very valuable commodities and too much oxidisation can, effectively ruin the work.[4] A further reason because the removal of the oxide flakes can take away the carburised parts of the blade.[5] As already mentioned, one of the purposes of using flux is to help protect the piece from oxidising when heated.

The selection of suitable metals

There are many traces of other elements within the surviving sword blades, but the two most important elements that are included, besides iron, are carbon, which we have discussed in the chapter on *Iron and Steel*, and phosphorous, which we will consider now. The smiths of northern Europe often had to contend with iron that had a relatively high phosphorous content,[6] which when pattern-welding, can be useful but could bring disadvantages, too. One of the more valuable qualities of phosphoric iron is that it is corrosion resistant and will tend to remain shiny. Another is that it will also etch more slowly than phosphorous free material,[7] leaving the surface of phosphoric iron raised slightly above the surrounding metal. Phosphoric iron provides a good contrast to low carbon iron and steel, both of which will remain much darker than the phosphoric iron, thus emphasising the patterns built into the sword.[8]

Using the evidence from the analysis of recovered swords, Lang tells us that the smiths would, generally, select three types of ferrous metal to make a pattern-welded sword: low carbon iron, phosphoric iron and steel of up to 0.8% carbon.[9] This seems to indicate that smiths were perfectly capable of differentiating between various alloys of iron.[10] We did, briefly discuss the ability to tell metals apart in the chapter on *Iron and Steel*,

---

[1] Lankton, 1989. 12.

[2] Engstrom, Lankton and Lesher-Engstrom, 1990. 4. Presumably if Lankton had only made 4 rods and used the split-core method, he would have needed less.

[3] Lang, 2007. 31.

[4] Sim and Kaminski (2012. 27) suggest that up to 65 to 70% of the original material could be lost through oxidisation, presumably unless precautions were taken. Skilled bladesmiths are likely to lose far less than this, I believe.

[5] Gilmour, 1990. 65.

[6] Buchwald, 2005. 177 to 178. McDonnell, 1989. 375. Points out that most iron from the British Isles contains phosphorous.

[7] Buchwald, 2005. 171.

[8] Lang and Ager 1989. 110. Lang, 2012. 238 and 259. Buchwald, 2005. 173. Gilmour, 2007. 107. Gilmour, 2010. 60. Godfrey, Vizcaino and McDonnell, 2003. 191.

[9] Lang, 2012. 257 and 263.

[10] Lang, 2012. 257. Gilmour, 1990. 227.

but breaking up the hot bloom by submerging it in cold water and then testing the resultant lumps by appearance, scratching, resistance to hammering or filing, listening to the sound a piece would make when struck or dropped and even tasting, are all methods that have been employed by smiths to select suitable metals.[1] Modern Japanese swordsmiths, using traditional methods, test their *tamahagane* steel by forming it into small cakes and distinguishing the harder and softer pieces by their colour, feel and the sounds they make when hit or dropped;[2] the softer pieces are used for the core of the blade and the harder, the edge. Islamic smiths reportedly used various methods to tell different alloys apart including the type of hammer scale a metal produced and the colour of the rust that it produced.[3]

Phosphorous will, just like carbon, tend to harden iron,[4] and for the ancient smiths this could have brought advantages. Such metal was certainly used to make swords and Lang reports that many Romano-Barbarian pattern-welded swords utilised layers of phosphorous rich iron alternating with layers of steel.[5] Phosphoric iron may have been the hardest ferrous material employed in weapons and tools, from around 500 B.C. until close to the Roman period according to some commentators.[6] Buchwald explains that the hardening effect is actually quite complex and will produce different behaviours and characteristics under different circumstances.[7] Phosphoric iron can also be work-hardened when cool, apparently doubling its hardness.[8]

Williams and McDonnell among others say that phosphorous when present in iron discourages the absorption of carbon,[9] and it has also been suggested that such iron will not carburise and cannot be welded.[10] However, it has been found that small amounts of phosphorous, up to 0.2% will not alter the rate at which carbon is absorbed in iron,[11] and, Lang reports on an experiment where an iron bar with a high phosphorous content and up to 0.7% carbon was smithed and welded successfully.[12] She suggests that phosphoric iron could be more easily weldable than many other ferrous materials, because the melting point of the alloy is lower and she supplies an example where the Romans regularly used layers of phosphoric iron layered with wrought iron to make glass blowing tubes. 'Ternary' alloys[13] consisting of iron, phosphorous and carbon certainly do exist and in fact occur regularly in swords.[14] One other benefit of working with phosphoric iron is that it is pleasant to work because it is very malleable, easy to forge and twist in any way that the smith desires.[15]

The real problem with phosphorous is that, because of the increased hardness, it can have an 'embrittling' effect on iron rendering it susceptible to catastrophic failure when suffering shock or, perhaps just fatigue over time, particularly in colder climates.[16] This behaviour of phosphoric iron has been known and dreaded for centuries, but not at all understood until recent times; smiths of old were not aware of the element, 'phosphorous' nor of 'carbon' for that matter, it was only in the 18th century that observers began to realise

---

1 Lang 2012. 259. Of course, there are more methods than those listed here.

2 Loades, 2010. 215.

3 Gilmour and Hoyland, 2006. 162. These methods were reported by Muḥammad ibn Aḥmad al-Bīrūnī in the 10th century.

4 Ager, and Lang, 1989. 110. McDonnell, 1989. 375. Lang, 2012. 259. Buchwald, 2005. 172. Godfrey, Vizcaino and McDonnell, 2003. 191. Although it cannot be further hardened by heat treatment. Gansum, 2004. 42.

5 Lang, 2012. 259. Lang, 2006. 85.

6 Godfrey, Vizcaino and McDonnell, 2003. 191.

7 Buchwald, 2005. 172 to 173. Godfrey, Vizcaino and McDonnell, 2003. 191.

8 Tylecote and Gilmour, 1986. 9. Welton, 2016. 13.

9 Williams, 2012. 68. McDonnell, 1989. 375. Lang, 2006. 86. Points this out, too.

10 But not by Williams or McDonnell, for details see Lang 2012. 259.

11 Godfrey, Vizcaino and McDonnell, 2003. 192.

12 Lang is referring to work by Rolsch and Kuhn.

13 Buchwald, 2005. 173.

14 Lang, 2012. 259.

15 Buchwald, 2005. 177.

16 McDonnell, 1989. 375. Lang, 2012.259. Buchwald, 2005. 178.

that it may be an element within particular types of iron that was responsible for embrittlement. They found, too, that there was little to be done about it except to avoid using certain ores for particular tasks.[1]

It would appear that large items, such as massive chains, ship plates and anchors had real problems and could fracture without warning, especially at low temperatures. However, smaller items made from similar iron, nails and locks, for example performed perfectly even in the cold. The reason for this has been put down to the different lengths of time a large object, like an anchor will take to cool after being made. Large objects will lose their heat slowly and the resultant grains of phosphorous, within the microstructure of the metal, will be comparatively big and coarse and, therefore, prone to failure at the grain boundaries; with smaller items the grains will be much finer due to the effect of rapid cooling.[2] 'Smaller items' includes Iron Age and later sword blades.[3]

From the 18th century until now, the manufacture of large iron based objects such as engines, locomotives, bridges, buildings, railway lines and ships has become commonplace, and engineers avoid using ferrous based metals with a high phosphorous content.[4] Metals are subject to shock-loading tests and rated according to how and where they can be used. Levels of phosphorous are regulated and kept below 0.4% and for certain functions below 0.1%.[5]

Early smiths, it seems, did not regard phosphoric iron as a problem for sword blades and it was certainly sought out to be an added ingredient in the best pattern-welded blades.[6] There are indications that smiths tended to concentrate the phosphoric iron within the core of the sword most of the time and not usually along the edge.[7] There are a number of other factors that can mitigate possible disadvantages of phosphorous to a large degree, apart from those contributing to the visual impact of the pattern-welding. It seems that when pattern-welding, a high phosphoric iron will stop the carbon diffusing out across the welds during forging, thus keeping the boundaries between the different metals sharply defined.[8] Modern investigations have also shown that carbon present in the phosphorous rich areas of the blade tends to weaken and perhaps cancel out the embrittlement.[9] In fact in recent times phosphorous has been found to strengthen some alloys of iron without causing them to become brittle.[10] To what extent ancient smiths were aware of the advantages and disadvantages of phosphorous is not known, at present.[11] It is possible, too, that they had an engineering technique that could have helped to ameliorate the possibility of brittle parts within the blade. Vince Evans has noted that the twisted parts of the rods are much harder to file than the 'straights', it is not certain, but this is probably because the twisted structure presents more 'edges' to the file. By having two layers, or more, and by offsetting the 'straights' and 'twists' against each other on opposite sides of the blade, it is probable that the smiths were deliberately combining the qualities of the metals in order to provide greater strength and toughness and overcome any weaknesses within the amalgamated and laminated metals.

There is evidence that the production of phosphoric iron, and for that matter low carbon iron and steel, had become specialised and highly organised during the Iron Age and probably continued as such well into the

---

1 Buchwald, 2005. 177.

2 Buchwald, 2005. 178. Lang, 2012. 259.

3 Lang, 2006. 86.

4 McDonnell, 1989. 375. Lang, 2012. 259.

5 Buchwald, 2005. 178. Lang, 2006. 86.

6 Lang, 2012. 238.

7 Lang, 2012. 238 and 256. Gilmour, 2010. 66.

8 Gilmour, 2007. 107. Godfrey, Vizcaino and McDonnell, 2003. 192. Noted by Welton, (106. 16) too.

9 Lang, 2012. 259.

10 Buchwald, 2005. 178.

11 Ulrich Lehmann thinks that smiths, possibly even from Romans times, were able to control the amount of phosphorous within the metal and that phosphorous when it occurs in a blade was desired. Personal comment.

Middle Ages and beyond.[1] Gilmour explains that the phosphoric ferrous material found in early blades tends to be around 0.2 to 0.3%, whereas from the 6th century onwards it is between 0.5 to 1%. This suggests that whoever or wherever it was being produced the industry could adjust to the demand.[2] Perhaps an even more compelling hint at organisation is the ferrous currency bars that have been found in many parts of Northern Europe. These differ in size and shape, and, it seems, that these factors could show where the bars originated from and, or their quality.[3] The bars demonstrate that there was a market and a demand, and that the customers could be discerning.[4] Welton notes, too, that the partial forging of the bars would have helped to demonstrate their quality.[5]

It is possible that the smiths of Europe were far more sophisticated in their knowledge of the tasks that they were undertaking and their understanding of the materials that they were dealing with than has often been thought. The fact that they were aware of the various ferrous alloys available and the materials' capabilities and were able to seek them out and obtain what they required, is surely good evidence of this. In the surviving blades, the smiths have demonstrated a large number of skills and shown a very high level of proficiency. Surely these factors alone are a testament to a very complex and organised, and probably international, industry involving clients and merchants as well as craftsmen?[6] Even today, with much more technology available and a far more complete understanding of what is happening to the chemistry of the alloys used, the skills involved in making a pattern-welded sword are not easy for modern smiths to master and require an intimacy with, and understanding of the versatility and limits of the materials and tools available. Even for the more accomplished and experienced smiths there are ancient patterns that provide great challenges; perhaps examples of these are the serpent design of the West Heslerton sword, the very complex designs from the Illerup bog finds and the blade from Vehmaa in Finland?

Based on the evidence available to us, the quality of pattern-welding found does vary, and not all swords are of top quality, which could mean that not all smiths making the blades were either. There is the possibility that a smith, sometimes, during times of shortages of suitable materials or when trade in iron-alloys was disrupted, would have had to make do with what alloys he had available. It is also likely that he would have to adjust the standard of his work to the purse of the client. However, Welton demonstrates that sometimes smiths just mistook one alloy of iron for another as their methods of selection were not always perfected.[7]

How do we know the patterns were valued?

Several times now I have stated that the designs were prized or valued, and to us it is obvious that a patterned blade looks more interesting than a plain one, but how do we know that the peoples of northern Europe didn't just view the patterns as a necessary by-product of the manufacturing process, after all, aren't nearly all swords of the 5th to 7th centuries patterned? [8] Lang and Ager provide us with some statistics from their study of Anglo-Saxon swords and bladed weapons, from the British Museum collection, regarding the proportions that were pattern-welded: 44% from the 5th to 6th century, 77% from the 6th and 100% from the 7th century.[9]

---

1 Gilmour, 2007. 93. Lang, 2007. 31 to 33 and Lang, 2012. 263.

2 Gilmour, 2007. 107. Gilmour, 2010. 68. This, of course, ignores the problems of the actual dating of swords.

3 Buchwald, 2005. 80, 103, 127, 131, 173, 175 to 177. Lyngstrøm, 2003, 23. Lang, 2012. 253 and 254. See also Gilmour, 2006. 74 and figure 17. Gilmour includes pictures of Assyrian iron trade bars dating from the late 8th century B.C. and reports on trade bars in Islamic areas up to 14th century A.D.

4 For example, Sim and Kaminski (2012. 58, 64) conclude that there was an organised trade between Rome and India.

5 Welton 2016. 24.

6 Stenvik certainly thinks so. He suggests that there were, "...advanced systems of distribution." Stenvik, 2003. 80.

7 Welton, 2016. 13 and 16.

8 Lang and Ager, 1989. 92. I am specifically referring to swords of the fifth to seventh centuries and not later swords.

9 Lang and Ager, 1989. 107. It must be borne in mind that the dating of swords is fraught with difficulties.

From those statistics it looks as though there are some plain blades to compare the patterned ones to. There are a number of problems here that we need to think about. The first is that many of the blades found in England have not been x-rayed, and as nearly all of them were located within a grave, contained within their scabbards and not in the acid waters of a bog, any patterns on them are unlikely to have been recognised and will have gone unreported by their excavators. This was the case, generally, until well after the middle of 20th century. The second problem here is that we need to be clear what we mean by 'patterned' and 'plain' and it is not entirely straightforward. All surviving swords of the period seem to have been hammer or forge welded together from several pieces of metal and could well have had distinct lines on the blade, perhaps just providing a layered, grainy or watery appearance. It is highly unlikely that any swords in northern Europe were entirely constructed from one type of metal; I know of none that have been clearly identified until the late 'Ulfberht' swords. The least that seems to have occurred is that the sword would have a core of iron with a welded edge of different iron-alloy, usually containing some steel. It is possible that a sword might consist of a core which has been completely covered with a different ferrous alloy; the latter would give the appearance of a plain blade. Although there are seaxes made this way, at the time of writing, I am unaware of any sword blades with this form of construction from Anglo-Saxon England.[1]

Until relatively recently it was not possible to peer inside the structure of a blade without using destructive methods and of course there are many remains of swords that are so corroded that little is left of their blades, so whatever process was used with these they would reveal very little information. Nor must we forget that most swords in graves from this period are found with the blades in their scabbards.[2] There are swords that have been x-rayed, where no discernible pattern was seen, but that does not mean that they didn't have a pattern. Gilmour explains that, "(However)….the simple x-ray-based sorting of the sword blades into pattern-welded and non-pattern-welded has been found to be misleading. "[3] When Lang and Ager wrote their seminal paper in 1989, they stated that, after discussing swords from Kent, "On the other hand Mucking (Essex) yielded 5 swords none of which was patterned."[4] Subsequently, it has been found that Mucking actually produced eight swords and six of these (presumably, including some of the ones considered by Lang and Ager) possibly have patterned blades. Referring to the seven swords found within Cemetery II at Mucking, Gilmour and Edwards state that there is "…only one of which shows no sign at all of welds or pattern-welding."[5] Gilmour and Edwards are careful to point out that there are no definite examples in Cemetery II of pattern-welding either. However, the single remnant of a sword in Cemetery I has a very clear triple herringbone pattern on its blade.[6] This is a good example of what archaeologists are up against; the poor quality of many of the remains, the resources to properly examine all swords that come their way, and the possibility of overlooking the often, very subtle signs of welding.[7] Unfortunately, Lang and Ager tell us little about the construction methods involved in making the swords that they say were not pattern-welded,

---

[1] Buchwald, 205. 266. Provides a useful diagram showing some methods of making a blade from different metals, including this technique.

[2] Davidson, 1994. 33.

[3] Gilmour, 2010. 60. Lang, 2012. 238. Lang seems to recognise the limitations of the technique, too.

[4] Lang and Ager 1989. 108. On page 91 they report that 5th – 6th century sword from Mucking, grave 618, was made from three strips welded together side by side, but otherwise without a pattern.

[5] Gilmour and Edwards, 2009. 565. All the blades were too fragile for sampling.

[6] Hirst and Clark, 2009. 674.

[7] There are also limits to what can be achieved with techniques, such as radiography. See Lang and Ager, 1989. 87 and 88. See also Gilmour, 1990. 224. Gilmour tells us, "Uncorroded pattern-welded structures, such as Anstee's modem replica, will show up much less well on x-ray. The opposite situation may also be true, and buried pattern-welded blades which are totally corroded may show little or none of their pattern-welded structure on a radiograph,…". Lehmann, (2013. 4.) explains that the patterns only show up in x-rays because of the way that different alloys react to corrosion, phosphoric iron, for example, degrades more slowly than low carbon iron or steel.

although they do tell us that at least some of them were made from different layers of metal. Two of the swords from Saltwood, from graves 3799 and 4665 seem to have had blades with a non-regular grainy, watery pattern, harking back to the practice with earlier swords from the La Tène and the Roman period.[1] It might be that many other swords that did not seem to have been patterned had layered, stacked or banded blades; such blades are still complex blades. Lang and Ager define a pattern-welded sword as one that has 'twisted rods' welded together.[2] Gilmour seems to take a broader view and regards the early Iron Age, layered or banded blades as 'patterned' blades, some of these are what he calls, 'free form pattern-welded'[3] and, in this book I will follow his lead. Perhaps banded, layered or stacked blades were often sufficient for their Anglo-Saxon owners and would have had a particular significance for their owners.

Nevertheless, the fact remains that many smiths went to great lengths to try and ensure that not only were patterns included on many of the blades, but that they were precisely done and easy to see.[4] They certainly didn't need to do that, a banded or stacked blade would have sufficed, so the demand for more complex blades existed and it was met. We know that King Theoderic the Great, 456 to 526 A.D., liked the patterned blades that were sent to him, possibly by King Thrasamund of the Vandals,[5] because he had a letter of appreciation written and sent to the Vandal. The extract referring to the swords, read,

"You have shown genuine brotherly love in sending me swords, made of iron more costly than gold. The polished blade shines so brightly that you can use it as a mirror – In the middle of the blade are beautiful incisions like winding worms, and there are shadows numberless so you believe that the bright metal has been interwoven with many different colours."[6]

Gilmour is probably correct when he states that the patterns were the real point of the pattern-welded sword and were meant to be seen. Certainly he gets some support from Old English literature, including *Beowulf*, where the sword, *Hrunting*, gleams with 'twigs of venom' and the poem, *Elene* tell us:

> …heardecg cwacaþ
> beofaþ brogdenmæl 7 bleom wrixleð
> grapum gryrefæst.
>
> Hard-edge quakes
> Trembles the patterned mark and in colours changes
> Gore-fast in grips.[7]

Unfortunately, the literature does not assist us in our understanding of the purpose of the patterns, nor why smiths would go to such lengths to include them. Lang and Ager make the point that, "It is quite possible that pattern-welding was thought to improve the properties of the swords, and it might be remembered that a

---

[1] Gilmour, 2010. 64, Figure 1. 67 to 68.

[2] Lang and Ager, 1989. 85.

[3] Gilmour, 2007. 94 and 95. Buchwald, 2005. 266. Both authors appear to regard 'piled' blades as pattern-welded. See *Patterns and Pattern Categorisation* in this volume for a brief discussion of the term and an attempt at standardisation.

[4] Gilmour, 2010. 65 and 69.

[5] Ulrich Lehman has pointed out to me that the letter may have been sent to the King of the Warni. Personal comment. We will return to this subject, elsewhere. See also, Mäder, 2001. 214.

[6] Quote taken from Buchwald, 2005. p.282 citing Magnus, 1555, Book 7, chapter 3.

[7] Gilmour, 2007. p.95 and 103. Gilmour uses this excerpt, but I have used a different and possibly more accurate translation by Stephen Pollington. Gilmour provides other examples from Old English including the words *sceadenmael* (branching patterns) *hringmael* (curving patterns) and *brogdenmael* (woven patterns).

smith with the skill to produce fine pattern-welding might be likely to produce a good quality sword anyway."[1] Were the patterns included merely to demonstrate the virtuosity of the smith, or show off the expense of the blade so that the owner could display his weapon to other aficionados? Was it akin to owning a very rare and expensive car today, or could there be some other purpose behind the patterning?

Gilmour suggests that there is, and that it had a long tradition. He observes that the patterns, especially of the early layered or banded blades, resemble flowing water.[2] On some of the Iron Age swords, although the entire blade, including the edge, is made from the same watery-grainy structure, prior to carrying out the etching process to reveal the pattern, the edges of the blade were covered in wax, grease or clay to protect them from the acid and then the blades have been heavily etched. According to Gilmour the finished effect was to bring to mind the idea of water flowing between two river banks.[3] Obviously, the smith did not have to protect the edges from the acid, it would not have affected the mechanical performance of the sword in any way; it is purely so that the overall effect can be seen. From the 2nd century, at least, patterned swords have separate edges welded on, so that this effect of banks either side of a flow is created by the welding itself.

Gilmour goes further and says, "This would appear to indicate that the patterns relate to specific pagan symbolism."[4] He suggests that what we continue to see in the swords from the 5th to 7th centuries is, in the main, a stylised version of this idea of water flowing between two banks.[5] Gilmour again, "In this way pattern-welded swords might literally have been seen as belonging, in some way – perhaps as the death dealers of epic fame – to a liminal zone between this world and the next. Possibly the decorated watery surface and (unnecessary and impractical) great length of these weapons[6] may have had as much to do with their talismanic value and supposedly almost supernatural or magical properties, as alluded to in epic verse….."[7] There is, too, a curious reference to a 'wave sword with hard edge' in Beowulf, *wrætlic wægsweord heardecg*, which seems to literally reflect the movement of water within the blade.[8]

Mäder, too, thinks that the patterns have a meaning in the ideology of the pre-Christians, although he sees the patterns as being symbolic of serpents. Snakes can be either a beneficial or threatening creature, possibly a symbol of birth and rebirth but also of venom and death. Certainly *wyrms*, serpents or snakes occur frequently in designs depicted on many artefacts of the period. According to Mäder, this symbolism changes with the conversion because this ambivalence towards serpents does not exist in the Christian worldview, where the creature is consistently considered to be evil due to its connection to Satan.[9]

If the designs had as much to do with the pre-Christian world view as they did with display, then perhaps that could be why the later literature tells us nothing about the purpose of the patterns and also why the pattern-welding of swords seems to be on the wane after the conversions in the 7th century.[10] We might have expected

[1] Lang and Ager, 1989. 110. This idea has been echoed by personal comments from Owen Bush and Mick Maxen, and Paul Binns all experienced pattern-welding smiths.

[2] Gilmour, 2010. 69.

[3] Gilmour, 2007. 94 to 95. Gilmour uses pictures of the Orton Meadows sword to illustrate this technique. See also Lang, 2006. 85 to 114.

[4] Gilmour, 2010. 69.

[5] Gilmour 2007. 107.

[6] I feel that this observation is incorrect and reveals a misunderstanding about how swords were used. We will take this further elsewhere in this volume.

[7] Gilmour, 2007. 107.

[8] Davidson, 1962 (1994). 144.

[9] Mäder, 2001. 217 to 227.

[10] Gilmour, 2007. 107. Lang and Ager, 1989. 107.

a Christian king like Theoderic to disapprove of such an overtly pagan custom, but he was living at a time when many interests had to be considered, including the upholding and supporting of some old customs.[1]

That there is some connection between swords and water in the minds of the peoples of Northern Europe from the Iron Age until the late Middle Ages is difficult to dispute, Iron Age peoples were placing swords (and other bladed weapons plus other items) into rivers and lakes as votive deposits in Britain (and elsewhere) until Roman times.[2] Curiously, in England, the practice seems to almost cease during the 5th to 7th centuries but starts again during the 8th century, well after the conversion, and continues for centuries.[3] We should remember the watery deposition of many thousands of items, including numerous swords in the bogs of Scandinavia and Northern Germany from the 1st century A.D. until the 4th century. Legends such as that of King Arthur's sword Excalibur and its return to the lake after its owner's death will be familiar to almost everyone, although I am not sure how far back that particular part of the story goes, it is an intriguing and very late medieval connection of a sword with water. Perhaps the fact that Beowulf collected the sword that he killed Grendel's mother with under the lake surface is an echo of this connection? There is a later story, too, in which Grettir kills a giant under water with his sword.[4]

Quite what the connection is with water is a matter for speculation, but perhaps it has to do with the powers to give and take life, or maybe a particular deity or spirit whose role has been forgotten, or as Gilmour suggests the liminal aspect of water meant that the sword could inhabit the world of gods and men at one and the same time. It is possible that the significance of the patterns changed over time and from place to place or it could just have been an appeal for help from the supernatural.

There is, too, the possibility that bone ash or ash from cremations was used by smiths as a flux to make the slag run more freely or in the mix to add carbon to the iron; imagine if the ashes of a famous dead hero or a fierce animal were used, how this would affect the numinous quality of the blade and the way that it was perceived in the minds of its owner and others? Perhaps the animal ornament on some swords is an indication of the ingredients that went into the making of the weapon.[5]

In Europe, it would seem, few understood the 'mystical powers' of the smith and how he did what he did, not even King Theoderic. It is possible that the smith would deliberately keep his procedures secret, he would tend to live on the outskirts of a village, settlement or hall-complex, he kept his workplace quite dark, the heat, the strange liquids, oils, urine, that he would use to quench the metal, the fact that he could turn stone into objects of awe and splendour; is it any wonder that the swords he made would acquire 'magical' properties in the minds of others and maybe even that of the smith, himself?

We will return to some of these ideas in the chapters, *The Meaning of the Sword* and *The Mysterious Smith*.

---

[1] Thrasamund, too, was a Christian, but the king of the Warni may not have been.

[2] Stead, 2006. 26, 79. Stead points out that this type of deposition continued until Roman times and then began again in the 8th century. Lang and Ager, 1989. 114.

[3] Stead, 2006. 79. We will return to this concept again in the chapter called *The Meaning of the Sword,* where we will refer to a few examples from 5th to 7th century England. As we shall see, in the chapter, *The Riddle of the Pyramids*, some sword parts and paraphernalia may have been deposited on dry land instead.

[4] Davidson comments on the strange coincidences contained in both Beowulf's encounter with Grendel's dam and Grettir's struggle with the giant. Davidson, 1962 (1994) 139 to 140. Brunning, too, comments on watery deposition and seems to agree that it is a ritual act. Brunning, 2014. 37.

[5] Gansum, 2004.

Where were swords made during the 5th to 7th centuries?

As suggested above, it is generally thought that weapons, including swords were being made, under Roman control, in the Rhineland,[1] at least from the 2nd century onwards. It is likely that organised manufacture continued in those areas for quite a while after the Empire faded away, perhaps even into the late Middle Ages as indicated in the meta-analysis by Frank Nikulka.[2] It is possible that many of the swords found in Northern Europe that date from the 5th to 7th centuries, were manufactured in that region, and perhaps other regions, too. However, whether swords were made in England during the period has not been established with any certainty.

Davidson said of pattern-welded swords, "...as yet there is no evidence that they were ever produced in England or Scandinavia, though there seems no convincing reason why they should not have been."[3] Lang and Ager also consider the question of whether the swords were made within the British Isles and say that it is a difficult one to answer but suggest that the technique of surface removal, that is when adding a fuller, or fullers to a blade, or to modify the pattern, is a continental one and was probably not practiced by smiths in England.[4] Fullers in Anglo-Saxon England, they feel, tended to be hammered in using fuller blocks so as not to alter the patterns. Gilmour has noticed that most 'herringbone' pattern swords found in England that have received a full analysis, would appear to have undistorted patterns; it seems to have been important to reproduce precise work as a mark of excellence, he feels that different criteria may have had to be met with in Frankish areas.[5] Lang and Ager ask the question, "If Anglo-Saxon swords were made in England what was the source of the iron, and where were they fabricated."[6] As we have seen in the chapter on *Iron and Steel,* there are many iron deposits within the lands that make up Britain and, as discussed there, much of the more specialised iron may well have been traded abroad in exchange for other alloys from elsewhere, perhaps from Scandinavia, the Continent, or even from other parts of Britain, or Ireland or bought from itinerant merchants, as required by the smith.

As for where within the islands swords were being made is difficult to answer, but as research into metal working continues and new discoveries of slag and ferrous metal working remains are being discovered perhaps it is a question that may be answered, to some extent, soon. It is improbable that during the 5th to 7th centuries every sword carried by an Anglo-Saxon warrior was made on the continent; some enterprising smith(s) during those three hundred years would have been able to spot the gap in the market and fill it. It is unlikely that a great chief or king would have been without his weapon smiths and facilities to help arm himself and his warriors. Lang and Ager appear to come to the same conclusions.[7] In my opinion they were probably being made everywhere that there was a significant settlement.

It must be said that at the time of writing, we really do not know of any region in Europe that definitely hosted smiths who made swords during the 5th to 7th centuries. Yet they were being made! I am sure that

---

[1] Wiesbaden, Mainz, Trier, Lorch, Niederbieber, Köln and Neuss. See also, Nicolay, 2007. 1 to 6, just to get an idea of the size of the Roman military presence in some of the Rhineland areas. In this case, they are Batavians. It is probable, too, that production continued in parts of what had been Gaul. However, Gilmour feels that they would have been made in other places, too. Personal comment.

[2] Nikulka, 2003.

[3] Davidson, 1994. 34.

[4] Lang and Ager, 1989. 112. Gilmour seems to agree. Gilmour, 2010. 65. Gilmour, 2007. Lang, 2012. 253. We will discuss the fuller, its function or its absence in the chapter entitled, *Blade Shapes*.

[5] Gilmour, 2007. 101. To a limited extent, this suggestion is borne out by the work of Lehmann, (2013. 13.) who found that curving patterns on swords are common on swords found in Westphalia Germany.

[6] Lang and Ager, 1989. 112.

[7] Lang and Ager, 1989. 113. Birch's paper, (Birch, 2016. 9 to 10) possibly hints at an answer with his 'utmark/innmark (or outfield/infield)' model which means looking just outside the accepted 'edges' of the settlement for evidence.

weapon smiths were located in many of the areas of Anglo-Saxon England; it is highly improbable that such militarised societies that existed during that time would have left themselves dependant on imports from abroad.

As for whether the perceived differences as discussed above between Continental and Anglo-Saxon swords are real or not needs further discussion and certainly more research, 'herringbone' pattern sword blades exist in both areas and in Scandinavia, too; many more blades needs to be considered and a catalogue created so that genuine comparisons can be made.[1] I have tried with Brian Gilmour's aid to create a catalogue of blade patterns from Anglo-Saxon England, but as far as I am aware nothing similar has been attempted for any other region. One other consideration is that we know that some members of the aristocratic warrior class and their retinue tended to travel in their search for glory and renown when young; presumably they took their weapons with them. This phenomenon would complicate the matter of the origin of swords and where they may have ended up.

Why did the pattern-welding of swords decline?

Based on current evidence, the production of patterned swords reached its peak during the 6th and 7th centuries then appears to decline and it has been suggested that pattern-welded blades after the 7th century are constructed from fewer parts with less complex patterns. I am not sure that evidence really supports that view that patterns became simpler.[2] We have, briefly, considered the idea of pagan symbolism being an important part of the display and the fact of the conversion during the 7th century England may have meant that attitudes to pattern-welding changed such that it was no longer acceptable in its old forms. It must be admitted that there is a problem with the evidence, as burials with grave goods cease during the 7th century and no swords have been definitively identified from the 8th century at all in England. Ironically, many finds from the 9th century onwards are from rivers and watery places. The Vikings briefly re-introduce burial with grave goods during the 9th century, thus increasing the sword count for the later period a little.

However, there is no doubt that what statistics we have indicate that the frequency of such swords does diminish, according to Lang and Ager the proportion of patterned swords in the 9th and 10th centuries falls to 45%.[3] By then the famed, all steel, Ulfberht swords and their imitators (not all steel) are being used, and some of these were probably superior to many patterned swords as far as performance in the field is concerned.[4] Although Ulfberht swords could be part of the reason why patterned blades became less common, they don't really appear until the 9th century and they are likely to have been just as expensive and not too easy to obtain. Intriguingly, the names on the sword blade are often in-laid with pattern-welded letters.[5]

Davidson suggests that by the 9th century better ores were available as conditions became more settled and that the design of furnaces had improved, so, presumably patterned blades didn't need to be made so often anymore, although existing ones continued in use.[6] One assumption here is that patterned swords were only made because the smith did not have sufficient material to make other types of blades. That may have been the case originally, but the fact that more and more complex patterns were developed over time, shows that the patterns were used for other than purely practical reasons by the 2nd century and beyond. Warfare after the

[1] We will consider the patterns and their structures in the chapter on *Patterns and Pattern Categorisation.*
[2] It is difficult to be definitive, partly due to the lack of examples in the later period; it is certainly true that all of the known most complex blades (four rods or more) come from 5th to 7th century contexts. See *Appendix 7 Pattern database.*
[3] Lang and Ager, 1989. 107.
[4] Apparently the true Ulfberht swords were wholly made of steel, many of the imitators were not, see Williams for details. As for their supposed 'superiority', it is a really difficult notion to test scientifically.
[5] Williams, 2012. 116 to 183.
[6] Davidson, 1994. 32. Unfortunately Davidson doesn't provide details about improved furnace design and how she knows they were better.

7th century was still endemic, and may have been conducted on an even larger scale; as we know, the Carolingians were particularly active in seeking quarrels and the Vikings were aggressive and very active as was the local response in many parts of northern Europe, including England. All European societies continued to have highly developed martial cultures.

A reasonable possibility discussed in Lang and Ager is that smiths could no longer spend so much time making a blade that took weeks to make, when other more simple weapons like seaxes could be made more quickly and cheaply, although it is difficult to understand how a seax could always replace a sword as they can perform quite different functions.[1] Gilmour does suggest that to some extent patterning shifts from swords to other weapons like spearheads and seaxes, he says, "Already simple forms of pattern-welding were being used for some spearheads, especially larger examples, from the mid-7th century..."[2] However, patterned spears had been found from earlier periods, eight out of the sixty five spearheads from Mucking were found to have patterned blades and so were three of the one hundred and sixty-eight knives so this tradition may well have started before the decline of patterned sword blades but has just not been recognised.[3]

Possible, too, is that during the later period more swords were needed than before in order to engage with the Vikings, etc., and that cheaper and less time consuming methods were required in order to cope with the demand.[4]

The use and effectiveness of pattern-welded blades as weapons[5]

It has been suggested that pattern-welded sword were primarily for decoration and display and not really for use as weapons, Gilmour asks, "Are we really so confident about what swords meant in Anglo-Saxon society as well as the manner in which they were used, how they were made and why?"[6] Gilmour, goes on to say that they, "...have long been known to have been important symbols of rank and cultural icon in early Europe and elsewhere."[7] His point is a good one: swords in modern armies have a purely ceremonial role and are no longer used offensively. I would add that swords could also act as a medal, a sign of achievement, indicate social status and perhaps a few more possibilities that we will return to in the chapter on the *Meaning of the Sword*. We will also think about the way Anglo-Saxon swords perform as weapons in more detail in the chapter, called *The sword as a weapon*. However in order to try and establish that patterned swords were used in warfare it should be remembered that the 5th to 7th centuries in Northern Europe were times when warfare was embedded within the fabric of the existing societies; Davidson, says, "Warfare permeated the social structure of North-western Europe, influencing art, symbolism and literature; it played an important part in the pre-Christian religion and left its mark on popular beliefs and customs."[8] In such circumstances it is

---

[1] Lang and Ager, 1989. 113.

[2] Gilmour, 2007. 107. See Buchwald (2005. 291), who reports on some early patterned spearheads.

[3] Hirst and Clark, 2009 (i). 424 and 425. Some of the finds are possibly as early as the 5th century; there is a strong correlation with these and Swanton type H spearheads (pages 425 and 426). Two of the knives are only possibly patterned. There are patterned spearheads from the 5th to 7th century cemetery at Wasperton. Carver, Hills, Scheschkewitz, 2009. 70. Grave 1 at Coddenham contained a patterned seax. Penn, 2011. 62. This might be 7th century.

[4] Personal comment, L. Kerr.

[5] Different authors have used different methods of labelling the hardness of a metal. I have kept each commentator's method and have not attempted to standardise them.

[6] Gilmour, 2007. 91 and 93. See also Lang and Ager, 1989. 109. Where they report Tylecote saying something similar. Tylecote goes on to say, "If it were not for the fact that we know such swords were used for fighting (Beowulf, etc.) we would have supposed that its purpose was like the ceremonial sword of today." Ager again suggested that some such swords had a 'ceremonial function' in 2015; see his work on the swords from Tranmer House (Fern, 2015. 123).

[7] Gilmour, 2007. 93.

[8] Davidson, 1989, 11. Halsall, 2003. 2 to 5. Mortimer, 2011. 186 to 187.

difficult to imagine a warrior encumbering himself with a weapon that was not of some use, especially in the competitive atmosphere of the meadhall, where an offensive remark or challenge may require an answer.

James was certain about efficacy of pattern-welded swords[1] Davidson was in little doubt either, and says, "It is no longer difficult to understand the popularity of the pattern-welded blade in North-Western Europe for so many centuries."[2] However we need to see if the structure of the blades precludes their practical use.

We shall report briefly on two metallurgic studies of some pattern-welded swords and consider what they tell us. Firstly, we will look at the work of Vagn Fabritius Buchwald, the metallurgist, who has examined a number of the swords from Nydam (200 to 400 A.D.). Their owners liked them because of their patterns, but Buchwald wanted to know if they were better than or as good as blades with other forms of construction. Of the one hundred and six swords found at Nydam, ninety-three were pattern-welded[3] and Buchwald analysed eight of these swords. Unfortunately, none of the swords that Buchwald examined had complete blades all had been broken or cut at some stage and had suffered corrosion to a greater or lesser extent. He found that, although many of them were not as hard or as ductile as they could be, most of them were able to perform very well and the pattern-welding had been executed perfectly, something that Gilmour has also noted. However, two of the blades were described as 'rather miserable' as their edges were only as hard as the point of an ordinary nail. He did point out that, on these fairly early pattern-welded swords, none had been quench-hardened (rapidly cooled in water or oil) and that this process would have turned most of them into excellent blades.[4] Although there had been an attempt to harden at least one blade's edge (Nydam No. 4154) Buchwald could not say what process had been used for certain but he suggests that when hot, the edge was drawn through wet clay.[5] Perhaps Williams has the answer when he says that, "Many smiths evidently (and understandably) preferred a blade of moderate hardness but reliable behaviour to one of greater hardness but possible brittleness."? [6] Buchwald also examined a much simpler blade from Vimose, which just had hammer-welded edges and no pattern-welding, and he found the sword itself to be extremely hard, although the edges were not very good.[7]

In summary Buchwald says that the blades represented a variety of qualities, structures, width and all were 'surprisingly' free of slag inclusions.[8] The joining of the different metals, phosphorous and carbon iron alloys, had been accomplished successfully and two of the swords had excellent martensite edges of 676 VPH and four others were acceptable.[9]

Buchwald goes on to say that, "It is difficult to maintain that the pattern-welded sword was just a prestige weapon. The examination shows that it was indeed of high coherence and quality and sometimes was hardened (the Vimose sword) or had martensite-hardened edges over a tough nucleus. It would be difficult to understand that a company of warriors, in casu (*sic*) Nydam, would enter serious combat, armed with 90 pattern-welded and 16 average swords, if they did not appreciate the combat value of their pattern-welded swords."[10]

We will turn to a more recent study of some of the swords that have been recovered from the cemetery at Dover Buckland, in Kent. These swords are from the period that this book is attempting to cover, the 5th to

---

[1] James, 2011. 184.
[2] Davidson, 1994. 30.
[3] Gilmour, 2007. 96.
[4] Buchwald, 2005. 283 to 291.
[5] Buchwald, 2005. 287.
[6] Williams, 2003. 11.
[7] Buchwald, 2005. 265.
[8] This echoes the findings of Sim and Kaminski (2012. 18) with their consideration of Roman iron and steel.
[9] Buchwald, 2005. 290 to 291.
[10] Buchwald, 2005. 291. Buchwald is specifically replying to the comments of Tylecote, 1987, here.

7th centuries, so the findings from these swords will be more germane to our period than the earlier Nydam swords. Seven swords have been found in the more recent excavations at Dover Buckland and all were double-edged and pattern-welded.[1] Five of the blades were examined by Janet Lang using various techniques. None of the extraneous materials adhering to the blade, such as scabbard remains or mineral corrosion was removed from the blade prior to X-radiography; it is likely that some details were obscured because of this. Approximately 20 mm sections of metal were taken from each blade, these were then mounted in resin and had their surfaces ground and polished. Care was taken at all stages to avoid unintentional heat treatment occurring to the samples. The polished samples were examined in an electron microscope and later were tested for hardness.[2] I will briefly report on some of the characteristics of each blade and then try to summarise Lang's findings.[3]

The sword from grave 264 had excellent martensite edges, which were lightly tempered. [4] Its hardness was +900 $HV_{0.1}$.[5] The blade has three main components in its central core with patterned layers parallel to edges on either side.[6] Lang judged this a very good weapon.

Grave 265B provided a sword that had cutting edges which looked as though they may have been deliberately carburised although there was no attempt at tempering. The blade consisted of two layers with four twist segments and possible curving elements.[7] Hardness maximum was 226 $HV_{0.1}$ and this was not at the edge.[8] This was not as effective a weapon as 264.[9]

The sword from Grave 346 had unusual amounts of nickel in its composition and Lang discusses possible reasons and parallels.[10] It has a broad ferritic core layer parallel to surfaces in the centre, and on either side of that there are sections with twists and straights. It was judged to be not outstanding and not very hard throughout.[11]

The sword from grave 375 has two central layers, each having two segments of twists. The edges are considerably harder than the centre, over 900 $HV_{0.1}$. The blade has been quenched, and was an effective weapon.[12]

Finally, the blade from grave 437 consisted of a two layer centre with twists. The cutting edges had been made of a carbon-rich iron sandwiched between two phosphorous-rich layers. The blade had been quenched and some martensite was present on the edges. The hardness along the edge is of 440 $HV_{0.1}$ to 520 $HV_{0.1}$ and Lang considers it a good weapon.[13]

The swords from graves 264 and 375 were probably auto-tempered (slack quenched) others seem to have been air cooled. The average carbon content at the edges was not high and was around 0.4% but at that level

---

1 Ager, 2012. 49. Lang, 2012. 237.

2 Lang, 2102. 238. For full details consult Lang. During earlier excavations at the same site many more swords were discovered. For full details see Evison, 1987.

3 For full details see Lang 2012, where she provides very comprehensive reports on each sword.

4 Lang, 2012. 242.

5 Lang, 2012. 243.

6 Lang, 2012. 242.

7 Lang, 2012. 245.

8 Lang, 2012. 245.

9 Lang, 2012. 246.

10 Lang,2102. 261.

11 Lang, 2012. 248.

12 Lang, 2012. 250.

13 Lang, 2012. 252 to 253.

it is still possible to harden via quenching.[1] Lang makes the point that heat treatment is usually regarded as a technical advance but its absence does not indicate a lack of skill and it is possible that some smiths preferred not to quench their sword because they were afraid of inducing brittleness.[2] Phosphorous was usually higher within the central sectors but was diffused generally throughout most blades. There was little difference in hardness between the phosphoric iron and the low carbon iron, which would probably mean that both would stand up to wear equally well. The quality of the swords shows some variety in manufacture and the capabilities of the smiths, but all of the blades would have performed as weapons, but some would have been better to use than others. Unfortunately, Lang does not provide reconstructed pictures of the patterns included within the blades.

It would appear, then, that both Lang and Buchwald felt that patterned sword were perfectly capable of performing as weapons, even though some were 'better' than others.

There have been two recent reconstruction attempts at trying to work out why pattern-welding was so popular and whether it had any advantages over more simply constructed blades. The first is a study undertaken for a MA degree at the University of Groningen by Sebastiaan Pelsmaeker. Pelsmaeker set out to answer the question, 'Were pattern-welded blades of a higher functionality than mono-steel swords or did the patterns serve a purely decorative purpose?'[3] He tried to use bloomery iron to construct the blades but explains how this part of the experiment was not able to proceed, mainly due to running out of time, a lack of appropriate tools and experience with the material so had to use a modern material, S235 steel.[4]

Once he had constructed the swords he subjected them to a series of tests to compare cutting abilities, edge retention and the flexibility of the blades.[5] His findings were that pattern-welded blades were stiffer and less flexible than less complex blades and could, he believes be made thinner and therefore lighter than other types of blade.

The second study is by Thomas Birch and took place at Department of Archaeology, School of Geosciences, University of Aberdeen, during 2013. Birch set out to answer a similar question, 'Does pattern-welding make Anglo-Saxon swords stronger?'[6]

Birch dismisses Pelsmaeker's attempts with, "The subjective nature of Pelsmaeker's (2010) testing methods, however, prevents any objective comparison."[7] Birch is correct and it would be difficult to replicate Pelsmaeker's experiments exactly. However, it was a brave, if naive attempt and involved a great deal of effort, time and thought. How valid Pelsmaeker's results are is difficult to say, especially as he only used one type of steel in his blades and he assumed that the alternative to a patterned blade was a mono-steel one. Birch is correct, too, in that more work is required and he sets out to try and address the problem. Birch had two samples made up by the English smith, Hector Cole. One was pattern welded while the other was made of a single piece of iron. Both samples were then cut into three equally sized pieces.[8] The materials chosen broadly reflect the composition of Anglo-Saxon swords in that the pattern-welded piece was constructed from a low-carbon iron (0.2% C) and Victorian phosphoric wrought iron. However, the non-patterned piece was made

---

[1] Lang, 2012. 265.

[2] Lang, 2012. 265.

[3] Pelsmaeker, 2010. 3.

[4] Pelsmaeker, 2010. For an account of the attempt at forging the bloom see page 38 to 48. The decision to use one type of steel to forge a 'mono-steel' blade and a patterned one is explained on page 51.

[5] Pelsmaeker, 2010. 67 to 81.

[6] Birch, 2013. 127.

[7] Birch, 2013.128.

[8] Birch, 2013. 128.

from a single piece of the same low-carbon iron used to make the more complex piece. Both patterned and non-patterned samples were made to the same dimensions.[1]

The pieces were subjected to x-radiography to examine their crystalline structure and to see how successfully the welding had been accomplished, apparently there were small faults. The pieces were then subjected to a series of standardised and accepted test methods to examine the comparative performance of the samples. Test methods looked at hardness, impact toughness and tensile strength.[2] Birch's conclusions were that "NPW (none pattern-welded) values for proof-stress and ultimate tensile strength (maximum stress) were marginally greater than PW (pattern-welded), indicating that NPW specimens were slightly stronger. The clearest difference between NPW and PW in tensile testing was the reduction in area, where NPW was deemed to be more ductile. Although both exhibited increased toughness at 5°C, Charpy testing confirmed NPW to have a greater toughness than PW specimens."[3] So it would appear that the none pattern-welded pieces were slightly superior to the patterned, except that it is not that clear cut and he found that the fracture performance of the patterned samples were better, the reason, it seems is due to the 'fibrous quality' of the patterned sample which promote 'crack closure'. He likens the structure to the thin veneers of plywood, which will not crack, easily, across the grain.[4] He feels that this is why pattern swords were preferred.

Birch's experiments are intriguing but cannot provide a full answer regarding the relative performance of swords, as we have already seen; very few swords, if any, were ever made from one piece of homogenous iron, at least in the period under focus; even non-patterned swords had complex edge constructions, unlike Birch's non-patterned sample. Another problem with Birch's experiment is that the materials used to make the samples were of necessity modern and that will have affected the outcome and, testing just two samples anyway can only be of limited value, if any.

In many ways the question posed by both Pelsmaeker and Birch sets up a false dichotomy and both fail to take into account the way that swords may have actually been used during the period; the things that they were trying to test probably did not matter that much to a swordsman of the period.

Limitations of the test methods

There are limitations with all the analytical methods that have been used; perhaps the most important is that the tests cannot tell us what is missing. If a blade is corroded, and all the above were to a greater or lesser degree, then no information can be gained concerning the missing sections, and some parts of a blade are more vulnerable than others, the thin, steel cutting edges for example, which can be very thin. From each of the blades, Lang cut one section, taken from varying positions wherever metal still remained. She did not take complete sections across the sword, just from the cutting edge to the centre of the blade so that she was never able to look at a complete cross-section of a blade.[5] Nevertheless, she was able to show that the swords the blades came from were, in the main, creditable weapons.

There is scope for much more research to be carried out regarding the merits of patterned swords, perhaps even involving whole swords made from authentic materials taking into consideration possible fighting methods used during the early Middle Ages?

Again it must be said that looking for advantages of patterned swords over non-patterned examples may not really be a worthwhile exercise as it may be missing the real purpose(s) of the weapon.

---

[1] Birch, 2013. 129.
[2] Birch, 2013. 131 to 132.
[3] Birch, 2013. 133.
[4] Birch, 2013. 133.
[5] Lang, 2012. 238.

What none of the tests tell us is how they handle as swords. We will consider this aspect later.[1] I am not clear, either, that any of those commentating on the efficacy of patterned swords, have really thought through the way in which such swords were used in combat; in short I think we have all seen to many Hollywood movies!

However, it is reasonable to assume that the owners of such swords believed that pattern-welding made a better weapon for them, even if its merits lay in the supernatural world rather than the practical. They were determined to equip themselves with what they regarded as the best technology available by whatever means, and that 'technology' may have been, at least partly supernatural. In fact, warriors may sometimes have gone so far as to take them from burial mounds; for example, Midfjord-Skegge opened the tomb of Hrolf Kraki to get the sword Skøfning.[2] Indeed, pattern-welding could be taken as a sign of quality just because the ores making up the sword could visibly be seen to have withstood the stresses of layering and twisting.[3]

Mass production?

An interesting, if flawed, paper from 1997, raises some questions about how many swords could be made in a given period and how smiths could have met the demands of the Carolingian army.[4] There are many major difficulties in what Conti, Bowlus and Braithwaite write; for instance they seem to believe that such swords were invented by the Franks, "Moreover the technology that lay behind Frankish pattern-welded sword manufacturing diffused rapidly beyond Charlemagne's empire to Anglo-Saxon England and Scandinavia, whence Varangians brought these techniques into the land of the Rus. (Ellis-Davidson, 1994, p. 114f.)"[5] They appear to be blissfully unaware of the fact that pattern-welded swords were being used in England, Scandinavia and elsewhere hundreds of years before Charlemagne was born. However, they are prepared to accept that Charlemagne's armies were very large indeed, taking some time to explain how such numbers were possible and go so far to say, "It is confirmed that Charlemagne fielded in excess of 30,000 men."[6] However, the idea of Charlemagne engaging, 'vast numbers of smiths is just dismissed as 'implausible' without further discussion.[7] I suspect that this is largely because they believe that they have come up with a 'new' and labour saving idea about how the swords were made and just want to tell us about it.

They seem to assume that none of Charlemagne's 30,000+ men had any swords to begin with and that each every one of them would need a new one. Based on that idea they suggest that it would take ten smiths with two assistants each, twenty years to make 5,000 swords.[8] They calculate that it would take 6.66 million man hours to produce enough for 30,000 men! They get their number of man-hours to make a sword from the experiments carried out by the smith J.W. Anstee, who made a complete sword, grip, scabbard and belt and buckle in seventy-three and a half hours. The full report is included as an appendix in Davidson.[9]

---

[1] Lang, 2007. 108 to 113 looks at the handling of swords and tries to discover scientific models to work out how they would perform and how effective they would be. In my view she fails completely, partly due to a poor understanding of how such weapons would have been employed in actual combat. We will discuss this aspect of the sword in the chapter entitled, *The Sword as a Weapon*.

[2] Grettir's Saga.

[3] Personal comment from Paul Binns, sword-smith. The reader will recall that similar sentiments were expressed by Owen Bush and Mick Maxen.

[4] Conti, Bowlus, Charle and Braithwaite, 1997. *Questioning the Accepted Techniques for Sword Forging in Anglo-Saxon England and Frankish Europe*. In Journal of Arkansas Academy of Science Volume 51. Little Rock.

[5] Conti, Bowlus, Charle and Braithwaite, 1997. 6. They do take Davidson's writing completely out of context.

[6] Conti, Bowlus, Charle and Braithwaite, 1997. 4 and 12. Of course, no such confirmation can exist.

[7] Conti, Bowlus, Charle and Braithwaite, 1997. 12.

[8] Conti, Bowlus, Charle and Braithwaite, 1997. 2.

[9] Davidson, 1962 (1994). 217 to 224. We have seen, that some more recent work on sword making could mean that the time taken to make a decent sword was more than seventy-three and a half hours.

They go on to propose a method of production, involving a division of labour between the master smith and his assistants and a production line. I am not sure that there is anything particularly radical in their suggestions but they feel that the production could be simplified to the following stages which would require a large, charcoal-fired forge:

> a skilled smith welds two or three bars[1] together at one end,
> which are then passed to assistants who heat the bar to red or white heat,
> withdraw the same,
> once coated in oxides – clamp it firmly together and replace in the fire,
> left in for 'some time' 'during the process of heating in excess of four hundred and fifty degrees the carbon from charcoal and heat of the fire should reduce the scale to pure iron,
> new assistants take the bar and twist it,
> then passed on to be forged square – then reheated to white heat,
> bar returned to master who welds it to two other bars (core now complete) this allows many bars to be in the forge at the same time,
> master smith then welds the edges,
> blank is now forged,
> filed and,
> polished,
> smith will chase out any failed welds with a chisel and sift in some powdered scale or oxide,
> put back in the fire so that oxides are reduced back to pure iron and fill the failed welds.[2]

Basically, what they suggest is that the master smith is free to carry out the more demanding jobs while a group of trained assistants perform the 'easy' manual tasks and that failed welds are remedied at the end by melting scale into them. I am not sure how effective this would be and have grave doubts about its efficacy. Conti, et al, had intended to carry out some experiments themselves, but I have been unable to find any evidence that they ever did.

However, leaving aside all the problems, the paper is in its own way, quite an interesting one partly due to the ideas of a division of labour, a production line and the possibility of a sophisticated arms industry. Taken together with the evidence for a highly developed trade in different qualities of metals, it is an idea that would be intriguing to take further and seems to resemble what we know of the earlier Roman *fabricae*.[3] Possibly not exactly the system that Conti, *et al* envisage but something similar could have meant that there were plenty of weapons, not just swords, around with which to equip armies.

Were swords as rare in Anglo-Saxon England as many commentators suggest?

Conti, et al.'s work does bring us to this question quite neatly. If we judge the numbers of swords in circulation by the numbers left in graves, then swords were very rare objects indeed, but the recent finds in Staffordshire of mostly damaged weapon parts has provided good evidence that this may not have been the

---

[1] These 'bars' are the rods of laminated metals that will form the core of the sword and will carry the patterns. Presumably, these are made by the master smith.

[2] Conti, Bowlus, Charle and Braithwaite, 1997. 14.

[3] Incidentally, La Salvia, 1998(21 to 23) and 2007 (44 to 71) provides considerable evidence for such workshops in contemporary Lombard Italy and suggests that there was continuity from the earlier Roman period to the time of the Lombards.

case; eighty-five pommels and pommel caps1 were found within the Hoard, all of them very high status objects of gold, or garnet and gold or silver. That number represents a really significant addition to the corpus of such finds from England and has to alter our perception of the numbers of swords available during the 5th to 7th centuries. In my survey of around ten thousand graves, there were one hundred and sixty-nine swords recorded, most of these had hilts with no metal parts of very simple ones, so it is pretty certain that many more 'ordinary' swords were around. It must be admitted that it is always a risk to use percentages from cemeteries as a reflection of real-life, as Härke tells us, "Burials are not 'mirrors of life': if anything, they are a 'hall of mirrors of life' providing distorted reflections of the past."[2]

I would like to conduct a little thought experiment. We don't really know the population of Anglo-Saxon England during the 5th to 7th centuries; the only time that we get enough data to make any kind of informed guess is after the Domesday Book was compiled in 1087, and it is thought that there may have been as many as one point eight million people making up the population.[3]

Let us suppose that there were half a million people living in England during the 6th century. We know that the society was highly militarised, so there would have been a demand for weapon smiths.[4]

I would suggest that one hundred weapon smiths in a land largely geared up for warfare would be a very low number, but for the sake of this exercise we shall just go with that number. That would make one smith for every five thousand head of the population (1:5,000 or 0.02%). I think to provide each smith with a couple of assistants would be perfectly feasible and still keep the numbers of people involved in making swords very low.

It doesn't seem unreasonable that each smithy would produce ten swords each per year, so for any given year there would be ten swords times one hundred smithies, making one thousand swords being made each year. In ten years there could be ten thousand swords produced, in twenty years twenty-thousand, thirty years thirty thousand and so on. Of course there would be some attrition, swords can get damaged, lost or buried, but generally they would be looked after and kept and stored safely, they were expensive items after all. Let's assume 10% attrition; after fifty years there could be potentially fifty thousand swords minus the five thousand lost or destroyed, still leaving forty-five thousand possible swords.[5]

The potential for there to have been large numbers of swords in circulation was certainly there, so were swords as rare as some would have us believe? We will return to this idea elsewhere within this book. Personally, I feel that anyone who saw their business as war, even if only part-time would have been able to equip himself with at least one sword, although he may not have 'owned' it himself.

---

[1] Cool, 2015. 87 to 88. Cool points out that more swords might actually be represented and Pieta Greaves has recently confirmed that around two hundred swords could be represented in the find. Personal comment.

[2] Härke, 1997c. 25. Personally, I would suggest that burials are not 'mirrors' in any sense, funerals are, and were, very special events outside of most day-to-day activities, Härke's quote may reveal that he agrees..

[3] Broadberry, Campbell. Bruce and van Leeuwen, 2010.

[4] I have no idea what the population size actually was and in some ways it does not matter; one hundred smiths capable of making swords in England at any time, I think is a very low estimate. Incidentally, various scholars have used equations and calculations that produce greatly varying figures up to several million.

[5] Halsall too, feels that swords were not rare, although they could be 'expensive'. Halsall, 2010. 2.

6A. A Different Method for Forging Sword "Fullers"

Lee Sauder

Over the past few years, I've occasionally puzzled about the fullers in Viking era swords. Modern smiths recreating these swords either grind them, or forge them in with spring fullers, neither of which methods fits with what we know about the Viking toolkit.

Recently, through dumb luck and serendipity, I came up with an idea that the fuller shape could be created by upsetting the edges of the core bar before the steel edged are welded on. That is, starting with a bar that represents the thinnest part of the blade profile, rather than working back from the thickest one.

I found this idea attractive because it would require only a hammer and anvil (plus of course, some practice and skill). It could explain why the edges of the "fuller" are usually so nicely concurrent with the edge of the core bar. And it would explain how the form arose, as an artifact of the construction method of steel edges on a pattern welded core, that was more and more accentuated over time, from the flat cores of the Anglo-Saxon era to the shallow fullers of earlier Viking era swords. As the form was defined, its advantages of lightness and stiffness would be recognized, and retained and exaggerated. As the construction methods of the blade changed from pattern-welded to all steel blades, this would suggest why the fullers tended to get narrower and deeper over time.

So first I'll explain the method, and then I'll come back to point out a few things in favor of this hypothesis.

Trial #1 Here is the first somewhat successful attempt at doing this. The early trials that didn't work failed from improper choice of starting sizes. This case is a small dagger sized version of the basis sword shape. (Both of the following example are simply forged from mild steel, disregarding for now the nuances I'm sure will arise from using a pattern-welded core and steel edges.)

Starting with a core bar of 3/16" by 1½", the point is forged, and then the edges are upset using a small hammer and light snappy blows.

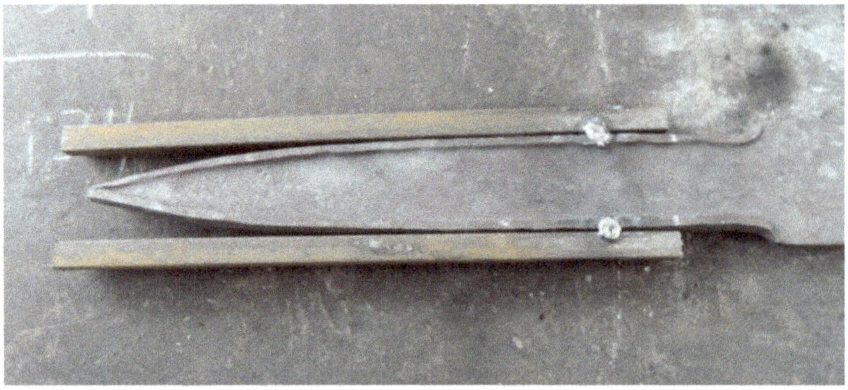

The edge bars of 5/16" square are tacked on (with an arc welder for convenience and clarity).

Then the bars are welded on (I think this is showing halfway through the second of two passes of welding)

And here the bevels are beginning to be defined:

Then the rough shape is forged. This could have been further defined by hammer, but I deemed it good enough for the moment.

Trial # 2 With this successful experiment in hand I tried again, trying to choose sizes that would end up close to real Viking sword dimensions. Here I used a core bar of 3/16" x 2", with 3/8" square bars for the edge.

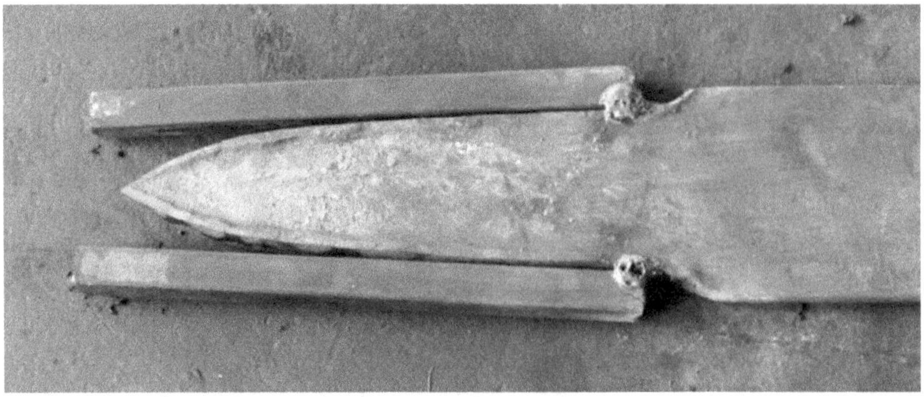

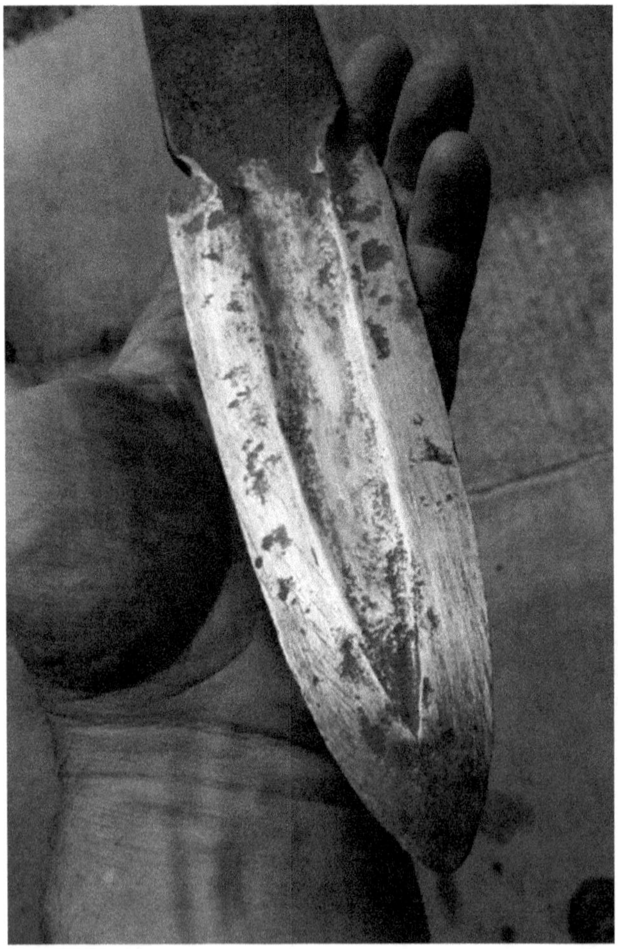

And here's the result, after spending ten minutes with a file on the edges and rubbing the "fuller" with a crumbly piece of sandstone.

So here are a few practical advantages of doing things in the way described, and arguments in favor of this hypothesis as a historical method:

1) I think this is actually a rather quick, easy, and accurate way to create this form. It's all done by one man, with small hammers, with small bars, and short welding heats in a solid fuel fire. The second example above took me 85 minutes, and it's only my second attempt.

2) The design and proportions of the blade all derive from the shape and size of the core, and the size and cross-section of the edge bars. Like so many forging problems, if you get the setup shapes right, the final shape is almost automatic. The coincidence of the bevel with the core bar is now also automatic, without the struggle this would cause when grinding or hammering in a fuller.

3) With the core bar defining both the fuller and the overall outline, tapered blades and tapered fullers are now a much simpler proposition.

4) So many historical blades show a simple chevron pattern, which indicates very minimal stock removal in the fuller. This method allows that.

5) It would seem to be a much simpler thing to add inlays of lettering and symbols to the flat bar before upsetting and welding, rather than adding such a tricky little detail to a curved surface after you've already forged the whole blade!

I'm hoping some of you who are more knowledgeable and blade-obsessed than I am will pitch in to working out this idea. The only way we'll really know if this (or a variation of this) is how it was done is to make lots of them. Looking for characteristic pattern distortions in the core (or lack thereof), and other signatures that might arise, and then looking at the originals to compare. Already, as I've educated myself more on the historic examples, I have seen a few details of shape (rather than pattern) that encourage me, especially in the blade-to-tang transitions, and in the taper and termination of the fullers.

And to preach a few seconds- this can't be answered by the smiths or the academics alone - an answer will require honest work by both!

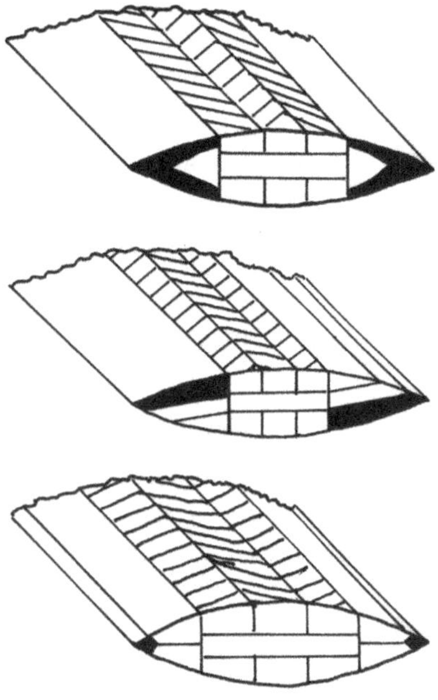

7.1. (not to scale) The cross-sections illustrate how complex the blade construction can be. The top drawing has steel confined to a wrap around the edge core. This sword is S43 from Gilmour's thesis (1999) and was found in the RiverThames.

The middle illustration shows a sword from Sarre, grave 34. Again the edge construction is complex and the harder metal, low carbon steel, occurs on only one side of each edge. S17 in Gilmour's thesis.

The bottom drawing is from Mitcham, Surrey (Museum of London 56:106/14). The fairly low carbon steel (0.4 to 0.5%) is confined to a very thin strip along the edge. S10 in Gilmour's thesis.

All drawings, although re-drawn, are used with the kind permission of Brian Gilmour.

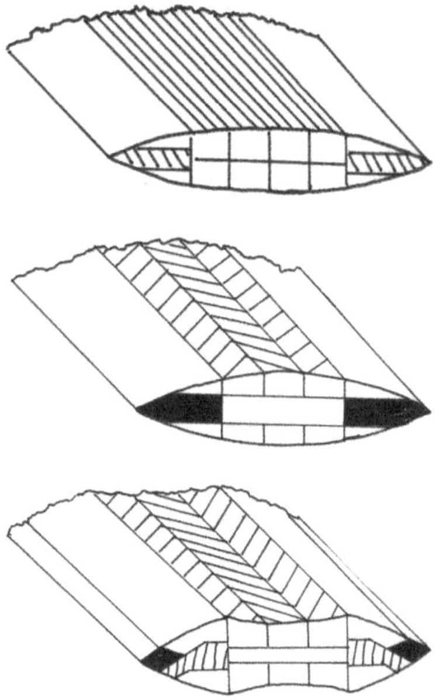

7.2. The top illustration is of a sword from Chesterton, Cambridgeshire. This sword has a low carbon steel insert in the centre of the edge on both sides. S40 in Gilmour's study.

The middle sword is from Wickhamreux in Kent and, again has a relatively low carbon steel insert in the edge. S19 in Gilmour's study.

Bottom shows a cross-section of a sword from Caversham, Berkshire. This sword has a fuller and each edge is made from four different pieces of metal positioned in such a way that the edge tip is made of steel. The date given is 7th to 9th centuries and it is S31 in Gilmour's study.

All drawings, although re-drawn, are used with the kind permission of Brian Gilmour.

7. Sword blade patterns

Paul Mortimer

As we have seen the patterning of blades seems to have been a deliberate attempt to add interest and meaning to the blade and was meant to be noticed and appreciated by the observer. It has been noted that smiths appeared to be trying out ideas and seeing how far they could push the boundaries during the 2nd to 4th centuries, but that by the 5th century a set of more standard, but still complex, patterns tended to be used. Most of these patterns are variations of alternately twisted and straight rods in different arrangements but not all. It has been suggested that sometime after the 7th century the patterns became less complex and that they eventually disappear from swords altogether, probably during the 11th century.

The work of Brian Gilmour has provided an opportunity to look at the patterns in many of the more than three hundred blades from Anglo-Saxon England that he examined. Gilmour's study provides insights into the ways in which the sword blades were constructed and the patterns achieved. His survey was the first to show that there is often more going on within the blades than just the patterns in the centre of the blade, edge construction varies considerably and can be very complex. Edge construction may well have been designed to make the most of limited amounts of steel, but some of the lengths that the smiths seem to have gone to can appear unnecessarily complicated; perhaps this too complemented the pattern in the core of the blade in ways yet to be explored. The diagrams show some of the examples illustrated in Gilmour's thesis. (figure 7.1, 7.2, 7.3, 7.4, 7.5) It is likely that the edge constructions, especially when etched, would have left visible signs on the edge surface, over and above those of folding and grain. Few modern smiths, if any, have gone as far as creating these structures and usually use a single piece of metal, and several swords from our period do exactly the same, including, it seems, the Sutton Hoo Mound 1 blade. One blade from Saltwood (3944 and 6th century)[1] has a 'serpent' along the edge; another from Stratfield Saye[2] has an edge described as having 'twisted pieces' in the edge. A second sword seems to have a similar edge to the Stratfield Saye example, this time from Droxford[3] and is dated to the 5th or 6th centuries.

The underlying structure of the core of the blade can be elaborate too; there can be up to three layers present in the centre. The middle layer, when present, tends to be made from softer iron. Occasionally, the patterned layers can be quite thin, practically just a very thin coat. The sword fragment from Bamburgh is reported to have a very thin 'veneer' of highly complex patterning on each side for example.[4] Almost all swords in this sample are constructed from at least two layers but there are five swords which are composed of only a single layer.

---

[1] Gilmour, 2010. 62.

[2] Attributed to the 9th or 10th centuries. Tylecote and Gilmour, 1986. 235.

[3] Tylecote and Gilmour, 1986. 267.

[4] Starley, 2006. 2. It is possible that the smith cut the rods very thinly and used the pattern slices on more than one sword, although the patterns will alter as the slices go through the blade.

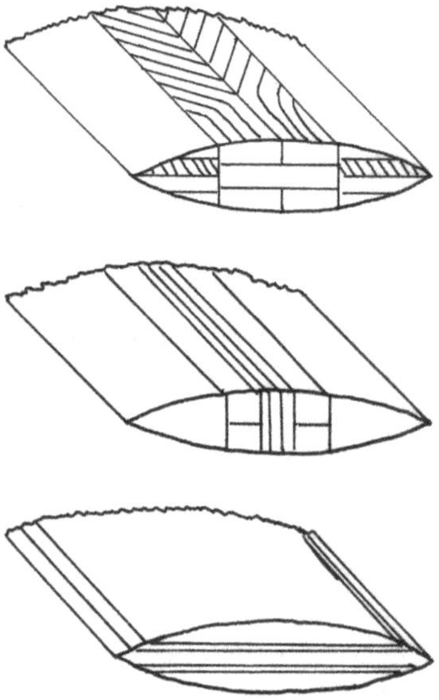

7.3. The top drawing shows a sword from Eastry in Kent which, again, exhibits a complex edge structure. It is dated to 6th/7th centuries. S14 in Gilmour's study.

The middle sword is from Aylesford in Kent and was found in a gravel pit in 1924. The edge is very simple consisiting of one piece of metal per side. S18 in Gilmour's study.

Below is shown a non-patterned sword, which has stacked plates one on top of the other, five in all. The harder metal at the edge protrudes from the sides of the blade. It comes from Sarre, Kent, grave 64 and was excavated in 1864. S51 in Gilmour's study.

All drawings, although re-drawn, are used with the kind permission of Brian Gilmour.

Summary of layers in sword pattern sample

5 (2.3%)	1 blade layer
84 (40%)	2 blade layers
55 (25.3%)	3 blade layers
73 (33.6%)	Unknown, unclear or unrecorded –
217 Total	

The sample

Gilmour's data is found in the appendix to his thesis,[1] he examined the blades using radiography, and by analysing some samples metallurgically and with microscopy. Most of the blades in his appendix are from the Anglo-Saxon period, but the study does include some outliers from earlier periods and a few from the continent, one from Ireland and another from the Hebrides. Gilmour included spearheads and seax blades but only the sword data is relevant to this study and has, therefore been extracted to form a more specialised database. I have not used all the information on the sword blades because some are not from the Anglo-Saxon period (400 to 1066 A.D. approximately) but I have included a very few blades from 2nd and 4th centuries because these, Gilmour felt, were used by 'Germanic' soldiers[2] and are part of the story of the sword in Early England. To Gilmour's data I have added the swords that he reported on from the Saltwood cemetery in Kent. The information on the sword blade fragment from Bamburgh, Northumberland, which was examined by David Starley[3] brings the sword count up to two hundred and eight, those from Susan Tyler's report[4] on Prittlewell add six more patterned blades to the sample and the three swords from the recent excavations at Sutton Hoo (Tranmer House)[5] raise the total to two hundred and seventeen.

---

[1] Gilmour, 1990. Volume 1, 235 to 274. It can be found in Tylecote and Gilmour, 1986, too.

[2] Gilmour, 2010. 60.

[3] Starley, 2006.

[4] Tyler, 1988.

[5] Fern, 2015.

One hundred and sixty-three swords in the sample come from 5th to 7th graves, while forty-five are post 7th century, four swords are pre-5th century and two blades are undated. In all one hundred and seventy-two swords were patterned, although of these, some twenty had patterns that could not be identified, mainly because of severe corrosion. No patterns at all were discernible on thirty-five swords and while some of the earlier period swords undoubtedly did not have patterns, generally the later the given date for the blade, the more likely they were to be plain looking, at least on the surface as even plain blades employ complex structures and will be made from several different pieces of metal and could have had grain structures or other features showing on the blade. Some plain swords could turn out to be patterned when looked at using new technologies. Some of the later blades in Gilmour's study contain inlays including examples of 'Ulfberh+t' and 'Ingelrii', but as these are not part of our present focus, they will not be commented upon further here.

The complexity of the patterns on most blades depend on the number of rods showing on each face and in the sample the number of rods varies from one to six.[1] It was not always possible to identify the exact numbers of rods showing on a blade, but where this could be achieved the statistics are as follows:

one rod	4 blades
two rods	30 blades
three rods	74 blades
four	31 blades
five rods	11 blades
six rods	4 blades

As reported elsewhere, the more complex blades are generally finds from the 5th to 7th centuries,[2] with more than 93% (twenty-nine of thirty-one) of four rod swords, 91% (ten of eleven) of five rods swords[3] and all the six rod swords being found in 5th to 7th century graves. This might be evidence of the swords becoming less intricate during the later period as has been suggested, although there are many fewer blades available to us from the 8th to 11th centuries as depositing them in graves had generally ceased during the 7th century.[4] Nevertheless, it is a fact that fewer swords with highly complex blades are found from the later period and it is possible that the ones that have been attributed to a later date are 'heirloom' weapons and were very old when they were left in rivers and streams. However, the earlier period also saw plenty of less complex swords from the 5th to 7th centuries being made, with 85% (sixty-one of seventy-two) blades with three rods, 79% (twenty-three of twenty-nine) swords with only two rods, 75% (three of four) with one rod coming from that era. It is possible that very sophisticated blades were still being made later, alongside the more straightforward examples which were constructed throughout the Anglo-Saxon period, but that these are just not available to us. In the case studies we have included a report by Jesus Hernandez about his reconstruction of, perhaps the most extravagantly patterned blade found from the 5th to 10th centuries, the Vehmaa sword from Finland. The find is dated to the 8th century,[5] but the blade could be much earlier. After the 8th century, there does seem to be a move away from using patterns and a tendency towards adding inlays within the blade, perhaps

[1] It is possible that one of the swords found with the two murdered Roman soldiers (4th century - Gilmour, 2007. 92, 93 and 96) had seven rods laid side by side – pattern AS+P. I have only attributed six rods to it in this study. It is now in Canterbury Museum – museum number 77509.

[2] The information contained in this section can be seen in table 1, below.

[3] The single sword of five rods not dated to the 5th and 7th is dated 7th to 9th, so could be from the earlier period.

[4] All the dating conventions used are those from the original reports; I did not attempt to reconcile the differing styles of dating.

[5] Peirce, 2002. 148 to 149. The sword comes from Vehmaa, Lahdinko, Huolila, Finland. Some unkind souls have suggested that the complexity of the blade is the result of the smith just using up bits from old or failed swords!

to satisfy the requirements of display and all such swords from this sample are dated to the 9th to 11th centuries. of the swords from the 5th to 7th centuries within this collection seem to have had a fuller, while only eight of the post 7th century swords were fullered, although the cross-sectional shape of the blade can vary from lenticular to almost perfectly flat. The possibly inaccurate and inconsistent dating of many of the blades should also be borne in mind.

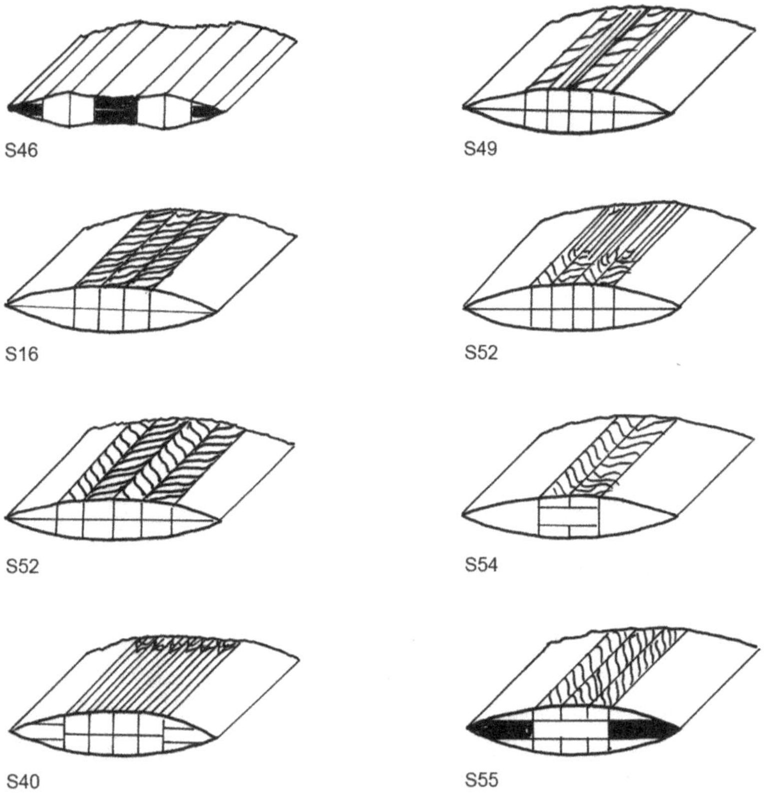

7.4. **S46** shows the construction of a fullered blade from the River Thames at Brentford. The dates range from 7th to 9th centuries. (Gilmour, 1990. 134). **S16.** is from the cemetery at Sarre, Kent, probably grave 264 and is from the 6th to 7th centuries. Excavated during the 19th century. (Gilmour, 1990. 128). **S45** is from Mitcham, Surrey and has been dated to 6th century (by Wheeler 1935,175) (Gilmour, 1990. 125). **S40** was found by chance in a ploughed field near Chesterton, Cambridgeshire during 1979. Dating is 5th to 7th centuries. (Gilmour, 1990. 118). **S49** probably from Bifrons, Kent. Possibly excavated during the 19th century and part of the Tomlinson bequest to the Kent Archaeological Society (Gilmour, 1990. 79). **S52** grave 113, Sarre, Kent. Excavated 1864. Dated to the 6th century. (Gilmour, 1990. 82). **S54** either from Sarre or Bifrons. Dated to 6th century. (Gilmour, 1990. 86). **S55** again from Sarre or Bifrons and dated to the 6th century. (Gilmour, 1990. 88).

All drawings, although re-drawn, are used with the kind permission of Brian Gilmour.

One further pattern illustrated here is of a sword from Germany, Krefeld-Gellep grave 1782 dated to the 5th to 6th centuries.[1] (figure 7.6) Like most swords in our sample, it has patterns on both sides that are different to each other, even if that is just due to the reverse of the blade being offset when compared to the obverse. On one side of the grave 1782 blade it shows a centre made from four rods continually twisted, except for a section where the two centre rods have interrupted straights. The twists have been ground down to form aligned circles. However, the other side of the blade is completely different to any other in our sample or elsewhere, as far as I am aware. It is made from straight and ground down twisted rods parallel to each other, but the whole has been laid at about a twenty degree angle to the blade.[2] How the pattern was made is not precisely known, but Lehmann thinks that it would have been made in the normal way that the core of a blade in constructed first, then heated and while hot, wrapped around a cylinder and then later cut to obtain the required angle.[3] During the 19th century pattern-welded shotgun barrels were made in this way. Lehmann suggests that the pattern would give the impression of something winding up and down the blade,[4] and I am sure that he is correct.

Not only is this a remarkable pattern, it was discovered by Ulrich Lehmann using digital computed x-ray tomography. He describes CT scanning thus, "The main procedure is similar to conventional radiography: the object is scanned by x-rays. But in contrast the find is mounted on a turning plate so that during the examination it rotates slowly around 360°. The resolution of the imaging and the measuring of the scanned section can be adjusted due to the position of the object. A smaller distance between find and x-ray source reduces the extent of the scanned section but also leads to better resolutions. The results are displayed as a three-dimensional model of the object.....Like a digital picture that consists of pixels, this model is built up of cubes, the so-called voxels – which is short for volume pixel. The grey-scale value of these voxels correlates with the density at the given point in the scanned object."[5] Once complete, digital sections can be taken from any part of the scanned blade or a moving image going through the structures can be created. It is a very powerful tool and we can expect such techniques to reveal more in the future, and maybe some new patterns. It has already shown that some 'plain' blades were actually patterned.[6] I wonder how many other 'plain' blades lay in museums and elsewhere that are actually patterned? Another benefit of this non-destructive technique is that it has been able to illustrate the crafting techniques hidden within the blade structures in ways that would have been hard to visualise with any other non-destructive method.[7] One other benefit of CT scanning appears to be that the surface image of the blade can be imaged more accurately than when using other radiography methods.

---

[1] Lehmann, 2013. Lehmann, 2016. 118, 159, 193 to 199, 414 to 416 and 525 to 528 (Tafel 99 to Tafel 102).
[2] Lehmann, 2013. 11.
[3] Lehmann, 2016. 226.
[4] Lehmann, 2013. 12.
[5] Lehmann, 2013. 5. Lehmann describes, in great detail, two different radiography techniques that he has used in his book; 3-D radiography and 3-D neutron computer tomography in his book. See Lehmann, 2016. 34 to 44.
[6] Lehmann, 2013. 12.
[7] Lehmann, 2013. 14.

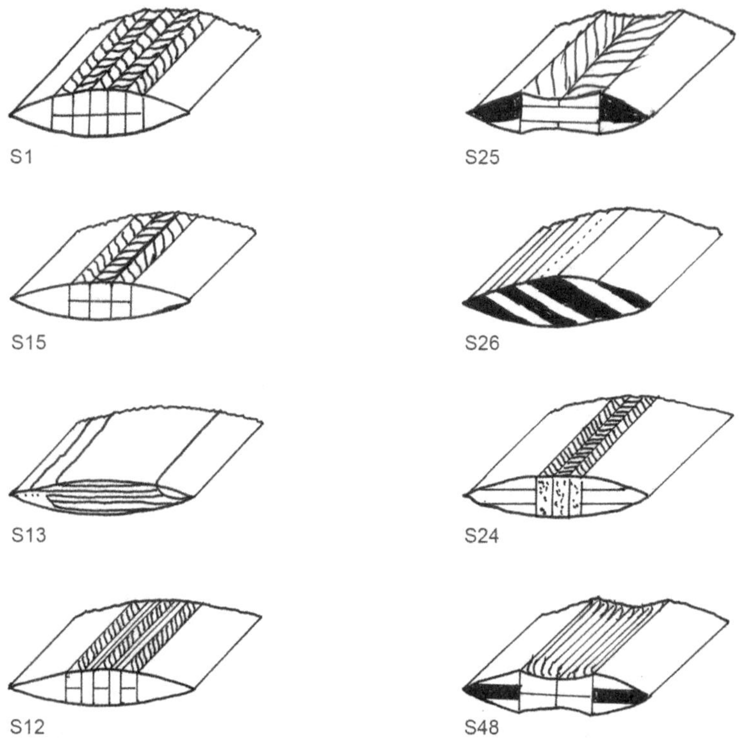

7.5. **S1**, grave 31, Loveden Hill, Lincolnshire (also known as Hough-on-the-Hill). The sword was not buried in its scabbard and had been bent double in two places. 6th to 7th centuries. Excavated in 1956. (Gilmour, 1990. 115). **S15** came from the cemetery at Holborough, Kent which seems to be 7th century in date. (Gilmour, 1990. 102). **S13**, Is possibly from Bifrons, Kent. It is part of the Tomlisnon bequest and is from the 5th to 7th centuries. This is a 'stacked' blade. (Gilmour, 1990. 96). **S12** was found in a gravel quarry in 1886. Near Wickhambreux, Kent. 6th to 7th centuries. (Gilmour, 1990. 93). **S25**, was found in 1936 at Reading in the Oxford Road at Tilehurst Street but no further burial circumstances are known. It is fullered and has been dated to the 10th or 11th centuries by Wheeler on the basis of its pommel guard. (Gilmour, 1990. 161). **S26** is a sword from the River Thames at Wallingford Bridge. The blade has a very pronounced convex shape and is probably very late – 12th to 14th centuries. It is shown here for comparative purposes. (Gilmour, 1990. 167). **S24**, From the River Thames at Cleeve, Goring-on-Thames, Oxfordshire. Probably after the 7th century based on the remains of the lower guard. (Gilmour, 1990. 154). **S48**: "It was found by chance near the junction of Watts Avenue and St Margaret's Street in Rochester, Kent, and evidently came from a grave,…" (Gilmour, 1990. 145). The suggested date is 9th to 10th centuries, but this is long after the practice of furnished burial was thought to have ceased. All drawings, although re-drawn, are used with the kind permission of Brian Gilmour.

A similar work to Gilmour's on Continental or Scandinavian swords could throw a great deal of light on what smiths were doing in particular regions and help to decide whether there actually were characteristics that showed whether the blade came from a 'Frankish', 'Anglo-Saxon' or other region with some degree of certainty, however I could find no such references to use as a comparison. Perhaps such studies would be even more revealing if CT scanning were used in conjunction with other methods.

7.6. Details of the blade from Krefeld-Gellep, grave 1782. Both sides of the blade are shown together with a cross-section at the bottom. Lehmann feels that the completed core may have been wrapped around a cylinder or rod and then cut length ways in one straight cut. It would then have been flattened and welded to the other side. Drawing used with the kind permission Ulrich Lehmann.

The patterns (figures 7.7 to 7.11)

When considering the patterns on the blades within the sample, it must be remembered that for most of them just part of the sword was surveyed and often only quite small fragments were available to be examined. It could be that, in some cases, the pattern seen and recorded did not cover the whole blade. I am sure, too, that there were other patterns that are not recorded here that were being used; some of these could be discovered in the future, so the catalogue that I have compiled will certainly require amending in the future. Some of the patterns in the register are quite similar to others and further work might mean that the number of patterns that I believe that I have observed will diminish. I have attempted to standardise the way that I have drawn them and in some cases the design has been compressed in order to get the information into the standard space available. I have concentrated on the interactions of the patterns rather than showing how much width the pattern took compared to how wide the sword is; anyone interested in exploring further should look at the original source. I have tried to keep the nomenclature that I have used fairly simple, but I am aware that there are some inconsistencies. I am sure too, that the way that I have built the catalogue and the designations that I have used will be improved by others in due course. This work should be regarded just as part of the discussion on patterns; there is much further to go.

As mentioned above, many of the patterns are variations on a theme, some, though are quite radical and demonstrate that much was going on in the world of the swordsmith. The sample shows, too, that at least one smith actually added a patterns to the edge, which was a surprise to me, as hitherto I had always thought that the edge was kept relatively plain, despite the fact that it might show grain or laminations due to complex constructions. Interestingly, Ulrich Lehmann reports another example from Beckum II, grave 24 in Germany that has patterned edges.[1]

Another surprise was the number of swords with four or more rods per side, (thirty with four, eleven with five and four with six) which makes the Mound 1 Sutton Hoo blade fairly ordinary by comparison.

I have identified more than thirty different patterns and I have endeavoured to classify them and provide names for the patterns. I should point out, that however complex or simple the pattern, the sword blade construction underlying the patterns could be very elaborate, but I have made no attempt at commenting on that particular aspect of the sword in this chapter, other than to report the number of layers within the core, where that has been recorded and that information can be found in *Appendix 5 Patterns Database*.

There is, I think, some evidence which suggest that the patterns may have been further enhanced with the selective use of different etchants being applied to differing areas of the blade. When carried out successfully this could add extra colour to the blade. King Theoderic the Great's letter, mentioned in the previous chapter on pattern-welding, says,

"…– In the middle of the blade are beautiful incisions like winding worms, and there are shadows numberless so you believe that the bright metal has been interwoven with many different colours."[2]

This mention of colours has always interested me and in discussing this with Brian Gilmour, we agreed that it is a relatively simple matter to add extra hues to the blade by selectively using different etchants in separate areas of the blade surface; different solutions would react with the metal and colourise it in particular ways. Of course, such colouration would not survive the grave or being submerged in a bog or river for centuries and so, if it was present, can no longer be seen in the archaeology. Vince Evans, too, reports that he has observed that some later Indian and Persian pattern-welded blades had been deliberately coloured by using controlled rusting, to make certain areas of the blade brown.[3] Mike Loades has an Anglo-Saxon style blade made by the smith Hector Cole, who used urine from an in-season mare as an etchant which gave the blade a 'golden patina'.[4] Loades reports, too, that tannic acid would create a bluish black finish.[5] I have seen a patterned blade that had been left in acid that had bronze dissolved in it and some of the bronze had adhered to parts of the pattern in a very attractive way. In this case, it is possible that over time the bronze could cause the blade to corrode through a galvanic reaction, though.[6] However, partially sealing different parts of the blade at different times, perhaps with clay, and then immersing it in a selection of etchants could mean that the smith might have been able to make a sword blade that could show a variety of coloured hues. It is an intriguing idea, but one that cannot be proven archaeologically.

Of the more than thirty patterns observed and recorded here, many of them do not appear very often, for instance twenty-four patterns only occur once within the sample. The most common pattern is the continuous herringbone (Hcon) which appears on blades from the 4th to 11th centuries and in almost all areas where swords have been recovered. There are sixty examples in the sample and perhaps this is because

---

[1] Lehmann, 2013. 9. Lehmann, 2016. 381 to 382 and 448 to 450 (Tafel 22 to 24).

[2] Quote taken from Buchwald, 2005. p.282 citing Magnus, 1555, Book 7, chapter 3.

[3] Vince Evans. Personal comment. Brunning has similar ideas, Brunning, 2014.

[4] Loades, 2010. 83 to 84.

[5] Loades, 2010. 84.

[6] Personal comment, Owen Bush.

the pattern can occur on swords with any combination of two or more rods. The next most numerous pattern is the alternate herringbone and straight pattern (AHS), which once again is found in most areas and on swords from the 5th to 9th centuries. There are thirty-seven instances of AHS in the sample and again the number of rods is not important in achieving the pattern. The third most numerous pattern is similar to AHS and seven of these are recorded and it is one that I have called, 'alternate twists in both directions and straights, offset' (ATS2O). This particular pattern, too, has a long duration being found on swords dated from the 5th to the 11th centuries, but only, so far as our sample is concerned, in London, Kent and Bedfordshire. A very similar pattern, but with twists in only one direction (ATS1O) occurs four times and has only been found in graves in Kent from the 5th to 7th centuries. A third variation of this theme, ATS2O°, has been found on two occasions on the Isle of Wight and in Yorkshire but the attributed dates are far apart, one sample being from the 5th to 6th centuries and the other 9th to 10th. A variation of the AHS, alternate herringbone with short straights (AHsS) has been found five times in Oxfordshire, Kent and County Durham on blades from the 6th, 7th and 9th centuries.

Six swords have completely different patterns or numbers of rods on each side. Gilmour's catalogue number 67 dated to the 6th to 7th centuries and found in grave 256 at Sarre, Kent, has three rods on one side and six on the reverse, both patterns are Hcon. A sword from Mitcham, Surrey, again dated to the 6th and 7th centuries (cat. No. 123) has three rods on each side and Hcon on the front, and ST1 + CT1 (see below for explanations of the abbreviations) on the reverse. An earlier blade, 5th to 6th centuries, from Lyminge in Kent, (cat. No. 249) has three rods per side and T1con on one face and ATS1O on the other. A fourth double patterned sword comes from Barlaston, Staffordshire, dated to 6th and 7th centuries (cat. No. 277) and has five rods on each side with Hcon on one and AHS on the rear. Howletts, Kent provides another sword of this type from 5th to 6th centuries (cat. No. 318); it has three rods on each side with Hcon at the front face and AHS + ladder on the reverse.[1] The final example was found in a 6th century grave at Saltwood, Kent and again has three rods per side and has H con on one and AHS on the other.

Vince Evans has commented on some of the more unusual patterns and suggested ways in which a particular design could have been achieved. T1+Geo, he feels was carried out by exposing the cores of two of the twist rods the Ladder was created using short pieces of twist and straight rods transversely.[2] Owen Bush has duplicated the pattern Fla+TE on a blade and achieved it by zig-zagging a long flat rod back and forth. In his experiment he used one long rod to make all the loops. This particular blade is one of the five blades consisting of just one layer.

One way of modifying the pattern when the rods have been twisted is to grind the rods down, as the grinder or file cuts deeper the pattern will change, until almost half way through the rod semi circles will appear.[3] If the rods are aligned perfectly with each other, complete circles can be seen in the twisted sections of the rods. (See figure 6.9) Aligning the patterns perfectly, as many of the swordsmiths were able to do, requires great skill and precision.[4] If the rod is ground further the half-circles will disappear and the direction of the twist will be reversed.[5] Making the layers within the rods thicker or thinner or doing the same to the rods themselves can add extra features.

---

[1] Putting the rods across the blade is probably a bad idea, as any poor welds could lead to fractures across the blade; the pattern AHS was, I think necessary to make this sword serviceable.

[2] Vince Evans, Personal communication.

[3] Lehmann, 2013. 3.

[4] Lehmann, 2013. 13 to 14.

[5] Thanks to Jesus Hernandez for this insight.

'Piled' blades

'Piled' blades are mentioned fairly frequently in the literature and it is often not clear what is meant by the term, particularly as different authors use it to describe different things that have happened in the construction of a blade. I have avoided using the term. Where plates of metal have been laid on top of each other and welded together I have called that technique, 'stacked' (See figure 7.9). A blade where layers of metal have been laid next each other and welded, I have called 'banded' although it could be 'loose' or 'tight' and this could be due to the smith using different techniques. A banded blade could be further processed by folding, perhaps many times and the result could resemble a 'watered' finish, much like some of the finishes on true damascened blades. Some blades with what appears to be a plain iron core might exhibit the 'grain' of the metal that makes it up. All such 'plain' blades that I have found evidence for so far, have separate edges welded to them, usually of a harder metal and sometimes steel. Details of these types of non-patterned blades can be found in the pattern data base.

The patterns are listed below together with frequencies and the county names where they were found. As with all information regarding swords and Anglo-Saxon England, the sample maybe skewed a little by the comparatively large number of swords found in Kentish graves. Other areas might not have been so well excavated or recorded and a lot of pilfering of graves occurred during the 19th century and earlier.

What follows are some notes on the nomenclature and abbreviations that I have used. All the patterns can be seen in figures 7.5 to 7.9.

More details can be seen in *Appendix 5 Patterns Database*.

Register of patterns

Nomenclature and abbreviations:

T – twists
T1 – Twists in one direction only
T2 – twists in two directions
T1 T-1 – twist and counter twist
S – Straight
Con – continuous
O – offset
L – lined
H – herringbone
A – alternate
Geo – geometric

E – edge
C – central
P – plain
s – short
B – bordered
Geo – geometric
E – edge
1, 2, 3, 4, 5, 6 – number of rods – n if it occurs on blades with different numbers of rods
° -- distortion – probably due to grinding
+ between different rods

Pattern names:

Hcon – herringbone continuous
AHS = alternate herringbone and straight
CS+T1 – central twist plus rods of twist in one direction
AHS + ½ AHS + AHS
AS+P – alternate straight and plain
T1+CS+T1 – rod with a twist in one direction + central rod straight + another rod with twist in same direction as the first
AST1 alternate straight and twisted – one direction
T1con – twisted in one direction only – only pattern on blade
AHsS – alternate herringbone and short straight
T2S – rods with twist in both directions separated by a straight
T1 + geo – twists and rods with a geometric look
ST1 – rods with straight and twists, always in the same direction
ATS2O – alternate twists in both directions and straights, offset
Hcon1 + CS + H con -1 - herringbone facing up + central straight + herringbone facing down
ST1 + CT-1 + ST1 – two outer rods with straights and twists in one direction, separated by a central rod with continuous twists in the other direction
Scon – straight continuous
T1 + CATS-1 + T1 – two outer rods with twists in the same direction, separated by a rod that is straight and twisted in the other direction
H spear – spear shape formed from herringbone patterned double rods
ATS1O – alternate twists and straights – twist all in the same direction
HLcon – Herringbone lined – continuous herringbone separated by a central straight rod and bordered by two more
Scon (grain) – straight rods – look grainy
H + CS – Herringbone separated by a central straight
Fla + Te – central rods forming small flat loops, offset, and what appear to be edges that have been twisted in opposition to each other
Serpent – serpent pattern running along the whole blade
AHS + ladder – alternate herringbone and straight for part of the blade – second part looks like rungs on a ladder
ST2O – rods that are alternately twisted and straight, offset against each other – adjacent rods twisted in different directions
Circles – small circles, probably achieved by grinding twisted rods
Hcon + serpent (e) – herringbone continuous plus serpent design within the edges
AHSO – alternate herringbone and straight double rods, offset against each other
Hcon + S – continuous herringbone bordered by two straight rods
ATS2O° -- alternate twists in both directions and straights, offset and distorted in places by grinding
ST2OA – a slightly different configuration to ST2O, hence ST2OA – alternate.
T1+C+T1+C – very similar to 6 – T1+CS+T1

Frequencies and locations[1]

1. Hcon, sixty-four examples -- 4th to 11th centuries – Buckinghamshire, Bedfordshire, Cambridgeshire, Hampshire, Nottinghamshire, Isle of Wight, Warwickshire, Staffordshire, Kent, London, Middlesex, Oxfordshire, Lincolnshire, Norfolk, Berkshire, Surrey, Hebrides, Europe[2]
2. AHS, forty-two examples – 5th to 9th centuries – Buckinghamshire, Kent, Hertfordshire, Lincolnshire, Yorkshire, Oxfordshire, Essex, Berkshire, Isle of Wight, Middlesex, Staffordshire, Europe[3]
3. CS + T1, one example – 6th to 7th centuries -- Kent
4. AHS + ½ AHS + AHS, one example – 1 6th to 7th century - Kent
5. AS + P, one example – 4th century -- Kent
6. T1 + CS + T1, two examples – 5th to 7th centuries --Kent
7. AST1, two examples – 6th to 7th and 9th to 10th centuries – Kent, Derbyshire
8. T1con, four examples – 5th to 10th centuries – Kent, Oxfordshire, Isle of Wight[4]
9. AHsS, five examples – 6th to 7th and 9th centuries – Kent, Oxfordshire, County Durham[5]
10. T2S, one example – 6th to 7th centuries -- Kent
11. T1 + Geo, one example – 6th to 7th -- Kent
12. ST1, three examples – 5th to 7th centuries -- Kent
13. ATS2O, eight examples – 5th to 7th and 10th to 11th – Kent, London, Bedfordshire[6]
14. Hcon1 + CS + Hcon -1, one example -- Middlesex
15. ST1 + CT -1 + ST1, one example – 6th to 7th centuries -- Surrey[7]
16. S con, four examples – 5th to 7th centuries – Kent, Hampshire
17. T1 + CATS + T1, one example – 9th to 10th centuries -- Berkshire
18. H spear, one example -- Norfolk
19. ATS1O, five examples – 6th to 7th centuries -- Norfolk[8]
20. HLcon, one example – 5th century -- Bedfordshire
21. Scon (grain), one example – 6th century -- Kent[9]
22. H + CS, one example – 5th to 7th centuries -- ?
23. Fla + TE, one example – 10th century -- Berkshire
24. Serpent, one example – 6th to 7th centuries -- Yorkshire
25. AHS + ladder, one example – 5th to 6th centuries -- Kent
26. ST2O, one example – 6th to 7th centuries -- Kent[10]
27. Circles, one example – 6th to 7th centuries -- Kent

---

[1] All numbers in the footnotes in this section refer to the number that of the sword in Appendix 5. Those that have one, two or three entries are listed in a note at the end of this chapter.

[2] 7, 8, 13, 16, 18, 20, 22,2 4, 26, 28, 39, 45, 47, 50, 57, 62, 63, 64, 65, 66, 67, 78, 71, 86, 95, 96, 99, 100, 103, 108, 109, 113, 121, 134, 147, 150, 151, 152, 154, 156, 158, 160, 161, 171, 173, 178, 180, 181, 183, 185, 186, 187, 188, 191, 193, 197, 198, 203, 204, 205, 212, 216.

[3] 10, 11, 19, 21, 27, 30, 34, 36, 40, 44, 49, 51, 56, 144, 76, 78, 80, 83, 87, 89, 97, 105, 117, 125, 136, 132, 144, 145, 146, 160, 162, 167, 170, 174, 175, 179, 184, 192, 199, 201, 217, 218.

[4] 31, 88, 110, 136.

[5] 6, 46, 128, 136, 195.

[6] 43, 85, 112, 133, 164, 172, 182, 194.

[7] 63.

[8] 6, 46, 128, 136, 195.

[9] 84, 126, 153, 168.

[10] 199.

28. H con + serpent, one example – 6th century -- Kent
29. AHSO , one example – 7th century -- Northumberland
30. Hcon + S, one example – 6th to 7th centuries -- Kent
31. ATS2O°, one example – 9th to 10th – Yorkshire
32. ST2OA, three examples – all from Prittlewell, Essex
33. T1+C+T1+C, one example Bifrons, Kent
34. Stacked[1]

Table 7i. Number of rods in sword blades by century in the pattern sample[2]

Century	1 rod	2 rods	3 rods	4 rods	5 rods	6 rods
4th		1				1
5th			1			
5th to 6th	1	5	19	3		
5th to 7th		1	6	3	1	
6th	1	2	10	4	1	
6th to 7th	1	14	20	15	7	1
7th			3	3	1	2
7th to 9th			1		1	
9th		1	1			
8th to 9th			2			
9th to 10th		5	6	2		
10th	1					
9th to 11th			2			
10th to 11th		1	3			
Totals	4	30	74	30	11	4

Note. The following appear in Appendix 5 once only: CS+T1 (14), AHS+1/2AHS+AHS (15), AS+P (17), T2S (35), T1+Geo (41), Hcon+CS+Hcon-1 (60), ST1+CT+ST1 (63), T1+CATS+ST1 (90), Hspear (124), Hlcon (140), H+Cs (157), FlaTe (166), Serpent (189), AHS+ ladder (197), Circles (200), Hcon+ serpent (198), AHSO (209), Hcon+S (176), ATS2O° (190), T1+C+T1+C (38). Appearing twice: T1+CS+T1 (23, 25), AST1 (29, 115). Three times: ST1 (42, 129, 196), ST2OA (210,213, 214).

[1] 32, 37, 48, 54, 207, 208.

[2] The rather strange datings given in the date column are due to the dates for each sword provided by the original report being retained.

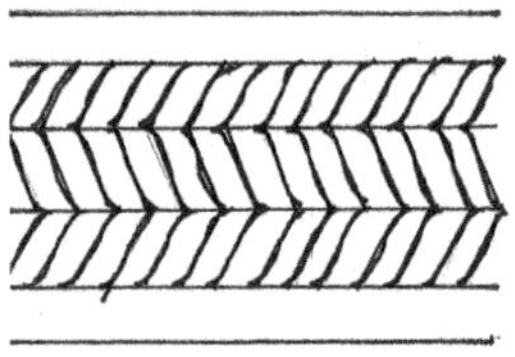
1 Hcon

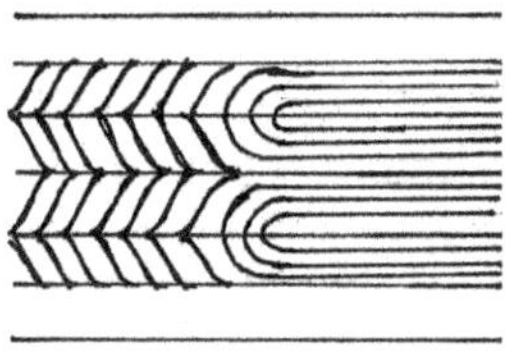
2 AHS

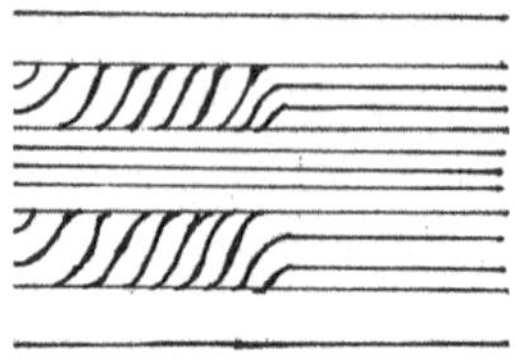
3 CS+T1

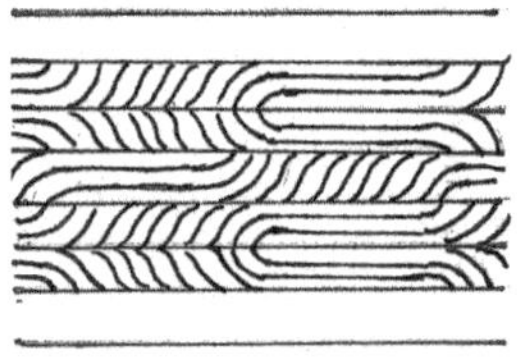
4 AHS+½AHS

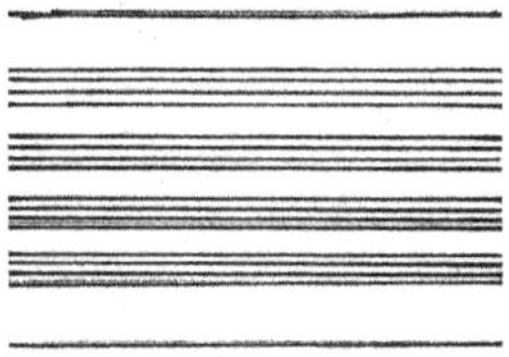
5 AS+P

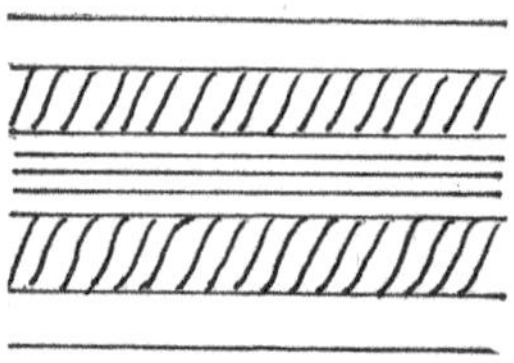
6 T1+CS+T1

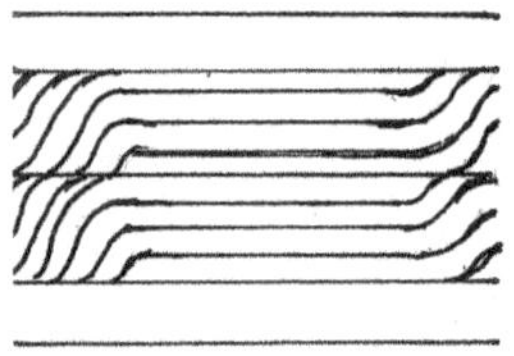
7 AST1

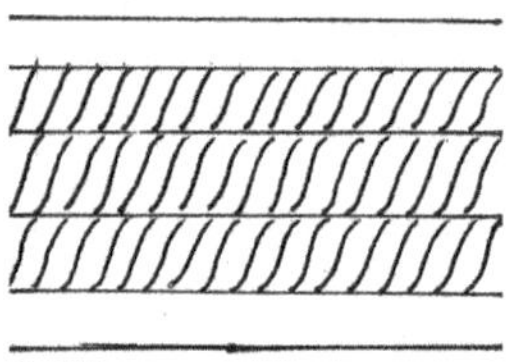
8 T1 con

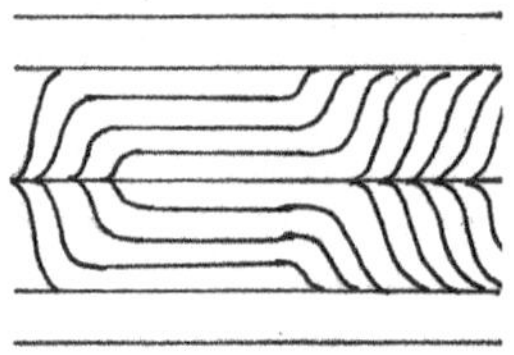
9 HsS

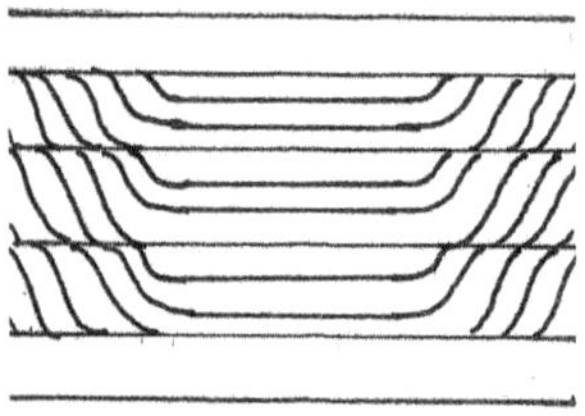
10 T2S

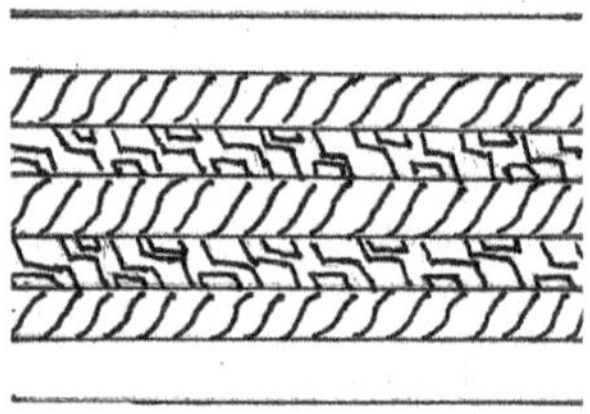
11 T1+Geo

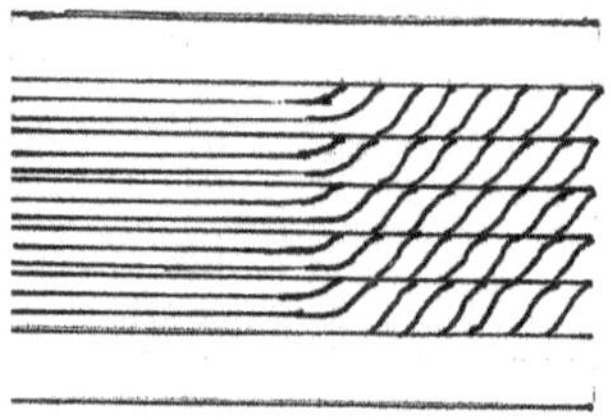
12 ST1

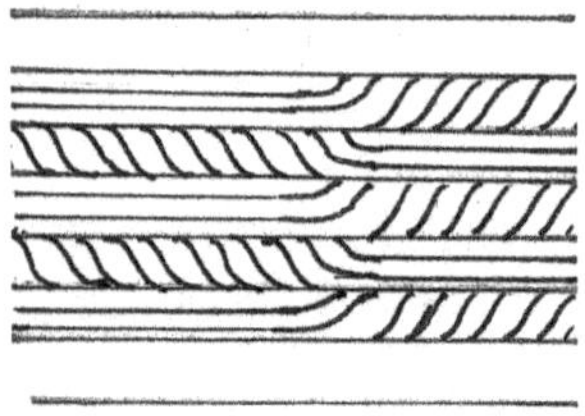
13 ATS20

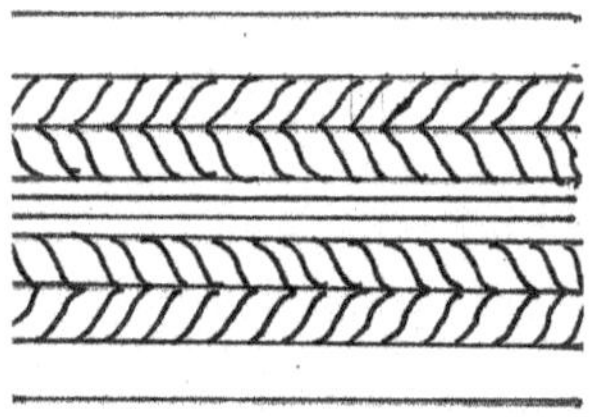
14 Hcon+cs+Hcon-1

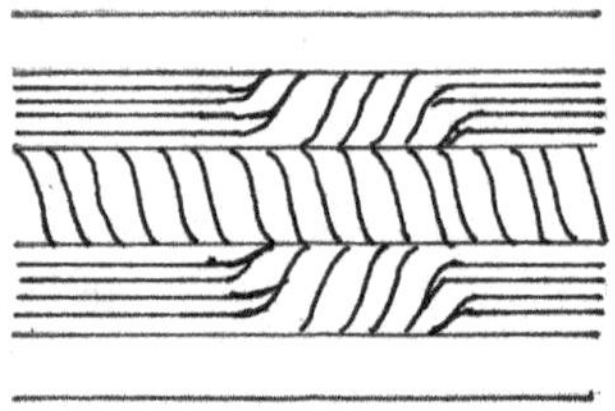
15 ST1+CT-!

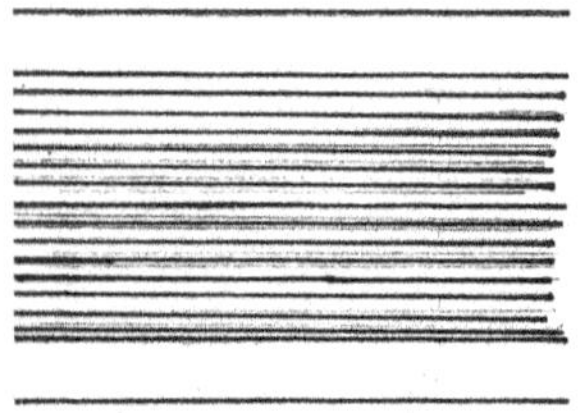
16 Scon

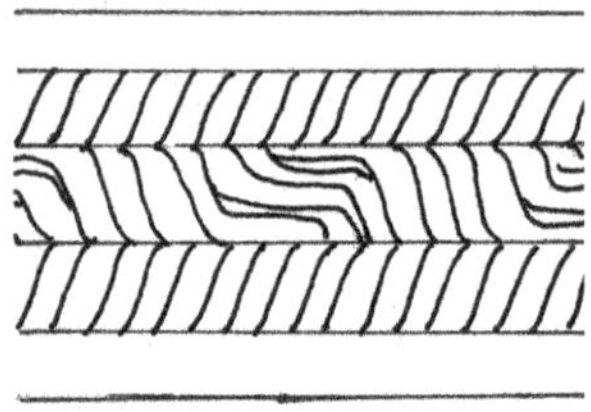
17 T1+CATS-1+T1

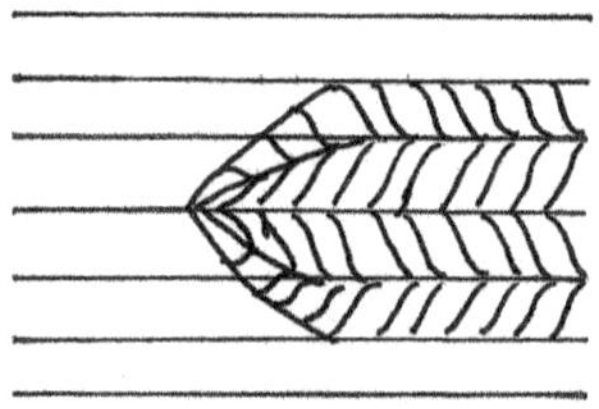
18 H spear

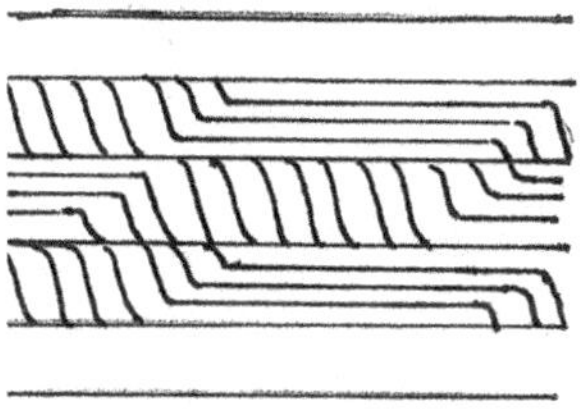

19 ATS10

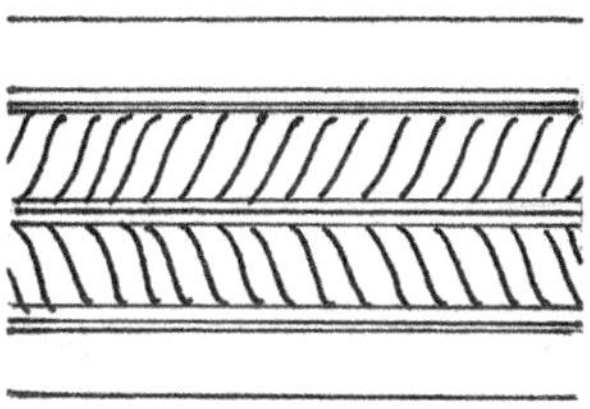

20 HL con

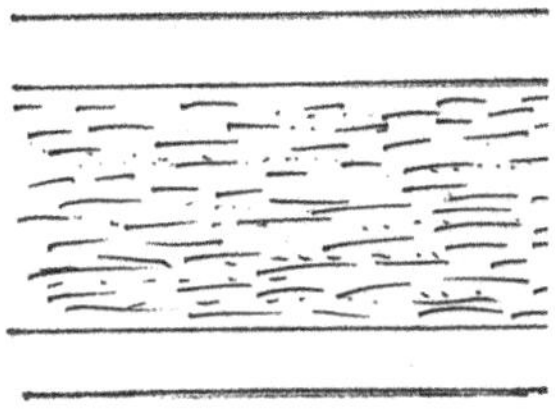

21 Scon(grain)

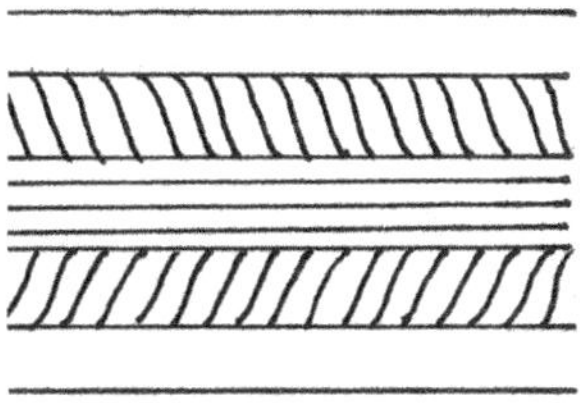

22 H+CS

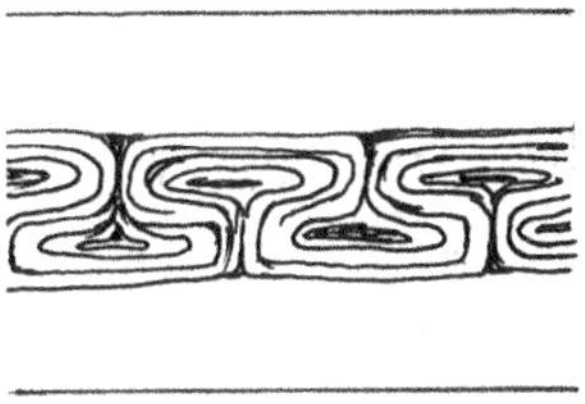

23 Flat Te

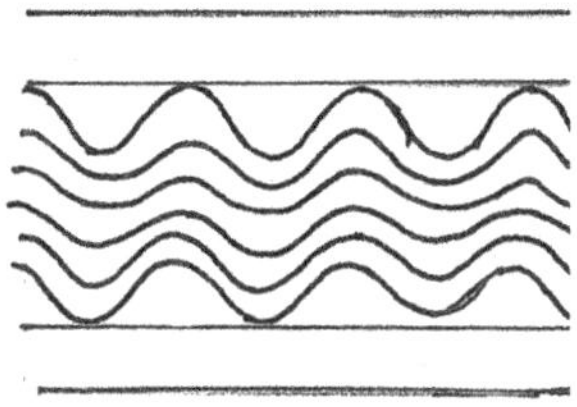

24 Serpent

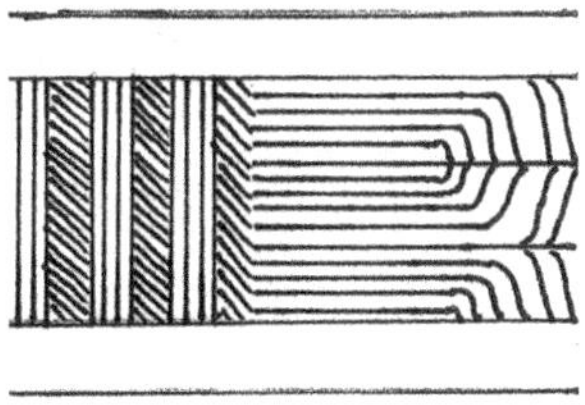

25 AHS+Ladder

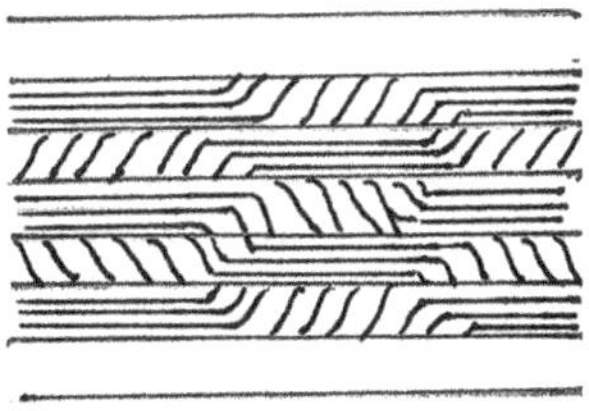

26 ST20

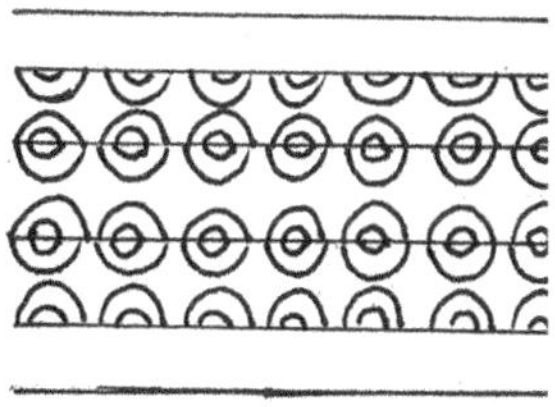

27 circles

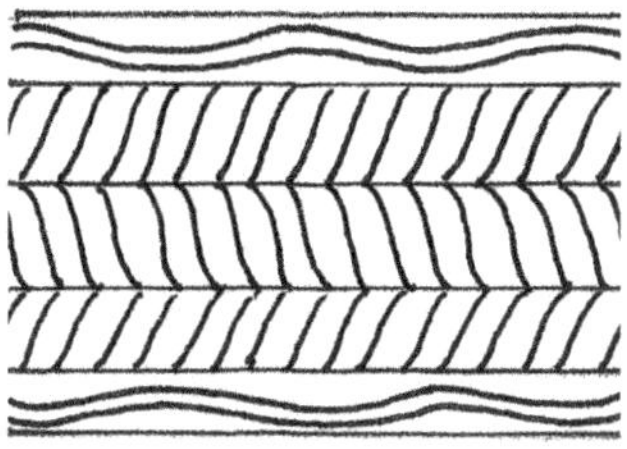

28 Hcon+serpent (e)

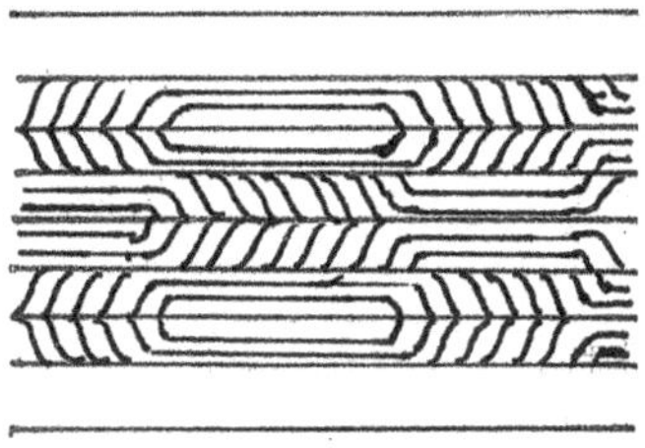

29 AHSO

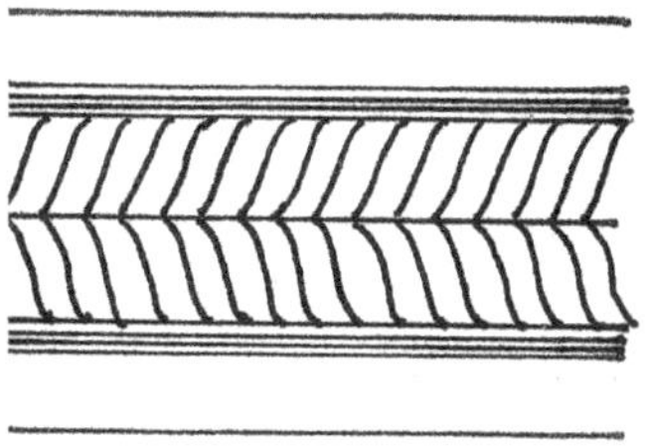

30 HCCR+S

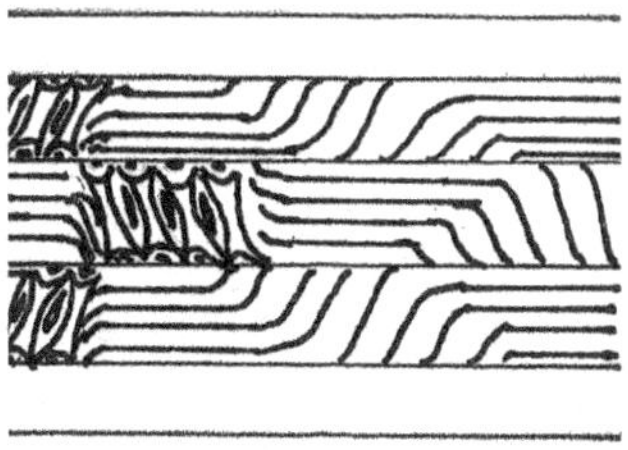

31 ATS20°

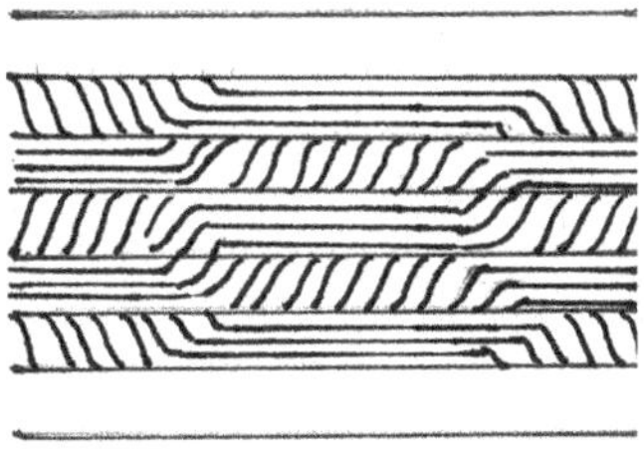

32 ST20A

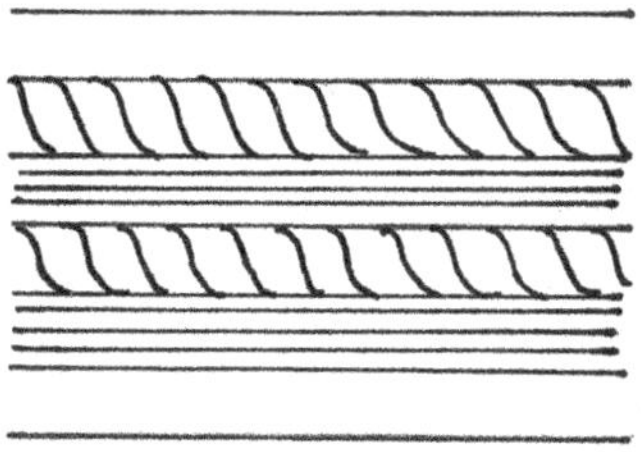

33 T1+C+T1+C

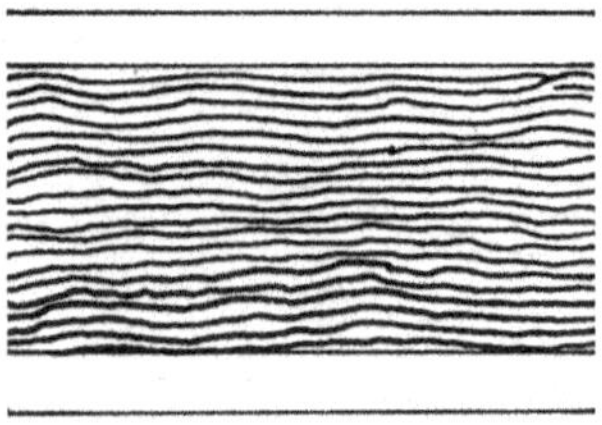

Banded loose

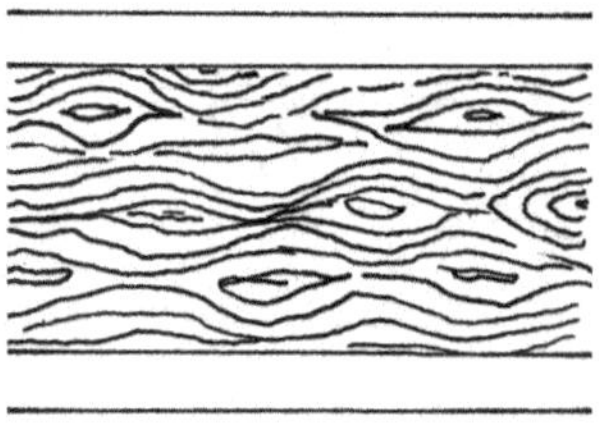

Watery grain

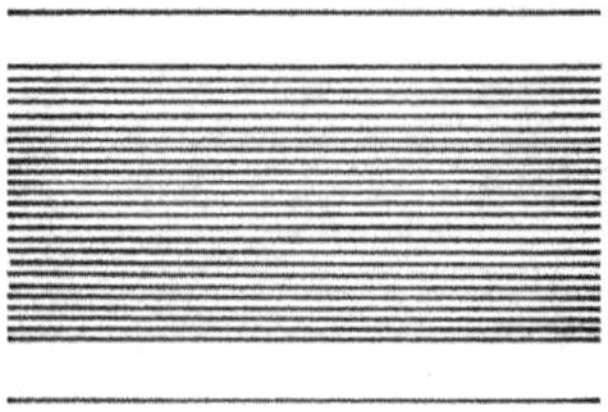

Banded tight

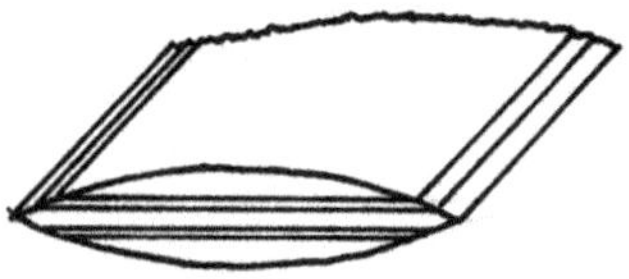

Stacked

8. The mysterious smith?[1]

Paul Mortimer

The art of the smith is a mystery that invests him with menace and awe. He is in command of the supernatural forces of fire and metal: a nature distinct from that of other men. Ursula Dronke.[2]

We have considered the materials that went into the making of a sword and discussed many of the processes. Several times, in previous chapters, there have been hints that there could have been something a little special and perhaps alien about the maker of the sword and it is to that final ingredient involved in making iron and steel, the human one, the smith, to whom we will now turn. We may find, too, that the smith was not the only human being to have been involved, very intimately, in the process of making steel.

Mitchell suggests that smiths were regarded as not like ordinary folk, or even craftsmen such as carpenters, for a number of reasons; the awe inspiring nature of their work, the sparks produced by virtually all phases of fabricating metal, their association with thunder-like noise and through that to the thunder gods and the heavens, and finally the ancient relationship between the concepts of stone and firmament.[3] It appears that the smith's residence and workplace were set apart from other dwellings, often on the outskirts or even some distance from a village. Perhaps even close to a cemetery.[4] At first this separation might just seem to be a sensible precaution; as the smith is dealing with fire and there would always be the risk of setting light to nearby domiciles, so it makes sense for his workplace to be located elsewhere. However, there is much more to it than that, besides the evidence suggests that some smithies could be in use for a considerable period, perhaps even generations, and they were probably less likely to catch fire than more ordinary houses as the smith's furnace or hearth would require constant tending and vigilance in order for it not to be overly wasteful of resources and to be capable of functioning when needed.

As Mitchell suggests there were probably quite a few reasons for the liminal position of the smith within society as well as the physical placement of his workplace. He was the possessor of secret and arcane knowledge[5] that even other craftsmen were not party to; his smithy was a place of darkness, fire, heat, sparks, smoke, smells and noise. His work could have involved rituals and rites that were forbidding and frightening

---

[1] 'Smith' in most of this chapter can refer to both the smelter of metal and the smith who turned the metal into an artefact. In this chapter there will be no attempt at distinguishing between different types of iron smiths e.g. smelters, tool makers, sword makers, finishers, etc. See Hinton, 1998 for possibilities and discussion. It is possible that many smiths did not specialise, we just do not know for certain.

[2] Dronke, 1997. 260.

[3] Mitchell, 1985. 12. Mitchell demonstrates that there are several Indo-European words that are cognate both with 'stone' and 'sky' and that this relationship is ancient.

[4] Gansum, 2004. 44 to 45. Gansum says, too, that the "...production of iron was sometimes situated at cemeteries." We will return to this possibility below. Hinton, 1998. 14. Wright notes that at Yeavering, Northumberland and at Lake Tisso, West Zealand, Denmark, the smith's hearth was part of the complex but on the periphery. Wright, 2010. 133 to 134.

[5] Gansum, 2004. 50. Brian Gilmour feels that there was nothing 'secret' about the work of the smith, in that the smith did not need to keep his processes from others, just as computer technicians today are not deliberately preventing us from learning their knowledge; most of us are just not in a position to master and use those skills. Even if Gilmour is right, this does not mean that others from outside the profession did not find their work 'magical' or awe-inspiring.

to members of the wider community and he worked magic, turning the rough and unprepossessing raw materials into useable artefacts; turning stone to shiny, hard metal![1]

Gansum says that, "In studies by ethno-archaeologists and anthropologists there is substantial evidence that supports the idea of the production of iron as ritualised and regulated by taboos......Many of the taboos and ritual performances link up with ideology and religion."[2] He goes on to report that it is possible that Iron Age Scandinavians did the same and then considers material from the *Poetic Edda* to look for clues regarding how rituals, technology and religious belief might have been linked to smithing. "The point of departure is based on the conviction that people did not separate actions, thoughts and technique: they were all interlinked in cosmology and revealed by rituals that gave meaning to the material world."[3]

Bone coal

Gansum did not just consult the literature; he also conducted experiments with a Norwegian smith, Hans Johnny Hansen, to find out more about the processes involved. Hansen informed him that when he was learning to be a smith, during the 1960s, he had been trained in using 'bone coal'.[4] Gansum notes that the knowledge of the use of bone coal in smithing is not widely known amongst archaeologists and explains that osteologists tend not to be able to tell the difference between burnt bone and bone coal and so could have missed the significance of many finds.[5] Bone coal is created by roughly crushing the bones and putting them into a container with the open end of the container facing the earth; a fire is lit on top of the vessel and left to burn for a considerable time; Hansen and Gansum kept their fire going for three hours when they made bone coal. Using this method very little oxygen is available to the bones, which means that the resultant bone coal will have very different characteristics to most bones burnt openly on a pyre, for example. Bone coal can, when ready, be used together with charcoal to get carbon, and perhaps phosphorous into the iron but it could be that what was most important was to get something of the deceased into the metal. Gansum and Hansen crushed the resultant bone coal and charcoal together and wrapped this mixture around a piece of iron and put the combination into an iron box; Gansum notes that in the past smiths would have used clay to seal the mixture rather than an iron box. The box was put into a furnace and kept at 900°C for more than six hours; they used forty-four kg of charcoal during that time. At 720°C the carbon starts to transfer from the bones and charcoal to the iron in the mixture and will penetrate as far as 3mms into the surface.[6] Gansum and Hansen used ox bones for their experiment, but any type of bones, including human bones, could have been used and this method of getting carbon into iron has been in continuous use since ancient times.[7]

Analogies from traditional iron production

Before we look at what Gansum has to say about the literature, it is instructive to consider his findings from observing practices among traditional iron producing societies in Asia and Africa and then consider how he connects that to what we know of Scandinavian smiths and iron production. Gansum reports that in such iron producing societies the process is governed by strict taboos and ritual behaviour, that furnaces could have female attributes, perhaps breasts or openings shaped as vulvas and that the whole process is bound up with

---

[1] Hedeager, 2011. 139, too comments on this.

[2] Gansum, 2004. 41. Gansum supplies many references to support this statement. Hinton comments on this aspect of the smith, too. Hinton, 1998. 15.

[3] Gansum, 2004. 42. For further discussion of this idea that everything in life was part of the belief system, see, for instance, Sundqvist, 2002 (12) and Winterbourne, 2004 (22).

[4] Gansum, 2004. 42.

[5] Gansum, 2004. 42.

[6] Gansum, 2004. 42 and 43.

[7] Gansum, 2004. 52.

notions of fertility. For instance, in Nepal, no-one is allowed to watch when the smelter places the phallic-shaped tuyère into the furnace and while doing this the smelter has to be totally naked in front of furnace; "…his nakedness illustrating the intimacy between husband and wife."[1] The Nepali name for the tuyère is *tora*, a modified version of their word for penis. The smelter is the husband, the furnace his wife, the pumping of the bellows represents the heavy breathing during intercourse and the bloom of iron is the 'child' delivered by the furnace.[2]

Motz sees the smith a little differently, "Since metals are conceived as embryos which grow and ripen in the womb of mother earth, the metallurgist who 'delivers' them also has the aspect of an obstetrician, and since he 'delivers' them before they would naturally have been discharged, he again demonstrates his ability to intervene successfully in the processes of life."[3] Pollington, *et al.* echoes this concept when he suggests that in many cultures the smith is seen as similar to a midwife in that he "...oversees the coming-into-being of a new entity."[4] Whatever, midwife or husband of the furnace, the smith was a powerful and awesome figure transgressing the normal boundaries of society, including, perhaps even crossing those of gender.

The ancient literature

To see if there is any remnant of any of these ideas in Scandinavian literature, Gansum turns to the *Poetic Edda* and looks at one particular poem, *Vǫluspá*, (Prophecy of the *Völva*[5]) stanza 40 where the poem looks at the build up towards Ragnarǫk, the end of things, where the wolf, Fenrir, will kill Óðínn. The offspring of Fenrir are mentioned as being born of an old giantess in a wood called Jarnviði, the 'iron woods'. Gansum considers the language in the poem and reasons that the 'old giantess' is a reference to the earth itself, acting as a furnace in this case, and that the 'offspring' are iron and steel weapons for use against the gods. He also believes that the name of the wood is no coincidence either, as iron and wood are both so essential to the production of weapons.[6]

Gansum points out, too, that the dwarfs are linked through the mythology to the technology of production and quotes Lotte Motz, "We must view them [dwarfs] as the mythical representatives of a profession, paralleling the craftsmensmiths (*sic*) of early society, who were, indeed, endowed with ritual importance."[7] According to Motz, therefore, dwarfs function as symbolic representations of the secret knowledge of the smiths in the mythology and also have close connections to earth and stones, as do swords.[8] Dwarf smiths are mentioned in the stories as the makers of powerful swords; one example comes from the *Hervarar saga* where Durin and Dvalin are forced to make a sword for Svafrlami. The sword is called Tyrving and, in revenge, they curse the sword so that it will always bring misfortune to its owner, which it does.

---

[1] Gansum, 2004. 48.

[2] Hedeager, 2011. 139, Hedeager remarks on this aspect of the smith's work, too.

[3] Motz, 1983. 4.

[4] Pollington, Kerr and Hammond, 2010. 124.

[5] A *völva* is a seeress. In the poem the *Völva* is relating the story of the creation and the end of the world to *Óðínn*.

[6] Gansum, 2004. 46.

[7] Gansum, 2004. 47. The quote is from Motz, Lotte, 1993. 84. *The Host of Dvalin. Thoughts on Some Dwarf-names in Old Icelandic.* Collegium Medievale 1.

[8] Gansum, 2004. 51.

The art

A particular art style that Gansum feels is linked to the use of animal parts in the smithing process is one that is usually referred to as Salin's Style I animal art.[1] Typically only parts of animals and parts of humans are depicted in this style and it is usually encountered on metal items, such as sword, scabbard and belt parts as well as on brooches and drinking horns, often in gold or gilded metal. Sometimes what appears to be breathing is emphasised on the designs or animals are depicted coming out of the mouths of human heads.[2] The thinking is that the icons represent stories and myths and only those who understood the 'code', could 'read' or recognise the stories and that the artefacts, themselves display status and ideologically link their wearer to the stories and their ancestors.[3] Gansum goes further, "The idea is that the mixing of bones makes it possible for the smith to transform humans and animals into weapons with all the symbolic meaning such actions may bring. There is a parallelism between bloomery iron (soft; newborn) on one hand with steel (hard; mature) on the other." Is the art style then, a representation of what the smith could have been doing; mixing the body parts of humans and animals in order to incorporate them into the blade of a sword? Were there myths that referred to this phenomenon? The fact is that we do not know for sure, but the possibility is there and it is an enticing one, as Gansum speculates that it could easily form part of the cult of the ancestor and having "...the old warrior transformed into the symbol of power, the sword, and by doing so passes on his strength, spirit and luck,..."[4] through the use of some of his bones to his son or comrade would have been a powerful and magical concept that would provide added significance and biography to the weapon. Hedeager has recently, possibly unknowingly, supported Gansum's suggestions in her paper on 'Split Bodies'.[5] Her writing is about the integrity of the body during life and after death and she links more than just Style I art symbolism to the idea that bodies could be split up and even hybridised with animals, her additional evidence includes images on golden bracteates,[6] the *pressbleches* designs on the Vendel XIV helmet and some belt buckle images,[7] as well as symbolism on textile remains and the goldgubber.[8] She remarks, "Thus, in the Scandinavian Iron Age, the end of a human lifecycle did not necessarily involve the maintenance of bodily integrity. Through a process of deconstruction, skeleton remains achieved an afterlife and thus outlived the living person in a variety of contexts. Bodily remains were imbued with agency and a biography of their own:..."[9] Back Danielsson has a different perspective on how the mingling of body parts with iron could have been perceived and suggests that the coming together of the ancestral human or animal parts with the iron might have been seen as a type of intercourse that resulted in the birth of a new material with characteristics that had been chosen by the maker.[10]

Gansum provides some evidence that the smith could have carried out some of his work within or close to cemeteries[11] and he believes that it is possible that some burial mounds that were robbed in antiquity might have been entered by smiths to retrieve a few bones of a dead hero or powerful figure. One of the dwarf

[1] Gansum, 2004. 51. For an excellent summary of the development of Style I and for a list of attempts to call it by other names, see Pollington, Kerr and Hammond, 2010. 101 to 107. There are some notes on Style I and II in the glossary. Hedeager seems to support Gansum's thesis, see Hedeager, 2011. 140.

[2] Gansum, 2004. 52.

[3] Gansum, 2004. 51. I am not so sure that the 'code' was as exclusive as Gansum implies, as most people within the community would know the stories, it is likely that everyone would understand the significance of the decoration.

[4] Gansum, 2004. 51.

[5] Hedeager, 2010.

[6] Hedeager, 2010. 111.

[7] Hedeager, 2010. 112.

[8] Hedeager, 2010. 113. 'Goldgubber' (Danish and Norwegian) – Guldgubbar (Swedish).

[9] Hedeager, 2010. 111.

[10] Back Danielsson, 2007. 248.

[11] Gansum, 2004. 50.

names, Haugspori, "one who treads on mounds" fits this idea well, Gansum suggests.[1] Hedeager also noted in her essay that, "In some cases skeletal remains were removed after burial."[2] It is possible, then, that smiths were able to make bones the medium of birth and re-birth as far as the iron and steel artefacts that they made were concerned, they were able to transform the dead hero into a weapon where he would have a new life! The naming of a weapon under such circumstances would take on a new importance.[3] Gansum wonders whether different mixtures of types of animal would have had different meanings, for instance mixing elk, horse and human bone could have had a different significance to that of bear, boar and dog;[4] perhaps Hedeager's recognition of the hybrid beings in some of the art-work is germane to this idea? It is possible that some of the animal imagery on metal objects is there because parts of that animal have been incorporated into the metal with all the magical power and symbolism that the animal represented.

A curious echo of idea of the use of bone coal can be found in modern day Taiwan, where at least one smith who makes Chinese swords, sometimes uses human bones in the process. The smith, Kuo Chang-hsi, throws human bone into the furnace while forging. He says that in ancient China swordsmiths threw a man into the fire because it "...helps to melt the metal and creates the pure iron that we can further hammer." Kuo claims that the burning bone adds phosphorous to the metal, a combination of iron (presumably low carbon) and steel that he uses. He gets his bone from cemetery relocations and from relatives of the deceased who ask him to make a sword in their relative's memory. Apparently, the human bone will add 'spirit' to the sword.[5]

Kuo does not claim a continuous practice from ancient times, just that he started experimenting with human bone and found it to be successful. He does say that the old stories tell of the practice.

How relevant this is to what might have been happening in Europe more than 1,300 years ago is debatable, but the ideas are interesting and using bone could only add to the mystique of the smith.

Transformation, travel and suspicion

The ability to transform ordinary materials such as earth, rock, wood, charcoal and bone into a symbol of authority and martial power or insignia that displayed the ideology of the group, may have given the smith an ambiguous place within society. This ability to take raw materials and make precious artefacts in some ways put the smith on a par with the gods who could make plants, animals or men out of the ordinary materials to hand. The smith was a creator, too, whose craft was surrounded by mystery; he could transform materials from the natural world and embed them within his culture,[6] his work united, "Cosmology, magic and materiality",[7] and provided his finished items with the qualities, "...that ensured the efficacy of rituals and the legitimacy of the elite."[8] His symbolic working of iron could have been envisaged as a magical act which benefited the community. As Hedeager points out, "Blacksmiths and jewellers are usually associated with power in traditional societies because they forge the instruments by which the natural and social world may be dominated. Furthermore, they create objects that mediate between mankind and the supernatural."[9]

---

[1] Gansum, 2004. 45. The reopening of graves is something that we will discuss again in *The meaning of the sword.*

[2] Hedeager, 2010. 111. See also Gardela, 2009, for 204 to 206 for his perceptive comments with regard to the metal included with *seiðr* staffs in Scandinavia. We will return to this idea of the deliberate removal items from graves in the chapter, *The Meaning of the Sword.*

[3] Hedeager, 2011. 140.

[4] Gansum, 2004. 51. However, the particular combinations that I have suggested are not suggested by Gansum.

[5] The National Geographic video of this story can be found here: https://www.youtube.com/watch?v=uzRYO98lQD8 -- apparently Kuo made swords for the film, 'Crouching Tiger, Hidden Dragon'.

[6] Hedeager, 2011. 139. Hinton, 1998. 16.

[7] Hedeager, 2011. 140.

[8] Hedeager, 2011. 140.

[9] Hedeager, 2011. 139.

However, the smith might have belonged to, "...the realm outside human society..."[1] and this position would have been amplified if he had no permanent dwelling in any one settlement and was peripatetic.[2] The suggestion is that many smiths, if not all, could have been, "long distance travellers; they mediated between the settled heartland of human society and the dangerous outside world."[3] Hedeager explains that traders and artisans from outside the community would often have been thought of as superior just because they were from somewhere different,[4] somewhere exotic. This would, nevertheless add to the idea that they were from beyond normal society.

Dronke tells us that in some parts of the world smiths would have great prestige but that in others they were treated with great suspicion, were controlled by taboos, sometimes hated, and would have the status of slaves and that these differing attitudes might have been reflected in the treatment of smiths in the north.[5] Certainly a smith's abilities and skills could be viewed as a threat to those in power, despite the fact that the elite needed the weapons that he could produce in order to keep them in power, his very mastery of the materials that he used, might have given him a 'superhuman stature' but could have caused others to regard him as a 'dangerous and ill-intentioned sorcerer.'[6]

Some of these possibilities surface in the stories.

Wayland/ Vǫlundr

We will consider the story of the smith Wayland (Wēland OE Velent OHG, Vǫlundr ON, also, Vølund *or* Völundr and sometimes written as Wieland, Weyland and Weland) who appears in a number of different sources, not all of them literary.[7]

The most complete version of the story is contained in the poem from the *Poetic Edda*, the *Vǫlundarkviða* and has been recently studied in depth by Dronke,[8] who does not believe that the story originated in Scandinavia but that there was a German version in, at least, the seventh century based on evidence of the names of some of the characters.[9] However, the earliest surviving evidence of the story comes from Anglo-Saxon England and is carved into the Franks Casket, which is thought to have been made during the early eighth century.[10]

---

[1] Hedeager, 2011. 143.

[2] Pollington, Kerr and Hammond, 2010. 124. The possibility of travelling smiths is also discussed by Hinton, 1998.

[3] Hedeager, 2011. 144. And Pollington, Kerr and Hammond, 2011. 124. La Salvia, 2007. 64.

[4] Hedeager, 2011. 140.

[5] Dronke, 1997. 260. Motz agrees. Motz, 1983. 2.

[6] Motz, 1983. 14.

[7] A summary of the most important sources referring to Wayland would include an OE charter of King Eaðreð. This Anglo-Saxon ruler bestowed in 955 A.D. a plot of land on one of his retainers. The charter which grants the gift specifies that the property, in the Berkshire Downs, is to reach in one of its extensions to the landmark known as Wēlandes *Smiððe* (see Dronke, 1997. 259 and Motz, 1983. 68.) a Neolithic barrow now in Oxfordshire (originally Berkshire), pictures carved into the Franks Casket, constructed from whalebone, the OE poems of *Deor*, *Waldere* and *Beowulf* – the Poetic Edda or Elder Edda (*Vǫlundarkviða*) *Velents Þáttr* from the *Þiðrekssaga af Bern*, and the picture stone found in Gotland at Ardre, stone VIII. For a fuller discussion of the sources and of others not mentioned here, see Dronke, 1997. 269 to 290. Wright has also listed the Anglo-Saxon sources that mention the Wayland story. Wright, 2010. 131.

[8] Dronke, 1997.

[9] Dronke, 1997. 269 to 270 and 280. However, Hall believes the tale originates in Anglo-Saxon England, see Hall, 2007. 39-41. Bradley, 1987, believes that far too much has been invested in the pre-Christian and Scandinavian connection of 'Weland', he feels that those associations weren't really current in mid- to late Saxon England, and that Weland had become thoroughly Christianised. However, he does see connections between Weland and the Greek legend of Daedalus. See Bradley, 1987, 140 and 158 to 160.

[10] Dronke, 1997. 270.

Although the various versions of the story contain different elements there is a great deal of agreement between them. The brief account here is based mainly on the version contained in the *Vǫlundarkviða*.

Vǫlundr had two brothers, Egill and Slagfiðr; they were sons of the king of the Lapps. The brothers made their home in a place called Wolfdales, where there was a lake called Wolf Lake. One morning the three brothers met three women by the lake shore who had swan garments with them, they were the Valkyries; Hlaðguðr Svanhvít, Hervǫr Alvítr and Ǫlrún. Each brother then lived with one of the women for eight winters, Vǫlundr lived with Hervǫr Alvítr. When the ninth winter was reached, the Valkyries divorced their lovers and left to seek battles. Egil and Slagfiðr were so desperate at their loss that they went to try and find their women but never returned. Vǫlundr stayed put and believed that his love would return, he was a consummate craftsman and when not hunting for food he spent his time making jewellery for his lover and crafted seven hundred rings for her.

One night he discovers that one of the rings is missing and he thinks that Hervǫr Alvítr has returned and is playing a game with him. He waits but eventually falls asleep. When he awakes he discovers that he is bound by strong bonds and is now the captive of Níðuðr, lord of the Niárar who with men, fully armed and mail clad, had tied up Vǫlundr while he slept. It was Níðuðr who had taken the ring and he, later, gives it to his daughter. Vǫlundr is imprisoned on the island of Sævarstǫð after being hamstrung so that he could not run away and is forced to make items for king Níðuðr.

Dronke makes the point that the great smiths of legend are all avaricious and hate to see their precious items in the hands of others and Vǫlundr is no exception,[1] although as he had been so badly treated, it is understandable that he could not bear seeing his work being worn by his captors. In revenge, Vǫlundr kills the king's sons when they visit him in secret, fashions goblets from their skulls, jewels from their eyes, and a brooch from their teeth; their bodies he buries beneath his forge well. He sends the goblets to the king, the jewels to the queen and the brooch to the king's daughter. When Bǫðvildr the king's daughter, takes her ring (the one that Níðuðr stole from Vǫlundr) to Vǫlundr to be mended, he takes the ring and rapes her, fathering a son and then escapes on wings he had made.

Three times during the story Vǫlundr is referred to as 'lord or commander or prince of elves', (álfar lióði, vísi álfa)[2] That he was not a normal human being is demonstrated by the several features of his story, that a Valkyrie was his wife, that he lived outside normal human society, that his level of craft skill was so high and that he could make incredible jewels, named swords and wings; there is also his ability to locate gold where none was thought to be which provides Níðuðr with the excuse to capture the smith as Níðuðr claimed it was his gold. Another indication that Vǫlundr is far from ordinary is that the armed men of the king would only approach him when he was asleep; despite their numbers and their weapons they knew they could not take him at any other time. It should be remembered that both Vǫlundr and his brothers were sons of the King of the Lapps, a region that is often connected to magic in the tales. However the association with elves is unusual in that smiths in the myths are usually linked to dwarfs but Dronke feels that this may "...relate to an older tradition in which álfar were subtle smiths, before the popularity of the dwarfs as underground metal works made the elves' title forgotten."[3] She suggests, too, "...that he should be identified as an elvish being, an elf man, as his wife is a swan woman." That he could be a mixed being; part man part elf is perfectly plausible as Dronke explains such beliefs have had wide currency in human cultures, even down to the present day.[4] Whatever, the most intriguing aspects of the Wayland story as far as our study of smiths goes is that he was

---

[1] Dronke, 1997. 257.
[2] Dronke, 1997. 261.
[3] Dronke, 1997. 262.
[4] Dronke, 1997. 262.

not quite human or part of human society, he was feared, captured and made to work for the king and that, when the time came he was quite capable of taking his revenge.

The version of the tale in *Þiðrekssaga af Bern* depicts Velent going to extraordinary lengths to make the sword Mímungr. He has promised to make a sword for the king and forges one. He is not pleased with it and files it into a heap of filings which he then mixes with feed for farmyard fowls. He feeds the mix to the fowl that he has starved for three days and then collects their droppings and refines them in his forge and makes a new sword which is not quite as long as the first. He repeats the process and finally forges Mímungr, which is razor sharp and of perfect balance. He hides the sword under his bellows and makes the king a more ordinary sword; he keeps Mímungr for his son.[1] The process that is described, laborious as it is, is unlikely to have produced a superior metal, according to Gilmour, as, "A few hours of exposure to dilute hydrochloric acid in the stomach of a bird would etch the metal to some degree, and might dissolve a little slag near the surface, but any more than a fairly minimal effect seems unlikely."[2] However, La Salvia points out that bird faeces are an organic compound and are rich in carbon and nitrogen and could assist in the process of carburisation.[3]

This story of cutting metal up and feeding it to birds is also reported by the Islamic writer al-Bīrūnī, only in his version it is ostriches that eat the metal;[4] the two versions were separated by 200 years and 5,000 km. al-Bīrūnī reported that the Russians and Slavs do the same with ducks,[5] and there are also Chinese versions of a similar episode, again with ostriches. The story resurfaces in the writing of the Italian Renaissance master craftsman, Biringuccio in 1540 A.D., this time geese are the hungry birds.[6] Gilmour goes on to speculate about the origin of the story and how it spread.

Reginn

Reginn's story appears in a number of sources, apparently written down in the 13th century but certainly referring to much older traditions, *The Poetic Edda* (*Reginsmál*) and the *Volsunga Saga* and *Þiðrekssaga af Bern*.[7]

Reginn (*Regin*) like Vǫlundr had two brothers; Reginn's were called Fafnir and Ótr, and they were the sons of Hreiðmarr, the avaricious king of the Dwarves. Three gods of the Æsir were travelling through the world; they were Óðínn, Hoenir and Loki. While near a river, Loki saw an otter catching a salmon, so he throws a stone at it and kills it. After they have gained lodging for the night with Hreiðmarr the three gods discover that the otter was none other than Ótr, the son of their host. Hreiðmarr and his two remaining sons bind the gods and will not release them until they have paid compensation for Ótr's death and demanded sufficient gold to completely fill and cover the skin of the otter. Loki was sent to collect the ransom and captures a dwarf named Andvari and makes him give up all his treasures, including a special ring, Andvarinaut that can increase wealth. Andvari has cursed the ring and says that it will wreak destruction on anyone that possesses it. Óðínn would have liked to have kept the ring but it was needed to cover up one of Ótr's whiskers and so was left with Hreiðmarr. The curse soon worked as Fafnir and Reginn kill their father for the ring but Fafnir decided he wanted Andvarinaut for himself, so he turned into a dragon and drove Reginn away.

---

[1] Dronke. 1997. 275.

[2] Gilmour and Hoyland, 2006. 159, footnote 56.

[3] La Salvia, 1998. 38 to 39. La Salvia points out, too, "...that many analyses of Merovingian swords have yielded the evidence of nitrogen."

[4] Gilmour and Hoyland, 2006. 158.

[5] Gilmour and Hoyland, 2006. 150.

[6] Gilmour and Hoyland, 2006. 160 t0 161.

[7] The story line of this the tale appears in a rock carving from the Viking Age in Sweden at Ramsundsberget. Hedeager, 2011. 142. Many of the tales refer to events during the 2nd to 5th centuries – the Migration Period. Bycock, 1999. 11 to 29. *The Nibelunenglied* which has another version of Siegfried's story was written around 1200. Bycock, 1999. 3.

Reginn travels and, like human smiths,[1] takes jobs as smith to kings and eventually becomes the foster-father of the young soon-to-be hero, Sigurd the Volsung. He teaches Sigurd many things, including languages, sports, chess, and runes. Reginn is the only one who is able to forge a sword that will kill Fafnir and he makes it from the broken sword that had belonged to Sigurd's father, Sigmund. The sword was called Gram and it had magic woven into it. Reginn encourages Sigurd to kill Fafnir and explains that it can only be done if Sigurd digs a pit and waits in it until the monster goes to the river to get a drink. When the dragon passes over Sigurd he must stab the dragon from underneath and kill it. All of which the hero does, but Fafnir does not die before he discusses with Sigurd doubts concerning Reginn's intentions towards the young hero. After drinking Fafnir's blood Reginn tells Sigurd to cut out the dragon's heart and cook it so that he, Reginn, can eat it after he has had a sleep. However, while cooking the heart Sigurd burnt his thumb on the morsel and sucked it to cool it down. As he absorbed the blood of the dragon he found that he could magically understand the comments that some nearby nuthatches were making. The birds were saying that Reginn planned to kill him, so Sigurd cut off Reginn's head before he awoke, ate Fafnir's heart and took all the treasures for himself.[2]

In this tale there is no doubt that Reginn was a dwarf who was very knowledgeable and gifted with magical skills and was a smith of great talent.[3] Once again, he is strange and stands outside of normal human society. He is devious and vengeful and will use Sigurd to obtain his ends, which is to possess Andvarinaut and the rest of the dragon's treasure. However, Motz casts Reginn in the role of the initiation master who reveals to the young the secret knowledge of their tribe and leads them to their goal of manhood and feels that he was probably a priest or elder.[4] The problem with this is that Sigurd murders Reginn, which is not the behaviour that would be expected of an initiate towards his guide but the story probably does have echoes of an older tradition within it and there are certainly examples of situations where boys are removed from the world of women at an early age to be trained by elders in Anglo-Saxon literature.[5]

Vǫlundr is possibly a slightly more sympathetic and honourable character than *Reginn* but is no less skilled.

Dwarfs

"The magician-craftsmen and artisans par excellence in Old Norse mythology are the dwarves, who produce objects through which the cosmic order is sustained; for example, Óðínn's spear Gungnir and his golden ring Draupnir, Thor's hammer Mjollnir, Freya's golden necklace Brísingammen (flaming necklace), and Frey's boar Gullinbursti (Golden bristles) and his magical ship Skíðblaðnir. They brewed the mead – the drink of wisdom, poetry and inspiration that constituted Óðínn's supremacy."[6] Following Motz, Hedeager goes on to remark that the dwarfs were teachers, chanters, wise men, magicians and ritual officials (priests) and makers of mighty

---

1 Hedeager, 2011. 142.

2 There are a number of surviving pictorial references to elements of this story from the Viking period – late 8th to 11th centuries. Davidson, 1969. (101.) lists examples from the Isle of Man, Lancashire, Norway and Sweden and says, "Although the style and treatment of the different carvings vary greatly, the same incidents are singled out: the stabbing of the dragon from below with a sword, the roasting of the heart, the birds on a tree, the beheaded smith and the horse with loaded treasure." She does mention other elements that are shown on other stones, too.

3 Motz, 1983. 96. Motz says that the adjective *fjölkunnigr*, which is used to describe him means 'knowing the arts of magic'.

4 Motz, 1983. 97 and 98.

5 Mainly in the training of warriors. See Mortimer, 2011. 186 to 188. Davidson, 1989. 11 and 20. Kershaw, 2000.p 24 to 25. See also the chapter in this book, *The Sword as a Weapon*.

6 Hedeager, 2011. 140.

weapons.[1] They are also identified with craft and with the making of metal artefacts. Like smiths they were linked to the heavens and many of the dwarf names that have come down to us show that connection clearly.[2]

The dwarfs do not have any women within their race and although they have sons, it is not always clear how they get them.[3] Dwarfs mostly dwell underground, or in mountains or stones[4] and the gods are dependent on them for their ability to make magical objects that guarantee the cosmic order but dwarfs do not always do this work willingly and can be resentful and seek vengeance. However, it is the dwarfs, not the gods, who are in charge of forging; there is no smith god in this set of beliefs.

The word 'dwarf' cognate with Old Icelandic *dvergr*, German *Zwerg*, OE *dweorg*, OHG *twerc*, does not convey the meaning of 'short in size'. The name has been related to Old Indian *dhvara* - a demonic being, or to an Indo-European root **dhver-* which means 'damage' and is also found in an Avestan word for a physical defect.[5] In the descriptions of dwarfs in the literature the emphasis seems to be on ugliness or deformity rather than physical size and this could be related to an initiation process that the dwarf might have had to endure in order to gain his knowledge, as Motz says, "If the word *dvergr* is indeed traceable to an Indo-European root meaning 'damage' we may understand it to reflect the physical deformity which had marked dwarfs as the 'damaged ones', those who had paid the price for their magical endowment."[6] Dwarfs are sometimes referred to as 'short' but they are also shape changers, occasionally having the appearance of men and once or twice called 'giants' in the poetry; indeed four dwarfs are reputed to hold up the sky which is a task one would assume not entrusted to a stunted being.[7] The dwarf Fafnir, it must be remembered could hardly have been small once he became a dragon! It seems that dwarfs might have been all shapes and sizes when it suited them or the needs of the story.

To add to the idea of smiths/dwarves and damage comes this description from a 17th century Polish 'iron-master' of his work quoted in Bradley might be appropriate:

"Look how my body is broiling! Is it little pain that we have in this fire? The flame and sparks fly into the eyes like ashes. I am constantly roasting in the fire on every side! If your shirt catches fire, part of your body is burned, and sometimes it won't heal for half a year. I am half-deafened by the crash of the hammers! I must perpetually toil under such wretched conditions. Our life is miserable. When forging, we must beware of injury at all times. We must take heed, for the hammer has no pity; he who forges carelessly will soon get a bruise! Even Vulcan himself, though a skilled master in his craft, was nevertheless lamed in one leg. Some say it was his mother Juno herself who crippled him in his childhood, but I do not believe Juno to have been the cause of his injury. It is more likely he was lamed by a blow of the hammer when he was busy with the craft of ironwork on the isle of Lemnos. This is why so many of us are deaf and lame, and I doubt if there is one of us who has a whole body. Look how my skin has stuck to the bones from the great heat of the fire!"[8]

---

[1] Motz, 1983. 7, 96 and 98.

[2] Hedeager, 2011. 141.

[3] Motz, 1983. 91. Motz quotes ""Then the gods went to their seats of rock and took counsel about who should be the ruler of the dwarfs" More dwarfs were formed from clay by the first members of the race, or grew, according to other versions, in 'earth and mould', or in the flesh and blood of the murdered giant Ymir." The first part of the quote is from *Vǫluspá* 9. Found within the Poetic Edda.

[4] Motz, 1983. The book is an exploration of the dwarfs and smiths who live in stones and mountains and often near water. Motz has amassed a huge amount of evidence in folktales of the connection s between the earth and dwarfs mainly in German speaking areas, but also some from other parts of northern Europe.

[5] Motz, 1983. 89 and 118.

[6] Motz, 1983. 118.

[7] Motz, 1983. 116 to 118. Motz treats the appearance of dwarfs in full.

[8] Bradley, 1987. 148. The quote comes from, Walenty Roździeński, Officina Ferraria: A Polish Poem of 1612 Describing the Noble Craft of Ironwork, pp. 99-100.

Summing up

As Gansum says concerning swords, "Iron and steel represent and define power functionally and ornaments in gold underline the symbolic value of the weapons. Both gold and pattern-welded blades are connected to great expense, whether it concerns the price of the metal or the expense of production."[1] The man who made such weapons must have had a status all of his own, the fact that his methods could have been considered secret and perhaps magical, involving ritual, death, gods and was full of symbolism would have meant that he was to some extent not part of 'normal' society, that he would have been regarded as something other, someone on the boundaries of more than one world. He could not be completely shunned as his products were in demand and of great value, but he must always have been looked upon with suspicion and perhaps envy, rather like dwarfs are regarded in the stories. As Hedeager explains, "...his ambiguous position, half technician and half-magician, could also place him at a disadvantage in society. He then had to live a separate and secluded life, regarded with disdain by the elite in order to hide the fact that he possessed the technical mastery which was necessary for any ruler in the mediation between himself and those he ruled."[2] She says, too, "The smith's work requires esoteric knowledge that enables him to manipulate the dangerous forces unleashed in the process of transforming shapeless metal into a finished product;..."[3] Perhaps the story of Volundr remaking the skulls of the sons of the king is a reflection of some of the ritual and magical practices that a smith would use in his craft, or something from the practice of shamanism where an apprentice has to be taken apart, ritually and remade in order to go through initiation.[4] So too with the burial of the boys' bodies within his forge, Dronke suggests that this could be an echo of a practice where humans were sacrificed to the hearth, she explains that there are accounts of such practices in the legends of Japan and that foetuses were buried beneath forges in Africa.[5] Gansum also reports that graves have been found inside furnaces in Scandinavia.[6]

It is little wonder that the smith/smelter occupied a special place in the society in which he lived, part magician, possibly part priest and living on the edge. However, what is absolutely certain is that the products he made were indispensable in an iron using community, and not just to the military elite but to just about everyone within it.

A sword made by a famous smith would be a valued sword indeed!

[1] Gansum, 2004. 51.

[2] Hedeager, 2011. 139.

[3] Hedeager, 2011. 139. Wright, 2010. 132 says something very similar.

[4] Dronke, 1997. 267.

[5] Dronke, 1997. 268. Dronke does not tell us whether the foetuses were human or not but the association with the burial of the boys in her explanation implies that they were.

[6] Gansum, 2004. 45.

9. Cemetery/sword ratio; a brief summary with a list of sword finds recorded

Paul Mortimer

Burials are not 'mirrors of life': if anything, they are a 'hall of mirrors of life' providing distorted reflections of the past. Heinrich Härke.[1]

In this work I have surveyed the reports of a total of two hundred and thirty-one burial places. At some of the sites only one grave was found while others had hundreds of graves recorded and investigated. The standards of the excavation and reporting varied enormously which means that completely accurate figures are not possible.[2] However I believe that there is enough data to be able to come to the conclusion that swords in graves are very special additions and not a normal object to be put into a grave during the 5th to 7th centuries; this is not a new observation. We will explore that idea again elsewhere in this book. One hundred and sixty-nine swords were recorded from the sites.

The total number of counted graves recorded in the two hundred and thirty-one burial sites is 9,818; the actual count would probably be more than 10,000 if better reporting had occurred at some of the sites. I have not included cremations in the count, although one cremation at Combe (Woodnesborough), Kent was accompanied by one ring sword and that sword is included in the one hundred and sixty-nine as is a second sword that has been subsequently lost.[3]

Of the two hundred and thirty-one burial sites, fifteen of them do not have any number of burials recorded so I have no idea how many graves they contained, at least one I assume, but I have not recorded any number from the fifteen in the count.

With another twenty-seven grave fields the numbers are not reliably reported so I have used the lowest figure provided in each case.

The reasons for the lack of precision with the grave count are due to a number of factors. Many of the reports are isolated and accidental finds where no further investigation was possible, some were just partial excavations, presumably due to time, money and other prevailing circumstances and as mentioned above, there is a serious lack of standardisation in reporting methods which adds to the problems when attempting an analysis.

Judging by the above figures perhaps 1.7% of graves can be expected to hold a sword or to put it another way seventeen swords per thousand burials.[4] Of course if all grave sites in the sample had accurate numbers of graves recorded that percentage may change, but I suspect not by much. The ratio of swords-to-graves could have been a little different if it were not for the very high proportion of cemeteries included from Kent, which

---

1 Härke. !997. 25.

2 Something noted by Brunning, too. Brunning, 2014. 116 to 117.

3 The grave was excavated by William Wakeling Boreham, 'an amateur collector of antiquities' around 1845. An account of its discovery was not published until 1853. (Davidson and Webster, 1967. 1 to 2 and 9 to 13). There is every possibility that many important details concerning the context of the find(s) were not reported and there must be some doubt as to whether it really was a just a cremation or whether it had become confused with a nearby inhumation or two, despite the confidence expressed by Davidson and Webster..

4 I would suggest a plus or minus figure of three swords.

amount to 72.5% of the sites, although not necessarily the same proportion of graves. There are some grave fields where a relatively high proportion of swords were collected, such as Saltwood, Kent, where eleven swords from two hundred and seventeen graves (5%), and a similar ratio at Mitcham, Surrey, twelve swords to two hundred and thirty-eight graves (5%). Dover Buckland yielded a high number of swords with twenty-four swords (6%), but there were four hundred and four graves on that site. The site where the most swords were found though is Sarre, again in Kent, where twenty-seven swords were recorded from four hundred graves, which gives a percentage of almost 7% (6.75%). The grave site that yielded the highest ratio of swords, Gilton, Kent, five swords to six graves is unlikely to have been fully excavated; in fact many of the cemeteries surveyed for this project have only been partially, if in some cases extensively, explored.

From our sample eight ring-swords were reported and one of those was recorded as being found in a cremation (Combe/Woodnesborough). It would seem that such swords, at least as far as being used in a funeral was concerned, were very rare indeed and although ring-swords represent 4.7% (8/169) of the swords in our sample, as far as graves are concerned only about 0.08% (8/9,818)of graves can be expected to contain one, perhaps less, as two of that number were not found in a burial. These figures can be contrasted with the number of finds of sword-rings or sword-ring parts and sword pommels located by detectorist as the examples in the appendices might indicate; eight possible sword-rings to seventy-nine sword pommels (8/79 = 10%).[1]

The frequency of swords found within graves could be also adversely affected by the grave robberies of the late 6th and 7th centuries noted by Klevnäs and others,[2] where swords were a particular target. Although it is difficult to estimate how that would change the figures, perhaps the ratio would improve by up to 10%?

It would have been useful to be able to differentiate male from female burials, but this is often not possible due to the excavators themselves not being able to tell, or the remains being so poorly preserved that sexing was not viable and sometimes such information was just not recorded. I suspect that if we were able to reliably distinguish adult male graves from all others, including children, then the deposition rate of swords would rise to 3% - 7% in male graves.

I realised too late that it would have been of benefit to collect a count of the total numbers of other weapon graves in the cemeteries and to compare that with the number of sword burials; unfortunately I did not do that, and it was only when the project was quite advanced that I realised that I had could have improved the study, but by then much of the reference material was no longer available to me.[3]

---

[1] See Appendix 10 for sword pommels and Appendix 11 for sword-rings. Some of the pommels may have been part of burials.

[2] Klevnäs, 2013.

[3] Härke, 2000. 387. Reports that around 18% of all inhumation burials contained weapons and 1% to 2% of cremations. Härke, 1992. 156. Reports that of 291 weapon burials only 19 contained swords; that is 6.5% of weapon burials.

Table 9i. Sword graves.

2.[1] Abingdon 42	50. Dover Hill 10	98. Murston 1	134. Sarre 57
3. Alton 1	51. Edix Hill	99. Orpington 36	135. Sarre 64
4. Alton 16	52. Finglesham 22	100. Ozengell A	146. Sarre 189
5. Alton 2	53. Finglesham 204	101. Prittlewell Grave group 17 sword point 173	147. Sarre 190
6. Alton 42	54. Ford Laverstock 1 (17)	102. Prittlewell Grave group 17 sword 171	148. Sarre 250
7. Apple Down 12A	55. Gillingham 1	103. Prittlewell Grave group 17 sword 172	149. Sarre 264
8. Barfriston 47	56. Gilton 10	104. Prittlewell Grave group 18 sword 179/45	150. Sarre 275
9. Barham Down 33	57. Gilton 23	105. Prittlewell Grave group 25 sword 179/1	151. Short Bros. Works 1
10. Berg Apton 19	58. Gilton 56	106. Prittlewell Grave group 29 sword 425/1	152. Sibertswold 98
11. Bifrons 24	59. Gilton 66	107. Ringwould with Kingsdown 1	153. Sibertswold 140
12. Bifrons 39	60. Gilton 89	108. Risley 20	154. Sibertswold 58
13. Bifrons 62	61. Great Chesterford 65	109. Risley 19	155. Snape 47
14. Bifrons 72	62. Great Chesterford 122	110. Risley 87	156. Springhead 2645+2643
15. Bishop's Bourne 10	63. Holborough 2	111. Risley 75	157. Supperton Farm 1
16. Blacknall Field 1	64. Holborough 7	112. Risley 86	158. Sutton Hoo 2
17. Blacknall Field 22	65. Howletts 3	113. Saltwood C1163	159. Sutton Hoo 1
18. Blacknall Field 47	66. Howletts 20	114. Saltwood C1145	160. Sutton Hoo 17
19. Blacknall Field 70	67. Howletts 1	115. Saltwood C3779	161. Temple Farm 3
20. Boughton-under-Blean 1	68. Kempston? (March 14)	116. Saltwood C4665	162. Wanlip?
21. Bromeswell 21	69. Kempston? (Feb 2)	117. Saltwood C6653	163. Watt's Avenue 17
22. Bromeswell 27	70. Kingston Down 208	118. Saltwood C3944	164. West Heslerton G74
23. Bromeswell unassociated	71. Kingston Down 16	119. Saltwood W1767	165. Westgarth Gardens 51
24. Buckland 71	72. Mill Hill 40	120. Saltwood C3826	166. Westgarth Gardens 66
25. Buckland U/A8	73. Mill Hill 81	121. Saltwood C3885	167. Wigber Low 4
26. Buckland 360	74. Mill Hill 91	122. Saltwood C1048	168. Woodnesborough cremation
27. Buckland C	75. Mill Hill 93	123. Saltwood C1081	169. Worthy Park 96
28. Buckland 22	76. Mitcham 27	124. Sarre 211	170. Combe/Woodnesborough[1]

[1] Number 1 is a blank.

29. Buckland 27	77. Mitcham 73	125. Sarre 3	
30. Buckland 33	78. Mitcham 93	126. Sarre 6	
31. Buckland 41	79. Mitcham 163	127. Sarre 8	
32. Buckland 56	80. Mitcham 181 to 184	128. Sarre 11	
33. Buckland 91	81. Mitcham 186	129. Sarre 12	
34. Buckland 93	82. Mitcham 188	130. Sarre 15	
35. Buckland 94b	83. Mitcham 192	131. Sarre 18	
36. Buckland 96a	84. Mitcham 65	132. Sarre 39a+b	
37. Buckland 96b	85. Mitcham 89	133. Sarre 54	
38. Buckland 98	86. Mitcham 193		
39. Buckland 131	87. Mitcham 211		
40. Buckland U/47	88. Morning Thorpe 218	136. Sarre 68	
41. Buckland 249c	89. Mucking Unstratified I	137. Sarre 71	
42. Buckland 264	90. Mucking 557 II	138. Sarre 86	
43. Buckland 265b	91. Mucking 600 II	139. Sarre 88	
44. Buckland 346	92. Mucking 618 II	140. Sarre 91	
45. Buckland 375	93. Mucking 682 II	141. Sarre 103	
46. Buckland 414	94. Mucking 766 II	142. Sarre 104	
47. Buckland 437	95. Mucking 769 II	143. Sarre 113	
48. Castledyke South CS 39 II	96. Mucking 883 II	144. Sarre 156	
49. Castledyke South 179	97. Mucking 933 II	145. Sarre 160+161	

Table 9ii. Number of swords per sword-yielding site:

Sarre Kent x27	Howletts Kent x3	Abingdon Oxon	Orpington Kent
Buckland Dover K x24	Sibertswold Kent x3	Apple Down Sussex	Ozengell Kent
Mitcham Surrey x12	Sutton Hoo S/folk x3	Barfriston Kent	Ringwould Kent
Saltwood Kent x11	Tranmer House S x3	Barham Down Kent	Short Bros Kent
Mucking Essex x9	Castledyke Lincs x2	Berg Apton Norfolk	Snape Suffolk
Prittlewell Essex x6	Finglesham Kent x2	Bishops Bourne Kent	Springhead Kent
Gilton Kent x5	Great Chesterford E x2	Boughton-under Kent	Supperton Farm Kent
Risley Kent x5	Holborough Kent x2	Dover Hill Kent	Temple Farm Kent
Alton Hants x4	Kingston Down K x2	Edix Hill Cambs	Wanlip Leicester
Bifrons Kent x4	Kempston Beds x2	Ford Laverstock Wilts	Watts Avenue Kent
Blacknall Field Wilts x4	Westgarth Suffolk x2	Gillingham Kent	West Heslerton Yorks
Mill Hill Kent x4	Woodnesborough x2	Morning Thorpe N/folk	Wigber Low Derby
		Murston Kent	Worthy Park Hants

[1] The second sword from the 'same' cremation, now lost, but reported in Davidson and Webster, 1967. 1.

Cemeteries/burials

The name of the burial site is followed by a set of brackets, the first number shows how many swords came from the site, the second, the number of burials; cremations are given afterwards where present. The second set of brackets indicates the time range of the site given in the report.

Total 231 burial places

Total known graves 9,818 cremations not included

Fifteen grave fields – number of burials not known so not included within the count.

Twenty-seven grave fields actual number of burials not known, so lowest possible figure used

Seventy-three cemeteries contain swords (31%) of these forty-six (63% 46/73) have only one or two swords (it is difficult to be absolutely precise as in many cemeteries accurate data is not available)

Abingdon, Oxfordshire (2/119 [44 male] +82 crem)

Adisham Down,*[1] Bekesbourne, Kent (?/46) (7th to 8th)

Alton, Hampshire (4/47)(5th /6th)

Alwalton, Cambridgeshire (0/34 and 30 cremations) (5th 6th)

Apple Down, Sussex(1/28 male graves) (5th to 7th)

Ashford,* Kent (?/?) (6th to 7th)

Barfriston, Kent (1/48) (7th to 8th)

Barham Down,* Kent (1/22) (6th to 7th)

Bay Hill,* St Margaret's -at –Cliffe, Kent (Y?/8) (?)

Beacon Hill,* Deal, Kent (?/200) (?)

Bekesbourne, Kent (0/45) ()

Beckford,[2] Worcestershire (0/130+4 crem) (5th to 6th)

Belmont Park,* Throwley, Kent (0/1) (6th)

Berg Apton, Norfolk. (1/63) (6th to 7th)

Betsham,* Southfleet, Kent (?/?) (?)

Berinsfield, Oxfordshire (0/114) (5th to 7th)

Bifrons,* Patrixbourne, Kent (4/110) (5th to 7th)

Bishop's Bourne, Kent (0/9) ()

Bishop's Cleve, Gloucestershire (0/24) (5th to 8th)

Blacknall Field, Pewsey, Wiltshire (4/104 or 105) (5th 6th)

Bloodmoor Hill, Carlton Colville, Suffolk (0/29) (settlement 6th to 8th - cemetery 7th to 8th)

Boss Hall, Ipswich, Suffolk (0/29?) (5th to 8th)possibly pw seax in grave 150

Boughton Aluph,* Kent (0/2 at least) (6th to 7th?)

Boughton-under-Blean,* Kent (1/1) (6th to 7th?)

Bourne Park,* Bishopsbourne, Kent (1/110) (?)

Brabourne Lees,* Kent (?/2) (?)

Bradford Platt,* Wrotham, Kent (0/3) (7th to 8th)

Bradstow School,* Broadstairs, Kent (Y?/135) (7th)

Brandon Road, Thetford, Norfolk (0/1) (5th 6th)

Breach Down,* Barham, Kent (Y?/103) (7th to 8th)

---

[1] *Taken from Richardson, 2000 – often no details of swords available or information sparse.

[2] Includes both Beckford A and B.

Brickfields,* Milton Regis, Kent (2/4) (5th to 6th)
Broomfield,* Reculver, Kent (?/1) (5th to 6th)
Bromeswell, Sutton Hoo, Suffolk (3/19) (6th)
Buckland, Dover Kent (24/404) (5th to 7th)
Butler's Field, Lechlade, Gloucestershire (x2) (0/219)
Buttermarket, Ipswich, Suffolk (0/73) (5th to 8th) very large seax grave 1306 pw seax grave 4344
Caistor, Norfolk (0/39)
Castledyke South, Barton-on-Humber, Lincolnshire (2/)
Chadlington, Oxfordshire (0/16) (7th [?])
Chalkwell,* Sittingbourne, Kent (0/2) (7th to 8th)
Chatham Lines,* Kent () ()
Chartham Down,* Kent (0/80) (6th to 8th?)
Cherry Garden Hill,* Folkestone, Kent (?/?) (?)
Chislet,* Kent (1/1at least) (7th to 8th)
Claylane Wood,* Shorne, Kent (?/?) (?)
Cleatham, North Lincolnshire 1,200 cremations and 62 inhumations. (pommel cap)
Cliffe,* Kent (?/1 at least) (6th)
Coach and Horses,* Strood Hill, Kent (?/2) (?)
Cook's Lane,* Milton Regis, Kent (?) (7th to 8th)
Cop Street,* Richborough, Kent (?/?) (5th to 6th)
Court Lodge Farm* Harrietsham, Kent (0/1) (7th)
Cowslip Wood,* Aerodrome, Kent (0/43) (6th to 7th)
Crayford,* Kent (?/1 at least) (?)
Croydon, Surrey(0/)
Crundale Down, Kent (1/79) (6th to 7th)
Cuxton, Kent (0/36) (6th to 7th)
Darenth,* Kent (0/13) (5th to 7th)
Dartford,* Powder Mills, Kent (Y?/2) (?)
Didcot, Oxfordshire (0/17) (6th to 7th)
Dover Hill,* Folkestone, Kent (1/52+) (6th to 7th?)
Durham Hill,* Dover, Kent (0/4) (?)
East Brabourne,* Kent (0/3) (6th)
Eastry,* Kent (Y?/6+) (5th to 6th)
Eastry House,* Eastry, Kent (0/1) (7th)
Eastry Mill,* Eastry, Kent (0/7) (?)
Eccles,* Aylesford, Kent (0/215 at least) (7th to 9th?)Mostly late without goods
Edix Hill, Cambridgeshire (0/149) (5th to 7th)
Empingham II, Rutland (0/)
Eynsford,* Kent (?/?) (?)
Eythorne,* Kent (?) (?)
Farningham,* Kent (0/10) (7th to 8th)
Faversham Church Yard,* Kent (0/1) (?)
Finglesham, Kent (2/216) (6th to 8th)

Ford Laverstock, Wiltshire (1/2) (6th to 7th)
Freedown,* Ringwould with Kingsdown, Kent (0/3) (6th to 7th)
Gamlingay, Cambridgeshire (0/118) (5th to 7th)
Gas Alley,* Margate, Kent (?/4) (?)
Gillingham,* Kent (0/1) (?)
Gilton, Kent (5/6)
Godmersham,* Kent (0/1) (?)
Great Chesterford, Essex (2/117) (450-600)
Greenbank, Darlington, Durham (2/6) (6th 7th)
Greenwich,* Kent (?/40) (7th to 8th)
Gunthorpe, Peterborough, Cambridgeshire (0/36 approx, 1 cremation)(6th)
Guston,*Kent (1/1+) (7th to 8th)+disc brooch
Half-Mile Ride,* Margate, Kent (0/34) (7th to 8th)
Harford Farm, Caistor St Edmund, Suffolk (0/31) (7th)
Harrietsham,* Kent (/1+) (6th to 7th)
Harwell, Berkshire (0/6) (5th to 6th)
High Meadow,* Dover, Kent (0/1) (8th)
Holborough,* Snodland, Kent (?/44) (7th to 8th)
Hoo Farm,* Minster, Kent (0/8) (?)
Hoo Junction,* Shorne, Kent (0/4) (5th to 6th)
Horton Kirby,* (South Darenth) Kent (0?/84) (5th to 6th)
Howletts,* Littlebourne, Kent (3/42+) (5th to 7th)
Huggin's Fields,* Milton Regis, Kent (Y?/50 (7th to 8th)
Hythe,* Kent (0/1) (6th to 7th)
Kempston, Bedfordshire (2/92?) () -Collecteana
Kemsing,* Kent (0/2) (?)
Kingsdown,* Ringwould with Kingsdown, Kent (1/1+) (?)
King's Field,* Faversham, Kent (?/?) (5th to 7th)
Kingston Down, Kent (2/308) (7th to 8th)
Lackford, Suffolk (0/cremations) (5th to 6th)
Lad's Farm,* Holborough, Kent (?/?) (?)
Laundry Hill,* Minster, Kent (0/14) (7th to 8th)
Leeds,* Kent (1/3) (?)
Lindenthorpe Road,* Broadstairs, Kent (0/2) (?)
Littlebrook,* Dartford, Kent (0/8) (?)
Little Chart,* Kent (0/3) (?)
Lower Halstow,* Kent (1/1) (?)
Lutterworth, Leicestershire (0/1) (6th to 7th)
Lullingstone,* Eynsford, Kent (0/1) (7th)
Lydden,* Kent (?/2) (?)
Lyminge I,* Kent (Y?/2+) (6th)
Lyminge II,* Kent (?/64) (5th to 7th)
Lympne,* Kent (?/?) (7th)

Maidstone,* Kent (1/4+) (7th)
Market Lavington, Wiltshire (0/42) (5th 6th)
Martello Dairy Farm,* Folkestone, Kent (0/1) (7th)
Martyr's Field,* Canterbury, Kent (0/1) (5th)
Mersham,* Kent (0/3) (6th)
Milky Down,* Newington, Kent (0/3) (?)
Millbank,* Hoath, Kent (0/1) (7th to 8th)
Mill Hill, Deal, Kent (4/76) (6th)
Mill Hill II,* Deal, Kent (0/3) (7th)
Mill Service Station,* Ripple, Kent (0/1) (?)
Minnis Bay,* Birchington, Kent (0/3) (?)
Minster Churchyard,* Kent (0/2) (?)
Mitcham, Surrey (13/238) (5th to 7th)
Monkton,* Kent (0/35) (6th to 7th)
Morning Thorpe, Norfolk (1/365+9 crem) (5th to 6th?)
Mount Pleasant,* Minster, Kent (0/18) (7th to 8th)
Mucking, Essex (7/311) (5th 6th)
Murston,* Sittingbourne, Kent (1/1) (?)
Nethercourt Estate West,* Ramsgate, Kent – no data
Nethercourt Farm,* Ramsgate, Kent (1/4) (?)
Northdown,* Margate, Kent (0/1) (?)
Northfleet,* Kent (2+/26) (5th to 7th)
North Leigh, Oxfordshire (0/5) (7th [?])
Norton, Cleveland (0/120) (6th)
Norton/Ospringe/Eastling,* Kent (?/2+) (?)
Olantigh Mount,* Wye, Kent (0/3) (?)
Old Park,* Whitfield, Kent (0/6) (7th to 8th)
Old Tilt Yard,* Greenwich, Kent (0/1+) (7th to 8th)
Orpington,* Kent (1/84) (5th to 7th)
Orsett, Essex (0/4) (7th or 8th)
Otford,* Kent (?/1) (6th to 7th)
Oxborough, West Norfolk (0/10) (6th)
Ozengell,* Ramsgate, Kent (1at least/710) (5th to 8th)
Parrington's Lane,* Snodland, Kent (2 at least/?) (?)
Pilgrims' Way,* Harrietsham, Kent (0/1+) (?)
Polhill,* Dunton Green, Kent (0/200+) (7th ---)
Preston Hall,* Aylesford, Kent (1/4) (6th to 7th?)
Priory Hill,* Dover, Kent (Y?/8) (7th)
Prittlewell, Essex, (6/16 [+11 possible]) (6th to late 7th)
Puddlehill, near Dunstable, Bedfordshire (0/10) (6th)
Quarrington, near Sleaford, Lincolnshire (0/15) (5th to 6th)
Ramsgate III,* Kent (?/4) (?)
Rayleigh, Essex (0/145 crems) (5th to 6th)

Reculver Foreshore,* Kent (/1+) (5th to 6th)
Redsul Avenue,* Deal, Kent (0/1) (?)
Richborough,* Kent (0/at least 2) (7th?)
Ringwould with Kingsdown,* Kent (1/3) (6th to 7th)
Risley,* Horton Kirby, Kent (5/118) (5th to 7th)
Rochester Cathedral,* Kent (?/2) (?)
Rondeau Estate,* Milton Regis, Kent (1+/39) (7th)
Sand Hills,* Deal, Kent (0/1) (?)
Salisbury Road,* St Margaret's -at -Cliffe, Kent (0/2) (?)
Saltwood, Kent (11/217) (5th to 7th)
Sarre,* Kent (27/400) (5th to 8th)
Saxby, Leicestershire (0/?) (6th?)
Sewerby, Yorkshire (0/50) (5th to 7th)
Shelford Farm,* Hackington, Kent (0/1+) (6th to 7th)
Short Bros. Works,* Rochester, Kent (1/1) (?)
Shrubland Hall Quarry, Coddenham Suffolk (0/50) (7th to 8th)
Shudy Camps, Cambridgeshire (0/142) (7th?)
Sibertswold, Kent (3/181) (7th 8th)
Snape, Suffolk (1/47) (430 to 670)
Springfield Lyons, Essex (0/282[inc crems]) (450-700a. d.)horse head burial
Springhead, Southfleet, Kent (1/10?) (6th)
Standlake Down, Oxfordshire (0/6+[58?]) (7th)
Station Approach,* Ramsgate, Kent (0/1) (7th to 8th)
Star Hill,* Rochester, Kent (0/20) (6th to 7th)
St Augustine's Churchyard,* Canterbury, Kent (?/19) (?)
St Lawrence,* Ramsgate, Kent (0/2) (?)
St Margaret's -at -Cliffe,* Kent (0/30) (7th to 9th)
St Martin's Churchyard,* Canterbury, Kent (0/1) (?)
Stodmarsh,* Wickhambreaux, Kent (Y?/2) (6th to 7th)
Stone,*Kent, (0/1+) (7th to 8th)
Stour Street,* Canterbury, Kent (0/4) (5th)
Stowting,* Kent (2/66) (6th to 7th)
Stowting Green,* Kent (0/1) (?)
St Peter's Tip,* Broadstairs, Kent (?/405) (6th to 7th)
Strood,* Kent (?/1) (?)
Supperton Farm,* Wickhambreaux, Kent (1 at least/3) (7th)
Sutton Hoo, Suffolk (3/18) (6th to 7th)
Swan Lane,* Crayford, Kent (?/?) (?)
Tallington, Lincolnshire (0/12) (6th)
Temple Ewell,* Kent (?/?) (?)
Temple Farm,* Strood, Kent (1/3) (6th to 7th)
Temple Hill,* Dartford, Kent (0/2) (?)
Teynham I,* Kent (2+/4) (6th to 7th)

Teynham II,* Kent (0/1) (7th)
The Brent,* Dartford, Kent (0/1) (?)
Thorne Farm,*Minster, Kent (0/3) (6th to 7th)
Thurnham Friars,* Kent (0/2) (7th to 8th)
Tremworth Down,* Crundale, Kent (/23) (5th to 6th)
Upchurch,* Kent (0/2) (7th)
Updown, Eastry, Kent (0/35) (7th)
Verulamium, St Albans, Hertfordshire (King Harry Lane)(0/39) (7th to 8th)
Wanlip, Leicestershire (1/?) (5th to 6th)
Wasperton, Warwickshire (0/115 [24 c and 91 i])[1] (5th to 7th)
Watchfield, Oxfordshire (2/?)
Watersend,* Temple Ewell, Kent (1/20) (?)
Watt's Avenue, Rochester, Kent (1/38) (7th to 8th)
Westbere,* Kent (0/60) (5th to 7th)
West Cliff,* Ramsgate, Kent (Y?/3) (?)
Westgarth Gardens, Suffolk (2/65+4 crems) (5th to 7th)
West Heslerton, Yorkshire (x2)(1/201)(1/185) (late 5th to 7th)
Westwell,* Kent (0/1) (7th to 8th)
Whiteheath,* Hollingbourne, Kent (1/4+) (7th)
Whitfield II,* Kent (?/?) (?)
White Fields,* Wrotham, Kent (0/4) ()
Wickhambreaux Church,* Kent (?) (?)
Wigber Low, Derbyshire (2[2]/5) (7th)
Wigston Magna, Leicestershire (1/?)(6^{5th} to 7th)
Willesborough,* Kent (?/?) (?)
Wingham,* Kent (0/10) (7th)
Woodnesborough (Coombe) Kent (2/1 cremation) (6th)
Worth,* Kent (?) (?)
Worthy Park, Hampshire (1/?)
Wye,* Kent (0/23) (7th to 8th)

Table 9iii. Summary of number of cemeteries surveyed by county:

Bedfordshire 2	Essex 6	Lincolnshire 4	Sussex 2
Berkshire 1	Gloucestershire 2	Norfolk 5	Warwickshire 1
Cambridgeshire 5	Hampshire 2	Oxfordshire 7	Worcester 1
Cleveland 1	Hertfordshire 1	Rutland 1	Wiltshire 3
Derbyshire 1	Kent 165	Suffolk 10	Yorkshire 2
Durham 1	Leicestershire 4	Surrey 2	

[1] 24 Anglo-Saxon cremations from 5th century and 53 inhumations 6th to 7th. Carver, Hills and Scheschkewitz, 2009. Ix.and 49. Earlier, 4th century non- Anglo-Saxon burials take the total to 203.
[2] One found along with other items in 19th century – all now lost.

10. Analysis of selected data from cemetery survey

Paul Mortimer

> The challenge for the future is to remember Härke's definition of the interpretation of mortuary data as a 'hall of mirrors', investigating the reflections that we see, the distortions that have affected these images in the past as well as the process of how archaeologists interpret these visions into stories for people today.
>
> Howard Williams and Duncan Sayer.[1]

Placement of swords (see Appendix 4 – Placement, age and date)

Placement noted in one hundred and eight instances

Left-side = seventy-one (71/169 = 42%) or (71/108 = 66%)

Right-side = ten of 5.9% of all swords (10/169) or 9.3% (10/108)

Behind back = 2 1.2% (2/169) or 1.9% (2/108)

Across body = 5 2.9% (5/169) or 4.6% (5/108)

The vast majority of swords were placed on the left hand side, as they would have been carried by most, right-handed men in life whether on a baldric or a belt. In this data set it is difficult to detect any shift from left to right during the 5th to 7th centuries. This is roughly in line with Brunning's findings for early Anglo-Saxon sword graves and contemporary ones from Sweden.[2] It is not possible to explain why swords were occasionally placed elsewhere, such as across the body or at the back of the deceased. It should be noted that the deceased was unlikely to be actually wearing the sword in the grave, as it appears from a closer inspection of the available evidence that the sword was placed by the mourners in the grave after the body had been deposited.

Comment. It has been suggested that the placement of the sword could indicate whether the owner was right or left-handed. This is, of course, possible, but what then are we to make of placement across the body or behind it? Twelve percent of the 'handed' graves (71 plus 10 = 81 – 10/81 = 12%) have the sword placed on the right side, which could correlate to the proportion of left-handed men within society. It has been very difficult to obtain any accurate data on left-handedness in modern society and, even if found, modern populations might differ to early medieval ones in this respect. Many swordsmen, I suspect, would have been able to use the sword in either hand depending on the circumstances.

---

[1] Sayer and Williams, 2009. 22.

[2] Brunning, 2014. 147.

Pommels (see Appendix 1 – Hilt data)

Swords with significant metal parts to the pommel from the cemetery survey

Alton 1 iron curved top first half 7th

Alton 16 iron curved early 7th

Bifrons 39 cocked hat? – copper-alloy and silver-gilt – swastika one side ring-sword*[1]

Bifrons 62 cocked hat – silver niello ring-sword*

Blacknall Fields 22 – cocked hat copper-alloy – ill cut groove*

Blacknall Field 47 small t-bar iron

Bromeswell (Tranmer House) boat-shaped iron 6th

Buckland 71 boat-shaped iron 625 – 650

Buckland U/A8 trapezoid or ovoid copper-alloy 575 – 625*

Buckland C cocked hat silver-gilt 3 ring and dots one side – 2 scratched lines on other – ring-sword 575 – 625*

Buckland 27 boat-shaped iron 625 – 650

Buckland 33 boat-shaped iron 575 – 625

Buckland 93 boat-shaped iron 575 – 625

Buckland 96a boat-shaped iron 625 – 650

Buckland 96b washer (iron?) 625 – 650

Buckland 98 washer (iron?) 575 – 650

Buckland 131 probably boat-shaped 625 – 650

Buckland 264 square section bar iron 575 – 625

Buckland 375 truncated pyramid iron 580 – 650

Buckland 437 ovoid iron button

Castledyke South II iron cap

Finglesham 204 cocked hat silver-gilt and niello, 3 drilled dots one side and ring and dot and swastika the other ring-sword*

Gilton 56 curved cocked hat silver-gilt? Dots around edge one side*

Gilton 66 circular iron

Mitcham 93 cocked-hat copper-alloy*

Mitcham 163 boat-shaped iron

Mitcham 192 cocked-hat gilded copper-alloy*

Mitcham 211 gilded copper-alloy lower plate (?)*

Morning Thorpe 218 cocked-hat silver?*

Mucking 557 II lozenge iron

Mucking 600 II cocked-hat copper-alloy*

Mucking 618 II lozenge iron

Mucking 682 II thin lozenge iron

Mucking 766 II thin lozenge iron

Mucking 769 II thin lozenge iron

Mucking 833 II lozenge iron

Prittlewell Group 18 179/45 washer

Saltwood C1163 convex ovoid iron early 6th to late 7th

---

[1] An asterisk means that this sword pommel is included in the high status group.

Saltwood C1145 pyramidal copper-alloy early 5th to early 6th*
Saltwood C3779 convex ovoid iron late 5th to 7th
Saltwood C4665 short bar sub-rectangular iron late 5th to early 7th
Saltwood C6653 short bar rectangular iron 6th to early 7th
Saltwood C3944 cocked-hat silver-gilt inset line one side ring-sword mid to late 6th*
Saltwood W1767 convex ovoid early 6th to late 7th
Saltwood C3826 convex ovoid 6th to mid 7th
Saltwood C3885 convex ovoid iron 6th to mid 7th
Saltwood C1048 long convex iron late 6th to mid 7th
Saltwood C1081 cocked-hat silver niello different zoomorphs each side 6th to mid 7th*
Sarre 88 cocked-hat silver niello ring-sword*
Sarre 91 cocked-hat copper-alloy runes on one side*
Sarre 275 pommel missing but ring attachment present
Sutton Hoo Mound 1 cocked-hat gold and garnet cloisonné*
Westgarth Gardens 51 cocked-hat silver punched circles around edges*
Westgarth Gardens 66 stud?
Wigber Low 4 'small humped' iron
Woodnesborough (Coombe) cocked-hat silver-gilt and niello zoomorphs one side ziz-zag and dots the other ring-sword*

Total number of swords from the cemetery survey is one hundred and sixty-nine swords including the second sword at Woodnesborough.

Fifty-six of the above have significant, surviving metal parts to the pommels, with at least a washer to reinforce peening (56/169 = 33%).

Eight of the swords were ring-swords and all of those were of the 'cocked-hat' type, with the possible exception of Sarre 275 where the pommel was missing (8/169 = 4.7%).

Twenty swords (including the ring-swords) have a pommel of copper-alloy, gilt copper-alloy, silver-gilt silver and niello, and one of gold garnet cloisonné, mostly of the 'cocked-hat' type (20/169 = 12%).

Twenty-seven swords have 'washers', 'curved top', 't-bar', 'square section bar', ovoid button, 'cap', 'circular', 'lozenge', 'convex ovoid', 'short bar sub-rectangular', 'long convex', a 'stud' (material not reported), 'small humped' pieces of iron to supplement the, presumably organic pommel (27/169 = 16%).

One sword has a pommel described as a 'truncated pyramid' of iron (1/169 = 0.6%).

Eight swords have 'boat-shaped' pommels, seven of which are reported to be of iron, and the material of the eighth is not recorded (8/169 = 4.7%).

Adding the last three groups together makes thirty-six fairly ordinary swords with only iron fixtures on the pommel (36/169 = 21%).

Summary

Considering pommels only, the status of all one hundred and sixty-nine pommels can be divided into the 'ordinary' pommels without surviving metal parts which amounts to one hundred and thirteen (169 – 56 = 113). If you add to them the pommels with only, seemingly, undecorated iron parts from the fifty-six which amounts to thirty-six, you arrive at a figure of one hundred and forty-nine (36 + 113 = 149 149/169 = 88%). Some nine of those with iron pommels have been crafted into a 'boat-shape'; and, perhaps these pommels could be considered a slightly 'higher status'? That would leave 82% (140) pommels in the lower category.

The 'high status' pommels amount to twenty, (20/169 = 12%) and this includes all eight ring-swords and the Sutton Hoo Mound 1 sword. Only one of these, the last one named, is of solid gold. Another eleven[1] have silver-gilt, silver or silver and niello pommels, the rest have copper-alloy, sometimes gilded.

Pommel shape

Seventeen pommels are described as being of the 'cocked hat' design, all of these are of gold, copper-alloy or silver-gilt which puts 10% (17/169) recorded pommels in this category.

Eight pommels (8/169 = 4.7%) are described as 'boat-shaped' mostly of iron.

Seven are recorded as 'ovoid' (7/169 = 4.1%).

Six are described as a 'lozenge' and are all from Mucking (6/169 = 3.6%).

Grip

Forty-six grips reported (46/169 = 27%) .

Thirty-two are stated to be of horn -- 70% (32/46).

Five are said to be of wood --11% (5/46).[2]

One of antler.

One recorded as being of horn or wood.

One was of Ivory or horn.

One said to be of ivory.

One sword had silver sleeves (collars) on the grip.

Four sword had horn grips packed with wood -- 4/46 = 9%.

Upper guards

Forty-six swords are reported as having upper guards (46/169 = 27%).

Twenty-four were of horn -- 24/46 = 52%.

Two were 'possibly ivory'.

Seven upper-guards were a sandwich of metal and organic materials -- 7/46 = 15%.

Six upper-guards were of wood -- 6/46 = 13%.

One is reported to be of wood or horn.

One stated to be of copper-alloy.

Five report traces of material only, presumably organic -- 5/46 = 11%.

---

[1] Bifrons 39, Bifrons 62, Buckland C, Finglesham 204, Gilton 56, Morning Thorpe 218, Saltwood C3944, Saltwood C1081, Sarre 88, Westgarth Gardens 51 and Woodnesborough.

[2] There is doubt about the use of wood. See Cameron, 2,000. 34. Cameron reports that she was unable to find one hilt that was constructed from wood. I think that she is referring to grips, rather than whole hilts.

Lower guards

Forty-six are reported as having lower guards (46/169 = 27%).

Twenty-four were of horn -- 24/46 = 52%.

Two were 'possibly ivory'.

Seven contained a sandwich of metal and organic materials -- 7/46 = 15%.

Four were of wood.

One of wood or horn.

One of copper-alloy.

One of iron.

Six report traces, presumably organic -- 6/46 = 13%.

Sword-beads

Sixteen swords were accompanied by sword-beads or are associated with beads – two swords had two each, although in the case of one of those swords, the excavators admitted the possibility that the beads were from a different grave and the grave at Saltwood C3885 had seven monochrome beads that do not appear to be connected to the sword.[1] 16/169 = 9.5%.

Sword-beads made from organic materials might well have not been noticed by many excavators in the past. Many could have degraded completely. To some extent, this is also true for glass beads and others made from other materials. An added problem is that it was very easy, particularly in digs from an earlier time, to overlook or dismiss the odd bead or two. Meaney makes this point very well.[2]

The graves that the beads come from seem to come from the entire date range that is the 5th to 7th centuries. Unfortunately dates were not given for all sixteen graves.

Alton grave 2 is dated to the first half of the 7th century, while Alton 42 is, apparently early 7th century. Mill Hill 40 dates from the 6th century; Apple Down 12A is dated 5th to 6th. Mitcham 192 carries a date of 5th to 7th centuries, but Buckland 264 is more precisely dated with a dating of 575 to 625 A.D. The graves from Blacknall Field (22), Howletts (20), Risley (86 and 87) and Westgarth Gardens (51) are undated, but come from the 5th to 7th centuries.

Scabbards (see Appendix 2 -- Scabbard data)

Twenty-five (25/169 = 15%) scabbards have tape, cord or both, somewhere on the carcass. The tape is most frequently found near the top of the scabbard. The tape can be made from linen, other textile, leather or bast; full details are in Appendix 2.

At least eighteen swords have throat bands and these are usually made from copper-alloy or silver, but might also be gilded. The bands are often swaged, ribbed or otherwise decorated. One is described as possessing an iron rim.

Only five scabbards are reported as having metal chapes and these are mostly made from copper-alloy, although one is of iron.

Just two scabbards in our sample might have had a slider to assist the sword suspension. The one definite slider is said to have been made from wood.[3]

---

[1] Riddler, McKinley and Skittrell, 2006. 63 and 64.

[2] Meaney, 1981.

[3] There are two more sliders that were found by detectorists in Appendix 11.

The material used to make scabbards from is wood and various types are reported, ash (nine times), alder (one), lime (one), maple (one) and willow or poplar (three). Sixty-three scabbards are reported as having been made from 'wood', and all of the swords seem to have been buried in their scabbards. Thirty-three scabbards have traces of lining and the most frequent material appears to be fleece from a sheep (nineteen examples), but fur is reported too, (twelve times) and it does appear that the two types of lining can be distinguished as both fleece and fur are noted by the same excavators in the same cemetery. A combination of fleece and flax occurred once as did fleece and textile.

Many scabbards had traces of a cover, usually of 'leather' (thirty-four times reported) but calf-skin is mentioned as well (twice), which, of course, is also leather. Most scabbards on discovery were too far degraded to distinguish any decoration, but four in my sample had signs of tooling or foundation moulding and one other had cord laid lengthwise along the scabbard, which was possibly part of a foundation moulded design.

Seventeen scabbards had some form of edging to strengthen them. This usually occurs in two strips either side of the scabbard where it would join the belt or baldric. The metal ones are made from copper-alloy (four occasions), silver (two times) and iron (three reported) but they can also be made from wood and there are seven of these in our sample. Another set is reported, but we are not told of what material it was made from. It is entirely probable that many more scabbards were supplied with edge strengtheners of organic materials but that these were either not recognised as such or had decomposed so much that they left no sign.

Twenty scabbards had various types of metal attachment or decoration, from buckles to gold embossed foil and bits of horse harness. Other types of adornment, from sword-beads, bosses and pyramids are discussed elsewhere.

Spears, shields and knives (see Appendix 3 -- Spears, shields and knives)

One hundred and thirty sword graves (130/169 = 77%) had other weapons with them in the grave.

Thirty-nine (39/169 = 23%) swords were the only weapon in the grave.[1]

Fifty-eight (58/169 = 34.3%) swords were accompanied by a spear, shield and knife of those six have at least two knives, six, two or more spears and one more than one shield.

Twenty-six (26/169 = 15.3%) have a combination of spear and shield only, six of these have more than one spear and 1 has more than 1 shield.

Fifteen (15/169 = 8.9%) have a spear and a knife, three of those have two or more knives.

Ten sword graves (10/169 = 5.9%) have a shield and a knife, and one has more than one knife.

Seven sword graves (4/169 = 4.1%) have just a spear, but two of those have more than one spear in the grave.

Five sword graves (5/169 = 3%) have just a shield.

Eight sword graves (8/169 = 4.7%) have just a knife, but one of those has two knives.

One grave contained a fauchard and a knife, Buckland Dover grave 437.

Härke wrote that the weapon sets found in graves were not practical weapon sets; if I understand him correctly, he meant that, in most cases, a warrior would not go out equipped with all the weapons in the grave on any given occasion.[2] That could be true, but it is perfectly possible to wear a sword, carry a spear, or two, and a shield and have a knife fixed to the belt and fight. The wealthy warrior would have had several spears, knives, etc., and he would not use them all at once, but he might be buried with extras. I think that Härke's

[1] This includes the two Coombe/Woodnesborough swords from a cremation.

[2] Härke, 1989. 55. Härke, 1997. 119.

main motivation in raising this point was to draw attention to what he thought was a new and important finding; that not everyone equipped in the grave with weapons was able to use them in real life.[1] Härke could be correct about some weapon burials, but it does not seem to hold true for sword burials in Anglo-Saxon England at least.[2]

Age of swordsmen at death (see Appendix 4)

Only seventeen reports of sword graves give any indication of age and they are as follows:

Abingdon 42 -- adolescent
Blacknall Field 1 -- 22 -- 25 years
Blacknall Field 22 -- 45+ years
Blacknall Field 47 -- 40 – 45 years
Blacknall Field 70 -- 23 -- 25 years
Castledyke South 179 -- 14 –16 years
Mucking 600 II -- young adult
Saltwood C3779 -- 20 to 30 years
Saltwood C4665 -- 20 to 30 years
Saltwood C3944 – circa 18 years
Saltwood W1767 -- 20 –30 years
Saltwood C3826 -- 18 –20 years
Saltwood C3885 -- 25 -- 35 years
Saltwood C1081 -- 30 – 40 years
Sarre 275 -- 25 –30 years
West Heslerton G74 -- 25 – 35 years
Wigber Low 4 30 years and 45 – 50 years
This is a very small sample, far too small to draw definitive conclusions from. However this is the only sample that we have and if we put them into age ranges we can tentatively say something about the ages of the swordsmen in this sample. In each case I have taken the highest age suggested to fit into the range. I have assumed that 'adolescent' and 'young adult' are below twenty years old.

Below twenty years – four graves (4/17 = 23.5%), twenty to thirty years – seven graves (7/17 = 41%), thirty to forty years – three graves (3/17 = 17.6%), forty to fifty years – three graves (3/17 = 17.6%).

If this sample can be trusted, which is asking a great deal from such a small sample, it would seem that men with swords, tended to die quite young, with 65% of them fitting into the first two, younger age ranges. How this compares to the age at which men who were not accompanied with swords died, is not possible to say as I did not collect that information.

Additional items found within the graves (se Appendix 7)

In the following I have included some simple statistics on the more common items found within the sword graves in my sample, I have not included every item, but they are listed in Appendix 7. Some of the items found in the graves might be parts of shields or scabbards. It would have been useful to be able to make a comparison between the items found in sword and non-sword graves but I did not set out to collect that

[1] Härke, 1992. 153, 155 and 156. Härke, 1997. 120. Härke, (1992.157) seems to concede the point as far as swords are concerned when he says that "…swords seaxes and axes not found with weakly built men…"

[2] Härke, 1992. 156 and 157. This idea is fully discussed in the chapter entitled, *The Meaning of the Sword.* and Härke's suggestion, at least in relation to sword graves, is dismissed.

information, nevertheless it may have revealed some interesting trends and could be considered for a future study. Swordsmen's graves were not always the most highly furnished in any given cemetery, but it must be borne in mind that there might have been elaborate organic, including textile, items placed in the graves that have subsequently left little or no traces.

One of the most common items to be found within a sword grave is a bucket, and at least nineteen, possibly twenty graves contained at least one (19/169 = 11.2%). These vary in the metal parts used to construct them, from being bound with copper-alloy fittings to iron or just being made of wood. Two buckets, Blacknall Field, grave 22 and Mucking 600 II, were decorated with metal vandykes. Whatever, they seem to have been an important item to include in many of the graves and are not uncommon in non-sword grave either.

Vessels associated with drinking are not infrequently included and within the sample some fifteen drinking vessels are specifically mentioned; two are drinking horns, nine are made from pottery and one of wood. Remains of metal rims form the rest of the collection and two of these were of copper-alloy while the remaining one was made from silver. Added to these should be the seven glass beakers that are included in the sample; four are claw beakers, one was made in the form of a bell and two were cone shaped. The inclusion of drinking vessels in a fighting man's grave perhaps should not surprise us, but at least two of the graves have cauldrons within them but both of these come from Sutton Hoo, Mounds 1 and 17. Fifteen graves contained bowls of various types and these could be made from copper (twice) or copper-alloy (seven), one was of glass, two were Coptic bowls and Mound 1 at Sutton Hoo contained several made of silver and a hanging bowl; Castledyke South, grave 179 also had a hanging bowl.

As for more personal items, twelve graves contained tweezers, four had purses, but only four have combs or traces of combs. One of those graves, Mound 1 at Sutton Hoo, contained three combs.

Three graves contained brooches; the occupant in grave 264 at Buckland Dover was buried with a copper-alloy Roman crossbow brooch, the man in Finglesham, grave 22, was accompanied to his final resting place with parts (at least) of a Roman cast copper alloy brooch, and the swordsman buried in Mill Hill, grave 40, possessed a Roman ovoid copper- alloy brooch decorated with blue glass. Another five graves had coins, at least four of which were Roman. Included within grave 66 at Gilton was a pair of 'goldsmith's' scales together with eighteen copper-alloy weights and a touchstone; this grave also included one of the Roman coins and it was from the reign of Julius Flavius. Grave 265b at Dover Buckland also contained a set of scales.[1]

Three of the graves contained a large part of or a complete horse, and these were Snape grave 47, Mound 17 at Sutton Hoo and the grave recorded at Wanlip.

Buckles were included in sixty-five graves, and some have more than one. Not all of these are belt buckles, some might have been part of other connection systems or even straps for shoes, there are examples of the latter in Mound 1 at Sutton Hoo. In some ways it is surprising that all the graves did not contain a belt buckle as it is, at least to my modern mind, an obvious way of attaching a sword to a person. However, as noted before, it is unlikely that any of the bodies were actually wearing the swords when they were committed to the grave, and many of the swords could have been worn on a baldric, or bandoleer, which might not have required a buckle.

Buckles can be made from a number of different materials including organic ones such as bone or ivory, none of which are recorded in our sample, but have been found elsewhere. In our sample twenty-eight are recorded as being made from copper-alloy and these are supplemented by another three that have gems embedded, mostly garnets. One buckle is listed as of silver, one more as silver-gilt, another example as silver plated and

[1] Parfitt and Anderson, 2012. 405 to 407 and 476. A swordsman's grave in Hegykő, Hungary (grave 34) also contained a set of scales, La Salvia, 1998. 47.

two others as 'white metal'. Two graves have buckles made of gold, Sutton Hoo Mound 1 has several, including the Great Gold Buckle, and Supperton Farm 1 has another single example. There are three graves that contained buckles but the material they are made from is not recorded.

Buckles also come in quite a variety of shapes. At least three in our sample are horse-head shaped and these are likely to be belt buckles, D-shaped buckles are more likely to come from other strapping mechanisms and there are at least seven D-shaped buckles in our sample graves. Other shapes, such as oval, or ovoid could be either from belts or straps depending on their size and level of decoration. The ten shield-on-tongue buckles are almost certainly from belts, especially those that have been plated or contain chip-carved decoration. Grave 73 at Mitcham, included a buckle described as a 'boar's head' buckle and studs of 'bronze' with which the belt was probably decorated.[1]

Seaxes

Unfortunately, when recording the items found within graves, I recorded all knives located within the graves as 'seaxes'; some of them may just have been utility knives and I should have differentiated between fighting knives and smaller examples wherever possible. Therefore caution needs to be exercised when making any interpretations of this aspect of the finds. However, ninety-four swordsmen's graves contained knives and ten of these, all in Kent, had two each and they are, Kingston Down grave 208, Dover Hill 10, Gilton 10 and Gilton 66, Ozengell A, Ringwould with Kingsdown 1, Sarre 39 a and b, Sarre 190 and Sibertswold 140. A further two Kentish graves, Buckland Dover 93 and Gilton 56, had a seax with two subsidiary knives attached.

Females and sword burials

Within the survey there are five burials that have been associated with females in some way and they are Castledyke South, grave CS 39 II, Buckland Dover 360, Buckland Dover 93, Buckland Dover 96b and Buckland Dover 33.

The Castledyke burial contains a workbox and some beads on a slipknot ring[2] and these have been interpreted as female items. It is not impossible that the 'sword', despite being pattern-welded, was a weaving batten, but in form, to me, it does look like a sword in form. The sword is badly degraded. It is also possible that the 'female' items were deposited by a sentimental loved one putting something valued into the grave with their dearly departed. The human remains are in such poor condition that the authors say, "…no osteological data are available for this burial."[3]

Buckland Dover grave 33 contains a sword, a seax and a spear, along with a buckle.[4] The grave layout and goods seem to fit the conventions of a male swordsman's grave but in one of two reports on the skeletal remains, they were reported as female.[5]

Incidentally, Buckland Dover 93 is a grave that has a full weapon set, including a shield, a seax, and a sword-bead.[6]

Grave 96b at Buckland Dover is a double burial, (with 96a) and would both appear to be swordsmen. The skeletal remains of 96b have some female attributes.[1]

---

[1] For more on buckles, see Marzinzik, 2003.
[2] Drinkall and Foreman, 1998. 95 and 207.
[3] Drinkall and Foreman, 1998. 248.
[4] Evison, 1987. 225 to 226, 291 and 341.
[5] Evison, 1998. 123 and 125.
[6] Evison, 1998. 238, 314 and 347.

Evison's report says, of graves 33 and 96b, "Where, therefore, a skeleton is provided with grave goods exclusively attributable to one sex, e.g. brooches and beads for a woman and a sword or spear for a man, it is reasonable here to assume that the grave goods are a true indication of sex, and that the bone evidence may indicate an intermediate stage not sufficiently marked to manifest itself to any perceptible extent in the life of the individual."[2] The final comment, this time referring to 33, 93 and 96b, "Of graves 33, 93 and 96b which are considered to be both male and female in different parts of the final report, the contents make it certain that they are all male."[3]

Despite the very poor condition of the skeletal remains in Buckland Dover grave 360, the occupant was probably female.[4] Most of the items are those associated with a woman, a set of keys and seventy-five beads. The sword is represented by a bronze sword pommel only; there never was a complete sword in the grave. I would suggest that the occupant was not considered a swords woman but that the pommel was put there by a loved one as a gift to the departed, or was buried with the woman because it connected to a son, husband, brother, sweetheart, etc.

---

[1] Evison, 1998. 123, 124, 126, 317 (96a 316), 348. Brunning, 2013 writes in some detail regarding the graves 96a and 96b. (150 and 152).

[2] Evison, 1998. 123. Written by Mr D. Brothwell of the British Museum.

[3] Evison, 1998. 126.

[4] Parfitt and Anderson, 2012. 428, 496 and 554.

11. Adorning the Sword

Matt Bunker

Introduction

In this chapter, we shall consider the hilt components of swords found in England during the 5th to 7th centuries. As a whole, the hilts main function is to cover the tang of the blade and to provide the sword's wielder with a secure grip. It also provides some counterbalance to the weight of the blade, may provide some protection to the hand of the wielder and, as with the scabbard, provides a ground for decoration and individualisation of the sword ensemble. How the form of the various hilt elements served these functions and how they changed over our period of interest is discussed in the relevant section in this chapter.

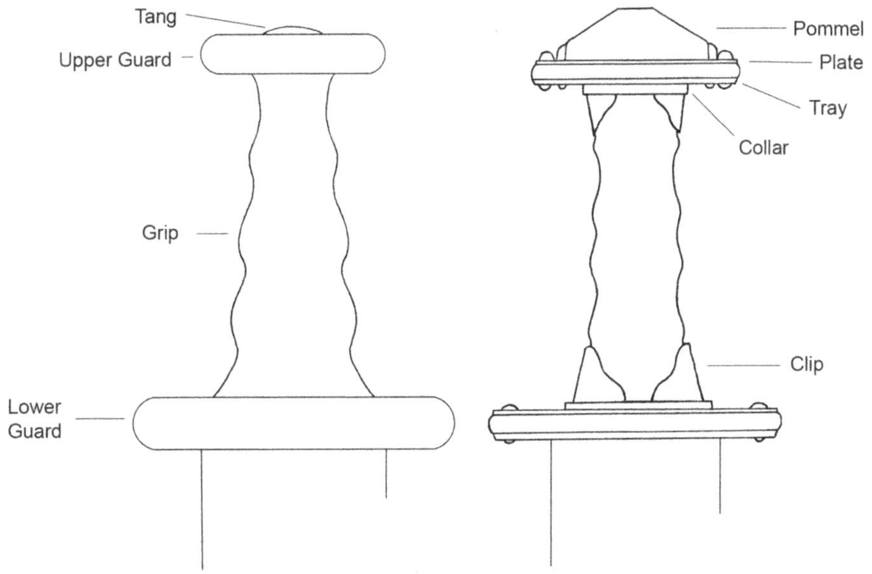

Figure 11.1 illustrates the two basic forms of hilt assemblages for this period and is labelled with the terms used to describe the elements within this chapter.

In general terms, the elements comprising hilts found on swords of this period are thought to change very little over time. As far as can be established, all consist of a lower guard, a grip and an upper guard which, when combined and viewed from the front of the sword, form an approximate 'H' shape (with the upper guard shorter than the lower). Whilst this is generally accepted, it must be stated that there is a degree of assumption here, as the majority of swords recovered from early Anglo-Saxon graves bear little or no indication as to the nature of their hilt elements, let alone the precise form which they took.

Of the 169 swords found in burials which form the basis of the blade study for this book, only 46 are reported as showing any evidence of guards and grip[1] and only a tiny number of those, in combination with some separate elements, give further evidence as to the actual shape of these components. There are only two full sets of organic hilts which survive from English contexts; those from the small boat burial in Grave 47 at Snape in Suffolk and a set in the possession of the British Museum, purchased in the 19th century and said to come from Cumberland. In addition to these, there is an upper guard, pommel, and hilt collars which come from a male burial excavated in 1861 on Crundale Down (also in the British Museum), a very well-preserved pair of guards on a sword from Grave 96a at Dover Buckland and a substantial amount of organic material, particularly the pommel, on the sword from Mound 17 at Sutton Hoo. To these can be added some metallic elements which have survived where the organic components have not and these too can contribute to our understanding of the form which these missing elements took. In addition, there are some contemporary continental examples to draw upon. All of these will be discussed in more detail in this chapter.

The majority of the 169 swords in the study, 73.4% of them, showed no evidence of any hilt fittings other than the three basic elements of guards and grip and, in every case where evidence of fittings survives, they are made from organic material (albeit with additional metal elements in a few cases). Horn, assumed to be bovine, is the most commonly identified material for all hilt elements with wood being the second most common. However, the identification of wooden elements is called into doubt by Esther Cameron, who states that she is "unaware of any sword hilts of the fifth to seventh century in England which have been verified as of wood"[2]. It is easy to mistake the slight remains and legacy marks of centuries old horn for wood, the former having an appearance not unlike the grain of the latter and it is perhaps telling that, in recent years, where enough material survives for a positive identification to be made, it is horn rather than wood which is identified. One exception to this is a recent study of the Cumberland hilt carried out by the British Museum in which it is stated that (an unspecified) part of the assembly is made from wood. However, it also states that the wooden part "might be interpreted as more recent repairs, as the wood itself does not appear in a state expected after years of burial".[3] Another reported exception is that from Grave 10 at Temple Hill, Dartford. In the published report, it is stated that "both guards were undoubtedly of wood" but that due to conservation "species identification is impossible"[4]. An inspection of the finds by the author failed to identify anything which could be identified as the remains of guards, wooden or otherwise. Unfortunately, Cameron reports that the sword from Dover Buckland 96a, which was recorded by Evison as having well preserved wooden guards[5], was "missing" in 1995 when she attempted to study it.[6]

There are also some surviving hilt elements which are made from antler. As well as the grip of the Feltwell sword (see below), there is a very well preserved upper guard made of antler from Caistor St Edmund which, is tentatively dated to the very early 5th century (now in the collection of Norfolk Museums Service) and another (decorated with four horizontal grooves) is recorded as having been found at Lakenheath[7], Suffolk, in association with a sword blade.

[1] See Appendix 10 for details
[2] Cameron (2000)
[3] Blakelock et al (2013)
[4] O'Brien (2015)
[5] Evison (1987)
[6] Cameron (2000)
[7] West (1998)

Predecessors

It is unfortunate and frustrating that there is a near total lack of swords from British contexts dating to the period immediately preceding our period of interest, which makes it difficult to establish which hilt forms were already being fitted to blades in Britain prior to the late 5th century. It would seem reasonable to assume that some if not all would be similar to those used in other parts of the Romanised western Europe at the time but there's no direct evidence to support this. Unfortunately, a sword described as "horn-hilted" which was found at Richborough and which may have dated to the late 4th century was neither preserved nor recorded in detail.[1]

11.2 Hilt of the sword from Grave 21, Hemmingen, Germany in the collection of the Landesmuseum Württemberg, Stuttgart.

There is one example which is thought to date to the early 5th century, the sword from the late Roman villa hypocaust at Feltwell.[2] A study of its features indicates that this is a sword of continental rather than native origin. The grip, which is of deer antler and which has survived very well, is hexagonal in cross section, a form which has its closest continental parallels in some of the early/mid-5th century swords found primarily in Southern Germany. Included in swords of this type are early examples of the "goldgriffspathen", swords whose grips are partially or total sheathed in gold (discussed in more detail below) so it is perhaps ironic that the only example of a hilt form of this type to be found in Britain should be the Feltwell. Its apparent lack of pattern welded blade and poor quality scabbard fittings puts the Feltwell sword at the opposite end of the spectrum in terms of quality when compared to the gold-hilted Frankish swords. Nothing remains of the Feltwell sword's other hilt components (indicating that they were made of an organic material other than the antler used for the grip, possibly wood), although the length of the blade's tang indicates that there was at least one other element. It is impossible to tell whether the tang itself is complete or if some of it has been lost to corrosion. A sword from Grave 21 at Hemmingen[3] (which also has a hexagonal grip, made from bone or ivory) has a low, oval iron pommel which is thought to have sat upon a wooden upper guard, with a wooden lower guard to match (Figure 11.2). Other Frankish swords with similar grips have been shown to have had very short guards, with the lower being not much wider than the mouth of the scabbard (e.g. the sword from Blučina).

It is possible that this "Germanic" hilt style was a feature of other 5th century English swords, especially given the use of Germanic forces as part of both the late and post Roman military structure in Britain. However, this form is not evident in any of the other examples from England which have substantial hilt remains. For the origins of those we have to look further afield, into the Northern Barbaricum and beyond.

[1] Hawkes (1961)

[2] Gurney (1986)

[3] Müller (1976)

11.3 Hilt assembly no. 445 from Nydam, Denmark in the collection of the National Museum, Copenhagen.

Students of the evolution of Northern European weaponry are fortunate to have the great number of weapons deposits from Southern Denmark to study. These bog deposits contain weapons from the 3rd to the 5th centuries, including some sword hilts in remarkable states of preservation. Of particular interest are some of the finds from Kragehul, Nydam and Thorsberg which demonstrate that totally organic hilts, some of which are very similar in form to those from Snape and Cumberland, were in use in Denmark and possibly other North Western European countries from at least the early 4th century. At Nydam in particular, we find hilt forms which not only exhibit the same configuration of organic hilt components as the two complete sets that we have from England but which have features which mark them out as very similar, particularly with regards to the way that the grip was formed (Figure 11.3).[1]

Prior to this (based upon the large numbers of hilt elements recovered from Illerup Ådal), the hilts for two edged swords of the 3rd century in the region were quite different, either featuring rounded guard and pommel assemblies which are reminiscent (and presumably influenced) by Roman hilt forms of the time or which are complex assemblages with multiple metallic elements which are probably local evolutions of earlier Iron Age types.[2] This latter form does have something of the 'H' profile when viewed from the front although, based on the surviving metal elements which embellished the guards, the guard elements were much more rounded in form than the flattened ovals seen on weapons of the 4th and 5th centuries

Of course, we are basing these observations upon the tiny number of surviving hilts from England. It is entirely possible that some of the other 4th and 5th century hilt styles found on the continent (such as the distinctive hour-glass shaped hilts found in Scandinavia) were in use in England at the same time, adorning blades which are now devoid of any clues as to the form they took. In fact, given the discovery in England of at least three cast metal pommels of the Holmegard/Kragehul variety (Menghin's Type 1d, discussed below, a type most commonly found in Denmark in early 5th century contexts[3]) outside of burial contexts, it seems very probable that some of the earliest "English" swords featured metal pommels and other non-organic hilt elements, just as their contemporary continental counterparts did. Whether such swords are still in the ground waiting to be discovered remains to be seen. If there are, and they emerge, they will help us fill gaps in our knowledge.

[1] Bemmenn and Bemmenn (1998)

[2] For examples see Matešić (2015) and Biborski & Ilkjær (2006)

[3] Menghin 1983

The Elements of a Hilt

Guards

The upper guard ensures that the sword hilt cannot slip through the hand of the wielder. It can also provide the surface over which the end of the blade's tang can be peened and, because it can serve this function, the upper guard is sometimes referred to as a "pommel", "pommel-bar" or "pommel-cap". However, for the sake of clarity, in this chapter we shall only use the term "pommel" when referring to an additional element which sits on top of the upper guard. It is usually the peening over of the tang which holds the whole hilt assembly together and secures it to the blade. It is possible that in some instances the tang may have been peened over a metal washer placed on the surface of the upper guard. It is very difficult to discern evidence for this washer, masked as it would be by the corroded remains of the peened tang and there are no traces of such a washer on any of the remaining hilt sets from England, although there is some evidence for their use on Continental remains (see below in the section on the Snape hilt).

The lower guard serves primarily to prevent the wielders hand from sliding onto the blade and would also afford some protection to the hand from an opponent's weapon, although the relatively short width of these guards, and their organic nature means that this aspect should not be over emphasised. Even with those examples where there is a metal element to the lower face of the guard, the metal is too thin and soft to have afforded much protection (a possible exception to this being the sword from Dover Buckland 96a discussed below). Where the remains allow, it can be observed that the bottom face of the lower guard is not only pierced to allow the tang of the blade to pass through it, but that the entrance hole is cut in at an angle so as to allow the shoulders of the blade to sit within the guard. This ensures a secure fit between guard and blade and for the hilt assembly as a whole once assembled.

In some instances, both upper and lower guards have additional metal elements; a thin plate covering the top of the guard and a thicker, walled tray into which the guard sits. This sandwich of elements is held together by a dome-headed rivet at each end (a hilt assembly discovered by a detectorist at Market Rasen in Lincolnshire in 2002 has the rivet heads covered by garnets). There are examples (e.g. the early 6th century Grave 76 from Chessell Down on the Isle of Wight[1]) where only the lower element is present, but whether this represents the original state of the fittings or is due to decay or even loss during excavation is unknown.

11.4 Hilt of the Goldgriffspatha from Pleidelsheim, Germany. Landesmuseum Württemberg, Stuttgart.

[1] Arnold (1982)

Whilst the addition of these plates would no doubt provide some additional strength to the guards, the materials used (often silver or gold) and their placement (the bottom face of the upper guard is not an area likely to need strengthening) make it obvious that these additions are primarily decorative. Whilst no examples have been found in England where the faces of the plates themselves have been decorated (a feature seen on some examples from the late 5th century contexts at Nydam[1]), the addition of these metal elements would enhance the appearance of the horn hilt and make it more distinctive.Until relatively recently, the majority of the finds with these additional guard elements were from Kentish cemeteries, leading some to conclude that perhaps this form was Frankish in origin[2]. Whilst it is true that some of the "Frankish" swords identified as being 5th century in origin have either a single plate on the lower guard (Pouan[3]) or a lower guard of sandwich construction (Blučina[4]), there are also examples of 5th century swords with guards of a metal and organic sandwich construction from the Danish bog deposits[5].

Whilst it is impossible to pinpoint the earliest example of this form of a fully developed set of guards, the sword from Dover Buckland Grave C discussed later in this chapter must rate as one of them, dating as it does (prior to its modification) to the early 6th century. As a form, it remained in use throughout the 6th and well into the 7th century if not beyond and can be found in graves throughout the sphere of Scando-Germanic cultural influence, from Scandinavia to Italy.

Where it is possible to discern the shape of these guards, either from surviving organic remains or the remains of the additional metal elements, both upper and lower guards are ellipsoid, with the upper guard being blunter and rounder at the ends than the lower but material thickness for both seems to be roughly equal. As stated previously, horn is the most commonly identified material. Where it can be discerned, upper guards seem to be shorter than the lower ones but that observation is based upon a small amount of surviving evidence. It is entirely possible that some swords had guards of equal length, like that seen on the very early 6th century sword from Mound 2 at Högom[6], something which is also a feature of some hilts seen on 4th and 5th century Danish swords (most notably those from Nydam[7]).

Grip and Grip Fittings

The grip covers the tang of the blade, providing the wielder with a secure, comfortable grip upon the sword. We have two surviving grips from English contexts (Snape and Cumberland – both discussed in detail below), both of which are made from a single piece of horn, with a hole running all the way through for the tang of the blade. Both are ribbed, the shaping of the grip providing grooves for all the fingers and both are oval in cross-section. In all of this, they are similar to contemporary Scandinavian swords and follow a tradition seen in an earlier example from Nydam.

---

[1] Jørgensen & Vang Peterson (2003)
[2] Evison (1987)
[3] Menghin (1983)
[4] Menghin (1983)
[5] Ørsnes (1988)
[6] Ramqvist (1992)
[7] Bemmenn and Bemmen (1998)

11.5 Pommel, guard plate and upper hilt collar from Market Rasen, Lincolnshire. British Museum.

11.6 (far left) Hilt from grave 23, Nocera Umbra, Italy in the collection of the Museo Nazionale dell'Alto Medioevo, Rome.

11.7 (near left) Hilt from grave 76, Chessell Down, Isle of Wight in the collection of the British Museum, London.

As with guards, we can also learn something about the shape of grips from the metal embellishments that they are sometimes adorned with. Some swords have thin bands or wide collars which encircle one or both ends of the grip and these can indicate not only the dimensions of the grip at these points but also the cross-sectional shape. Whilst some of these are oval, at least one shows that the grip was a flattened hexagon (see the section on Dover Buckland C, below). This is interesting as hexagonal grips are a feature of many contemporary Frankish swords, as well as matching that found on the earlier Feltwell sword. The form of these Frankish grips is well understood, thanks to the excellent condition of the organic example from Hemmingen[1] and also because the shape is preserved in more than 20 examples of what are termed "Goldgriffspatha". These are Frankish swords, dated to the late 5th and 6th centuries, with grips that were partially covered in thin gold foil, found in graves spread between Belgium and the Czech Republic, with the highest concentration (nearly 50%) being from the Baden-Württemberg region of south-west Germany[2]. Whilst in most cases, the organic element beneath has decayed, the gold foil (which usually only covers one side of the grip) retains its form and shows that, as with the example from Hemmingen, these grips were hexagonal, straight sided, tapering from bottom to top and, whilst they were divided into four fields by either incised lines or a slight raised rib running around the hilt, there was no attempt to create grooves for the fingers as with the Scandinavian examples. Figure 11.4 shows an example from Pleidelsheim.

Whilst impossible to prove, the hexagonal shape of some of the collars and rings found in English graves and as stray finds may be an indication that hilts of this type might have been used on some of the swords. The collars which adorned some grips can be quite plain, as seen with the examples from Grave 16 at Alton, which are quite large but which are of copper alloy and are undecorated[3], whereas others are fine examples of the art of the metalworkers of the period. The collars from the previously mentioned set of hilt fittings from Market Rasen are particularly noteworthy, each being made from gold sheet covered with gold filigree zoomorphic interlace decoration (Figure 11.5)[4].

Another embellishment sometimes found on swords are mounts which sit on the side of the grip at both top and bottom, referred to as 'clips'. There are two English examples of hilts with these clips found in situ, one being from Mound 1 at Sutton Hoo, which only has a pair on one side of the grip and the other being the Cumberland hilt which has one pair at the bottom of the grip only. However, there are continental examples (e.g. Grave 32 at Nocera Umbra[5]) which have four clips, two at the top and two at the bottom (Figure 11.6) and the Staffordshire Hoard contains many silver fragments and one complete pair of silver clips[6].

The final element that we know adorned the grips of some swords of the period can be classed as "applied decoration" and they came in a variety of forms. The Cumberland hilt (discussed in detail below), has 10 separate elements on its grip, six of which are in the form of rectangular panels of gold with gold filigree decoration arranged around the grip. Whilst rare, these applied decorations do appear on other grips. The sword from Grave 76 at Chessell Down has one (Figure 11.7), again of gold, rhomboidal in shape and decorated with filigree circles which (if the location in which it is currently displayed is correct) was mounted about half-way up the grip. A very similar piece forms part of the hilt decoration of a sword from Saint-Dizier in France, although in this case the rhomboidal plate was fixed to the front of the lower guard.[7]. The

---

[1] Müller (1976)

[2] Ament (1998)

[3] Evison (1988)

[4] See the 2002 Treasure Annual Report published by the Department for Culture, Media and Sport

[5] Rupp (2005)

[6] Fern (2015b)

[7] Truc (2012)

Staffordshire Hoard contains a number of items, including small zoomorphic gold and garnet cloisonné pieces and other pieces decorated with filigree scrollwork, which are thought to be embellishments of this type[1].

Study of the embellishments found on the grips of swords makes it obvious that they were decorative rather than functional and leads to questions as to their meaning and how they were acquired. This will be explored as part of the discussion at the end of this chapter.

Pommels

More than any other element of the hilt assemblies of swords in the early medieval period, pommels are the items which demonstrate the most individuality in terms of design and decoration, with no two of the hundreds of decorated examples from across Western Europe being identical.

From a practical perspective, the pommel serves one of two functions. Some are fully pierced, allowing the tang of the blade to pass through and to be peened over where it protrudes from the pommel, thus securing the blade within the hilt assembly. Others are hollow but not pierced and are riveted to the upper guard, covering the peened top of the tang.

11.8 Pommel Type 1c from grave 1, Selmeston, Sussex. Barbican House Museum, Lewes.

At this stage in the development of sword hilts, the pommel does not really have enough mass to provide any sort of counterbalance to the weight of the blade of the sword, as is the case with later medieval swords. Of course, being at the apex not only of the hilt but of the whole sword assembly, the pommel also served as an obvious ground for decoration. Whilst there are some examples that are of undecorated bronze, there are many which are made from silver (sometimes gilded) or gold which bear a wide array of decorative motifs, executed in all of the decorative techniques available to the craftsmen of the time including casting, filigree and cell cloisonné work as well as instances of carving applied post-manufacture (including the carving of runes, as seen on an example from Gilton[2]). Even those which are plain copper alloy may not all always have been so. There are examples of pommels decorated with gold filigree on gold backing plates which are supported by a copper alloy core (a particularly fine example, PAS reference BM-7C4457, was discovered at Middleham in North Yorkshire in 2014). A mark of the consideration that was given to the embellishment of pommels is the fact that the decoration is rarely the same on both faces.

The broadest study of pommel types to date is the one undertaken by Wilfred Menghin, published in his "Das Schwert im Frühen Mittelalter" in 1983. He identified eight types of pommel for his typology but within that

[1] Fern (2015b)

[2] Evison (1967)

there are three main physical forms, each of which contains sub-types.[1] Whilst it would be impossible to discuss all of the examples of pommels from English contexts, especially given the large number that have been discovered by detectorists and recorded with the Portable Antiquities Scheme (PAS) in recent years, we will detail some notable examples from each those of Menghin's categories which have been found in English contexts. As of the end of 2016, the PAS database records 53 pommels (not including those from the Staffordshire Hoard) which can be dated to our period of interest.

Type 1

Menghin's Type 1 are the "boat-shaped" pommels; long and low in form, with a curved back when viewed in profile, usually made from copper alloy and, in the case of the two sub-types found in England, pierced to allow the tang of the blade to be peened. Decoration is usually relatively simple although some examples have small animal heads at the ends as part of the casting.

Type 1 includes: -

Type 1c (Fralsegarden-Liebenau type), of which there are two recorded from burial contexts in England; Grave 30 at Fairford and Grave 1 at Selmeston (Figure11.8)

Type 1d (Holmegard-Kragehul type). Whilst Menghin gives no examples of this type in England, there are now five examples recorded in the PAS database. Three of these were found in East Anglia, one in Essex and one in Warwickshire. (Figure 11.9).

11.9 Pommel Type 1d from Compton Verney, Warwickshire (PAS WMID-916283)

These boat-shaped pommels almost certainly originate in Denmark. Examples of them are found in most of the Danish sacrificial bog deposits, many of them in association with the distinctive, metal sheathed hour-glass shaped hilts such as those found at Ejsbøl, dating to the early 5th century. Judging by the scabbard fittings associated with both the Fairford and Selmeston swords, it seems that these pommels were still being fitted to swords that were in circulation in England during the early 6th century or even later.

Whether they were re-used or were still fitted to swords which originated in Jutland in the 5th century is something that can never be known but both scenarios raise interesting possibilities which will be discussed later. In addition to all of these distinct copper alloy boat-shaped pommels, there are a few examples of low, iron pommels which are described as being "boat-shaped", with four being recorded at Buckland[2].

Type 2

Menghin's Type 2 are the "cocked-hat" (or, in his typology, "Pyramiden") pommels, by far the most common form of pommel found both in England and across the Scando-Germanic sphere of influence in the late 5th to 8th centuries. He divides these into seven sub-types, although the seventh, which Menghin calls "Sonderformen" (Special Forms) should really fall into a separate type. With the exception of the Sonderformen, all of the sub-types have a broadly similar shape, being roughly triangular with either a rounded or flattened apex and with sides which range from straight to strongly concave and were riveted to the upper guard, covering the tang. Materially, copper alloy, silver and gold are all used and a wide variety of decorative techniques were employed.

---

[1] Menghin (1987)

[2] Evison (1987)

The main characteristics of each sub-type are as follows: -

Type 2a - The holes for the rivets are at the ends of the main body of the pommel, rather than being in tabs or collars. Rivet configuration can be one at each end, pairs at each end or a single rivet at one end and a pair at the other. Materials are either bronze or silver.

11.10 Pommel Type 2a from Rockbourne, Hampshire (PAS WILT-5C2531)

Menghin reports one English example from Chessell Down, although it is atypical as Evison reports that it has been pierced to allow for a tang[1]. However, it is unclear whether the hole is deliberate or whether it is actually the result of damage to the top of the pommel. The jagged nature of the top of the pommel around the hole suggests the latter. Another type 2a, found at Rockbourne in Hampshire, is recorded within the PAS database (WILT-5C2531, see Figure 11.10).

Type 2b – The holes for the rivets are in flat lugs at the ends of the pommel. Rivet configuration can be one at each end as with Brighthampton Grave 31 (Figure 11.11), pairs at each end or a single rivet at one end and a pair at the other. Materials are bronze or silver.

11.11 Pommel Type 2b from grave 31, Brighthampton, Oxfordshire.. Ashmolean Museum, Oxford.

Menghin records eight pommels of this type in England with a further fifteen examples being detailed within the PAS database as of the end of 2016.

Menghin also includes the pommel from Grave 21 at Petersfinger in this type but it is really a hybrid type, having one rivet set in a lug on one side and the other coming out of the bottom of the other end of the pommel (Figure 11.12). Atkinson states that this pommel also has a pin running transversely through both the pommel and the tang inside[2] but in this he seems to have mistaken the circular design on the side of the pommel for the head of a pin. Both the sword from Petersfinger Grave 21 and Brighthampton Grave 31 are usually dated to the second half of the 5th century and described as being of Frankish origin[3], although there are more pommels of this type recorded in England than anywhere else. When the fifteen examples recorded in the PAS database are added to Menghin's eight, Type 2b becomes the commonest form eighty-one pommels reviewed for this study.

[1] Evison (1967)

[2] Atkinson (1953)

[3] Evison (1967)

Type 2c – The holes for rivets are in sleeves at the ends of the pommel, either one, two or three per side, with some instances of asymmetry. The two faces of the pommel are near perpendicular to the base. Materials are usually silver, silver-gilt or gold (over a copper alloy core) and are usually highly decorated, often with cell cloisonné work, filigree and sometimes combinations of the two on opposite faces. Although Menghin records no examples in England, the PAS database now contains a fine (if incomplete) gilded silver example found at Mautby in Norfolk (NMS-C91286 – Figure11.13).

11.12 Pommel Type 2b from grave 21, Petersfinger, Wiltshire in the collection of the Salisbury Museum, Salisbury.

Type 2d - The holes for rivets are in sleeves at the ends of the pommel, either one, two or three per side, with some instances of asymmetry. An example from Gloucestershire (PAS reference GLO-67083F) has three sleeves on one side with a dummy rivet set between two functioning rivets, but only two sleeves on the other. The sides of the pommel are often strongly concave and the faces angled inwards. The ends of the pommel are cut off rather than tapering towards the rivet sleeves. Materials are bronze and silver. This form of pommel is the commonest type recorded in England by Menghin, with the fourteen examples accounting for 50% of his English pommels. When we add in the six examples contained in the PAS database, this pommel type makes up 24.7% of the eighty-one pommels, making it the second most common. It is also worth noting that 75% of the type 2d pommels in Menghin's sample were found in England, which might be taken as evidence that the form originated here. (Figure 11.14)

11.13 Pommel Type 2c from Mauntby, Norfolk (PAS NMS-C91286)

Type 2e – The holes for the rivets are in tall sleeves at the ends of the pommel. Strongly concave sides and faces that are near or actually perpendicular to the base. Materials used are bronze, silver and gold. The use of high quality and complex decoration are the distinguishing features of this type with gold filigree and cloisonné being usual on English examples. Some continental examples feature the "Tauschierung" inlay technique and others have complex cast decoration (a good example of which is the pommel from Crundale Down, which is discussed in more detail below). The pommel from the Mound 1 Sutton Hoo sword is the most well-known English example of this type. (Figure 11.15)

In Menghin, Type 2e is most commonly recorded in the Uppland region of Sweden amongst the Vendel and Valsgärde burials and on Gotland, with a cluster in South Western Finland but with outliers in Northern Italy. Menghin only records two examples of this type in England but more have since been discovered (e.g. PAS reference WILT-B5EE27). Many of the pommels in the Staffordshire Hoard are of this type.

Type 2f. The distinguishing features of this type are that they have animal head terminals and that they have no obvious rivets. Materials are bronze, silver, silver gilt and gold cloisonné. This is an unusual form, with only five recorded by Menghin, none of which are from England, although he does overlook an example from Sarre in Kent, which does seem to match the criteria.

Type 2g. These are Menghin's "Sonderformen"; a catch-all group for those pommels which although still being vaguely "pyramiden", don't fall into any of his other categories. The only English example that Menghin records is the one from Grave 76 at Chessell Down which, whilst pyramidal in overall form, has a cruciform shape when viewed from above and is pierced for the tang. There are two other instances of this form in the PAS database, both from East Anglia with very close continental parallels in Denmark, including one from Kragehul (22453a[1]) from a late 5th century context. As with the Type 1 boat-shaped pommels, this raises the possibility that the pommels of this type found in England are Danish in origin.

11.14 Pommel Type 2d from Great Thurlow, Suffolk (PAS ESS-4BC242)

The other pommels which we could put into this category are those where there is no obvious method for attaching the pommel to the rest of the sword (see PAS reference HAMP-FA0ADC from Cheriton in Hampshire for an example, Figure 11.16). It has been suggested that the end of the tang might have been embedded in a lump of lead or solder which also filled the pommel [2] but it is difficult to imagine how this could be achieved in the final stages of assembling the sword hilt or indeed how this would be sufficient to secure the pommel, given that the inside surface offers no overhangs to prevent the lead from coming loose (something that Atkinson acknowledges). An example from Cherwell in Oxfordshire (PAS reference BERK-0B2EA7) may give some clue, featuring as it does a lead core into which the tang is embedded, with a sheet of copper alloy bent around the lead to form a rough pommel shape. As the tang passes through the sheet before entering the lead, the whole piece has some integrity but it still unclear as to how a pommel would sit securely on top. Whilst some of these pommels are pyramidal/cocked hat in shape, others such as PAS reference YORYM-4B4B66 from Wilberfoss in Yorkshire have curved backs and are ovoid in plan (Figure 11.17)

[1] Birch Iverson (2010)

[2] Atkinson (1953)

Type 3

Menghin's Type 3 are his "Trapezoide" pommels. These are hollow cast, often with slightly concave sides and a flattened apex. They are usually plain and have no rivets for fixing to the upper guard. Instead, they are pierced to allow the tang to be peened over the top of the pommel. Materially, they are mostly copper alloy. These trapezoid pommels are widely spread across North Western Europe but the main concentrations are in Germany, particularly in Baden-Württemberg. Menghin only records one English example of this type (from Alfriston) but there are nine examples in the PAS database, widely spread, from Berkshire, Hampshire, Herefordshire, Lincolnshire, Norfolk, Suffolk and Kent (Figure 11.18)

Whilst most pommels found in England continue to fall into Menghin's categories, at least one new form has been identified amongst the 86 pommels from the Staffordshire Hoard[1]. Whilst it shares the low, rounded profile of the Type 1 boat-pommels, StH 294 has flat sides and two holes for rivets at each end, with interlace ornament on both faces, one with a zoomorphic garnet decoration. It is, to date, unique and is just one example of how the discovery of the Hoard has made necessary a re-evaluation of the typology, distribution, manufacture and chronology of pommels and of hilt fittings in general in an early medieval context (Figure 11.19).

11.15 (top) Pommel Type 2e from Mound 1, Sutton Hoo, Suffolk British Museum, London (inventory number 1939.1010).

11.16 (left) Pommel Type 2g from Cheriton, Hampshire (PAS HAMP-FA0ADC)

[1] Fischer and Soulat (2011)

11.17 (2 x top) Pommel Type 2g from Wilberfoss, Yorkshire (PAS YORYM-4B4B66)

11.18 (3 x left) Pommel Type 3 from North Rauceby, Lincolnshire (PAS WMID-F4BF02)

11.19 (bottom) Pommel StH 294 from the Staffordshire Hoard

Sword-rings.

The final sword hilt elements to discuss are the rings which are sometimes fitted to them. They take the form of two interlocking rings, usually placed to one side of the pommel with a rivet descending from the vertical ring that passes through the upper guard. When Evison wrote extensively on the subject of sword-rings in "The Dover Ring-Sword and Other Sword-Rings and Beads"[1], she identified eighteen examples in England and another seventy or so in Scandinavia and across North Western Europe. Since then, more have been discovered in England with at least seven partial or complete sword-rings in the PAS database and three in the Staffordshire Hoard[2].

11.20 Sword-ring Type 1a from Brighstone Parish, Isle of Wight (PAS IOW-FB5A95)

Evison created a typography for these sword-rings, identifying four types, the first three of which have been found in English contexts.

Type 1

The most obvious feature of the type 1 sword-rings is that the ring itself is free moving. Evison divided this category into two sub-types.

Type 1a – A complete ring mounted on a rivet, with a moveable, interlocked ring.

At the time that Evison was writing, only one example (from Faversham) was known from an English burial context but a second complete example has since been discovered during excavations of the cemetery at Temple Hill, Dartford (see below). In addition, a stray find of silver with some gilding, was discovered on the Isle of Wight in 2005 (PAS reference IOW-FB5A95 – see Figure 11.20). Although illustrations of the example from Faversham show the mounting ring directly adjacent to the pommel, Evison points out that this is an unlikely reconstruction. The lack of wear which would result from the free-moving ring rubbing against the pommel (as seen on pommels associated with rings of Type 1b) can be taken to indicate that they were not adjacent. Instead, Evison suggests that the sword-ring was mounted elsewhere on the hilt, perhaps in the same position as that seen on the hilt of the sword from Mound V at Snartemo, southern Norway. This Norwegian mound burial is that of a high-ranking warrior and is dated to the second quarter of the 6th century. The sword from the mound has a beautiful hilt, its grip sheathed with decorated gold foils and with finely worked guards and pommel. It also has a Type 1a sword-ring which is fitted to the bottom of the lower guard. If Evison's assessment of the Faversham sword-ring is correct then it is possible that some, if not all, of Type 1a were fitted to the lower guard. This would allow for the sword-ring to be mounted without having to modify the pommel (an issue with Type 1b sword-rings). If all Type 1a rings were fitted to the lower guard and if it was (or became) important that the sword-ring be in a prominent position then this may be the reason why Type 1a sword-rings are the least commonly found type, as they would not be as visible as one

---

[1] Evison (1967)

[2] Cool (2015)

mounted on the pommel when the sword was in its scabbard. This can perhaps be taken as evidence that it was important for the ring to be seen, something that we will come back to later in the chapter.

A recently published find which adds to the corpus of this type of sword-ring was discovered during excavations of an early Anglo-Saxon cemetery at Temple Hill, Dartford, in 2002[1]. Grave 10 contained a sword with what are described as the remains of wooden upper and lower guards (although, as previously mentioned, this description is problematic and seems to be impossible to verify) and with a copper-alloy Type 1a ring and rivet (Figure 11.21). The ring itself is very simply made, being a round section rod bent into a circle with the ends overlapping whilst the rivet takes the form of a split pin, the free ends of which were simply bent over to secure it into the guard (it could be argued that the use of this 'split-pin' rather than a solid rivet warrants the creation of a new sub-category in Evison's typology). The report states that this ring was fitted to the upper guard but, in the absence of a diagram or other reference material showing the finds in situ, it is unclear whether this is merely an assumption. A study by the author of unpublished material held in the excavation archive held at Maidstone Museum, which includes x-rays of the sword before removal of the soil block, has failed to clarify the situation. Regardless, this is a remarkable example of a seemingly crudely made ring fitting mounted on an early 6th century sword which has no other apparent embellishments and was fitted with a very simple, low iron pommel. Whilst it is a unique find, its survival demonstrates that the plainest of swords could be fitted with a sword-ring which (other than the ring itself) would leave no trace[2]

11.21 Sword-ring of Type 1a (although potentially a new form) from Grave 10, Temple Hill, Dartford (TEMP 2015 209)

Type 1b - The ring on the rivet is incomplete, forming what Evison refers to as a 'bow', the open end butting up against either the side of the pommel or against the base. The interlocked ring is still circular and separate (although in some instances, Dover Buckland Grave C for example, the gap between the two allows only for rotation, not full movement[3]).

In some instances, it is possible to determine that sword-rings of this type were added to the pommel and upper guard at some point after the hilt had been assembled. In most instances the evidence for this comes from the fact that the pommel has had to be modified in some way to accommodate the sword-ring, usually by modifying or removing part of the rivet sleeves at the ends of the pommel to allow the rivet ring to sit closely adjacent to the pommel. Other pommels demonstrate that their asymmetry was deliberate, designed to accommodate a sword-ring (e.g. Grave 39 at Bifrons – see Figure 11.22). More evidence for sword-rings being later additions comes from the analysis and comparison of the niello inlay on the sword from Grave C at Dover Buckland, as discussed below.

---

[1] O'Brien (2015)

[2] I am indebted to Pernille Richards, Collections Officer at Maidstone Museum & Bentil Art Gallery, for providing access to the excavation archive relating to the Temple Hill cemetery.

[3] Evison (1987)

11.22 Sword-ring Type 1b from Grave 39, Bifrons, Kent. Maidstone Museum.

11.23 Sword-ring Type 2 from Grave 88, Sarre, Kent. Maidstone Museum.

Type 2

The two rings are still made separately but they are now both so thick as to prevent any free movement. In some cases, the two rings are also soldered together (as observed by Evison in the example from Grave 88 at Sarre[1] – see Figure 11.23).

A rivet shank allows the sword-ring to be fixed to the upper guard, against the pommel. A find from the Staffordshire Hoard, STAFFS-EA9345, seems to be the mounted part of a Type 2. It has a groove into which the other ring would have sat and clearly demonstrates that it would not have been free moving. Again, some examples of these were fitted to swords after their hilts were assembled and at least one (from Grave 88 at Sarre may have been fitted to more than one sword (based on an observation by Evison[2]).

Type 3

The whole sword-ring is cast as one piece, with an integral rivet to allow it to be fixed to the upper guard, closely adjacent to the pommel. There is a good example of this type in the Staffordshire Hoard, PAS reference STAFFS-6C4F55, which has some decoration on the face of the upper 'ring' in the form of a bird's head (Figure 11.24). Some continental examples of this form are very large, with the ring assembly being at least as big as the pommel itself. In the example from Vallstenarum, Gotland, part of the ornate gold and garnet work on the pommel had to be removed in order to accommodate the ring.

[1] Evison (1967)

[2] Evison (1967)

Type 4

The whole sword-ring is cast as one piece with the pommel (Figure 11.25). These are often much larger than other types, with some being larger than the pommel itself. There are no examples of this type known from English contexts.

11.24 Sword-ring Type 3 from the Staffordshire Hoard (PAS STAFFS-6C4F55 – Image courtesy of the Staffordshire Hoard website

11.25 Sword-ring Type 4 from Grave 1, Nocera Umbra, Italy in the collection of the Museo Nazionale dell'Alto Medioevo, Rome.

11.26 Detail from an electrotype replica of the Gutenstein scabbard in the collection of the Landesmuseum Württemberg, Stuttgart (the original being in the Pushkin Museum, Moscow).

11.27 Sword-bearing warriors on a pressblech patrix from Björnhovda, Sweden

11.28 Possible evidence of a sword-ring mounted on a lower guard, Staffordshire Hoard K374

Whilst it seems that the two forms of Type 1 are the earliest in the development of the sword-rings, with the progression through types 2 to 4 being obvious, it is unclear whether Type 1a or 1b came first. Whilst a strong case can be made for the sword from Grave 21 at Petersfinger being of mid to late 5th century in date (based in part on the mouthband of its scabbard, which is of a type found on Frankish scabbards of that period), that does not prove that the sword-ring with which it was fitted (and which was subsequently removed prior to deposition) was added at that time rather than early in the 6th century. Given that the Mound V Snartemo burial is also dated to the early 6th century, as is the sword from Faversham and that from Grave 10 at Temple Hill, all of which have a Type 1a sword-ring, there is every chance that the two forms are contemporaneous. This may be the reason for them being found in differing locations on the hilt; the practice of adding rings to hilts having begun in the late 5th/early 6th century but the preferred location not having been decided upon until later in the 6th centuries.

As previously stated, the practice of mounting rings on the lower guard seems to fade away and mounting on the pommel becomes the norm. A possible explanation for this might be that, whatever the function or symbolism associated with these sword-rings, it was important that they be seen, rather than being tucked away under the guard. Evidence for this comes from imagery of the 6th century, where we see figures with swords fitted with sword-rings. One of the bronze patrices from Torslunda in Sweden, used for making pressblech foils to adorn helmets and other items, shows two warriors wearing boar-crested helmets. The warrior in front holds up a scabbarded sword with its sword-ring clearly showing and pointing forwards. A helmet decorated with pressblech foils from boat burial XIV at Vendel also shows warriors, some of whom bear swords and these are also hold aloft with their sword-rings to the fore. A third example, that of the pressblech panel decorating a scabbard from Gutenstein in Baden-Württemberg, shows what may be an Ulfhednar, a wolf-coat warrior, again holding a scabbarded sword aloft with the sword-ring pointing forward (Figure11.26). In every pictorial instance where we see swords clearly fitted with rings, they are being held up by warriors and clearly displayed.

It is, however, worth noting that, whilst the sword borne by the second warrior depicted on the Torslunda patrix clearly has no sword-ring upon its pommel, there is an object next to the lower guard, above the thumb of the warrior and pointing towards him, which might possibly be a ring mounted in the same way as that from Snartemo (Figure 11.27). Whilst even a close study makes it impossible to determine if the object is really a sword-ring, if it is then perhaps the fact that it is pointing towards the bearer rather than away from him is significant, the meaning or symbolism being subtly different.

One piece of evidence which perhaps challenges the accepted idea that the mounting of sword-rings on the upper guard becomes the norm comes from the Staffordshire Hoard. One of the gold and garnet trays from a lower guard assembly (item K374) has an interesting feature. As well as having the large hole that the rivet which held the guard assembly together passed through, it also has two small holes close to the end of the tray (Figure 11.28). Whilst there is no evidence as to what went into them, if a pair of holes were found in the same position on an upper guard fitting, the assumption would be that they were to accommodate a sword-ring, probably a Type 1b. It is difficult to imagine what else might warrant holes of this type in that position and so we are left with the possibility that sword-rings were still being mounted on the lower guards of some swords in the late 6th or early 7th century, something which might be depicted on the patrix from Torslunda. Whether the choice between mounting a sword-ring on an upper or lower guard was a purely aesthetic one or whether there was some significance to the positioning will probably remain unknown. Given, however, that sword-rings mounted on the lower guard are less obvious than those mounted near the pommel, the continuation of this practice into the 7th century (when hilts were becoming increasingly ornate) could perhaps be an indication that the purpose was a symbolic one.

What then, might be the purpose of these sword-rings? One set of theories which can be fairly readily dismissed are that they had an overtly practical function. The Icelandic sagas mention that swords were sometimes tied to the wrist by a thong (as described in chapter 58 of Egil's Saga) and also that swords were sometimes tied into their scabbards using friðbönd, peace-ties to prevent them from being drawn in anger (Króka-Refs Saga). Behmer discussed both of these possibilities in the context of sword-rings, as well as whether they may have been used to suspend something like a healing stone of the type mentioned in Laxdæla

Saga. However, practical aspects aside (a charm or amulet swinging from a sword when in combat would be distracting at the very least) not only does the fact that it is impossible to pass anything through the rings of a Type 2, 3 or 4 sword-ring make it unlikely that this was the original purpose but also we have the imagery mentioned above, none of which shows anything tied to or otherwise associated with them.

Both Evison[1] and Davidson[2] discuss the idea that the joining of two rings symbolises a bond between two parties. Davidson concludes that swords with sword-rings were owned by kings or leaders and that oaths of loyalty were sworn upon them. However, as Evison points out, the only sword from England that we can be reasonably confident is that of a king is that from Mound 1 at Sutton Hoo, which has no sword-ring (although there was a Type 2 ring in the burial, associated with the shield). She suggests rather that the sword-ring was given by a lord or king as either a reward for service or as a badge of office or rank, as well as being symbolic of the bond between them. This interpretation is widely accepted, but it is still problematic. Whilst it would certainly account for the fact that some (if not all) rings of Types 1-3 were added to swords that were already in use, it does not necessarily explain why many were removed prior to deposition. If it were the case that the sword-ring had to be removed and returned to the giver upon the death of the bearer, then we wouldn't expect to find any sword-rings in burials at all. It is possible that the removal of a ring indicates the end of whatever service or bond existed, or that the bearer had fallen out of favour and lost the rank or office which it symbolised. In addition, what are we to make of the sword-ring from Temple Hill, a rather crudely formed circle of copper-alloy? Hardly something you would expect to be a gift from a lord to one of his favoured men.

Mortimer suggests another possibility, that whilst they symbolise an oath, it was an oath between a warrior and his god to perform a specific feat or undertaking, to be removed upon its completion.[3] This would allow for examples like that from Temple Hill, something hastily made and with its value more in what it symbolised rather than in the materials used.

However, that still leavesu with the issue of the Type 4 sword-rings, which (regardless of the nature of the oath) couldn't be removed. It may be that Evison or Mortimer are correct as to the original purpose of the sword-rings but that by the time that Type 4 evolves in the 7th century, part or all of that meaning had been lost and that by this stage they were simply badges of rank (albeit ones which might still have been given by kings).

All of these theories now need to be re-evaluated in light of a new form of sword-ring which has emerged from the Staffordshire Hoard. For the first time, we have pommels which have two sword-rings fitted to them, making the pommel and sword-ring combination symmetrical. Three examples of this type have been identified in the Hoard (K291, K136 and K163), all seemingly of Type 3 but fitted to Type 2 pommels which were designed to accept them rather than being later additions. All three pairs are highly decorated in different styles, with silver, gold, gold filigree and cloisonné elements, as are the pommels with which they are associated and they presumably adorned the hilts of fine swords (Figure 11.29).

If one sword-ring symbolised an oath between a warrior and his king or god, what did two represent? Two separate oaths taken at the same time is, of course, a possibility. Perhaps the idea that they somehow served as badges of rank seems more likely and this may inform the debate on the purpose or meaning of the rings once they are fully published. More prosaically, the development of these twin-ringed assemblies may be nothing more than a desire for symmetry in hilt fittings, marking the triumph of the aesthetic over the symbolic. It is certainly possible to discern the vestiges of the shape of this new form in the tri-lobed pommel fittings we find on swords of the 9th century but this may be entirely coincidental. It is unlikely that we'll ever know which, if any of these theories is correct. Perhaps they all are.

---

[1] Evison (1967)

[2] Davidson (1994)

[3] Mortimer (2011)

The hilt fittings in the Staffordshire Hoard

The Staffordshire Hoard (discussed in Chapter 4) is hugely important to the field of Anglo-Saxon studies in many respects. The current view as to the dating of the hoard is that the "vast majority of the material is of the first half of the 7th century", although "a small amount is 'heirloom' material - fittings from old swords of the 6th century, still in circulation in the 7th" and nothing from "the very late 7th century or 8th century".[1]

The study of the many thousands of elements of which it consists has already given us new insights into the decorative arts, metallurgy and gold-smithing techniques of the period and raised many questions which will discussed for many years. One area upon which it has already had a huge impact is the study of sword hilt fittings. The Hoard contains more than five hundred pieces of hilt fittings, the overwhelming majority of them being made from gold. This includes roughly eighty pommels, sixty-three being made of gold (the actual numbers are still unclear at the time of writing as work is ongoing in matching the one hundred and fifty or so fragments together), one hundred and fifty-seven pieces of guard plate, one hundred and five hilt-collars (including examples decorated with cloisonné) and one hundred and thirty-six items classified as "Other sword mounts", which includes a large number of small gold and garnet cloisonné pieces which may have been mounted on sword grips in the same way as the triangular cloisonné piece on the Cumberland hilt is.[2] According to Chris Fern, (Lead Typology and Art History researcher and one of the Academic Editors for the Staffordshire Hoard), when considering the number of swords which the Hoard represents, "a sensible estimate might be of a minimum of circa one hundred and fifty"[3].

As the work of piecing together the fragments into which some of the elements of the Hoard had been rendered progressed, some of these hilt elements were consolidated and sets began to emerge. Although the final analyses of the team carrying out the conservation and consolidation work are yet to be published, based on the information already available, it is possible for us to draw some initial conclusions.

11.29 Double ring-pommel K136 from the Staffordshire Hoard

---

[1] Chris Fern (2016) pers. comm. 3rd October

[2] Cool (2015)

[3] Chris Fern (2016) pers. comm. 3rd October

Given the very high quality of many of the pommels, with fifty-five being decorated with either filigree (thirty-five), cloisonné (thirteen) or a combination of the two (seven), the number which fall into Menghin's Type 2e will be high and no doubt we will see that type's frequency increase from the current level of 8.6% of English pommels. At least one new pommel type has been identified in the Hoard and others may emerge, as may new sub-types. Two of the Type 2 pommels within the Hoard are decorated in what is described as an "Insular" style[1], a merging of Germanic and "Celtic" styles to produce a new artistic form which emerges in the early 7th century. As a result of these discoveries, a revised typology will have to be developed. The overall ratio of pommels to swords represented in the Hoard (approximately eighty to one hundred and fifty – 53%) is higher than the 36.6% ratio derived from Mortimer's study undertaken for this volume. From this we might conclude that swords of higher quality (there is no doubt that the hilt fittings contained within the Hoard are of a far higher quality than those found in the vast majority of burial contexts) were more likely to have pommels and other non-organic elements than average. Yet it could also be seen to demonstrate that a high proportion of swords in use in the early 7th century had some non-organic elements, most notably grip ornamentation, but no pommels and that therefore the Cumberland hilt is a unique surviving example of something that would not have been unusual amongst swords of the period.

However, it should be emphasised that, whatever speculative conclusion we might draw from the Hoard, now or in the future, we should not forget that it is both an exceptional and unusual assemblage which skews any statistical analysis to which it might be applied. With that caveat in mind, the Hoard does seem to demonstrate that swords of the quality of that found in Mound 1 at Sutton Hoo, whilst no doubt not the norm, were not as exceptionally rare in England as was previously thought. Why we do not find more hilt assemblages of this quality in burials in England and why the hilts found in the Hoard were stripped from their blades is something that we will consider in the conclusion to this chapter.

11.30 Hilt of the sword from grave 47, Snape.
Courtesy of Suffolk County Council
Archaeological Services

[1] Fern (2015b)

Snape Grave 47

The sword recovered from Grave 47 at Snape, Suffolk in 1991 (Figure 11.30) is one of the most important and overlooked weapons recovered from an early Anglo-Saxon context as it is the only example to date of a complete (bar the very tip of the blade) sword from that period. The burial, contained within a dugout log-boat coffin, is thought to be that of a juvenile and is dated (based on a combination of the carbon-dating of some skeletal remains and on the typographies of the spearheads and shield bosses which it contained) to the late 6th or early 7th century[1].

11.31 Close up of the lower guard of the sword from grave 47, Snape. Courtesy of Suffolk County Council Archaeological Services

All of the hilt elements are intact and all are made from horn (assumed to be bovine[2]), with no evidence that it was fitted with metal embellishments of any sort, although the upper face of the lower guard does bear what appears to be some simple carved decoration (Figure 11.31).

The upper guard is elliptical with little tapering, 62mm x 21mm and 13mm thick with a rectangular slot cut for the tang. The tang was peened over the upper guard, possibly over a flat metal washer, although it is not possible to establish this for certain as the top of the tang is obscured by a section of preserved textile, the remains of a cloth in which the sword and scabbard were wrapped prior to deposition. An example of this type of washer can be seen on another sword with an intact organic hilt which is in the collection of the Rijksmuseum van Oudheden in Leiden (inventory number M 1967/12.1). Although lacking in provenance (all that is known about it is that it came from a private collection in Cologne and was allegedly found in Germany), there is no obvious reason to doubt its authenticity and it gives us insight into the one part of the construction of hilts of this type which the Snape find does not (Figure 11.32).

11.32 Detail of diamond shaped iron washer from a sword in the collection of the Rijksmuseum van Oudheden (inventory number M 1967 / 12.1).

[1] Filmer-Sankey and Pestell (2001)

[2] Cameron and Filmer-Sankey (1993)

It is also possible that the upper guard of the Cumberland hilt bears the legacy mark of such a washer (see below), as does at least one of the surviving upper guards from Nydam (no. 445[1]).

The grip of the Snape sword is roughly oval in cross-section, 94mm in length, tapering from bottom to top and is shaped so as to accommodate the bearer's fingers in four grooves running around it. This ridged hand grip is a feature which can also be seen on horn hilts from Cumberland, Nydam, Högom Mound 2 and Evebø (as well as the above mentioned "Cologne" sword) and presumably served to both make the grip more comfortable and more secure when the sword was being wielded.

The lower guard is both wider and a more elongated ellipse than the upper in plan, 87mm x 25mm, although it is the same thickness of material. The rectangular perforation for the tang which runs through it has been widened in the lower face in order to accommodate the shoulders of the blade, ensuring a stable fit. In being wider than the upper guard, it is consistent with other English and Continental finds, but there are examples of weapons from the period where both guards are roughly the same size (Högom Mound 2 being the most complete example[2]).

Despite its lack of adornment or embellishment, the importance of the Snape sword cannot be overstated as it represents the form of hilt which, as far as we can establish, adorned the majority of swords buried in England between the late 5th and early 7th centuries.

Cumberland (Figures 11.33 & 11.34)

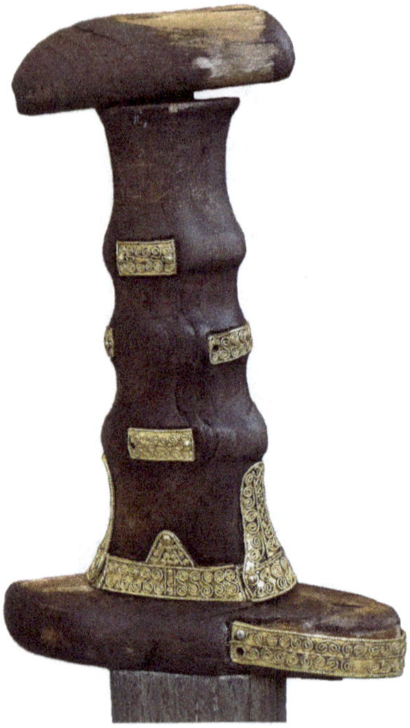

Until the discovery of the Snape sword in 1991, the best evidence that we had for the organic elements of early English sword hilts was the British Museum's item number "1876,0717.1"; the "Cumberland" hilt. Apparently discovered in the 19th century in the now defunct county of Cumberland and sold to the museum by Archibald Dodd in 1876, it consists of an upper guard, grip and lower guard. An analysis of the hilt carried out by the British Museum in 2013 states that these elements are made from horn and wood but fails to clarify further, other than to remark that the wooden elements "might be interpreted as more recent repairs, as the wood itself does not appear in a state expected after years of burial"[3]. When looking at the hilt, it is not obvious which these wooden elements might be, as all three parts appear very similar, mostly reddish-brown with a few spots of damage where a lighter colour shows through. However, there is one section of the lower guard which is distinctly different in appearance and which seems to be an obvious repair, so it seems reasonable to assume that this is the wooden element referred to.

---

[1] Bemmen and Bemmen (1998)

[2] Ramqvist 1992

[3] Blakelock, Mongiatti and La Niece. 2013

As with the Snape hilt, the upper and lower guards are both ellipsoid and of roughly equal thickness at approximately 14.4mm. The upper guard is fully pierced by a hole for the tang of the blade and there is no evidence that a pommel has ever been fitted. Presumably the tang was peened over the upper guard but the end of the tang is now missing (although much of the tang still remains within the hilt). There is a shallow, rectangular depression in the uppermost face, around the hole and it is possible that this is the legacy mark of a washer over which the tang was peened (Figure 11.35). The lower guard is 85mm wide and is designed to accommodate the shoulders of the blade and the grip is ribbed to accommodate the fingers. The overall length of the hilt is 127mm, 7mm longer than the Snape hilt but in all other respects of its construction it is effectively the same.

11.33 (opposite) and 11.34 (right). The "Cumberland" hilt. British Museum (inventory number 1876,0717.1) showing both sides.

Where it does differ is in its decoration. Unlike the Snape, the Cumberland hilt is fitted with eleven decorative pieces; ten gold filigree panels and one gold and garnet cloisonné triangle. One of the filigree panels is on the lower guard and all of the rest are mounted on the grip. These are all held in place by small round headed nails (although there are obviously several nails missing), all but two of which are gold and consistent with known early Anglo-Saxon manufacturing methods. The other two nails (used to fix the filigree strip which covers slightly less than half of the circumference of the lower guard) are brass, and of relatively modern manufacture. This piece coincides with the part of the lower guard that appears to have been repaired which might explain the use of these brass nails. Six of the pieces on the grip are set out in a pattern, with one on both the obverse and reverse of the upper and lower rib and another pair on the outside edges of the middle rib. Another panel encircles the base of the grip and over this is set a pair of mounts, one either side of the grip, extending from the base to the lowest rib. The triangular cloisonné panel (Figure 11.36) is set above the middle of this lower panel on the obverse side, mirrored on the reverse with a triangular filigree section which is part of the panel that encircles the grip. Stylistically, all of these decorative elements appear to date to the 7th century.

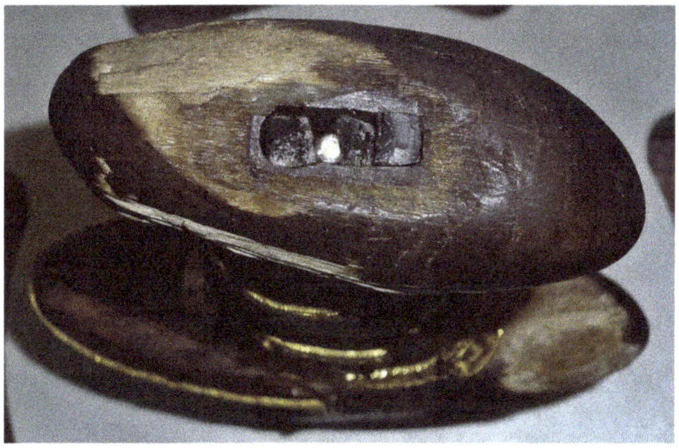

11.35 Top of the "Cumberland" hilt, showing possible legacy mark. British Museum (inventory number 1876,0717.1)

11.36 Detail of the "Cumberland" hilt. British Museum (inventory number 1876,0717.1)

There are no empty nail holes to indicate that any of the decoration is missing, although the overall symmetry of the piece is broken by the fact that the piece around the lower guard only covers one side. Davidson theorises that this might be because it was there to indicate the correct way to place the sword into the scabbard and it is true that if the "correct" way is with the cloisonné panel facing outwards then the filigree panel on the lower guard would face forwards when the sword was in its scabbard (assuming the wearer was right handed). However, rather than being there to indicate this, perhaps the reason for the filigree panel only covering that particular side of the guard is because that is the side which would have been seen, whether the sword was in its scabbard or being held aloft. Although no doubt beautiful, precious and valuable, these decorations still leave most of the hilt exposed and it seems that, if it is correct that none of the decoration is now missing, this can only be for one of two reasons. Either the owner could not afford (or was not gifted) more decoration (although one would think that a band which could completely encircle the lower guard would seem desirable if decoration alone was the paramount consideration) or, despite the enhancements and the statements that they made, it was important that the organic hilt should still be visible. There is an assumption that the hilt itself is 5th century (as there is with the Snape hilt) but there's no real basis for it other than a wider assumption that hilts of this type were only made in the first part of the early Anglo-Saxon period. A discussion of the problems with this assumption and the general issue of dating can be found further on in this chapter.

Dover Buckland Grave 96a (Figure 11.37)

Discovered at Buckland near Dover during urgent rescue excavations in advance of the construction of a new road in the early 1950s, the sword recovered from Grave 96a has a hilt which is a unique example of its type in England. The grip is almost completely decayed but appears (based on a visual inspection) to be horn, although in the absence of a proper analysis it's impossible to state that it's not wood. The report[1] does state that both the upper and lower guards are made from wood and that they were originally completely sheathed in metal. Both still have a thin outer band of iron running around their circumference, decorated with transverse bands of inlaid copper alloy. Each also has an iron rivet at both ends of the ellipsoid wooden guards, which (it is assumed) fastened metal plates to the upper and lower face of each guard, although these plates have not survived. Whilst it is true that the presence of rivets in a guard plate does not necessarily indicate a multi-layered construction (see the description of the Crundale hilt below) and that we cannot be certain that the guards were fully sheathed in metal as the report states, just the presence of these decorated, encircling bands of iron mark the Buckland hilt out and led Evison to conclude (based on continental examples such as the one from Hermsheimer Bösfeld in Mannheim) that the sword "must have been imported from the Rhineland"[2]. Making such a bold statement on the basis of such limited evidence does seem to be over-presumptive but there's no denying that there is nothing else like it known from an English context. The sword also has a long, low boat shaped pommel made from iron, fixed to the upper guard with two rivets. Grave 96a is classed as coming from Phase 4 of the cemetery, which equates to deposition in the second quarter of the 7th century although the sword itself (or at least its hilt) may have been of mid to late 6th century manufacture.

11.37 Reconstruction of the sword from grave 96a, Dover Buckland. Made by Tim Noyes at Heron Armoury.

[1] Evison (1987)

[2] Evison (1987)

Crundale (Figure 11.38)

Little seems to be known about the sword from Crundale Down, except that it was bought from F.W Wyatt by the British Museum in 1894[1]. The hilt elements are now mounted separately and consist of a pair of silver gilt collars, decorated with a simple interlace pattern, which were once fitted to either end of the grip, an organic upper guard which is pierced by a dome headed rivet at each end and a silver gilt pommel, decorated in a form of Style II animal art which is thought to date to the later 7th century. The Crundale hilt is interesting and worthy of more study for two reasons. Firstly, it's the only example that we have from England of a pommel in situ on an organic upper guard. Secondly, although the upper guard is covered in a thick layer of what appears to be conservator's wax, enough of the wax is missing to allow us to see the organic material within (which appears to be horn) and also to see that there do not appear to be any metal hilt plates on the guard. If this is the case, then it demonstrates that the two dome headed rivets which pass through it were used purely for decorative reasons rather than to hold together a sandwich of upper guard elements as is usually the case. The Crundale hilt is deserving of further study, not least to see what details an x-ray could reveal of the inside of the pommel and the peened tang beneath.

11.38 The pommel and upper guard from the "Crundale" hilt. British Museum (inventory number 1894,1103.1)

Sutton Hoo Mound 17

Excavated in 1991 by Annette Roe as part of Martin Carver's project to develop a deeper understanding of the burial ground at Sutton Hoo, Mound 17 proved to be the resting place of a young man and his horse[2]. His sword was placed at his side, to the right of the body which is unusual and might be taken as an indication of the handedness of the owner (see Chapter 10 in this book for a discussion on the position of swords in burials).

All three hilt elements of the sword are identified as being made of horn and, although the grip and lower guard have mostly decayed, a substantial amount of the upper guard remains, revealing two noteworthy features. Firstly, the underside of the guard is pierced by the grip rather than simply butting up against it as is the case with both the Snape and Cumberland hilts. Secondly, there is no sign that the tang of the blade fully penetrated the upper guard, giving us the possibility that some swords had hilts that were held together by some other means than simply peening the tang over the upper guard or pommel. It may be that these two unusual elements are related. If the tang were secured in place within the grip using an adhesive (perhaps something similar to 'cutler's resin', a combination of pitch, wax and either sawdust or charcoal, traditionally used by cutlers to fix knife blades into their handles) and wedges hammered into the top of the grip, this would serve to hold grip, lower guard and blade together securely. The upper guard could then be glued into place, with the seating of the end of the grip into a recess within the upper guard providing extra security (and

[1] Information taken from the British Museum's online record for the item: 1894,1103.1

[2] Evans (2005)

a larger surface area if glue were used). Whilst this is entirely speculative, the authors own experiments with adhesives made from hide have demonstrated that they are highly effective in glueing elements made from bovine horn together. Whether or not the resulting join would be durable enough to withstand the shock of combat is debatable (and opens up the question of whether the sword in its final state had ever been used in combat) but hide glues are used to fix organic elements of laminate bows together and certainly prove to be durable in that context.

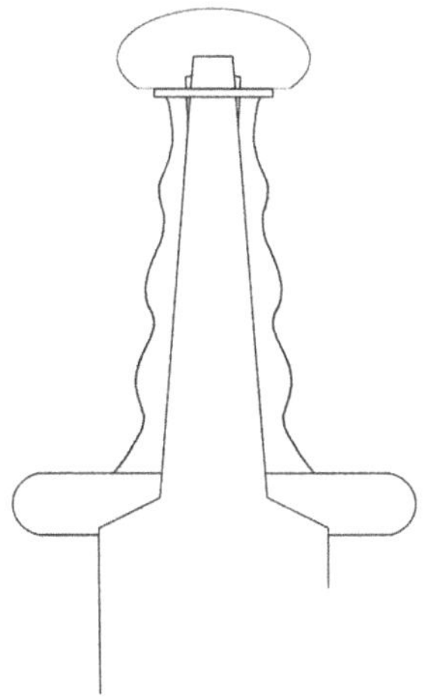

11.39* Cross section of the hilt of a sword from grave 2, Watchfield (drawn by the author, after Scull)

Of course, if all of the above is correct (and again, further investigation including x-ray analysis of the hilt elements would be useful here), then it leaves us with a question; why would you assemble a hilt in this manner rather than in the normal way? It is possible that it was necessary as there was not enough length in the tang to allow for the usual method of assembly. The grip of the Mound 17 sword is approximately 10cm in length but the tang only projects a few millimeters beyond it. It may be that the end of the tang had been broken off at some stage in its life and that this was the only method of re-hilting it without making the grip too short to be usable.

On the other hand, it may have been purely aesthetic, a method of assembling an organic hilt in such a way as to leave the upper guard unpierced and unblemished by a peened tang or additional pommel. The top of the Mound 17 upper guard appears to have been carved to a shape which bears a resemblance to a guard with a pommel on it, certainly something more complex than the simple forms of other surviving examples. There is some evidence to suggest that the method of construction was not restricted to this one example. In her analysis of the sword hilt from Grave 2 at Watchfield[1], Cameron observed that the horn pommel had been "separated from the grip by an oval copper-alloy washer" which had been fixed in place using two small wedges of an "amorphous substance...probably resin". The use of this washer, whilst adding a slight, almost imperceptible decorative element to the hilt assembly, would allow these wedges to be hammered in firmly, holding the tang securely inside the grip. There is no sign of the end of the tang of the Watchfield sword having been peened and, in the reconstruction drawing (drawn by Christopher Scull), the tang does not fully penetrate the upper guard (Figure 11.39). It should also be noted that a great many sword blades from early Anglo-Saxon contexts have tangs which show no obvious signs of having been peened. Whilst this might simply be the result of corrosion having removed the evidence, it is possible that the method used to assemble the Sutton Hoo Mound 17 and the Watchfield hilts was more widely used than previously thought.

[1] Cameron

Dover Buckland Grave C (Figure 11.40)

The last set of hilt fittings that we will discuss in detail are also from the Buckland cemetery near Dover. Grave "C" was excavated in 1951 by Mr. W. Stebbing in advance of Evison's main excavation[1] and the sword which it contained is one of the best examples that we have of a complex hilt assembly with metal elements from England. Although most of the organic material has decayed (that of the upper guard and grip being all but gone but with substantial remains of the lower guard still evident), the fact that nearly all of the metal elements are still in situ on the sword give us a very clear picture as to the overall appearance of a hilt of this type. It also contains some small but interesting features that are worth some attention.

11.40 The hilt assembly of the sword from grave C, Dover Buckland. Photograph courtesy of Lindsay Kerr

The decorated, silver-gilt cocked-hat pommel is fixed to the upper guard by two pairs of rivets, which pass all the way through to be peened on the lower face. One set is peened over an additional shaped washer whilst the other set are peened directly over the bottom plate. The reason for this difference is unclear but may be the result of the rivets on one side having been removed and replaced. One example from Gilton (6875.2) also has only one washer[2] and other examples from that cemetery have none at all. The sword from Mound 1 at Sutton Hoo has these washers in place for both sets of pommel rivets, as does one of the swords from Faversham. Their use is neither universal nor consistent.

The upper guard assembly consists of a very thin, flat silver-gilt upper plate, the main body of horn, which would have been 11mm thick, flat on its upper surface and chamfered on its lower, to fit into the silver-gilt lower tray. At one end of the guard, a rivet with domed heads at either end passes through the three elements. At the other there is a rivet whose upper part curves into roughly two thirds of a circle (the open end resting against one side of the pommel) to form the carrier for a decorated ring, the two parts forming a Type 1b sword-ring, according to Evison's typology[3].

The grip was adorned with two narrow, beaded, silver gilt bands (one of which went missing before the sword was received at the British Museum), encircling the horn at either end. The one surviving band sits at the bottom end of the grip and gives us an indication as to the original size of the grip at its base (approximately 40mm x 16mm) and also reveals that it may have been hexagonal rather than oval.[4]

The lower guard is again made up of three parts, the flat upper plate, the organic material and the dished lower plate, with a dome headed rivet at either end of both the materials (notably the different types of niello)

[1] Evison 1987

[2] Evison 1967

[3] Evison 1967

[4] Evison 1987

and the decoration on both the pommel and the ring, led Evison to conclude that whilst the hilt was early 6th century in origin, the sword-ring could not be dated to before the middle of that century, demonstrating that the sword-ring was a later addition to the assemblage, perhaps by as much as 50 years[1] (Figure 11.41). Not only is this intriguing in the wider context of trying to understand the purpose and meaning of these sword-rings (as discussed earlier in this chapter) but also makes it very likely that the person who owned the sword at the time that the ring was fitted was not the person for whom the sword had been made.

Conclusions and Discussion

The study of sword hilts found in English contexts that can be dated to the 5th to 7th centuries allows us to draw few definite conclusions other than obvious ones concerning their physical nature. It does however generate many avenues for discussion, not just with regard to hilts but also about the swords that they were part of, the nature of ownership and the possible relationship between a sword's furniture and how its original bearer (or those who came after him) wished to be perceived in life and in death.

11.41 Close-up of detail of the sword-ring from Grave C, Dover Buckland in the collection of the British Museum (inventory number 1963,1108).

It does seem safe to say that the styles of hilt fittings in use in England during this period were influenced by those found in both Scandinavia and Western Europe in the early 5th century with little evidence for any influence from British Insular forms (although, as discussed previously, we have no real idea what form such a 'British' sword might have taken) and form part of a body of similar hilt fittings that can be found throughout the sphere of Scando-Germanic influence during the earliest centuries of the Early Medieval period. What also seems clear is that, whilst some elements (most notably pommels) underwent changes in terms of form and decoration over time, the overall form remained, for the most part, unchanged both chronologically and regionally throughout the period. We could even create a rough chronology for the first use of particular pommel types and other non-organic hilt elements, although it would be of limited value; the point at which something is buried is not a great indicator as to the point at which it was created. We can do better with individual examples, based on the style of decoration employed and the results of the work being done on the Staffordshire Hoard will no doubt add to our understanding in that area.

[1] Evison 1987

Of course, what we cannot know is how long swords or, more specifically, hilt fittings actually stayed in circulation. Whilst the Snape sword has a hilt of the very simplest form, something that we might be tempted to date to the earlier part of our period, the grave in which it was buried contained items which date to the late 6th or early 7th century. Is it the sword fittings themselves which are old or is it just evidence that the form was long lived? Was Snape a relatively new sword with new organic fittings in a style that hadn't seen much change in over a century?

Likewise, with the Type 1c and 1d pommels which seem to originate in early/mid-5th century Scandinavia, did they remain fitted to blades which came from that time and place, or were they recycled, fitted to new blades and if so, why? Was it purely because somebody liked them aesthetically or was there a deeper symbolism in the act, reflecting attitudes towards ancestry and cultural identity? If the latter, had those fittings been passed down within the family? Some pommels certainly exhibit great wear to both their gilding and decoration and even to the structure of the pommel itself, which could be taken as evidence of them having been in circulation for a very long time. (for a detailed discussion on the subject of wear of hilt fittings and the possible implications, see Brunning 2013).

Even those hilts which feature pommel types that seem to emerge in the 6th century might have older organic fittings which were later modified by the addition of such a pommel. That is perhaps what we see with the remains of the Crundale hilt and might even explain the presence of what seem to be entirely decorative rivets, modernising an old hilt so that it more closely resembled the latest fashion without completely refurbishing it. The same can be said of other elements, such as the 7th century adornments found on the Cumberland hilt. The original organic elements could predate them by a century or more.This in turn raises another question which we can't know the answer to. If it is the case that the 'antique' forms of organic fittings which adorned some of the swords found in burials were old at the time of deposition, were these swords which were still in use at the time of deposition or were they heirlooms, preserved rather than maintained, which the family chose to bury with the deceased? The apparent plain nature of the vast majority of hilts on swords buried might lead one to conclude that these are old blades with old organic fittings, heirlooms placed in the grave by the family to act as a symbol of status or cultural association or even to serve them in the afterlife rather than being a sword that had been used by the individual, and that the sword actually used by the person in life was more 'stylish', adorned with the latest fittings and embellishments. If we take the evidence from burials at face value, the alternative to this would seem to be that most swords in use during the early Anglo-Saxon period were devoid of adornment and decoration, at least at the time that they were deposited. Whilst this is entirely possible, how do we square it with the fact that the hilt of a sword gives a prime opportunity for individualising the weapon, for adding embellishments to make it stand out in a culture where obvious displays of wealth and status seem to have played such an important role? We could even take this further and theorise that the hilts of some swords might have been so distinct as to mark them out as being recognisably the sword of an individual, swords of renown. How might these 'plain' swords, with their simple, organic hilts, have been individualised? If it was by the addition of pommels and other elements, then why do the majority of swords not have them? If the reason is (in some cases at least; there is no evidence that the sword from Snape had a pommel or any other form of embellishment at any time and we must conclude that Snape was not an isolated example) that pommels and other additions which marked the sword out as belonging to an individual were stripped from the sword prior to deposition for some reason, then that takes us into another area of theorising regarding the role played by hilt elements and how they were managed at the time of and after the death of the owner.

With so many swords from burials not having pommels and so many pommels having been discovered as stray finds rather than in burial contexts, there is a possibility that what we have is evidence for some form of practice whose significance we can only guess at. A high number of these isolated pommels show no signs of damage (unlike those contained in the Staffordshire Hoard, many of which bear witness to the force which was used to separate them from their blades) which could indicate that in some cases the sword pommel, possibly the most distinctive part of the whole hilt assembly if not the whole sword and scabbard, was deliberately and carefully removed from a sword and then specifically deposited on its own, rather than being melted down or otherwise recycled.

Why might a pommel be removed and deposited separately? Perhaps because it (like the sword-rings) represented something specific, a mark of office, a badge of rank or an oath undertaken, something that the deceased or his family didn't want to (or were not permitted to) put into the ground with the sword. Obviously, this theory assumes that there is any link between the absence of a pommel in some burials and the deposition of pommels outside of a funerary context.

It should be noted that, in addition to the large number of pommels dating to the 5th-7th centuries, there are a further fifty-one finds of pommels dating to the 8th-10th centuries recorded in the PAS database. Whilst a few of these have the remains of the tang within them, most do not. If the removal and deliberate deposition of pommels was a practice, it seems that it was a long lived one.

Another possible explanation for the isolated deposition of pommels is that it was part of a process that marked the handover of a sword to a new owner, a process which might involve stripping from it the identity of its previous owner and then carefully depositing those elements. Perhaps some swords of this period were badges of rank or status, gifted by a king or lord, which would have to be returned at the end of service. This is certainly something that we seem to have evidence for in the later Anglo-Saxon period. In a will dated between 946 and 951, written by Ælfgar, father of Æthelflæd (wife of King Edmund I) and Ælfflæd (wife to Byrhtnoth, Eolderman of Essex who fell at Maldon) there is mention of a sword which Ælfgar had already given to "his lord" (presumably King Eadred), a sword which Ælfgar had previously received from King Edmund and which had a value of "one hundred and twenty mancuses of gold and had four pounds of silver on the sheath"[1] (a mancus being worth thirty silver pennies, something in the region of £600 in modern terms). Although there is no more detail than that, we can speculate as to the reasons for Ælfgar's actions. It does not seem to be a simple case of the sword being part of his "*heregeat*", the war-gear that every man of means was required to bequeath to his lord in a will, as this document states that the sword had already been handed over at the time that the will was written. It also doesn't conform with other examples of heregeat detailed in wills of the time which consist of multiple items and sums of money (compare this with the will of Ælfhelm, whose will describes a heregeat of a hundred mancuses of gold, two swords, four shields, four spears and four horses[2]). As the sword was given by a previous monarch, perhaps it was a token of some office that Ælfgar held under Edmund but not under his successor and which therefore he was required to return. If this was a common and ongoing practice during the early medieval period, it may be that when such swords were subsequently bestowed upon a new incumbent, the recipient would want to mark the sword out as their own, or would be gifted a more precious fitting whilst the sword was in their possession but would also want to honour the previous owner. This may be what the embellishments on the Cumberland hilt and those contained within the Staffordshire Hoard represent, precious tokens given by a lord in recognition of some achievement or service or marking the bearer out as a member of a particular group of warriors, rather than just being decoration for decorations sake.

---

[1] Whitelock (1930)

[2] Ibid.

That pommels and other precious hilt elements were valued for more than just their material worth in the warrior culture of the period is something which the Staffordshire Hoard seems to indicate, although with the Hoard this is demonstrated in a negative fashion as the process of stripping the hilts from their blades may be connected with humiliation of the owner rather than honouring him. Little care was employed in the removal of the objects in the hoard and, in some cases, considerable violence seems to have been used. The process of conservation and restoration has re-united elements of single fittings which seem to have been deliberately destroyed in a manner reminiscent of the practice of ritually "killing" weapons by bending them and hacking them into pieces seen in many of the Danish bog weapons deposits.

11.42 Upper hilt assembly from Market Rasen, Lincolnshire. British Museum (inventory number 2006,1001.1.c)

We may never understand what the Hoard represents but perhaps it consist of items taken from the defeated in an act meant to destroy memory of their valour and fame. The process of tearing off the fabulous hilts and other adornments and then butchering them to leave the blade naked, stripped of anything which identified it as a specific sword, would be a symbolic act carried out on one of the most prized possessions of the vanquished warrior. It would remove their personality, their status and possibly their record of achievements from the sword. If this was common practice across England at the time it might go some way towards explaining the almost total lack of swords discovered in burial contexts with hilt fittings to rival those found in the Staffordshire Hoard.

There are, of course, more prosaic explanations for the absence of richly hilted swords in burials and for hilt furniture in general. Some have no doubt been lost over the centuries to a variety of causes. Some burial fields and mounds were ploughed out (Mound 17 at Sutton Hoo was so eroded by ploughing that it was barely visible[1]) or destroyed by other means and their contents either scattered (which may explain finds such as the Market Rasen hilt Figure 11.42, which was clearly part of an intact sword at the time it was deposited) or taken. Equally likely is that many of them were taken from graves and not recorded, either as part of 'excavations' by early antiquarians in the 17th and 18th centuries or during grave openings that took part in earlier centuries, perhaps within just a few years of the grave being first closed. A discussion on this subject, and on the ongoing work of "The Grave Reopening Research Group", can be found in Chapter 22 of this book. Hopefully, many more still lie in graves or isolated contexts, waiting to be discovered, to add to or perhaps modify our understanding of the importance and meaning of sword furniture in the early Anglo-Saxon period.

[1] Carver (2005)

12. Sword blades and sword blade shapes

Paul Mortimer

Working out accurate blade dimensions of swords in Anglo-Saxon England during the 5th to 7th centuries is fraught with difficulties as we shall see, and because of that it is only possible to come to some fairly general conclusions. Not all of the surviving swords have been x-rayed and had their overall measurements published, although this practice does tend to be followed with more recent finds, such as Dover Buckland 2 and Saltwood.

One of the main problems is that almost all swords found in burials were buried with the blade protected by its scabbard and during the centuries these components have decayed and become inseparable. Most measurements that have been taken include the remnant scabbard.

Another problem is that many of the surviving swords exist in a very fragmentary state with portions of the blade area completely decomposed. Of the one hundred and sixty-nine swords included in the finds database only sixty-five have are more or less complete and yield both width and length measurements. With quite a few of these the tang measurement has had to be worked out from the scale drawings as they were not provided in the reports, so we are reliant on the drawings being accurate.

Blade shape

The scabbard can obscure blade features and true dimensions but I have assumed that it reveals the general shape of the blade and in almost all cases the blades are long and described as either parallel or slightly tapering; that is, narrowing towards the point end. I believe that where blades have been described in the reports as 'virtually parallel', 'almost parallel' or 'slightly narrower at the tip end', or variations on this theme that the authors meant that the swords are slightly tapering toward the tip; that is certainly the case where scale drawings were provided. Proceeding on that basis, the sample contains thirty-seven sword blades described as tapering in some way and thirty-two as parallel. This number includes a few blades that do not have both width and length measurements included in the relevant report, and with some of those that do, I am afraid that the shape was not reported, so the two sets are not equivalent.

My reproduction of the Sutton Hoo Mound 1 sword has a blade width of 60 mm and a length of 720 mm, however the scabbard measures 85 mm by 750 mm. The width of the blade then, is 70.5% of the width of the scabbard; length is less affected as the blade is 96% of the scabbard length.[1] I do not have enough examples to hand to know if these blade/scabbard proportions are typical; we have tried to replicate the Sutton Hoo ensemble as accurately as we can, but what this simple exercise illustrates, is that the length measurements given in the cemetery reports are more reliable than those of the width as length is less affected with the addition of a scabbard. Matt Bunker, based on his extensive experience in building scabbards from the 5th to 7th, suggests that a complete scabbard would consistently be around 20mm wider on average than the blade in order to achieve the correct curve and accommodate the scabbard lining.[2]

The way a sword and scabbard decay together will vary subject to soil acidity and other local conditions so it is not possible to just subtract (or add) a given measurement, such as 20mm, from the width, because in some

[1] The difference in width is more than 41% of the width of the sword blade, whereas the difference in length is only around 4.1% of the original length.

[2] Matt Bunker, personal comment.

examples most of the scabbard has gone but much of the sword blade has survived its time in the grave, whereas in others the scabbard width might have been added to with corrosion products adhering to it. (figures 12.1 to 12.7)

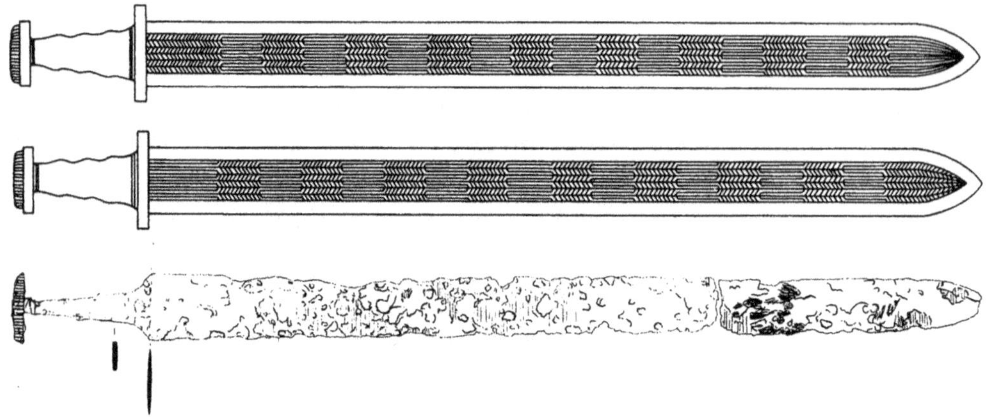

12.1 Acklam Wold sword from Yorkshire. The metal parts of the grip survive. The blade shape is fairly typical, virtually parallel and a fairly gently shaped tip that will be effective in the thrust. Drawing by Brian Gilmour.

Blade lengths[1]

As the discrepancy between the length of the blade and the blade in its scabbard can be largely disregarded, it is possible to make reasonable assumptions about the lengths of the sixty-four swords blades in the sample and not include the tang length; there is one sword, that from grave 179 at Castledyke South where I do not know the length of the tang, so that example has been excluded for now. The average blade length is 759.5 mm, but there is quite a wide range, the shortest being a sword from grave 27 at Dover Buckland at 520 mm (width 51mm)and the longest from Great Chesterford, grave 65, with a length of 891.5 mm (width 50 to 60 mm). Only two blades in the sample are actually below 600 mm, another five are in the 600 to 700 mm range, while forty-five have blades of 718 mm to 798 mm and twelve have blades of 800 mm or more. There are two modes – one at 760 mm and another at 780mm – both measurements have seven blades each at that particular length. 775.5 mm is the median length.[2]

Blade widths

As we have already seen there are difficulties in calculating widths, but it could be that something can be gleaned by analysing the reported figures from the cemetery reports. The narrowest sword is reported to be 39mm (grave 193, Mitcham – sword length 885mm with tang – 775mm without) and the widest is 76mm (grave 682 II, Mucking – sword length 952mm with tang, 812mm without). There are eight narrow swords (39mm to 45mm at the throat), fifty-one medium width sword blades (46mm to 60mm) and five wide swords (61mm to 76mm).

Despite the problems, putting all of the raw width data from our sixty-four swords together does produce some workable results, I would suggest. The average width of our sample is 52.625mm, the modal width is 51mm and the median is also 51mm.

---

[1] Table 2 at the end of this chapter is a scatter diagram showing length and width measurements.

[2] Cameron, 2000. 45.

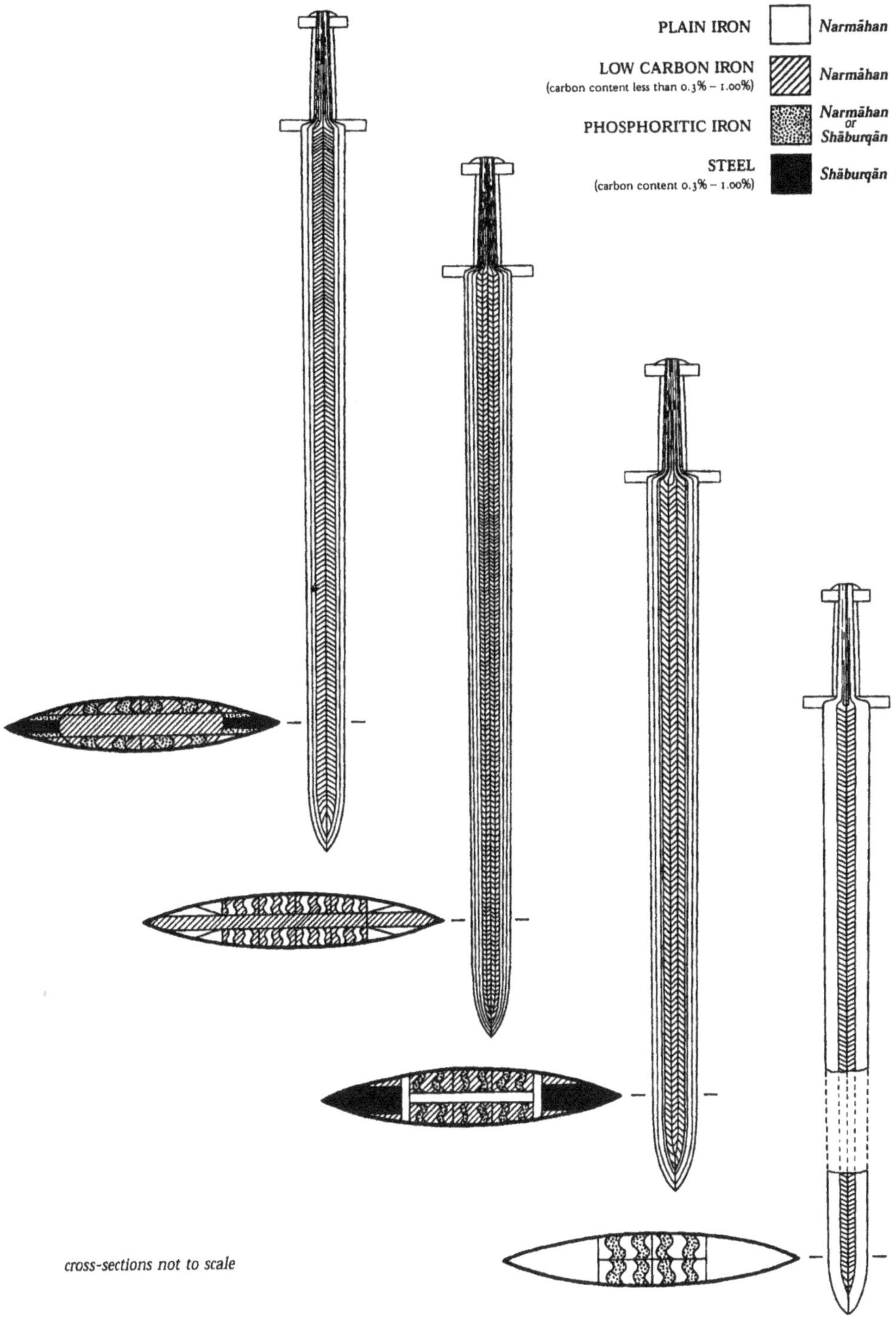

12.2 The Park Lane, Croydon swords. Again, the blade shapes are typical with a very slight taper towards the tip. Drawn by Brian Gilmour.

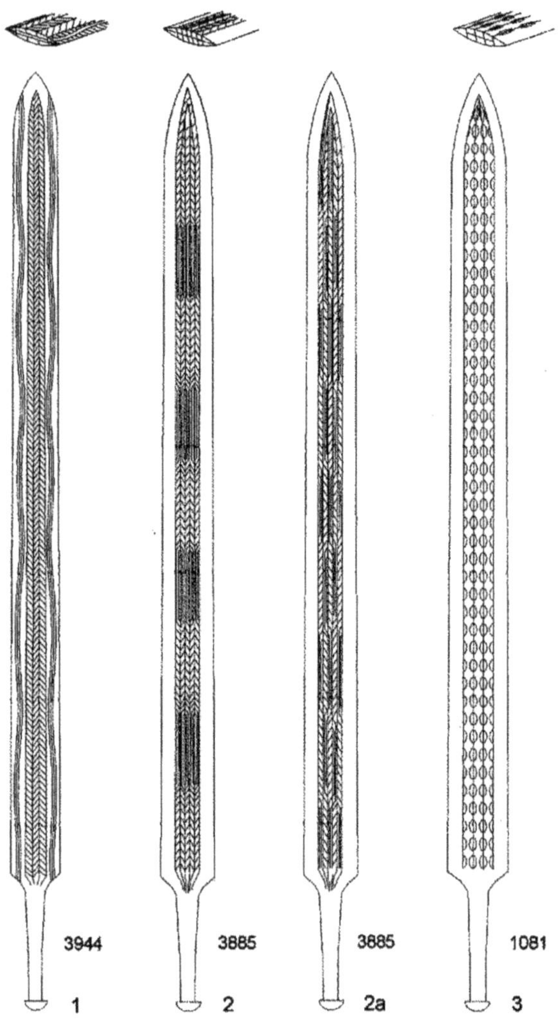

12.3 Three of the Saltwood swords. Graves 3944, 3885 (both sides of the blade) and 1081. The blade shapes are very similar to those previously illustrated.

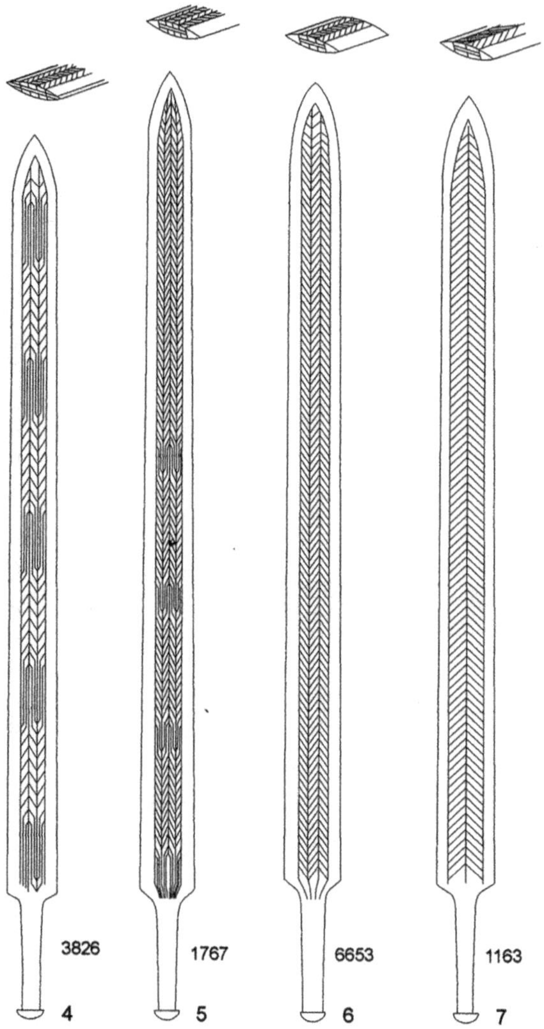

12.4 Four more swords from Saltwood, Kent. Graves 3826, 1767, 6653 and 1163.

All drawings by Brian Gilmour

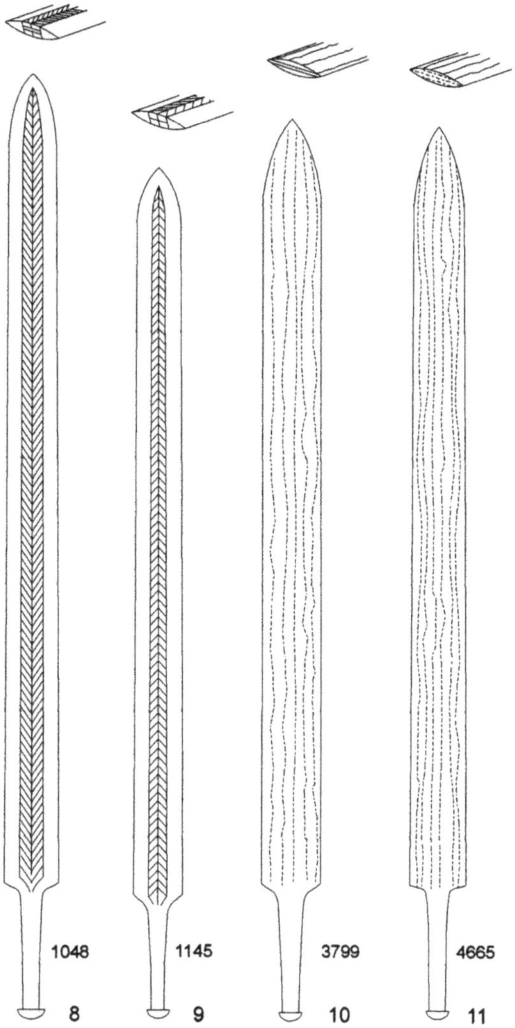

12.5 Another four sword blades from Saltwood. Graves 1048, 1145, 3799 (stacked) and 4665 (stacked).

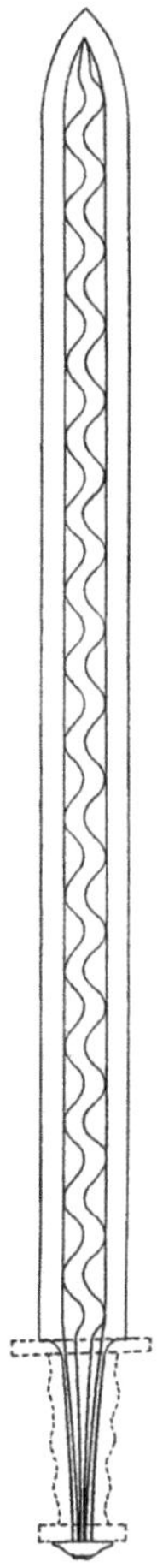

12.6 The West Heslerton 'serpent' sword.

All drawings by Brian Gilmour

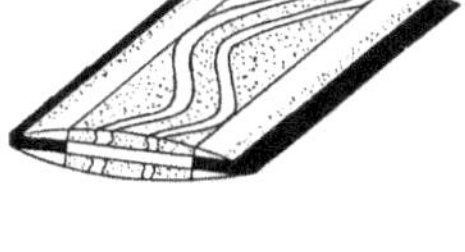

12.7 Cross-section of the West Heslerton sword.

Tangs

Tangs have to incorporate the whole of the grip, the lower and upper guards and the pommel, so they need to be lengthy enough to accommodate all those elements and strong enough to sustain the impacts that will arise from combat and still keep the grip assembly intact. They, too, vary in length within our sample; the longest is 160mm (Snape) and the shortest is 11.5mm (Great Chesterford, grave 65. As this is the sword with the longest blade, it is pretty certain that the tang is incomplete). It should be noted here that the tang on the Sutton Hoo, Mound 17 sword has a tang that only protrudes some 50% into the upper guard and might not have been peened over. If that is the case, then at the time of burial, the Mound 17 sword was not a serviceable weapon and would have failed in use. We will discuss this further, in *The Meaning of the Sword*.

There are sixty-four measurable tangs within our sample, some of the shorter ones, those that are less than 100mm are, presumably, damaged[1] and would be unable to fulfil the requirements of a tang which is to bind the blade to the hilt; I have, therefore, included two sets of statistics, those that include all the tangs and those that have tangs that are less than 100mm. For the all tangs, the mean is 114mm, the modes are 110mm and 114mm and the median is 121mm. The modified data with the shortest ten tangs excluded provides a perhaps, more realistic mean of 121.9mm. The modes remain the same, but the median would now be 124mm.

There are eighteen tangs in the 102mm to 118mm range, seventeen between 120mm and 128mm, eight from 130mm to 136mm, nine 140mm to 149mm and three over 150mm.

Blade measurements, a note:

Cameron provides the following average measurements from her study of 5t to 7th century swords in England; she was concerned to see whether there was any difference between swords with and without scabbard chapes:[2]

Table 12i. Average dimensions

	Total length	Blade length	Width
Average dimensions in mm of swords with chapes	908	804	56
Average dimensions in mm of swords without chapes	891	762	52

Apart from the fact that she found that swords with scabbard chapes tended to be longer and wider than those without, the other measurements accord well with what has been found within our own survey.

Blade cross-sections

It is not possible to provide anything like a definitive analysis of blade cross-sections for 5th to 7th century swords from our sample, although to some extent the work of Brian Gilmour does give considerable information regarding the cross-sectional structure of many swords. In the very small sample culled from the information contained within his 1999 thesis, of twenty-three swords with a possible 5th to 7th century date,

[1] Many of the others may be to a greater and lesser extent, so caution must be exercised with the tang lengths.

[2] Cameron, 2000. 45. See her Appendix 1 for full details.

only one is definitely fullered;[1] all the rest have lenticular blades. The shape, 'lenticular' (strictly, it should be 'double lenticular') is an oversimplification and might have had nuances which have not been looked for or detected within Gilmour's study. Several sword blades seem to have had flat centres, including the Mound 1 Sutton Hoo sword and it is interesting that Lehmann has found examples with a similar blade shape in his study of blades from Westphalia.[2]

Unfortunately, no measurements of how thick the blades actually were are given and precise drawings of the cross-sections are not provided.

A chart giving some information on cross-sections taken from Gilmour's work can be found below.

Table 12ii. Sword blade cross-sectional data
Taken from Gilmour (1999) Developments in Iron Smithing and Decorative Welding Techniques Found in Anglo-Saxon Swords and Related Edged Weapons, volume 2.

Gilmour's Catalogue Number	Location	Figure Number in Thesis	Cross-sectional shape	Dating
S32	Unknown	14	diamond	EIA
S27	Little Wittenham*	15	lenticular	LIA
S29	Little Wittenham*	16	lenticular	LIA
S39	Whittlesey	17	flattened	RB
S17	Sarre	18	lenticular	6th 7th
S49	Bifrons	19	lenticular	6th 7th
S51	Sarre	20	lenticular	6th 7th
S52	Sarre	21	lenticular	6th 7th
S53	Sarre	22	lenticular	6th 7th
S54	Sarre or Bifrons	23	lenticular	6th 7th
S55	Sarre or Bifrons	24	lenticular	6th 7th
S56	Sarre or Bifrons	25	lenticular	6th 7th
S12	Wickhambreaux	26	lenticular	6th 7th
S13	Bifrons?	27	lenticular	6th 7th
S14	Eastry	28	lenticular	6th 7th
S15	Holborough	29	lenticular	7th

[1] Gilmour's findings are not incorporated into my sample of swords from cemeteries. Although there are over two hundred swords taken from Gilmour's study, there are only twenty-three possibly from our period where he specifically mentions the shape of the blade cross-section.

[2] Lehmann, 2014. 9. Cameron, 2000 (34) also notes that swords from Anglo-Saxon England of this period, "...have a wide and flattened appearance..."

S18	Aylesford	30	lenticular	6th 7th
S19	Wickhambreaux	31	lenticular	6th 7th
S20	Lenham	32	lenticular	6th
S1	Loveden Hill	33	lenticular	6th 7th
S40	Chesterton	34	lenticular	5th 6th
S10	Mitcham	35	lenticular	6th
S45	Mitcham	36	lenticular	6th
S16	Sarre	37	lenticular	6th 7th
S31	Caversham	38	fullered	7th – 9th
S46	Brentford	39	fullered	7th – 9th
S7	Thames London*	40	lenticular	9th 10th
S8	Thames London*	41	fullered	9th 10th
S43	Thames London*	42	lenticular	10th 11th
S48	Rochester	43	fullered	9th 10th
S23	Thames Reading*	44	fullered	9th 10th
S47	Thames Waterloo*	45	fullered	10th
S24	Thames Cleeve*	46	lenticular	9th – 11th
S41	Thames Vauxhall*	47	fullered	8th – 10th
S25	Thames Reading*	48	fullered	10th 11th
S44	Thames? Brentford	49	fullered	11th
S26	Thames Wallingford*	50	lenticular	12th – 14th
S28	No provenance	51	lenticular	13th 14th
S30	Sunbury Weir Stream*	52	Lenticular with mid-rib	Late Medieval
S50	Barham Down	53	lenticular	6th – 8th

Blade tips

The way that sword tips are concealed within scabbards or corroded, means that it is difficult to come to firm general conclusions regarding sword tips, but they do seem not to have often been acutely pointed, if at all. Where the tip shape can be discerned, it can be rounded or relatively gently pointed.[1] This has led some authorities to assume that the swords of this period were only used for slashing but this is, probably, a mistaken view; experimental use with replica swords has indicated that thrusts can be devastating. Again, this will be discussed elsewhere in *The Sword as a Weapon*.

[1] Cameron, 2000. 34.

The way tips were formed is also somewhat obscure, whether edges were made in one piece and wrapped around the tip or made in two pieces (one per side) and both attached in turn is not certain. It is likely that different methods were used by different smiths and unless there is an intensive study of sword tips using x-ray or CT scanning it is unlikely that we will know any more than we do now.

Fullers

Fullers would appear to be quite rare features in swords from the 5th to 7th centuries included within Anglo-Saxon graves. There are, however, some occurrences within our sample of one hundred and sixty-nine swords.[1]

The Abingdon sword has fullers, in fact on one side it has single fuller, while on the other there are two. It would appear that one of the recent finds at Sutton Hoo, grave 21(called Bromeswell in my data base but now, it seems, called Tranmer House) and dated to the 6th century is fullered. Several swords from Dover Buckland had fullers or might have had; the sword from, 265b, (second half of 6th century) had one as did the late 6th to early 7th century grave 264, although that probably had a ground-in fuller. The sword from grave 414 (second half of the 6th century) has one, too, and it is possible that the swords from grave 346 (late 6th to early 7th century) and from grave 437 could have had them. Another possibility is the 7th century grave 179 from Castledyke South which contained the skeleton of a very young man, and the blade from grave 163 at Mitcham, Surrey might have been fullered. Visual inspection of the Watchfield swords indicates that it had a fuller.[2]

Fullers can be very shallow or hidden by corrosion and can, therefore, be easily be missed on inspection, so with our present state of knowledge it is wise not to be too dogmatic about how frequently they were used on swords of the 5th to 7th centuries.

As mentioned above, it would seem that some swords had flat centres, and this could have been common.

Conclusions

Within the sample, I think that there are indications that there were a variety of different sword blade types being used during the 5th to 7th centuries, perhaps they were made to suit different styles of fighting, or perhaps they had been tailored, originally, to suit an individual. As discussed elsewhere in this volume, many of the swords may have been very old at the time that they were buried and Sue Brunning has shown that old swords, especially ancient ones, were much more highly prized than new swords,[3] so it is likely, that the finds represent blade styles from quite a wide time range. It is also possible that only old, worn out swords were buried.[4]

We have seen that some sword blades were actually quite short, but lengths around 760mm to about 780mm would appear to be much more normal and some were obviously longer. When looking at widths and comparing them to the lengths, it can be seen from the scatter diagram, (table 2)below, that there is a very large cluster containing the vast majority of the blades around the 40mm to 60mm width range with lengths between 730mm to 800mm. If we break that down a little further, the majority of swords seem to be of two main types, narrow swords with widths at or below 51mm and wider, presumably heavier swords with widths of 52mm to around 60mms. There are a few that are wider and a few that are slimmer.

---

[1] La Salvia, (1998. 41) suggests that the typical 'Migration Period' sword had a fuller; this is a questionable claim in view of the evidence, at least from England.

[2] Matt Bunker, personal comment; this sword may actually have a double fuller.

[3] Brunning, 2013. 117 to 118, 137, 189, 235, 236,237, 238, 255 to 256, 260.

[4] See The Meaning of the Sword, in this volume.

Sword blade cross-sectional data, table 12iii.

This information below is taken from Gilmour (1999) *Developments in Iron Smithing and Decorative Welding Techniques Found in Anglo-Saxon Swords and Related Edged Weapons*, volume 2, and not from my cemetery survey. Samples marked with * are almost certainly river finds. For the sake of comparison I have included all the swords from whatever date that Gilmour uses. I have omitted all information on seaxes and spears.

Gilmour's Catalogue Number	Location	Figure Number in Thesis	Cross-sectional shape	Dating
S32	Unknown	14	diamond	EIA
S27	Little Wittenham*	15	lenticular	LIA
S29	Little Wittenham*	16	lenticular	LIA
S39	Whittlesey	17	flattened	RB
S17	Sarre	18	lenticular	6th 7th
S49	Bifrons	19	lenticular	6th 7th
S51	Sarre	20	lenticular	6th 7th
S52	Sarre	21	lenticular	6th 7th
S53	Sarre	22	lenticular	6th 7th
S54	Sarre or Bifrons	23	lenticular	6th 7th
S55	Sarre or Bifrons	24	lenticular	6th 7th
S56	Sarre or Bifrons	25	lenticular	6th 7th
S12	Wickhambreaux	26	lenticular	6th 7th
S13	Bifrons?	27	lenticular	6th 7th
S14	Eastry	28	lenticular	6th 7th
S15	Holborough	29	lenticular	7th
S18	Aylesford	30	lenticular	6th 7th
S19	Wickhambreaux	31	lenticular	6th 7th
S20	Lenham	32	lenticular	6th
S1	Loveden Hill	33	lenticular	6th 7th
S40	Chesterton	34	lenticular	5th 6th
S10	Mitcham	35	lenticular	6th
S45	Mitcham	36	lenticular	6th
S16	Sarre	37	lenticular	6th 7th
S31	Caversham	38	fullered	7th – 9th
S46	Brentford	39	fullered	7th – 9th
S7	Thames London*	40	lenticular	9th 10th

S8	Thames London*	41	fullered	9th 10th
S43	Thames London*	42	lenticular	10th 11th
S48	Rochester	43	fullered	9th 10th
S23	Thames Reading*	44	fullered	9th 10th
S47	Thames Waterloo*	45	fullered	10th
S24	Thames Cleeve*	46	lenticular	9th – 11th
S41	Thames Vauxhall*	47	fullered	8th – 10th
S25	Thames Reading*	48	fullered	10th 11th
S44	Thames? Brentford	49	fullered	11th
S26	Thames Wallingford*	50	lenticular	12th – 14th
S28	No provenance	51	lenticular	13th 14th
S30	Sunbury Weir Stream*	52	Lenticular with mid-rib	Late Medieval
S50	Barham Down	53	lenticular	6th – 8th

Table 12iv. Scatter-graph of blade lengths and widths

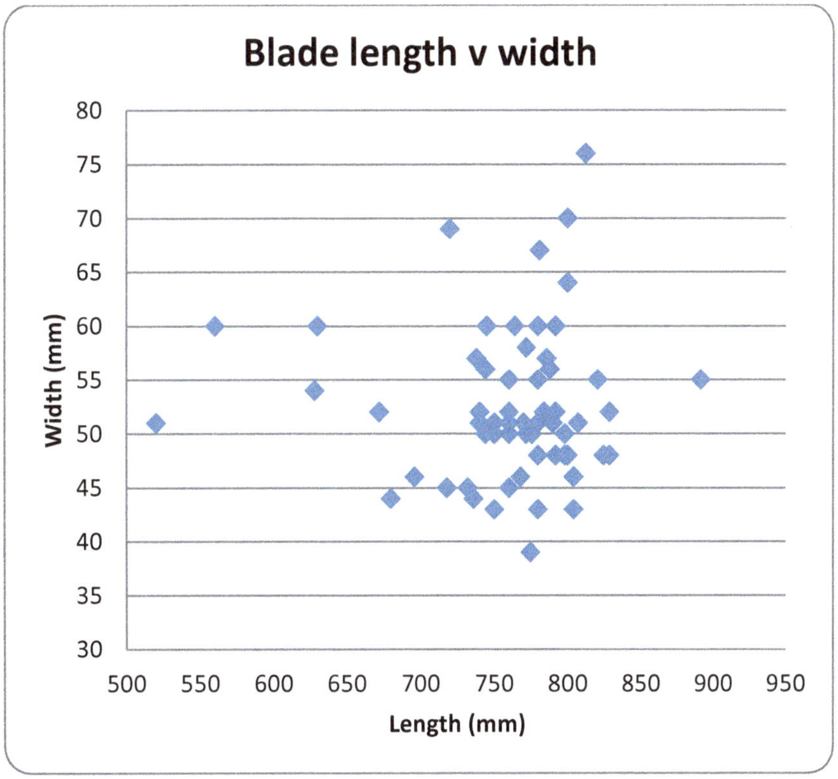

13. Clothing the sword; the scabbard and its fixed decoration

Matt Bunker

Introduction

In this chapter, we will look at the construction methods employed in the making of sword scabbards found in England in the early Anglo-Saxon period, their decoration and their fittings.

The primary functions of a sword's scabbard are to provide protection, both for and from the sharp blade, and also to provide a means for the weapon to be carried on the person. Due to its size, the scabbard also provides an excellent ground for displays of wealth and status by means of decoration and applied fittings. As the scabbard "was ordinarily the most visible part of the weapon, this decoration played an important role in communicating visual information about the weapon and about its bearer"[1].

English sword scabbards of the early Anglo-Saxon period were all (as far as we can tell from studying the archaeological evidence) made in a very similar fashion. Whilst decoration, fittings utilised and the precise materials used varied, a scabbard was essentially two laths of wood, usually lined with fleece or fur, bound together with some form of skin product to provide a cover for the blade. In this, English scabbards of the 5th to 7th centuries are following a North West European tradition, which goes back to at least the Bronze Age as evidenced by the example from Vapnö[2]. Other methods used in scabbard construction can be traced back to covers for bladed weapons of the late Neolithic period. An example of this comes in the form of a sheath for a flint knife from a bog near Wiepenkathen, Germany, dated to approximately 2400 BC. Not only does the well-preserved, moulded leather sheath bear traces of tooled surface decoration in the form of what has been described as a pine tree, but it also contains a lining of soft, alum-tawed skin product to protect the daggers cutting edges from abrasive wear and ensure a good fit in the sheath.[3]

A study of the tradition of scabbard making in Britain in the centuries preceding the 5th century shows that the method of constructing scabbards with a lined, covered wooden core does not appear to be a native one. British scabbards of the Iron Age are predominantly made of unlined iron or copper alloy.[4] Wooden scabbards from British Late Iron Age contexts have been found, but the increase in the use of wood in this period may be due to Roman influence (the scabbards of the long *spathae* employed by the Roman Army from the late 2nd/ early 3rd century onwards being predominantly wooden).[5] Neither the surviving native examples nor those thought to be of Roman origin show evidence of decoration being applied to the body of the scabbard, a regular feature of early English scabbards. In comparison, a Scando-Germanic tradition of decorated wooden sword scabbards existed from at least the Bronze Age.

One of the finest surviving examples of these North-West European Bronze Age scabbards was found inside an oak coffin in Muldbjerg, western Zealand, Denmark in 1883. It is dated to the middle of the 14th century BC and features two pairs of parallel lines executed in relief carving on the front face. Another Danish example, from Store Kongehøj is also dated to the 14th century BC but the decorative carving is much more complex, with multiple panels of geometric design set in numerous fields between two sets of parallel lines (fig 13.1). Other wooden Scando-Germanic scabbards of this period show that a thin layer of skin or leather was used to cover the wooden core, sometimes moulded over cord decoration or relief carving.

---

[1] Wells 2012

[2] Nyström

[3] Cassau

[4] Stead

[5] Southern & Dixon

This tradition of scabbard making continued through the Iron Age in the region, with many examples being preserved as part of the weapons assemblages which have been recovered from Danish bogs, most notably from Illerup Ådal (2nd-5th century AD)[1] and Kragehul (3rd-5th century AD).[2] We will refer back to these earlier Scando-Germanic examples, the materials and methods used in their construction, to demonstrate a continuity of tradition amongst the makers of the scabbards found in early Anglo-Saxon England.

13.1 A Bronze Age carved scabbard from Store Kongehøj, Denmark. National Museum, Copenhagen.

We will not be studying detailed typologies or addressing the chronology of scabbard fittings, partially because it is a subject which requires an entire volume to do it justice and also because that volume has already been written. Anybody interested in further study of that subject is directed towards Esther Cameron's excellent and essential "Sheaths and Scabbards in England AD400-1100" in which she details the remains of three hundred and fifty-one examples, with one hundred and thirty-seven of those being pertinent to the period under discussion. Also relevant is Wilfred Menghin's "Das Schwert im frühen Mittelalter", a wider study of swords in Early Medieval Europe which contains the typologies for fittings which Cameron draws upon for her study.

Construction

Lining

Applying a lining of fibre-bearing animal skin to the inside of the wooden laths was common. Although the fibre and skin itself seldom survives, its presence can be detected either by mineralised remains on the blade or inferred by a void between the wooden laths and the blade which the lining previously occupied. Whilst evidence for both textile and animal fur linings have been found (mirroring examples from the continent), a study of those English scabbards in which the lining could be identified (twenty-two examples out of a total survey of one hundred and seven scabbards, with eighty-one showing some evidence of having been lined), showed that the most commonly used material was sheepskin.[3]

A sheepskin or other animal fibre lining to a scabbard has ssome practical benefits but which (if any) were the most important at the time is unknown. One undeniable benefit is that it ensures a secure fit between sword and scabbard, due to the fibres being compressed when the blade is sheathed.

Some writers have theorised that the lanolin within the fleece is another reason for using sheepskin as a lining material, due to the waterproofing that it would impart to the blade. However, practical experience has shown that whilst there might be some benefit to this it is countered by the fact that a layer of lanolin on the surface of the blade helps to keep contaminants on it, which are in turn transferred to the inside of the scabbard.

[1] Biborski & Ilkjær

[2] Iversen

[3] Cameron 2000

It is worth remarking that the scabbard from Mound 1 at Sutton Hoo, the highest status burial known from England in this period, was lined not with sheepskin but with a "straight, soft type of animal fur, rather than sheep's wool",[1] possibly beaver. A similar lining, described as being of "fine fur" was used by the maker of the richly decorated 4th century 'HarkilaR' scabbard, thought to belong to the commander of the vanquished army whose equipment was deposited in the Nydam 'Pine Boat'.[2] Although it is only speculation, it is interesting that something other than the usual material was used to line both of these scabbards which belonged to people at the upper end of their respective societies. As the lining would never be seen, perhaps it was a property of the material itself which made it superior to sheepskin and worth the additional expense. This would seem to argue against sheepskins lanolin-bearing properties being the main reason for its use, as a fur lining would have no such properties.

Wooden laths

The pair of wooden laths at the core of a scabbard give it some rigidity and acts as a framework for the other elements. In her survey, Cameron found that 64% of identifiable lath remains were made from willow or poplar (the two woods being difficult to tell apart when mineralised),[3] with alder being the next most frequently used (at 17%), and laths of hazel, lime, beech, birch and ash making up the remainder. The same wood species were used in the construction of the scabbards recovered from Illerup Ådal although here, 53% of the one hundred and eight samples were made from lime with poplar being the second most frequent (32%) and with willow being one of the two least common types of wood to be used. The wooden remains recovered from the Illerup site are presumably in such good condition that it is possible to distinguish between poplar and willow. It seems reasonable to assume that wood species were chosen for use as scabbard laths for certain characteristics and this preference for poplar over willow at Illerup might be taken as evidence that, of the two, poplar was the more suitable. By extension, it would be reasonable to theorise that the same could be said for Anglo-Saxon scabbards and that, if it were possible to differentiate between the two, we might find that the majority of the willow/poplar remains are poplar. This contradicts the opinion expressed by Cameron who felt that "the superiority of willow over poplar is evident to craftsmen when the qualities of the two woods are compared".[4] However, Robin Wood, a woodworker who has studied and recreated early medieval items from simple bowls to the Oseberg ship states that both woods "have similar working properties" with "clean straight parallel grain that splits down nicely into thin laths. Both woods steam bend easily and both carve easily"[5].

As the laths were, on average only 2mm in thickness (with approximately 0.5mm variance),[6] being able to split it into thin sections may well have been one of the most important properties of the wood used. Whether the laths were cut into thicker sections and then worked down to the correct interior and exterior profile or whether they were cut to the required thickness and then curved (whilst green) to fit the profile of the blade is difficult to establish by eye.

Experimentation by the author has demonstrated that both methods produce the same end result in terms of form and both were certainly well within the abilities of the woodworkers of the period. However, they do produce scabbard cores with different mechanical properties. Laths which are split to

---

[1] Bruce-Mitford 1978
[2] Jorgensen & Vang Petersen
[3] Cameron 2006
[4] Cameron 2000
[5] R. Wood, 2015, pers. comms.
[6] Cameron 2000

the correct thickness and then curved to fit the profile of the blade are (once bound within their outer covering) under compression and have a degree of flexibility and strength which those carved down from thicker pieces seem to lack. A further possibility is that the timber was cut down into laths that were already curved in cross section, using a tool similar to the sort of curved froe used by coopers and others.

Without further microscopic study of the end grains of a wide sample of scabbard laths, it is impossible to say which of these methods was used by the early English scabbard maker. It may well be that all were employed, dependent on the wood variety and skills available to the individual makers. Regardless of the method used, the end result was invariably the same; a wooden core which followed the form of the blade it contained (albeit with a more rounded tip) so closely that, with the addition of the lining, it must have been a snug (if not tight) fit indeed.

Coverings

The wooden laths provide the core of the scabbard but, as has already been observed, these were always very thin. An outer covering was required to provide strength and, in England at least, this seems, usually, to have been made from skin product (as it is difficult if not impossible to ascertain whether animal skin has been tanned, tawed or cured in some other way after centuries in the ground, the use of the word 'leather' is not necessarily appropriate). On the continent, most notably in Rhineland contexts, birch bast was sometimes used to provide a binding cover for the whole scabbard (a particularly well preserved example which has been the subject of close analysis survives from Grave 61 of the cemetery at Wünnenberg-Fürstenberg)[1] . It is possible, given the very poor state of preservation of some scabbard remains, that this or a similar method may have been used in England (bast bindings have been found on English scabbards but so far they have been restricted to the scabbard mouth, over the covering – see below). However, to date, only coverings of skin product have been found.

The thickness of this covering varies but seems to fall into two distinct categories, one being a thickness in the region of 1mm and the other being between 1.5 and 2.5mm.[2] Cameron implies that the thinner group can all be dated to the 5th and 6th centuries whereas the thicker group belong to the 7th (although by her own admission, some of the ones which fall into the thicker group are of unknown date).

Where it can be identified, the seam closing the covering runs up the back of the scabbard either centrally or off centre, never along the outside edges. In the two samples identified by Cameron in which it was possible to determine the material used to close the seam, she describes it as 'thong',[3] with the stitching itself being serpentine. The same serpentine seams are seen on the back of two of the early Anglo-Saxon scabbards excavated from the Saltwood Tunnel cemeteries in Kent.[4]

The presence of a layer of corrosion products on top of the mineral-preserved wood of a scabbard recovered from grave 59 at Collingbourne Ducis might be evidence that an adhesive was used to glue the cover to the wooden core, either as part of the manufacturing process or to aid in forming of the cover over any decoration beneath.[5]

Decoration

With the exception of the shield, the scabbard presents the largest opportunity for decoration and display in the panoply of the early medieval warrior. As demonstrated in the introduction to this chapter, there is a

[1] Lehmann 2010
[2] Cameron 2000
[3] Cameron 2000
[4] Cameron 2006
[5] McCormick 2016

tradition of scabbard decoration that goes back to the Neolithic and in this section, we will look at the decorative techniques employed in our period of interest. Early medieval organic finds from England are, in general, in a very poor state of preservation and scabbards are no exception. There are, however, enough fragments surviving to allow us to identify some of the methods used and the patterns created.

Both of the decorative techniques which can be identified from archaeological remains are forms of foundation moulding, where the leather covering is dampened and then moulded over a pattern executed on the wooden face of the scabbard before being stitched closed. This pattern was created in one of two ways; either by carving the design into the wood itself or by applying it to the wood using cords made from fibres.

At its simplest, the pattern consists of a single ridge, running longitudinally for part of the length of the scabbard but variations on this are recorded (and it is possible that, in those examples where only a single ridge line has been identified, they are all that is left of a more complex arrangement).

The most frequently recorded of these variations involves the addition of two lines radiating from the central line, to create the shape of an arrow, pointing upwards towards the sword hilt. Three examples of this design were recorded by Cameron with two of them (from Clobb's Row, Broomfield in Essex and Grave 249C from Buckland, Dover) being cord-work and one (from West Garth Gardens, Bury St Edmunds, Suffolk) appearing to her eye to be carved in relief into the wooden core of the scabbard. Since the publication of Cameron's survey in 2000, at least one more English example of this 'arrow' pattern on a scabbard has been discovered in grave 21 at Tranmer House, Bromeswell, Suffolk, the pattern again being made up of cord-work.[1] Figure 13.2 shows a scabbard made by the author which has this pattern, executed in cord-work.

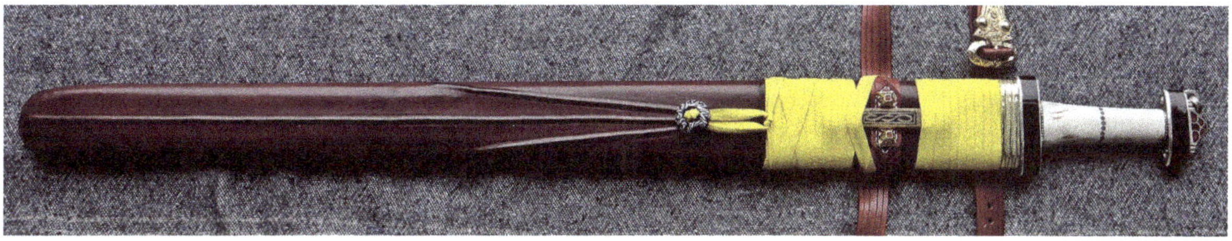

13.2 A scabbard made by the author, showing the 'Tir' rune pattern.

To these, we can add continental examples of scabbards which feature this same pattern of lines radiating from a central ridge. Some, like the example from grave 61 at Wunnenberg-Furstenberg in Germany[2] or that from grave 5 at Snartemo[3] are very similar to the English examples, the former being cord-work under its bast covering and the latter, Norwegian example, being carved from the wooden core. Others feature the radiating line pattern as part of a more detailed carved decoration, as can be seen on two of the scabbards recovered from Danish Nydam IV deposit.[4] Another Nydam scabbard (from the earlier 1d context, dated 375-475 AD) also features this design, but this time in the form of silver mounts fitted to the scabbard.

In the case of the two Nydam IV scabbards and the Snartemo scabbard, the radiating lines and the central line come together and then run parallel to each other, continuing over the suspension slide. It is possible that this

[1] Fern 2015

[2] Melzer 1991

[3] Rolfsen and Stylegar

[4] Jorgensen and Peterson 2003

may have been the case with some of the English examples. Unfortunately, nearly all scabbards recovered in England are in such poor condition that it is impossible to make out surface detail of this type.

This arrow shape is so ubiquitous that it is tempting to imagine that it has some significance, that its common use was not merely because of its simplicity and ease of execution but that it imbued the bearer with some sort of power or protection.

A possible explanation seems to be that it is a representation of the arrow shaped rune called 'Tir' in Old English. This rune is associated with the god Tyw, "the principle Germanic god of the early heathen period"[1] who Snorri Sturluson (writing in the 13th century) describes as "boldest and most courageous…For men of action he is good to invoke".[2] As such, it would seem to be an appropriate symbol for warriors to bear upon their equipment and there are examples of the Tir rune to be found on weapons of the period. At least three spearheads discovered in early Anglo-Saxon cemeteries have the rune inlaid on them; from grave 7 at the Anglo-Saxon cemetery in Holborough, Kent,[3] grave 192 of the excavation at The Meads, Sittingbourne in 2008 (fig 13.3) and grave 7049 of the cemetery at Wrotham in Kent[4]. The same rune can also be found at least four times on an Anglo-Saxon cocked hat pommel.[5] These examples bring to mind the passage from Sturlison's "*Sigrdrífumál*" in which the Valkyrie Sigrdrífa tells the hero Sigurd:-

"runes on thy sword-hilt write;
Some on the furrow,
and some on the flat,
And twice shalt thou call on Tyr"[6]

13.3 Detail of gold wire inlay 'Tir' rune on a spearhead from grave 192, Sittingbourne. Photograph from the 'CSI Sittingbourne' blog page.

[1] Pollington 2008
[2] Byock 2005
[3] Evison 1956
[4] Stoodley 2015
[5] Pollington 2011
[6] Bellows 2007

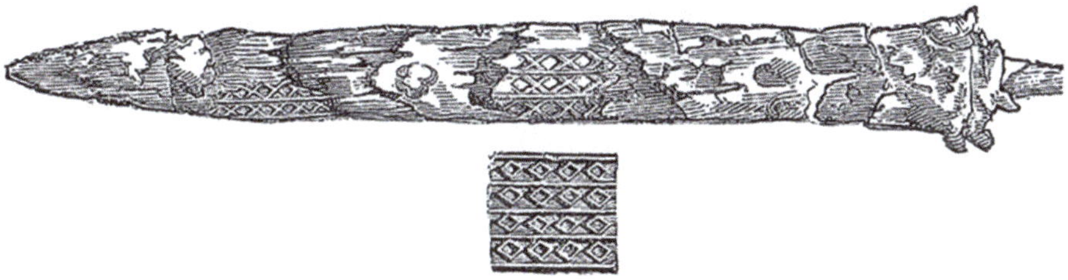

13.4* Line drawing showing the decoration found on the scabbard from Lapwing Hill when excavated. From Bateman's "Ten years' diggings in Celtic and Saxon grave hills", based on Llewellyn Jewitt's watercolour.

All of these examples, in conjunction with other uses of the Tir rune on cremation urns and amulets seem to support the idea that it was a rune that was used with what Pollington calls "apotropaic intent" and, as such, its use on a sword scabbard seems entirely appropriate.

Other than those featuring these simple geometric designs, we have very little to indicate the form that scabbard decoration might have taken. There are just two English scabbards which we can look to for evidence of something more complex and one of those requires us to study drawings which are now over one hundred and fifty years old.

When Thomas Bateman excavated a "finely shaped barrow, near Brushfield upon Lapwing Hill"[1] in the August of 1850, he discovered an Anglo-Saxon burial which included "a long and broad iron sword, enclosed in a sheath made of thin wood and covered with ornamented leather". Today, the sword (held at Sheffield's Weston Park Museum) is in a badly degraded, fragmentary condition, with traces of the scabbard all but lost but fortunately we still have the illustration made by Llewellyn Jewitt for Bateman's book *Ten Years Digging in Celtic and Saxon Grave Hills* (fig. 13.4). It shows the most extensive decoration known on a scabbard of the period, consisting of five longitudinal ridges which run for most, if not the whole length, of the front creating four long fields which each contain a continuous pattern of diamonds.

13.5 Detail of carved decoration on a scabbard from Charlton Plantation, Wiltshire, Salisbury Museum

[1] Bateman 1861

Whilst it is no longer possible to establish whether the pattern was created by carving or cordwork, Jewitt's watercolour is evidence that foundation moulding was used to create decoration more complex than the few lines which we normally find.

The other surviving example which demonstrates complex scabbard decoration in England comes from the Anglo-Saxon cemetery at Charlton Plantation in Wiltshire. (fig. 13.5) The find is unstratified as it was discovered by workmen during road-widening work which led to the excavation of the cemetery, but all of the finds that were handed in with it were early Anglo-Saxon in nature and there's no reason to suspect that SF124[1] is not from the same period. The scabbard remains adhering to the sword blade are extensive but, as usual, badly degraded. However, at the top end, very close to the mouth, there survives a lateral band and below that, a tantalising glimpse of some curvilinear decoration, consisting of three arcs which originate from a point at the edge of the scabbard and which then spread out across it. All of these features are carved into the wood and no trace of the skin covering remains. It is the only evidence that we have that early English scabbards may have been as extensively carved as those found in some Scandinavian contexts.

There is no reason to believe that some English scabbards were not decorated to the same extent as those found in the bog deposits of Jutland (fig. 13.6) or the boat graves of Uppland and it seems incredibly unlikely that the scabbards of the swords found in mounds 1 and 17 at Sutton Hoo would not be as beautifully carved as those from graves 6, 7 or 8 at Valsgärde, but time and the soil have left us with nothing but these thin shreds of evidence.

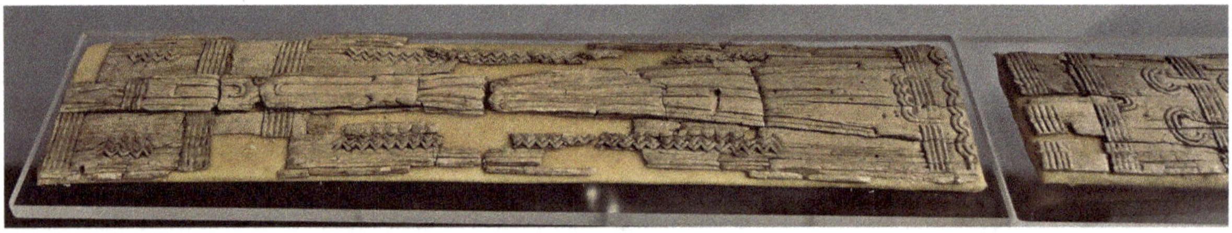

13.6 Detail of carved decoration on a scabbard from Nydam Mose, Denmark. Nationalmuseet, Copenhagen

St Cuthbert Gospel

Before moving on from the subject, any discussion of decorative leatherwork of the early medieval period must refer to the St Cuthbert (or Stonyhurst) Gospel, the earliest intact European book which has now been dated to the very late 7th or early 8th century.[2] This small book is bound with wooden boards, covered with thin, red leather and features complex designs on both the front and back covers.

The leather is interesting as it bears a close resemblance to the sumac-dyed goatskins which were imported into Europe from sub-Saharan Africa for bookbinding from the 15th century onwards.[3] If such fine, exotic leather was being imported into England in the 8th century for bookbinding, it's possible that it was also being used for other high quality leatherwork, including scabbards. Any scabbard with fine, complex decoration would have to be covered in a very fine skin in order to allow the detail of the design to be seen.

The decoration of the back cover is impressed, a geometric design made up of straight lines, created using a fine, smooth tool (possibly of bone, a material still used for such tools by leatherworkers today) or knife. The same method is employed on the borders of the front cover but its central panel features a much more

---

[1] Davies 1985

[2] Breay & Meehan 2015

[3] Pickwoad 2015

complex design; a raised, scrolling plant-like motif. Until recently, it was assumed that the design under the leather was made from either cord or leather but computed tomography scans taken in 2013 revealed something much more interesting. The design appears to have been executed using "an even textured, possibly clay-like substance"[1] found between the wooden board and the leather cover. The scans also reveal that this substance had been forced away to the sides almost completely in those areas where there is no raised design, indicating that the design was created by the application of a carved matrix to the leather covered 'clay'. Pickwood states that "the crispness of the embossing suggests the use of a metal matrix". We know that Anglo-Saxon craftsmen were capable of such work as evidenced by the patrices that they created for use in the production of decorative foils.

The discovery of an apparently new form of decorative technique opens up the possibility that something similar may have been used in the decoration of scabbards to create a complex or repeating design. Looking at Jewitt's watercolour of the scabbard from Lapwing Hill, we can see that it would have been easy to create the pattern using this process of stamping onto leather supported by a mouldable substrate. This is, of course, purely supposition but not an unreasonable one.

Fittings

The majority of scabbard remains found in England appear to have been devoid of any fittings at the time of deposition. Whether this is because they never had any, because they were made of materials which have not survived the centuries or because they were stripped prior to burial is impossible to know. In this section, we will summarise the types of fittings which are known from the archaeological record and consider their practical application.

13.7 Detail of the scabbard from Valsgärde 8, Sweden, showing a bast mouth-binding over a textile layer which overlaid the complex carving, in the collection of the Gustavianum Museum.

[1] Pickwood 2015

Organic Bindings

Some English scabbards (approximately 10% in Cameron/s study) were bound for some of their length with braids, presumably produced by the method known as tablet or card weaving. Examples of these woven braids made from both fine wool (e.g. Clobb's Row, Broomfield) and linen (e.g. Sutton Hoo Mound 1) have been identified, although such positive identification is often impossible due to the degraded nature of the finds. This degradation can also make it impossible to establish how much of the scabbard was bound with braid but the evidence points towards it being limited to the upper half of the scabbard, starting at or near the scabbard mouth. In the case of the scabbard from Clobb's Row, the woollen braid begins at the mouth and is wrapped tightly and densely (with enough overlap between the turns to increase the thickness of the scabbard by 6mm)[1] until it meets the suspension bridge, at which point it is bound over by fine cords which then continue down, passing under the arch of the bridge.

As well as these braid bindings, some scabbards (Cameron identified two in her study and another amongst the scabbards from Saltwood)[2] were bound with cord. The example from Mound 17 at Sutton Hoo was made from flax or hemp,[3] 0.3mm in diameter and covered the scabbard from its mouth downwards for at least 130mm.[4] One of the examples from Saltwood is bound at the mouth with cord made from bast, something mirrored on the continent by an example from Concevreux in Northern France.[5]

Although part of the purpose of these bindings may have been decorative (the braid binding of the scabbard in Sutton Hoo Mound 2 is described as being "not a plain twill but a diamond or chevron"),[6] it must also (given the plain nature of some of the materials used) have served a practical purpose, presumably to reinforce or possibly even repair the mouth of the scabbard. The overall lightness of scabbard structure means that the act of placing a sword into its scabbard can result in damage to the mouth or, if the sword is twisted as it enters, in the splitting of the laths themselves. Binding of some form would either reinforce the structure and prevent this from happening or would at least act as a repair which would hold the mouth of the scabbard together in the event of such damage. For evidence that such bindings served as a repair rather than an original feature, the scabbard from the Valsgärde 8 boat burial serves as an exemplar.[7] Here the mouth is bound by a woven textile which is in turn bound with a decorative band of woven bast fibres which is at least 45mm wide, obscuring some of the complex carving on the face of the scabbard. These bindings can only have been an addition to the scabbard to either reinforce it or, more likely to effect a repair (fig. 13.7).

When considering these bindings, what seems odd is that if they are providing reinforcement for what was clearly a common issue, why use organic materials rather than employing the sort of metal mouth bands employed on other scabbards of the period. We will return this question at the end of the next section.

Metal fittings

The two types of substantial metal components to be found on scabbards of this period are mouth bands and chapes. Neither are common finds in sword burials, with thirty-seven mouth bands and seventeen chapes in Cameron's study of one hundred and thirty-seven relevant examples of scabbard remains which in turn encompassed approximately two hundred and fifty sword burials. Metal mouth bands do greatly outnumber the examples of braid and cord bindings although this could simply be a result of the complete decay of the organic bindings. In the same study, we can observe that there are only five instances in which a chape was

---

[1] Cameron 2000

[2] Cameron 2006

[3] Walton Rogers 2005

[4] Cameron 2000

[5] Menghin 1983

[6] Bruce-Mitford 1975

[7] Arwidsson 1954

found in situ on a scabbard without a mouthband, leading us to conclude that a mouthband was either more popular (as a decorative element) or more necessary (as a structural one) than a chape.

Mouthbands

13.8 Simple, smooth copper alloy mouthband with punched decoration, from the Feltwell scabbard - Norwich Castle Museum

As well as being a decorative addition to a scabbard, the mouthband served the same practical purpose as organic binding; to reinforce a scabbards weak point at its mouth. At its simplest and crudest, the mouthband could be a strip of copper alloy, bent to fit around the mouth of the scabbard and closed with rivets, perhaps with some punched decoration. In fact, this is exactly what we find on the scabbard of the sword from Feltwell, recognised as being the earliest Germanic sword found in Britain to date.[1] Dating to the early 5th century and thought to have been brought to England from Southern Scandinavia, the Feltwell sword and scabbard are fascinating for many reasons, not least because of the absolute crudeness of the scabbard fittings. The punched unevenly spaced in lines which are anything but straight (fig. 13.8). It is possible of course that the mouthband was hurriedly fashioned to affect a repair, but the quality of the workmanship on the chape is no better. This form of 'smooth' mouth band continued in use until at least the mid-6th century, with the quality of the stamped decoration varying but never so poorly executed as that earliest example.

13.9 Gilded copper-alloy ridged mouthband from grave 303, Alton, Hampshire in the collection of the Curtis Museum, Alton

Another form of long lived design was the 'ridged' mouth band (fig.13.9). Rather than being a strip of worked metal, these were cast (usually in copper alloy which was then gilded) and featured parallel ridges running horizontally across the front of (and sometimes all the way around) the band. Some of these also feature simple stamped decoration. In some cases the entire band was cast as one piece but in some instances the ridged front only extends to the beginning of the back face, with the gap being filled by a separate plate which was soldered into place. A scabbard from the cemetery at Mucking had a ridged mouthband which was made of leather into which the ridges had been impressed, covered with a thin gold foil and fixed to a plain strip of copper alloy, which may have originally been part of a smooth mouth band.[2]

These ridged mouthbands were in use for the majority of the 6th century but Continental finds of exactly the same form (Merxleben grave 3 for example)[3] demonstrate that they were in use from at least the second half of the 5th century. In some instances, a small pierced lug depends from the bottom of the mouthband either as

---

[1] Gurney 1986

[2] Hirst & Clark 2009

[3] Miks 2007

part of the casting or as part of a plate soldered to the back (see Dover Buckland grave 96b).[1] The function of this lug is impossible to determine but it seems likely that it is to allow for something, presumably a bead or other object, to be suspended from the scabbard mouth.

Whilst the decoration of both the 'smooth' and 'ridged' form of scabbard mouth is relatively simple in style and execution, the third category feature much more complex cast designs. Cameron further divides these into three subcategories; The 'Abingdon' type with multiple rectangular fields containing geometric designs (fig. 13.10), the 'Brightampton' type featuring single rectangular fields containing spiral patterns (fig. 13.11) and lastly the 'Pewsey' type, the most complex form, with "Animal Style I" decoration either being contained in rectangular fields or being allowed to spread across the bottom of the band (fig. 13.12). This last type sometimes features additional decoration in the form of garnets, as seen in the examples from Alfriston (fig.13.13) and Selmeston (fig. 13.14), both to be found in the Barbican House Museum in Lewes, Surrey.

13.10 'Abingdon' type copper-alloy mouthband from grave 21, Petersfinger, Wiltshire in the collection of Salisbury Museum

Cameron states that all three of these types date to the 5th century and, in the case of the Abingdon and Brightampton types this doesn't seem unreasonable. The spiral decoration of the Brightampton type can be seen on scabbard fittings (Nydam and Kragehul) and jewellery (a large silver gilt fibula from Skjerne) from Denmark dating to the early/mid-5th century, whilst the decoration of the Abingdon style seems to exhibit a Late Roman influence and is mirrored in 5th century finds from the Lower Rhine region of Germany (for example, the find from grave 43 at Krefeld-Gellep).

Dating the Pewsey type to the 5th century is more problematic as, unlike the Abingdon and Brightampton styles, there are no continental precursors which can be securely dated to earlier than the 6th century. Dating of these finds then depends upon opinions as to the dating of "Style 1" art (for a discussion on Style 1 art, its origins and meanings, see Pollington's "Wayland's Work" p.101).[2] Inker sees Style 1 as a 6th century development,[3] as do others.

13.11 'Brightampton' type gilded copper-alloy mouthband from grave 31, Brightampton, Oxfordshire. Ashmolean Museum, Oxford

[1] Evison 1987

[2] Pollington, Lindsay & Hammond 2010

[3] Inker 2006

13.12 'Pewsey' type gilded copper-alloy mouthband from grave 22, Pewsey, Wiltshire. Wiltshire Museum, Devizes

13.13 'Pewsey' type gilded copper-alloy mouthband from Alfriston, Sussex. Barbican House Museum, Lewes

13.14 'Pewsey type gilded copper-alloy mouthband from Selmeston, Sussex. Barbican House Museum, Lewes

13.15 Quoit-brooch style copper-alloy scabbard mouthband from Cheriton, Hampshire (PAS SUR-029B13)

In addition to Cameron's three subcategories of complex cast designs there is a fourth which is unquestionably 5th century in origin. Discovered near Cheriton, Hampshire in May 2013, it is a cast copper alloy mount, partially inlaid with silver foil and decorated in the "Quoit Brooch" style, which combines animalistic forms with geometric designs and which emerges in the first half of the 5th century as a Germanic response to Imperial art of the late Roman period. The mount, recorded as SUR-029B13 in the Portable Antiquities Scheme database (fig. 13.15) was a stray find by a detectorist so little more can be said about it (indeed, it is impossible to establish for certain that it is from a scabbard) but it is, perhaps, contemporary with the Feltwell find, giving as an example of fine early 5th century work to stand alongside the Feltwell's crude one.

Chapes and chape mounts

At the other end of the scabbard we (rarely) find the chape. It seems obvious that the function of the chape is to reinforce the tip of the scabbard, presumably to protect against damage to the outside. If, however, such damage was common, we might expect chapes to be more numerous finds so it is entirely possible that, although a chape would serve to protect the scabbard, its primary function is decorative and is as much about displaying wealth and status as it is of any practical use. In Cameron's study, it was observed that the average size of those swords found with chapes exceeded that of swords without, which (following the thinking applied in Härke's study of early Anglo-Saxon knife blades[1]) suggests a link between the use of chapes and statements of status (or, more prosaically, that larger swords were more likely to suffer damage at the tip). It is also possible that the chape may have added weight to the bottom of the scabbard, helping to counterbalance the hilt fittings of the sword.

Chapes take to general forms; narrow edged and broad edged. Narrow edged chapes are typically u-shaped in cross section, presumably conforming to the shape of the scabbard edge, and are usually made from copper alloy (sometimes gilded) or silver.

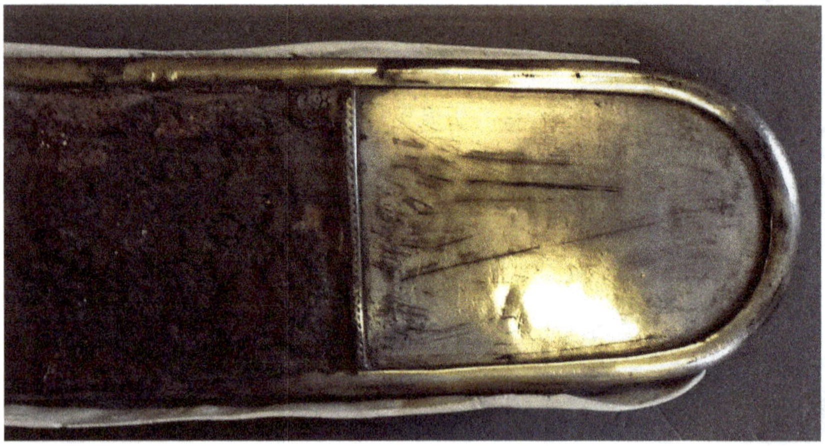

13.16 Silver asymmetrical narrow edged chape with gilding and niello decoration from King's Field, Faversham, Kent. British Museum (inventory number .951.70)

They are held in place by small rivets that go through the chape and scabbard edge. They can be equal armed or asymmetrical and can vary greatly in length. One silver asymmetrical example from Faversham extends up one edge of the scabbard for 480mm (fig.13.16).

13.17 Copper alloy equal armed, narrow edged chap from grave 42, Alton, Hampshire. Curtis Museum, Alton

---

1 Härke 1989

The chape from Alton grave 42[1] was the only equal armed, narrow edged chape from an Anglo-Saxon context recorded by Cameron. It has a bronze sheet which sits within the chape, covering the front of the scabbard, topped with a bronze strip with transverse line decoration (fig.13.17). Evison dated it to the 5th or 6th century but chapes of this form can be seen on swords from the Danish bog deposits of the 4th and 5th centuries.

Some asymmetrical narrow edged chapes also have this infill sheet and transverse band. The aforementioned example from Faversham (which is silver gilt) has an infill sheet of silver, topped with a band of silver which features niello decoration.[2]

Broad edged chapes are usually made from copper alloy and can also be subdivided into two main categories; long armed and short armed. Both forms are symmetrical and both offer a greater opportunity for decoration than the narrow-edged variety.

13.18 Copper alloy short armed, broad edged chape with gilding, from Kingston-by-Lewes, Sussex. Barbican House Museum, Lewes

Most of the continental examples of the short-armed variant are to be found in Scandinavia and many of these have fields upon them which contain additional decoration. A few examples also have a knob-like projection on the base and one of this type may have been found as part of the contents of a bag in a grave at Mucking, although the find is broken and its actual form is unclear. The small number of intact English examples (Cameron records only three) are relatively plain, with the only decoration being some transverse banding across the tops of the arms and a border around the inner front edge, as can be seen on the example from Kingston-by-Lewes (fig.13.18). This same form of decoration can be seen on several examples of the long-armed form.

The decoration of a long armed broad edged chape from Brightampton grave 31 is another interesting example of a scabbard fitting with 'Quoit Brooch' style decoration, in this case a line of gilded beasts and other figures running around the face of the chape (fig 13.19). The chape was obviously very old and worn, the detail on the decoration already obscured by wear, when it was buried. Even more interestingly, the mouthband fitted to the same scabbard was also of early/mid-5th century date, Cameron's "Brightampton" style featuring the spiral decorations more usually found on fittings from Denmark (see above). Again, this fitting was well worn, raising the possibility that these two objects, from two different cultures, may have remained on the same scabbard to be handed down at least once, perhaps twice, before being buried in the late 5th or early 6th century.

---

[1] Evison 1988

[2] Evison 1967

13.19 Quoit-brooch style gilded copper-alloy broad-edged chape from grave 31, Brightampton, Oxfordshire. Ashmolean Museum, Oxford

Cameron's study listed only two chapes as being made from iron rather than the usual copper alloy or silver (omitting a third, from Abingdon grave 49).[1] Whilst one of these (from Pewsey grave 22) seems to be a simple equal armed, narrow edged chape,[2] the other (Abingdon grave 42) whilst now fragmentary, supported an additional decorative element in the form of a cast copper alloy mount, fitted to the tip of the scabbard wrapping around from the front, underneath the chape to the back. The decoration on the front consists of a face, flanked by two birds whereas the back is plain. When Cameron compiled her corpus, this was the only known example of this form of mount from an Anglo-Saxon context, although there were several examples known from Germany and Belgium, (being one of the five styles of chape mount listed by Menghin in his typography of such fittings).[3] Since then, seven more of these mounts have been discovered by metal detectorists and recorded in the Portable Antiquities Scheme database. All of these feature the same design on the front, a human head between two inward facing bird heads. One example in the PAS database (HAMP-4CBF82) from Micheldever in Hampshire is so well preserved that it is clear that the human head is that of a male (fig.13.20). Obviously, this brings to mind the god Woden/Odin and his two ravens (who may be the character being depicted in the many Anglo-Saxon pendants which consist of a male face wearing a helmet bearing horns tipped with two birds heads – see PAS items HAMP2432 and LEIC-40DB05 for examples – fig.13.21) and perhaps the addition of this very distinctive form of mount marked the wearer out in some way, possibly as an Úlfheðinn; one of Woden's own 'wolf-warriors'.

---

[1] Leeds & Harden 1936

[2] Annable & Eagles 2010

[3] Menghin 1983

13.20 Copper alloy chape mount featuring a male face between two birds from Micheldever, Hampshire (PAS HAMP-4CBF82)

13.21 (left) Gilded copper-alloy 'Woden' pendant from Soberton, Hampshire (PAS HAMP2432)
13.22 (right) Copper alloy chape mount from Near Eastbourne, Sussex (PAS SUSS-427FF4)

The PAS database also contains examples of other forms of chape mount which conform to Menghin's types 3b and 3c, but there is also what appears to be a distinctly English variant which has no exact continental equivalent. This variant has little by way of decoration, with no faces or animal heads, but instead features three circular recessed settings through which the pins securing them to the scabbard passed (fig 13.22). It seems likely that these settings each held some sort of decorative material, glass or garnet possibly, which would conceal the rivet head and in one example (from Burnham Thorpe, Norfolk, PAS NMS-751713) the remains of a piece of glass are still in situ.

Another example of a similar type (but with simple holes for the pins rather than recesses) was discovered as part of a hoard of gold and silver at Patching, West Sussex in 1997, thought to have been deposited in the third quarter of the 5th century (approx. 461 AD).[1]

Given that the known number of these chape mounts has increased from one to at least a dozen as a result of work of metal detectors outside of funerary contexts, it seems reasonable to assume that they were only fixed to the body of the scabbard with fairly short pins rather than being attached in some way to the chape itself and, as such, were relatively easily lost (unless they were deliberately removed and deposited, see "The Riddle of the pyramids", "The Meaning of the sword" and "Adorning the Sword", all in this volume, for further discussions of this possibility). As mounts of this type seem to serve no practical purpose which the chape itself does not already serve, it seems that they must be primarily decorative or perhaps (in the case of the decorated variants) symbolic/apotropaic.

Edge Reinforcers

The final structural elements to consider are the edge reinforcers. These short lengths of metal are u-shaped in cross section, usually of copper alloy (sometimes gilded) and are riveted in pairs to the edges of the scabbard, in the upper half. Again, they are rare finds, with only fifteen instances recorded in Cameron's study. Some have simple decoration in the form of transverse ribs.

The purpose of these items seems to be to reinforce the edges of the scabbard at the point where a strap forming part of the suspension mechanism encircles the scabbard. Some of the reinforcers are fitted with pairs of disc-headed rivets, part of the scabbard's suspension system. This method of suspension will be further discussed in Chapter 14..

Before moving on from the subject of metal adornments, it is worth mentioning that one recently published find raises the possibility that some English scabbards were partially or fully sheathed in metal. The scabbard of a sword discovered at Temple Hill, Dartford is described as being "plated on each face with thin copper-alloy sheet"[2]. The report states that this sheathing is best preserved on the back of the scabbard at the lower end so this may simply be from the infill sheet of a long armed chape. It is unclear what evidence there is for a more extensive covering but it may be that these are fragmentary remains of decorative pressblech foils, fitted to parts of the scabbard in the same manner as those found on the example from Gutenstein, Baden-Württemberg (see fig 11.25). The sword itself seems to be ordinary with one notable exception; it is a ring-sword, possibly one of the oldest found in England. Whilst the meaning of these swords and the ring mounts which they bear is open to debate (see Chapter 11 – Adorning the Sword), there can be no doubt that they are of considerable significance so it is intriguing to think that the scabbard might bear images or decoration of equal significance.

Further observations relating to scabbards and scabbard decoration can be found in the concluding comments in Chapter 14 - "Wearing the Sword".

---

[1] White 1998

[2] O'Brien 2015

14. Wearing the sword

Matt Bunker

Introduction

As stated at the beginning of chapter 13, one of the functions of the scabbard is to allow swords to be carried on the person, suspended from a strap which could take the form of either a waist belt or a baldric. The actual method employed to suspend the scabbard is usually impossible to determine from the few remains found in early Anglo-Saxon inhumations, leading to the conclusion that it either relied on organic elements which are now lost to decomposition or that the material elements were removed prior to deposition. As swords seem, for the most part, to have been deposited in graves next to the body rather than being worn as they were in life, the latter is a distinct possibility. In this section, I shall consider those elements of suspension methods which do survive and how they functioned. For this, we shall look at what the archaeological record (for both English and continental contexts) can tell us but also what we can learn from the iconography of the period.

Suspension Fittings

Rivets

As mentioned in Chapter 13, some of the pairs of edge-reinforcers are fixed to the scabbard with disc headed rivets which stand proud of the reinforcers (see Pewsey Grave 22 for a particularly fine example) (fig. 14.1).[1] It is assumed that the heads of these rivets passed through small slits in the suspension strap, which must have formed a tightly fitting loop around the scabbard at this point, presumably by means of an additional short strap fixed to the main strap. It seems likely that these rivets served to prevent the scabbard from moving within the loop, with the friction between the loop in the strap and the metal edge-reinforcer being the primary means of bearing the weight of the scabbard.

14.1 Detail of the edge reinforcers with suspension rivets from grave 22, Pewsey, Wiltshire. Wiltshire Museum, Devizes

In some instances, the two straps which create the loop into which the scabbard fits were joined together by additional rivets (see, for example, Great Chesterford Grave 122),[2] providing a method of closure that was more decorative than simply stitching the straps together (which is what we must presume to be the case where no supplementary fittings are evident). All of the English examples of scabbards with these riveted closures also have edge reinforcers, although some continental examples such as the one from Krefeld-Gellep Grave 43 (which has the rivets but no reinforcers) demonstrate that this was not always necessary. The example from Brighthampton Grave 31

[1] Annable & Eagles 2010

[2] Evison 1994

features reinforcers with disc headed rivets, the additional rivets for creating the suspension loop and also a cruciform stud which may have been fixed to the strap which passed across the front of the scabbard, presumably serving as a decorative rather than structural element.[1]

Strap Slides

A strap slide is a bridge on the front of the scabbard under which the suspension strap passes. They seem to have originated as a method of scabbard suspension amongst the nomadic horseman of the southern Ural Steppe in the sixth or seventh century BC and from there moved both eastwards and westwards to be adopted by many different cultures[2].

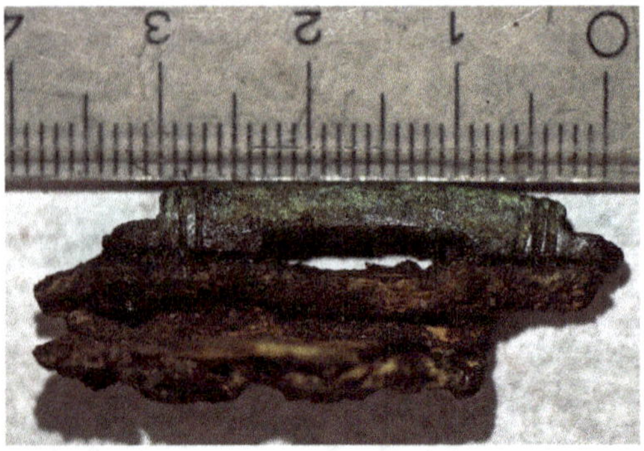

14.2 Copper alloy scabbard slide, one of a pair, with scabbard remains, from Grave 10, Temple Hill, Dartford (TEMP 2015 266)

Most of the surviving examples from England are wooden, either carved directly into the wooden core of the scabbard or applied to the core and would presumably have been concealed beneath the outer covering. The slide on the scabbard from Grave 5537, St Mary's Stadium Southampton[3] appears to have had some simple decoration carved into it which was probably displayed via the foundation moulding of the thin skin product which covered it, although it may have been that the covering was cut away over the slide in order to display the carved wood itself. Some of these slides have a waisted shape, splaying out at the ends (this is very noticeable on the example from St Mary's Stadium) but an example from Wickhambreaux in Kent is rectangular in shape.[4] Despite having apparently been covered, most surviving examples of these wooden slides appear to have been further secured by the type of braid bindings previously discussed, providing much needed additional reinforcement to the suspension at a point where it would be under the most stress. Unless they are merely decorative, the presence of these bindings can be taken as evidence that the slide was probably applied to the surface of the wooden core rather than being carved.

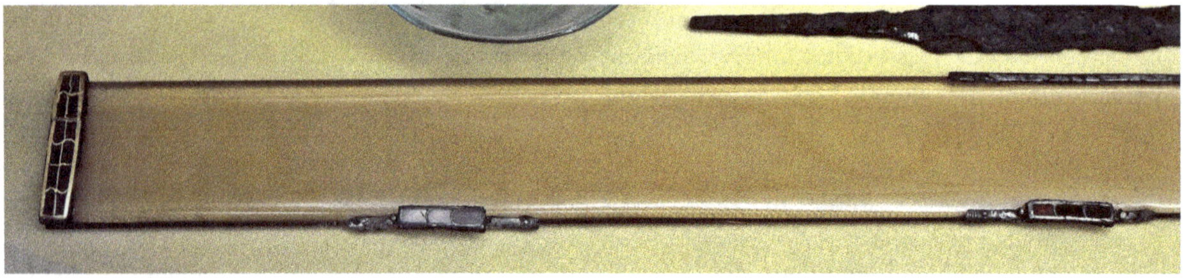

14.3 Scabbard slides mounted on the same side of the scabbard, creating a two-point suspension, from grave 319b, Lavoye. Musée d'Archéologie Nationale, Paris.

---

[1] Ackerman 1860
[2] Trousdale 1975
[3] Birbeck 2005
[4] Cameron 2000

Although metal slides are rarely found in England, continental evidence shows that they could be used singly (which is the norm for the many examples from Denmark between the third and fifth centuries) or in pairs (a frequently used method found on Continental examples of the fifth and sixth centuries). The few English examples seem mostly to be of the latter variety.

Cameron lists two pairs of slides found in situ in her study, from Alfriston Grave 26 (apparently undecorated iron) and Riseley Grave 86 (copper alloy with cross hatching which may have once held decorative inlay). A third pair was discovered in 2002 during excavations of an early Anglo-Saxon cemetery at Temple Hill, Dartford (figure 14.2). Although the excavation report only describes one slide being found in Grave 10[1], an inspection of the finds in the excavation archive held at Maidstone Museum by Dieter Huggins and I in August 2017, established that in fact a pair had been found. This small pair of copper alloy slides, approximately 35mm in length have three transverse grooves at either end and terminate in flat lugs through which pass rivets, three of four still being in situ. To each slide is still attached the mineralised remains of the wooden core of the scabbard to which it was attached, to a length of 37mm on one of the slides and 70mm on the other. Both of these pieces of scabbard are intact for the entire curve around the edge of the blade, which allows us to establish that these two fittings were rivetted to the upper face of the scabbard at the extreme edges rather than towards the middle as might have been expected based on a study of the position of such strap slides on Continental finds. Although the flat lugs would allow the use of a braid or cord binding over the ends of the slides, there are no visible remains or discolouration to indicate that this was the case. Given that elements of both the skin product covering and a textile within which the sword and scabbard were wrapped have survived, it doesn't seem unreasonable to conclude that the slides were secured by the rivets alone. Although this pair of slides can be categorised as belonging to Menghin's Type 2b[2] they are (as far as can be established) the first pair of this type which can be identified as having been fixed to opposite edges of the scabbard. The only other example in Menghin of Type 2b slides being fitted to the edge of a scabbard is that from Lavoye, France, Grave 319b (fig. 14.3). Here, both are fixed to the same edge of the scabbard, the left, with one slide being located near the top of the scabbard and the other about two thirds of the way towards the tip. This creates a basic two-point suspension system to which the open ends of the baldric would be fixed. As with the sword-ring from the same grave, discussed in Chapter 11 of this volume, the strap slides and associated scabbard remains from Temple Hill are unique finds, deserving of further study.

To these in-situ finds we can add a stray find from Barham, Kent, discovered in 2011 and recorded in the PAS as KENT-B9EB67 (fig.14.4). This fabulous piece is of gilded copper alloy, has a boar's head at each end and recesses along the back which may have held decorative stones of some sort. Both ends terminate with a low, projecting tab which would have slid into slits in the scabbard's covering, to be secured with an overlay of cord or braid bindings. Whilst it may have been used as a single slide, continental evidence suggests that it is more likely to have been one of a pair.

In addition to the slides which have survived, we have a very clear image, albeit Continental, of one in use. A sixth century pressblech patrix, one of four discovered in 1870 at Björnhovda, Sweden, depicts a figure wearing a scabbard suspended from a baldric which runs through a single rectangular slide (fig. 14.5). This is a rare instance where iconography of the period (albeit from Scandinavia) demonstrates scabbard suspensions in use.

---

[1] O'Brien 2015

[2] Menghin 1983

14.4 Copper-alloy scabbard slide with boar's head decoration from Barham, Kent (PAS KENT-B9EB67)

Bosses

Another piece of evidence for a form of scabbard suspension can be found on the helmet from boat burial XIV from Vendel in Uppland, Sweden.[1] One of the decorative pressblech panels depicts two helmeted warriors, each wearing a sword suspended from a baldric. The method of suspension is clearly depicted; the scabbard bears studs or bosses which pass through slits in the baldric. These bosses are somehow fixed to the scabbard and seem to bear its weight (fig. 14.6).

14.5 Pressblech patrix from Björnhovda, Sweden showing a scabbard with slide suspended from a baldric

Gold and garnet cloisonné bosses have been found, both singly (Wickhambreaux) and in pairs (Sutton Hoo Mound 1), in association with scabbards and there are several examples of finds from the continent. There are also stray finds recorded in the PAS database, including examples from Griston, Norfolk (NMS-Z32B24), Cotgrave, Nottinghamshire (DENO-2A0601) and the Staffordshire Hoard. One of the rare pieces of organic material from the Hoard has been shown to be a stone bead into which one of the bosses sat (fig. 14.7), something which is seen on Continental examples (see chapter 15i in this volume for further details). In addition, Evison[2] observed circular depressions on the front of scabbard remains from three of the swords from the Buckland cemetery, similar to the pair of circular depressions left in the front of the Sutton Hoo Mound 1 scabbard by its two bosses. She concluded that the depressions observed on the examples from Buckland had each been left by a single boss which had either been removed prior to deposition or made from an organic material which had subsequently decayed. There is plenty of evidence for the use of organic materials for scabbard fittings on the continent (most notably pyramids made of bone) which, if they were ever used, never survive in England.

Whilst it is impossible to know from the archaeological or iconographic evidence whether any of these examples were load-bearing parts of the scabbard suspension or were merely decorative elements, practical experimentation using reconstructions has shown that it is possible to use bosses of this type to recreate a functioning suspension system which matches that depicted on the Vendel helmet (fig. 14.8)

---

[1] Stolpe 1912

[2] Evison 1987

Pyramids

As well as its pair of bosses, Sutton Hoo Mound 1 also contained a pair of gold, garnet and glass cloisonné pyramids. Another pair, equally beautiful in workmanship were discovered at the same site in Mound 17 (fig.14.9). Other pyramids have been found in Anglo-Saxon burials (usually in pairs) and large numbers of them are recorded in the PAS database and elsewhere (see chapter 15ii in this volume for further detail). Not all of these are cloisonné pieces, some are cast copper alloy, some silver, some gilded. Amongst the large number of continental finds there are also examples made from organic material; wood, bone and antler (fig.14.10).

As with the bosses discussed above, it is impossible to know how many of these pyramids, so often found in association with scabbards, formed part of the suspension mechanism and which were either merely decorative or perhaps a part of the 'friđbond' or peace-bands described in later Norse sagas. The sheer volume of stray finds recorded in the PAS database could be taken as an indication that, whatever their usage (and it is by no means certain that they were only used in conjunction with scabbards or other martial equipment), their loss was not immediately noticeable and, as such, they were not load bearing elements but this is assuming that the finds outside of burial contexts represent accidental losses and not deliberate depositions. It is possible that they were fixed to the suspension strap either side of the slide to prevent the scabbard moving along the strap. Used like this, the loss of a pyramid would not be immediately noticeable. It is also possible that pyramids could have been used like bosses, passing through slits in a baldric and fixed to the body of the scabbard perhaps by stitching or tying them to a leather band encircling the scabbard. Lars Lüppes has done a great deal of work in reconstructing methods of suspensions using pyramid finds (including a form in which the pyramid is set onto a rectangular base-plate, something not yet found in England) and his paper on the subject is highly recommended.[1]

One definitive example which shows the way in which pyramids could be used as part of a suspension mechanism comes from Altdorf in Switzerland. This scabbard will be discussed in more detail below.

14.6 Pressblech from the helmet from the Vendel XIV boat grave, showing scabbards mounted with bosses, suspended from baldrics (Helmet reconstruction by Grzegorz Kulig, with thanks to Dieter Huggins)

14.7 Scabbard boss, one of a pair, and its stone bead mount, from the Staffordshire Hoard (K675 and K764)

[1] Lüppes 2010

14.8 Suspension reconstruction utilising scabbard bosses on organic mounts, emulating the method seen on the Vendel XIV pressblech.

14.9 (below) Reconstruction of the pyramids from Mound 17, Sutton Hoo, by Dave Roper, Ganderwick Creations.

14.10 Pyramids carved from bone from graves 441 and 712, Kletthamer cemetery, Erding. Erding Museum.

Two-point suspensions

Whilst both of the contemporary depictions that we have of scabbards being worn show only a single suspension strap, there is compelling evidence to demonstrate that this was not always the case.

As mentioned earlier in this chapter, the scabbard from Lavoye Grave 319b used slides in such a way as to create a two-point suspension, only really suitable for use with a baldric. The low placement of edge reinforces on some scabbards (Great Chesterford Grave 122 for example)[1] can be taken as evidence that two-point suspension systems were also in use in England, although it is unclear whether a baldric was employed or whether a waist-belt and additional lower connecting strap was used. It is possible that the extended arms on some of the longer, narrow walled chapes could have served the same purpose, reinforcing the scabbard at the point of connection.

A further evolution in two-point suspensions is the development of specialised fittings for the lower point of suspension. These can take the form either of a shaped plate, riveted to the lower strap to create a loop (as can be seen in the example from Altdorf) or a buckle with extended plate, curved to fit the body of the scabbard which, again, works to create a loop in the lower strap. This latter form is found in both Mound 1

[1] Evison 1994

and Mound 17 at Sutton Hoo (fig.14.11) and seems to be a late 6th or early 7th century development (although a buckle which may have served the same function was found with the scabbard from Basel-Kleinhüningen, Grave 63, which is dated to the late 5th or early 6th century). The lower buckle from Mound 1 (actually a false buckle, as is the example from Mound 17; the tongue does not move) is part of a larger suite of fittings which, when taken together are obviously from a sword belt rather than a baldric. There is even a remarkable articulated fitting which served to connect the belt to the lower strap; this would have no place on a baldric (fig.14.12).

14.11 Lower 'buckle' scabbard fitting from Mound 1, Sutton Hoo. British Museum

The utilisation of a two-point suspension system with a lower strap keeps the scabbard tilted forwards. In the Swedish depictions of scabbards being worn with seemingly single suspensions, all are tilted forward at an angle, although this may be an artistic convention to allow the hilt of the sword to be clearly displayed as, when such systems are reconstructed, they have a tendency to hang vertically or near vertically. Whilst this is not a particular encumbrance when on foot (especially as, if the depictions are to be taken at face value, scabbards were worn high up on the body with the hilt close to the armpit), it is a much more awkward position for someone on horseback. With the scabbard tilted forward by a secondary lower suspension point, particularly when worn lower down on the body with a waist belt, drawing the sword takes it diagonally across the body rather than vertically. A vertical draw might be better suited to the man on foot, particularly in the press of close combat, as it requires less space on either side than the cross-draw but it is more cumbersome to execute and exposes the wielder's underarm in the process. The young man buried in Mound 17 was accompanied by equestrian equipment as well as his horse. Although there is no corresponding evidence from Mound 1, it seems unlikely that a person of such high status would not also have been a mounted warrior and so, although it is not conclusive, we can perhaps draw a tenuous connection between equestrianism and this form of scabbard suspension. A study of other early English 'warrior' burials which contained remains of horses[1] didn't reveal any evidence of the suspension methods used but, as with the other methods discussed, it is possible to create the two-point suspension method using purely organic fittings or even with no additional fittings at all.

The Altdorf Scabbard

It seems appropriate to conclude this discussion on scabbard fittings by giving a brief description of an exceptional set of remains. Unfortunately, it is not an English find but it does contain many of the elements discussed above and demonstrates how they would have functioned together.

Discovered beneath a stone slab at St Martin's Church at Altdorf, Switzerland in 1969, the grave was that of a wealthy man in his 40s. He was buried with a sword in its scabbard, seax, knives, shield, bow and arrows.[2] The scabbard, made from beech and covered with goatskin, has a pair of edge reinforcers, a pair of relatively plain pyramids and a complex set of belt fittings which included fittings for a lower suspension strap. The grave also included stirrups. Crucially, as well as all of these fittings, the grave also contained considerable remains from the leather suspension strapping. This configuration clearly demonstrates that the pyramids acted as both a load bearing part of the suspension and as pivot points, connecting the two ends of the main belt strap with a strap that tightly encircled the body of the scabbard, which in turn passed beneath an integral wooden slide and around the two edge reinforcers. The Altdorf scabbard is the only surviving example which shows the relationship between all of these elements and how, when combined, they create an effective, adjustable means of suspending the scabbard. (fig 14.13)

14.12 Articulated connector strap fitting from Mound 1, Sutton Hoo. British Museum

[1] Fern 2005

[2] Marti 1995

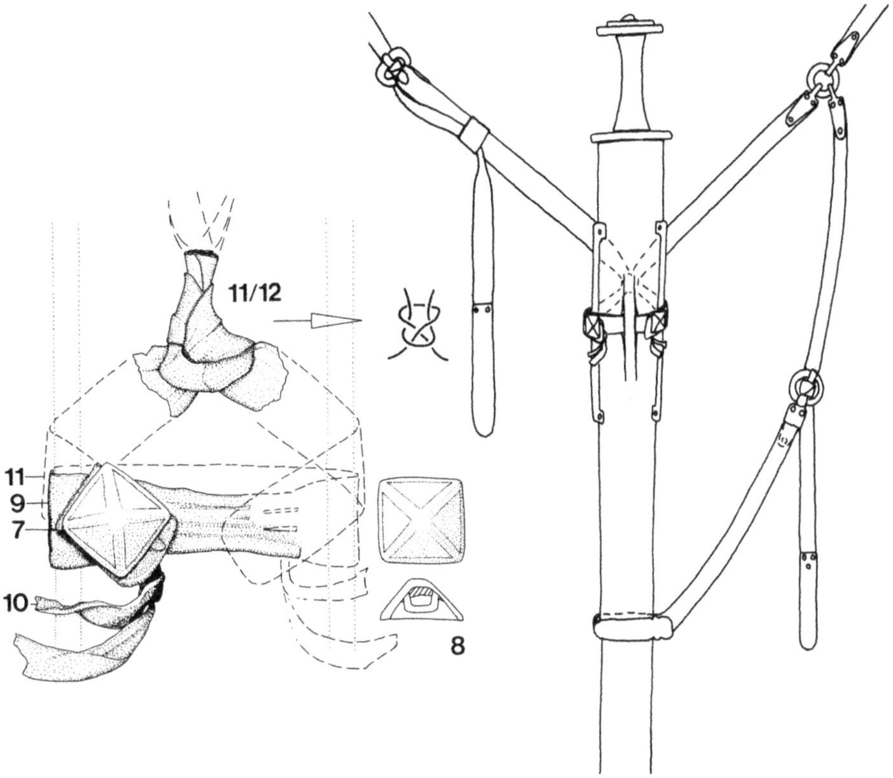

14.13* Detail of the surviving suspension and a reconstruction based on all the surviving elements from St. Martin's Church, Altdorf. Illustrations from "Das Grab eines wohlhabenden Alamannen in Altdorf UR, Pfarrkirche St. Martin", reprinted by kind permission of Reto Marti.

Closing thoughts on scabbards

As stated previously, most scabbards from early Anglo-Saxon burials show no sign of decoration and are devoid of fittings. Given that the scabbard is such a large opportunity for showing off wealth and status, this seems unusual (although the same could be said of the hilt fittings of swords, the majority of which were seemingly plain if the majority of finds can be taken at face value). The lack of obvious decoration can perhaps, for the most part, be put down to the poor preservation of organic material in English soils. The surfaces of most scabbards are so badly degraded that it is impossible to tell if there was ever any decoration present. However, that which is in evidence does seem to match the relatively simple decoration seen on the majority of scabbards from other late 5th-7th century Scando-Germanic contexts, none of which match the highly complex decorative carving seen on two of the late 5th century examples from context IV at Nydam[1] (fig.14.14). These do seem to be exceptional examples, perhaps those of the leaders of the armies whose weapons are deposited at the site, as other examples of carved scabbards from the Danish bog deposits exhibit work that seems to be on a par with the style of carving seen on the fragment from Charlton Plantation (fig.14.15).

[1] Jorgensen and Van Peterson 2005

14.14 Complex carving on a scabbard with surviving skin product from Nydam. National Museum, Copenhagen.

14.15 Simple carving on a scabbard from Nydam. National Museum, Copenhagen.

One decorative method not seen in England for which there is a continental example comes from Gutenstein, Germany in the form of decorative pressblech panels, executed in silver, fixed to the face of the scabbard using copper alloy edging and with stamped silver strips covering the joins between the panels. The uppermost panel shows a figure bearing a scabbarded ring sword in its left hand and a spear or staff in its right. The figure has the head of a wolf, leading to the obvious conclusion that this is a depiction of an Úlfheðinn, perhaps receiving his new sword as part of a ceremony celebrating him taking his place as one of Woden's warriors. The rest of the decoration shows animal heads and geometric designs which date the scabbard (or at least its decoration) to the 7th century (see fig. 11.25).

The lack of fittings on English scabbards is more curious, with two possibilities presenting themselves. It is possible that in some cases fittings which are no longer evident were made from organic materials. Slides, pyramids and bosses could be wood, bone or antler and continental examples still exist from a wide variety of geographical contexts. Mouth reinforcement could be of leather (as seen at Mucking) or cord binding and the protection provided by a chape could come from an extra layer of leather or cord at the scabbards tip (although there is no evidence that this took place during the period). As with surface decoration, all of these could easily decay over the centuries, leaving little or no trace. Another possibility is that some or all of the fittings and suspension straps were stripped from the scabbard prior to deposition, leaving just the leather covered core to protect the sword in the grave. This might have been a simple case of recovering valuable elements for re-use or perhaps it is more complex than that. It is possible that the fittings, which would presumably have been recognisable within the immediate community, were of symbolic value and needed to remain in circulation, to be fitted to the scabbard of the heir to demonstrate to all that he was the legitimate successor. It may be that some elements of the fittings (particularly the finer examples of the pyramids and bosses) were given to the bearer by a superior as badges of rank or importance which, in some cases, were returned upon the death of the bearer. Similar theories regarding elements of sword hilt fittings are explored in Chapter 11 of this volume.

Of course, it's also possible that, in many cases, there were no additional fittings. It is possible to create a functional suspension system for a sword without any additional elements, simply by stitching a short length of strap to the main strap to create a loop in which the scabbard sits or by creating two loops at either end of a baldric and seating the scabbard in those. Other fittings are mechanically unnecessary and may have been used more to satisfy fashion at a particular point in time rather than functional necessity. After all, the finest sword with the finest suspension fittings known from early Anglo-Saxon England, the one from Mound 1 at Sutton Hoo has neither chape nor mouthband whereas one of the poorest examples of 5th century work, the scabbard from Feltwell has both.

15i. Additions to the sword: beads, bosses and pyramids

Paul Mortimer

Sword-beads[1]

It is rare to find in Germanic graves men buried with beads or jewellery worn as amulets but less so if that jewellery is associated with a sword. The practice of wearing beads on a sword seems to have begun quite early and originated in the east. Meaney, quoting Werner, relates that it was probably copied from the Sarmatians in the first century BC.[2] The use of sword-beads seems to have travelled westwards, spread by the Huns and eventually reached the Rhineland. The Thorsbjerg bog deposit (100 to 500 A.D.) shows that the soldiers represented in the bog find were using them,[3] as at least nine beads have been found there in association with swords; these vary in size between 45 mm to 53 mm and are mostly plano-convex. Around twenty-three sword ornaments were located in the Nydam deposit and some of these are very elaborate and most are beads of glass, including millefiori.[4] A few have metal attachments and one is stone, triangular in shape and has runes scratched on to it. More sword-beads, at least thirteen, have been found in the Porskjær Mosefund,[5]and another five at Kragehul.[6] Perhaps even earlier are the beads from the Ejsbøl lake(1st to 5th century) where three sword-beads were recovered, one is a green glass bead, partially covered with a tapering gilt-silver collar.[7] Vera Evison believed that the habit of wearing sword-beads had travelled to England by the 5th century due to the influence of the Franks and the Alamanni.[8] Whatever, the custom of attaching amulets or beads to sword ensembles was established long before the traditional date given for the coming of the Angles, Saxons, Jutes, etc. to England. Evison draws our attention to two possible scabbard ornaments that are somewhat reminiscent of the triangular shaped bead from Nydam. The first is a perforated square horn bead,[9] about 2.5 cm in width, from grave 15 at Brighthampton, Oxfordshire which was found close to the pommel of the sword contained within the cut. That sword throat band has three decorative triangles, and a hole by which the bead may have been attached.[10] The second piece was found in grave 1 at Dorchester and is a perforated bone disc 4 cm diameter. The disc is decorated with six large ring-and-dot motifs and twelve small ones. Apparently, during the excavation several pieces of iron were just thrown in the river, which perhaps explains the absence of the sword which Evison felt the disc must have belonged to.[11]

---

[1] A fairly comprehensive list of glass sword-beads can be found in *Appendix 8 Sword-beads and Amulets*, compiled by Su and Mike Poole.

[2] Meaney, 1981. 196.

[3] They could have been employed at one time as Roman auxiliary troops, as, at least some of the weapons are Roman and from the 1st century A.D.

[4] Rau, 2010. 363 to 384 and Rau, 2010b. tafel 46, 47 and 48. The Nydam finds are from around 200 to 500 A.D.

[5] Nørgård Jørgensen, 2008. 54 to 57 and 177 to 178. Porskjær is dated 3rd to 7th centuries.

[6] Birch Iversen, 2010. 95 and tafel 59. Kragehul dates to 200 to 500 A.D.

[7] Ørsnes, 1998. Tafel 209. Revisited in Nørgård Jørgensen, Andersen, 2014. 165.

[8] Evison, 1976. 314.

[9] Meaney says that it is 'antler', see below.

[10] Evison, 1976. 309.

[11] Evison, 1976. 309.

15i.1 A sample of reproduction sword beads made by Tillerman Beads.

Glass is the most frequently used material for beads found with Germanic swords, although the colours and forms can be quite varied. It is possible that the use of beads could have been more common but that beads made from organic materials or ceramics might not have survived. (figure 15i.1)[1] Meaney lists three made from plain green glass: grave 49* at Abingdon, Berkshire which was 32 mm in diameter and dated to the late 5th century, grave 2* from Alton, Hampshire and grave 86, Horton Kirby, Kent.*[2] Another bead, this time from grave xx, at Howletts, Kent, was plain yellow glass.[3] The overwhelming majority of glass beads though, are multi-coloured. There is a large, late 5th century bead from Luffenham, Rutland* which is a flat disc with a zig-zag trail around it that looks like a 'sunflower'. Another from the same period was found at Fairford, Gloucestershire, again a disc, but was bluish-green with red trails 'cable fashion' around the circumference. A striking, cylindrical bead came from grave 93 at Dover (figure 15i.2); this is 28 mm in diameter and is composed of multi coloured layers, mainly red and green stripes running throughout in a spiral away from the centre. This has been dated to the 6th century. Another 28 mm disc was found at Petersfinger, Wiltshire, grave 20,* again, green but of a dark hue and translucent but with opaque, yellow zig-zag trails around it. The cemetery at Little Wilbraham, Cambridgeshire produced three beads, unfortunately undated. In grave 44 the bead was a 'translucent mid-green glass disc', of 35 mm, it has white zig-zag trails around it; grave 96* contained a black glass bead, one side is convex but the other is flat, it has white trails in a five-petal shape and is 43 mm;[4] grave 151 contained an 'oval black glass bead' with 'light blue crossing trails and red dots'; this bead was 25 mm in

15i.2 Multi-coloured cylindrical bead from grave 93 at Dover Buckland. Reproduction made by Tillerman Beads.

[1] Beads marked with * are represented in the reproduced beads in figure 15i.1.A list of known sword beads is contained within Appendix 8 compiled by Mike and Su Poole of Tillerman Beads.

[2] Meaney, 1981. 195. For a full list of sword-beads in the 169 sword graves (plus one cremation) in our data see chapter 10, *Analysis of selected data from spreadsheet.*

[3] Meaney, 1981. 195.

[4] Menghin, 1983. 209. fig 35.

diameter.[1] Two beads were found near the hilt of a sword in grave 44 from Brighthampton, Oxfordshire and in grave II at Selmeston, Sussex. A bead was located in a grave at Bishopstone, Buckinghamshire in 1875; it was dark in colour, had a white six-pointed star, about 27 mm in diameter, and appears to have been a slightly domed disc. A large bead of clear green glass with a wavy line of opaque white was uncovered, in 1932, at a site in Fetcham,* Surrey, unfortunately, it was shattered on excavation. A similar fate met another bead from a grave at Westgarth Gardens, Bury St Edmunds, Suffolk; at least the excavators were able to record that the bead was situated about 50 mm down from the scabbard throat.[2] Morning Thorpe 218 contained a blue bead that could have been associated with the sword. In our sample fifteen swords have beads associated with them.

It is possible that the colours and patterns of the beads had some significance; green does appear quite often, but it could be that they were just considered beautiful, lucky or protective in themselves. It is highly likely that there have been many other sword-beads but that they have not been recognised as such, and many stray finds of beads might also have been from swords when the bead has become detached; this may have happened with many sword pyramids, as we shall see.

In England the next most common material for sword-beads appears to be amber. Meaney reports that, up until her work was published, there had been about one hundred and fifty finds of amber beads in English graves from the beginning of the sixth century and continuing into the Christian period.[3] Of course, most of these amber beads were not sword-beads and some could actually have been playing pieces, but she does write that there is considerable evidence that the Anglo-Saxons treasured amber. She lists a number of finds from English graves that definitely were sword-beads. The earliest amber bead that she records is an example from grave 31, Brighthampton, Oxfordshire, which appears to come from the early 5th century. Grave XXI* at Petersfinger, Wiltshire, had an amber bead and the grave is dated to the second half of the 5th century, as were the beads from grave 42* at Alton, Hampshire, grave 87* at Riseley, Horton Kirby and grave LXXXVI, Sarre, Kent.[4] Amber beads are found on the continent too, particularly in Central and Eastern Europe and usually dated to the period 400 to 450 AD. Meaney cites examples at Čegem, North Caucasus, at Kerchin in the Crimea and provides examples of others from Hungary, the Alps to the Rhineland region. Amber is found naturally, from the Baltic to the shores of the North Sea, including England (mainly Essex and East Anglia) so the amber for the beads might have come from anywhere. Amber, when rubbed, has electrical powers of attraction, so it may have been regarded as magical for this property alone. However, it also is relatively warm to the touch - unlike most other 'stones' and is often transparent or translucent, it sometimes contains the remains of creatures or plants, and its colour is reminiscent of the sun; there is, too, a body of folklore which seems to indicate that it had healing properties.[5] Perhaps it should not be a surprise that some warriors chose to wear it on their swords?

Much less common are sword-beads of antler, in fact Meaney lists only one, the bead mentioned above by Evison. It was located near the pommel of a sword in grave 15 at Brighthampton, Oxfordshire; this was flat, 26 mm in diameter, unlike all other known beads in England, it was square with rounded corners and had two concentric circles around the central perforation.[6] It is possible that antler was chosen because stags renew them every year and so this material represents renewal and youth. A 40 mm diameter bead of bone was found at Dorchester, Oxfordshire; it had ring-and-dot patterns on one surface and was plain on the

[1] Meaney, 1981. 195.
[2] Meaney, 1981. 196.
[3] Meaney, 1981. 67.
[4] Meaney, 1981. 68. Also mentioned by Evison, 1976. 310.
[5] Meaney, 1981. 69 to 71.
[6] Meaney, 1981. 142.

other side. Meaney, following Salin, suggests that the ring-and-dot patterns are symbolic of the wheel and the sun.[1] There are a number of white materials that the northern European peoples used to decorate their jewellery and swords; among these are chalk, meerschaum, spar, cristobalite, gypsum, and ivory. Some of these materials are actually quite soft and could be deemed unsuitable for use where wear might be expected; it seems that it was the whiteness that was important, not the material. Exactly what the white colour signified is difficult to say. However, a ring-sword from Dover, Kent, was accompanied by three porosphaeras, white fossil creatures; they were unperforated and one was as large as 25 mm and they were associated with the sword in some way.[2] Fossil creatures, such as these and ammonites, were regarded as 'snake stones' and probably thought to have magical or protective powers. There will be more to say about white material when we discuss the 'eye in the sword' below.

15i.3 Silver and shell bead probably from a seax found in grave 76 at Kingston, Kent. (Drawn by Wayne Letting).

Meaney discusses one other possible sword-bead, or perhaps more accurately 'seax-bead', and this is of much more sophisticated manufacture than any of the others so far discussed. This bead is from grave 76 at Kingston, Kent; the grave contained a spearhead and a seax. The seax had a blade of about 10 inches long and 1½ inches broad. The bead is of silver with oblongs of shell set into its surface, and it has a central tube to accommodate a thong. (figure 15i.3) The dimensions of the bead are 30mm high and 30mm diameter and it could have originally formed a pommel were it not for the central tube and that part of the original suspension string attached to it, remained inside the tube. At least two other similar beads have been found in other graves but not, as far as we know, associated with weapons; they have all been dated to the early Christian period.[3] There have been other stray, detector finds which are listed in the Portable Antiquities Scheme that could be of sword-beads, some of them are of garnet and- gold, cylindrical with hollow centre tubes, and apart from that possibility they have no known purpose.

What was the purpose of sword-beads? We have mentioned that they might have been perceived to have had magical or protective powers but maybe there are other possibilities?[4] They could have been part of the attachment mechanism fixing the sword to the belt, certainly sword-bosses and some sword pyramids appear to have been employed in that way. Other uses have been proposed, perhaps as part of the 'peace-bands' mentioned in the later, Norse literature. Peace bands are recorded in several sagas, for example, the *Saga of Gisli* and the *Sturlunga Saga*. It is assumed that such bands were used to visibly show that the sword could not be drawn from its scabbard and therefor that its owner was at 'peace'.

The sword of Hrolf Kraki, Skofnung, once retrieved from its burial mound, was accompanied by a 'life stone'; this could heal wounds caused by the sword.[5] It seems that 'life stones' might not always have been attached to the sword itself, as in *Kormak's Saga*, one is carried in a little bag around the neck. Admittedly the sagas were written much later than our period, but they do contain much of relevance and the 'life stones' could be an echo of earlier practices. It must be admitted that sword-beads and 'life stones' could have had entirely different purposes, but if one evolved into the other, they had a very long period of use, indeed.

---

[1] Meaney, 1981. 142.
[2] Meaney, 1981. 115.
[3] Meaney, 1981. 129 to 130. Evison discusses one from Kempston, Bedfordshire and another from Kingston in Kent, however they do not seem to be directly associated with swords, Evison, 1976. 313.
[4] Meaney, 1981. 198.
[5] Meaney, 1981. 198. Hreinsson, 1997. Vol 5, 89.

Before we continue our discussion on other beads and sword accoutrements, it might be useful to consider briefly how they were attached to the sword. In some cases, beads are found near the hilt and in others on the scabbard; it is likely that there were several different methods of attaching them. Perhaps to simply tie a thong around the hilt and let the bead dangle from it would be quite practical, as when the sword was held in the hand and used the bead would not interfere with its use. Another method could be to attach the bead to the scabbard slide, where the means of attaching the belt or baldric was fixed to the scabbard; again, the bead would just be suspended from the scabbard and would not be an impediment to its use. Holes appear on a number of scabbard throat bands and beads could be suspended from such a position. These methods mean that the bead could be seen easily and its symbolic significance, if it had any, conveyed and understood by the observer. Other ways of carrying beads could have been utilised, possibly connected to the belt or some other equipment. (figure 15i.4)

The eye(s) in the sword?

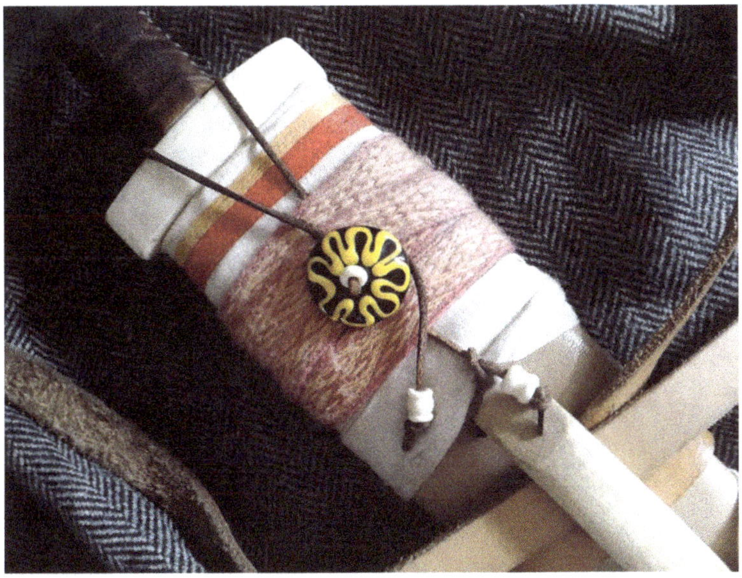

15i.4. Reproduction North Luffenham sword-bead mounted on a sword in a manner suggested by finds in some graves. The sword and ensemble put together by John Wills. (Photograph courtesy of John Wills).

At Selmeston, Sussex, grave 1 had a sword with a cristobalite bead of cylindrical shape and grave 192 at Mitcham, Surrey had a disc of spar as a sword-bead.[1] At Wickhambreaux, Kent a sword was located that had a circular, jewelled button with a rectangular loop at the back attached to a disc of white material. This is reminiscent of the Sutton Hoo, Mound 1 sword bosses and of several continental finds. A sword found at Great Chesterford, Essex, grave 122, had a semi-spherical boss attached to its scabbard, this is described as being covered in 'brass sheet' and it has a central hole for attachment purposes.[2] Grave 27 at the Buckland cemetery, Dover produced another sword with a circular shape on the scabbard in a similar place to that on the Great Chesterford, what this was is not known as the material has disappeared and left the distinct impression of a ring; the outer ring is 60 mm and the perforation, is 21 mm.[3] Evison presumed that it was a sword-bead. A magnificent, but damaged, gold and garnet hemispherical boss was found at Markle, East Linton, East Lothian, Scotland; it was a surface find but must once, have rivalled the Sutton Hoo bosses in quality.[4]

[1] Meaney, 1981. 96. Evison, 1976. 311.

[2] Evison, 1994. 108 and figure 47.

[3] Evison, 1987. 222 and figure 15.

[4] Proudfoot and Aliaga-Kelly, 1996. 4 to 5.

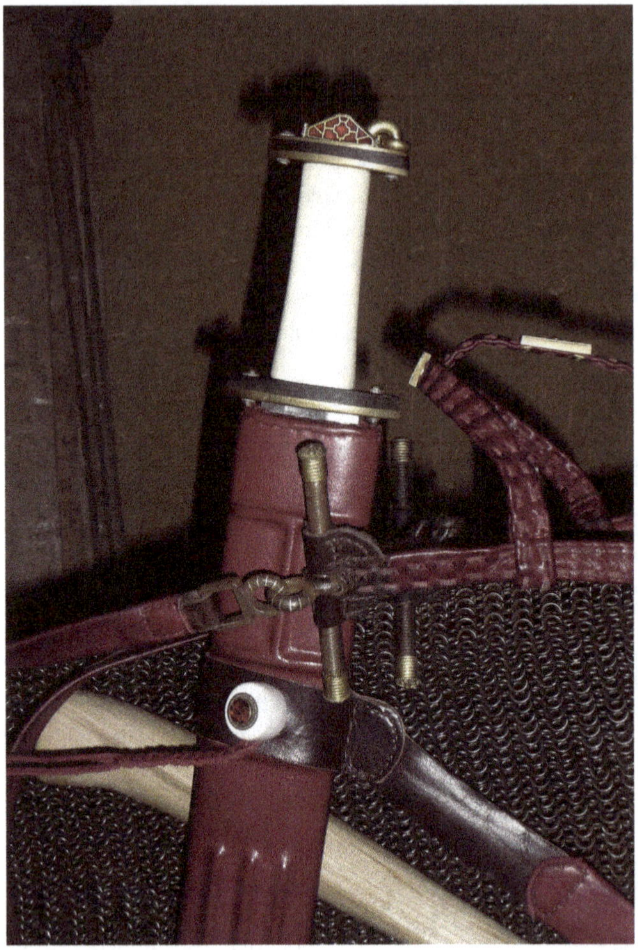

15i.5. Reproduction of ring-sword from Niederstotzingen, grave 9, owned by Arian Zilliox. The backing of the original gold and garnet cloisonné boss was a barrel-shaped cylinder of magnesite. Photograph, Paul Mortimer.

Several others have been reported to the PAS in recent years,[1] from Hampshire, the East Riding of Yorkshire, Northamptonshire, Staffordshire, Nottinghamshire and Norfolk; and, of course there could be more.

Meaney includes an illustration of another bead consisting of a silver-gilt knob, whose shank passed through a large amber bead and was fastened to the scabbard by the same shank. This was found at Kleinhüningen, Switzerland.[2] A sword from Taurapilis, Lithuania (grave 5) has a cylindrical bead of chalcedony surmounted by silver ornamented disc with 4 beast's heads dated to around the mid-5th century.[3] Menghin shows a sword with a gold and garnet cloisonné bead with a white surround from grave 20 at Chaouilley, France.[4] Several more of this type have been found in Germany and elsewhere; for instance, in grave 9 at Niederstotzingen, from Börstingen. (figure 15i.5) The latter was a metal dome with a staple at the back which went through a bead of and was mounted on the scabbard.[5] The dome consists of a garnet and gold cloisonné, mounted on a bulbous meerschaum bead; again the assembly is mounted on the sword by means of a looped staple protruding from the dome and out of the back of the bead. This bead and most of securing arrangements for attaching the sword to the belt or baldric. Another, similar bead is from Krefeld-Gellep, grave 1812,[6] and another from grave 1782;[7] both of these have meerschaum backing. (figure 15i.6) A complex bead with millefiori and gold and garnet work backed with a huge disc of white material was found at Bülach, Switzerland.[8] Menghin does list several others but the above should suffice to provide good examples and

[1] Unique identifying numbers: NMS-274C44, WMID-C02AF3, DENO-2A0601, HAMP-2DAB03, NLM-1A8B56 and GLO-4906E6.

[2] Meaney, 1981. 198.

[3] Menghin, 1983. 205. Dating sequences discussed on pages 58 to 60. The 'boss' button is almost identical to the recent find of a 'pyramid' from Sandby Borg, Ölund, Sweden. We will discuss these examples again in the next chapter. They can be seen in Appendix 9v.

[4] Menghin, 1983. 225. figure 57.

[5] Paulsen,1967. 93. Menghin, 1983. 253 to 254. figure 104.

[6] Menghin, 1983. 227. figure 60.

[7] Menghin, 1983. 239 to 240.

[8] Menghin, 1983. 258. figure 111.

we should note that this class of sword boss would appear to have had some antiquity as there are two examples that were found in the Nydam deposit.[1] One is particularly interesting as the metal part is a decorated conical, copper-alloy pyramid surmounted on a considerably larger, pale piece of wood. The pyramid, itself is about 15mm wide and 7mm high and has a simple whorl design around the base, the base is about 33mm wide and more markedly curved on top.[2] The other boss in the deposit is a similar shaped conical pyramid of silver, but plain, this time, it is not connected to a base any longer, but there is a disc shaped crystal with a centre hole in it and that is about 35mm wide.[3] As the Nydam deposits are from 200 A.D. to the beginning of the 5th century, it is possible that these are the earliest examples of the boss tradition, which probably evolves from the use of beads; it could be, too, that the conical pyramids are an early form of sword pyramid.

15i.6. Reproduction of the ring-sword from Krefeld-Gellep grave 1782. It has a gold and garnet sword-boss fixed to a the scabbard. Note that the boss is backed by a barrel-shaped block of white meerschaum. The sword is owned by Arian Zilliox. (Photograph, Paul Mortimer).

Apart from being decorative and attaching the sword to the belt there is no known purpose for this type of bead, except that, in some ways, they resemble an eye; this is particularly true of those with white material backing the jewel. Perhaps a reference to the one eyed god, Woden?[4] In that case what of the two eyes on the Sutton Hoo Mound 1 sword? We know from the Vendel XIV pressbleches that other swords must have had this arrangement but, as far I am aware, the only ones that are actually in pairs and were found with swords are a find from St. Dizier, France,[5] (figure 15i.7) a magnificent gold and garnet pair that reside in the museum at Berck-sur Mer, France,[6] and the ones from Sutton Hoo. It is interesting that with the Sutton Hoo the two bosses are subtly different from each other; the one on the wearer's left is smaller than the other.[7] Something similar is true of the St Dizier find where the beads are in the form of two thick but flat

---

[1] Rau, 2010b. tafel 48. They are 48, 6 and 48,5. These examples will be discussed more fully in the next chapter. They can be seen in Appendix 9v.

[2] Rau, 2010 b. 111 and tafel 48.

[3] Rau, 2010b 112 and tafel 48. The crystal bead is 48,4. There is also one other example from Nydam, similar to the one from Taurapilis, but we will discuss that in the chapter, *The Riddle of the Pyramids*.

[4] Or Wotan, Godan, Odin, etc.

[5] Fischer, 2008. 103 to 109. INRAP website – Institut National de Recherches Archéologiques Préventives. The discovery of the mid-6th century grave occurred in 2002.

[6] Musée Opal Sud Berck-sur-Mer 2008. Cover and 11.

[7] It is interesting to note that the smaller boss of the sword is on the wearer's left; the eyebrow without foils on the Sutton Hoo helmet is also on the wearer's left. Perhaps, as far as the occupant of the Mound 1 tomb was concerned, Woden's left eye was

discs, one of ivory and the other made from stone but surmounted by a double spiral of gold. The pair from Berck-sur- Mer are of different diameters, but the domes are of different in shapes, too. The smaller of the Berck-sur-Mer pair closely resembles another find of a gold and garnet single sword-bead from Wickham in Kent.[1] (figure 15i.8) It is just possible that these deliberately included differences are reference to the two differing eyes of Woden.[2] Of course, it may be nothing of the sort and was just the result of artists who wanted them to be different, if that is the case the Sutton Hoo jeweller went to a lot of trouble with the Sutton Hoo gold bosses because he could have made them identical but he chose not to.

15i.7. The ring-sword from St Dizier, France. The pair of bosses are each made of different materials and only one is surmounted by a double gold whorl. Courtesy of Musée de Saint-Dizier / Photo Claude Philippot.

There is one other seeming pair of bosses (K675 and K1425) and these were found in the Staffordshire Hoard, obviously detached from a weapon, or weapons. They are virtually identical and consist of a gold and garnet cloisonné dome with a twisted filigree surround. Compared to the ones from Sutton Hoo, they are quite small. A pale, stone, 'barrel shaped' bead (K764) is a perfect fit for either of the bosses,[3] and perhaps only one of them would have been mounted on such a bead?

Sword bosses do seem to have been attached in various ways to the scabbard, as mentioned above, some were connected via a strap to a scabbard slider, while others, it would appear, might have been employed directly or indirectly as a part of the suspension method between the scabbard and the belt or baldric.

---

the weaker of the two? However, it must be said that other figures, such as the one on the Torslunda die has the altered eye as the one on the right. For a full discussion see Mortimer, 2011.

[1] Åberg, 1926. 140.

[2] For a fuller discussion of this possibility see Mortimer, 2011. 26 to 28. Mortimer and Pollington, 2013 and Price and Mortimer 2013.

[3] Cool, 2015. 93.

15i.8. An illustration of the gold and garnet bosses from Berck-sur Mer, near Boulogne, France. Courtesy of Musée Opale Sud Berck-sur-Mer.

Sword pyramids

Sword pyramids are enigmatic objects, once thought to be very rare but with the use of metal detectors, it seems that they were actually comparatively common. As to why the shape of the pyramid is used it is difficult to say,[1] pyramids do occur in other contexts in Germanic graves, as decorations on horse mounts for example and on other types of scabbard mount, but why they became associated with swords can only be surmised. Curiously, the sword boss ensemble mentioned above from Nydam, has a round conical pyramid surmounting the organic base; is this a precursor for sword pyramids?

The majority of sword pyramid finds have square bases and are made of metal, bronze, silver or gold, or iron and silver *tauschieren*[2] work and occasionally just iron. Virtually all are hollow with a bar at the bottom to facilitate a connection to a strap; it is also been assumed that they should always be in pairs.[3] There are two basic forms of the square-based pyramids, flat topped and pointed, although recent finds have shown considerable variation and occasional moves away from square based pyramids to other polygonal and even circular bases. When Menghin wrote his book on swords, no pointed pyramids had been found in England and he recorded none in Scandinavia on his distribution maps.[4] That situation has changed and there are now several pointed pyramids that have been located in England and there are at least four known pyramid finds in Sweden; a circular and elaborate one from Sandby Borg,[5] the remains of a garnet and gold cloisonné one from the west mound, Odenshög, at Uppsala,[6] and an elaborate, small but tall pair were attached to one of the swords in the Valsgärde 7 grave,[7] and one was found in 2016 near Uppsala.[8] Recently, finds of sword pyramids have been made in Denmark and two are illustrated in a recent paper by Claus Feveile; one is a garnet cloisonné example from Vivegrøften and the other is plainer and from Gammel Hviding, another from Sandegård on the island of Bornholm is referred to.[9] Finds of sword pyramids are extremely rare finds in Scandinavia and this is possibly due to the legal restrictions against metal-detecting in the Nordic countries; most of the recorded finds that reach the public domain in England have been found by detectorists.

---

[1] This subject will be returned to in the Chapter, *The Riddle of the Pyramids.*

[2] *Tauschieren* is the name given to inserting wires of one type of metal into another in order to create designs.

[3] This assumption will be challenged in the next chapter.

[4] Menghin, 1983. see karte 22 and karte 24 for distribution maps of flat topped and pointed respectively.

[5] We will discuss the piece from Sandby Borg further in the chapter, *The Riddle of the Pyramids.*

[6] Lindqvist, 1936. 179. Bruce-Mitford, 1974, 41.

[7] Arwidsson, 1977.

[8] http://www.uppsalatidningen.se/aktuellt/guld-och-granater-i-stormannagraven-4245934.aspx

[9] Feveile, 2014. 84 and 85.

A sword pyramid of a completely new type has been found in the Salo cremation cemetery in south-west Finland.[1] It is possible that other examples of this type have been found elsewhere in Finland, too.

Menghin's distribution maps show that the flat topped pyramids are known from finds in England, the Rhineland, with some in France and Italy. It seems that the Franks, the Alamanni, the Lombards and the Thuringians used them. He shows three finds in the Black Sea area, indicating that they were known and worn in that region, too. Gilt or copper alloy cloisonné or gold cloisonné seem to be almost completely restricted to England,[2] with another find on the Dalmatian coast, and another in Germany plus the one from Uppsala. Most of the rest of the pyramids of this type that he lists are of relatively plain bronze or silver. A high status gold filigree and garnet example was found in Ezinge, Holland,[3] and this, in some ways resembles the pyramid from Danethorpe Hill Farm (the Collingham Jewel and number 7 in our non PAS corpus) that now resides within the British Museum.

The pointed iron and silver tauschieren pyramids follow a similar pattern except that they seem to be concentrated in a narrower band along the Rhine with one or two along the Danube.

15ii.37. Type 1ix. A pair of organic pyramids from Nocera Umbra (so far none recognised from England). Photograph Matt Bunker

The Sutton Hoo Mound 17 pyramids, which are of copper-alloy, with garnets and blue glass cell filling, are interesting in that there are others which have a very similar design. There is one pair that was found in an isolated mound burial at Coombe Bisset, Salisbury Race Course, Wiltshire[4] and another from near Snettisham, in Norfolk, this particular one, is made of copper alloy and garnet, like the Mound 17, but some of the cells have their fill missing, so it is unknown if it included any blue glass.[5] Another from Scunthorpe, Lincolnshire, did have blue glass as well as the garnets.[6] A pyramid with a similar design, another detector find from Chearsley, Aylesbury Vale in Buckinghamshire has been recorded by the PAS.[7] Intriguingly, the same design that is on the side of the Mound 17 and other pyramids, appears on both sides of the sword pommel of a ring sword from Laon, France.[8] (figure 15ii.37)

Ten pyramids were found in the Staffordshire Hoard making five pairs; all are either silver or gold, and are decorated in various ways.[9] At least two of those of garnet and gold pyramids also have garnet prisms along the edges, similar to the ones from Sutton Hoo, Mound 1.

---

[1] Mäntylä-Asplund and Storå, 2010. 56. The cemetery dates to A.D. 550/600–800. The authors attribute the pyramid to the Frankish weapon trade. It is illustrated in Appendix 9v.

[2] Unfortunately, no comprehensive study of sword pyramids from Continental Europe exists as far as I am aware; if it did the above statement may have to be revised.

[3] Nicolay, 2014. 96.

[4] Evison, 1963. 71. Swanton, 1973. 176. This pyramid is number 58 in our corpus.

[5] 25 in our corpus.

[6] Both the last two were sold by Timeline Originals to private buyers. The Scunthorpe pyramid is 26 in our corpus.

[7] Unique number, BUC-90E4A7, (P17 in our corpus). These will be discussed further in the chapter, *The Riddle of the Pyramids*. There are at least 7 of this type (M17), see pyramids P17, 10, 25, 26, 58, 70, 78 in appendices 9i, 9ii and 9iii.

[8] Evison, 1976. 303 and Plate LXIV. All known pyramids that resemble the ones from Mound 17 at Sutton Hoo, will be referred to in the next chapter.

[9] Cool, (2015. 93) seems to suggest that there are actually fourteen, but that does not appear to be the case.

Due to the proliferation of detector finds, we now know that sword pyramids were much more common than previously realised and, it seems could have been attached to more swords than previously thought as they were also made of perishable, organic materials. Grave 53 at Collegno, Italy contained a sword with one pyramid made from antler, while the other was of wood.[1] Other organic pyramids of bone or wood occur at the Santo Stefano cemetery in Pertica, Cividale in Northern Italy[2] and in Germany. It is likely that remains of many organic pyramids may not have been recognised for what they were, as if they survived at all they might just have been small, nondescript bits of matter due to the largely organic-hostile soils prevalent in England.

Pyramids, it seems, could be utilised in different ways. They have to have been attached to a textile or leather strap, thin enough to go under and round the bar at the base of the pyramid. In my reconstruction of the Sutton Hoo Mound 1 rig, I have used the pyramids as sliding buttons on 'peace bands'[3] made of tablet weave; it has to be a very fine weave to pass it under the bars. On this sword it is unlikely that the pyramids were used to attach the scabbard to the belt as the bosses fit into the position that fulfils this purpose. However, Menghin shows pyramids being used in such a way as to act as 'buttons' connecting the sword rig to the belt and there are several finds that seem to show that he was correct; one will suffice for the moment and that is the sword from grave 75 from Donzdorf, in the Rhineland, where the pyramids are clearly used for this purpose.[4] It is possible that they were used as suspension aids in Anglo-Saxon contexts, too. However, the general assumption, based on the comparatively large number of detector finds of 'stray' pyramids mainly from England,[5] is that they were probably worn dangling on the end of a band of textile or leather just before they became detached.[6] If pyramids were used as part of the method of connecting the sword to a belt or baldric, then the sword wearer would presumably have soon noticed them becoming detached and looked for them and not many would have actually been lost, although it must be admitted that in battle or on long and difficult journeys, it could have been possible for them to become detached and lost and not found until someone with a detector comes along.

We will discuss some of the aspects mentioned above further and more in the next chapter, *The Riddle of the pyramids*.

---

[1] My thanks to Lars Lüppes for sending me this information.

[2] Ahumada Silva, Lopreato, and Tagliare, 1990. 122. My thanks to Nini LaFara for this information.

[3] See report in The Sutton Hoo Mound 1 Sword and the Replica, in this volume.

[4] Neuffer, 1972. 101 and tafel 50. See also Marti, 1995. 88 to 92. Marti reports a find from Altdorf UR, Pfarrkirche St. Martin, Switzerland that has a reasonably intact set of pyramids and suspension system showing how they were used to connect the belt to the sword.

[5] The use of metal detectors in many European countries is subject to rigorous restrictions, if it is allowed at all.

[6] This assumption will be considered again in the next chapter.

15ii.The riddle of the pyramids: an attempt to unravell their meaning. A typology and comments on chronology

Paul Mortimer

Introduction

In the previous chapter we briefly discussed sword pyramids, sword-beads and sword-bosses. We will extend the discussion with a look at the evidence that I and others have been able to gather regarding sword pyramids, and to question some assumptions concerning these interesting and important objects. I will attempt a typology, something of a chronology and try to unravel what the significance of the pyramids may have been and where their origins lay in time and place. Very little work has been published concerning pyramids so this effort is just the start of a discussion about what seem to be significant objects. They might hold much information about the world-views of the inhabitants of the 5th to 8th centuries, and possibly even later, but to gain that information we need to reveal their secrets and their relationship to the sword.

I have compiled a sample of images of one hundred and ninety-seven examples of pyramids from England, and the most important, single source has been the PAS[1] website, which records finds by those who use metal detectors and report them. I have located others in books, museums, archaeological reports and on-line, particularly eBay, where many have been sold. Wherever possible, I have tried to get as much additional validation as I can, but for the most part, 'ordinary' sword pyramids of the type found for sale on-line do not fetch high prices and are not, therefore, usually faked,[2] the biggest risks, I think, are that in some cases later medieval pyramid mounts are labelled as 'Anglo-Saxon' sword mounts (I have tried to exclude these, but it is possible that one or two will have slipped through) and that unscrupulous dealers are passing off examples found on the Continent as originating from Anglo-Saxon England in order to be able sell them; laws regarding the use of metal detectors are much more restrictive in Scandinavia and Germany in particular. I am sure, too, that many more sword pyramids have been found in England, and either sold without being recorded or are hidden in drawers and unreported; legally only those with a significant gold or silver content have to be reported. I suspect that there are considerable numbers that have been re-buried in museum cupboards and await discovery anew.

When trying to locate new examples, I have been hampered by the fact that there is no uniformly accepted term to describe them, 'pyramid', 'pyramid mount', 'scabbard mount' 'sword pyramid', 'pyramidal mount', 'pyramidal ornament', 'Saxon scabbard pyramid' 'sword jewel' and others are all used, with or without hyphens; in this book, 'sword pyramid' is, I believe, always used.

My attempt at a typology, together with some notes on frequencies of various types, will be found below. It is very difficult to discuss sword pyramids without such a tool; being able to sort the pyramids into categories, consider their ornamentation, look at similarities and differences can provide clues about their manufacture, origin and meaning that might not be otherwise realised. There are of course dangers with such an approach and it is possible to make assumptions that will be found to be false, but it is at least the beginning of a conversation. A section that considers some of the results of this study follows the typology.

[1] Portable Antiquities Scheme https://finds.org.uk/

[2] This could be different for 'high status' examples, but even here the work and materials needed to make such fakes probably exceeds the price that would be achieved in many cases. The open sale of 'high status' pyramids, could be problematic and attract unwanted attention to the vendors.

The typology refers to all one hundred and ninety-seven examples of pyramids from our sample. The typology does not allow for structure or methods of fabrication, for example whether a particular pyramid is complete or just an outer casing that would have had an inner core, but it might be noted in the details of the pyramid provided in the relevant appendices. I have made no comments on the size of pyramids either, although I have tried to include dimensions wherever possible. The dimensions do vary quite a bit, but most of the highest status ones (quality A) tend to be larger than average. The pyramids vary considerably in quality and I have included a rather subjective quality analysis in the discussion that follows the typology. The notes in Appendix 9 provide varying amounts of information, mainly location of the find, where they are being kept and the way that they have been published, or not. Some further details, including images of all of the sword pyramids found for this study are provided in the sections of Appendix 9. Examples from other lands beyond the shores of England, are only used when a comparison is required; it would have been useful to compare 'Anglo-Saxon' sword pyramids with as many of their European counterparts as possible, but as far as I am aware no catalogue or serious study of them has yet been published.

In the discussion that follows the typology, I have included some general comments on distribution, although some of the find sites are fairly vague. In this sample I have not been able to locate examples from the English counties, Gloucestershire, Herefordshire, Rutland, Shropshire and Tyne and Wear where they could reasonably be expected. There are none from Cheshire, Cornwall, Cumbria, Devon, Lancashire or Merseyside, which is probably a little less surprising. There is, however, one from Southern Scotland (West Lothian) and one from Wales (Monmouthshire) to make up for the disappointment.[1]

The typology

The four main types of sword pyramid follow and there is some subdivision within most of the categories. Each type or sub-type lists all examples of that type that I have found. Type 5 pyramids are examples that may have been used as sword pyramids but are significantly different to most of them in some way.

1. Flat-topped square-based
2. Pointed-top square-based
3. More than 4 sides
4. Circular-based
5. Possible pyramids

Sub-categories and frequencies[2]

Type 1. Flat-topped square-based, is further subdivided into 10 sub-categories
1i. Plain: iron, cu-alloy, silver, silver-gilt cu-alloy gilt, gold (17 examples 8.6%)
1ii. Decoration consists mainly of a single gem on top (8 examples = 4%)
1iii. Decorated: cloisonné, complex (8 examples 4%)
1iv. Decorated: cloisonné, simple, often slabs of gems/glass on sides and top[3] (28 examples 14.2%)
1v. Decorated: filigree (8 examples 4%)
1vi. Decorated: chip-carved or cast, triangles dominate (59 examples 30%)
1vii Decorated: chip-carved or cast, other designs on sides (28 examples 14.2%)
1viii. Corrugated sides (3 examples 1.5%)
1ix. Organic pyramids (0 examples)

[1] Further details of the general distribution can be found below.
[2] Percentages will have been rounded slightly.
[3] Many have the slabs missing.

1x. Tauschieren decorated (1 example 0.5%)
1xi. Pierced top, round hole, no bar underneath (5 examples 2.5%)
1xii. Other square-based pyramids (2 examples 1%)

Type 2 Pointed-top square-based, is subdivided into 2 sub-categories
2.i. Non- tauschieren decoration (10 examples 5%)
2.ii. Tauschieren decoration (0 examples)
Type 3 More than four sides (3 examples 1.5%)
Type 4. Circular-based, Is further divided into 4 sub-categories
4i. Relatively plain (3 examples 1.5%)
4ii. Conical with flat sides (8 examples 4%)
4iii. Squat, button-like, may have Style I decoration or cloisonné (2 examples 1%)
4iv. Other (3 examples 1.5%)
5. Possible sword pyramids (2 examples 1%)

A list of all known pyramids follows at the end of the chapter and should be consulted in conjunction with Appendix 9 where images of the pyramids can be viewed.

Discussion

Before we begin to try and unravel what the significance is of different types of sword pyramid and what particular types of ornament could mean, I would like to challenge two assumptions that have often been made regarding them.

1. Casual loss

As mentioned above, I have found images of one hundred and ninety-seven[1] sword pyramids and I suspect that there are many more in museums and private collections and that two or three times that number have actually been found by detectorists, many of which will never have been formally recorded. Generally, such weapon ornaments would have been valuable items and must have represented a considerable loss to their ancient owners. It has always been assumed, by me and most others, that the losses must have occurred when they became accidentally detached from the sword scabbard during travelling, combat or just through wear and tear. Comparatively few pyramids have been located as grave finds;[2] the majority are probably detector finds, although the find status of many is difficult to determine.[3] Whether there are, one hundred and ninety-seven, two hundred, five hundred or more, the finds by detectorists would seem to indicate that their original owners had a very careless attitude to caring for such prized objects. The pyramids had to be carried on some form of tape, probably tablet woven braid, which is very strong, and unlikely to fall apart easily, although leather straps or cord could have been used too and that could have been less dependable, but it is unlikely that such large quantities of pyramids would just fall off and be lost. The seeming lackadaisical nature of their loss was brought home to me when I was re-reading Kevin Leahy's and Roger Bland's small book on the Staffordshire Hoard,[4] published remarkably soon after the discovery was made public. While discussing the possibilities concerning why the Hoard was deposited, Leahy says, "It could,

---

[1] With generous help from Lindsay Kerr, Sue Brunning, Andy Pilkington, Stephen Pollington, Matt Bunker and others.
[2] At least fourteen are from graves; 1, 10, 11, 23, 30, 35, 48, 58, 73, 79, 81, 89, 98 and 108. See addenda at the end of the chapter for full details.
[3] Some are listed as being from 'an old collection'. This is particularly true of those offered for sale by auction sites such as TimeLine Auctions. Others are stored in museums where documentation is poor or non-existent or are from private collections. Some of those sold on eBay are offered without much provenance. See addenda at the end of the chapter for full details.
[4] Leahy and Bland, 2009.

however, be some kind of ritual deposit, the Portable Antiquities Scheme has, over the last ten years, recorded nine pommels stripped from their blades and apparently abandoned. The Hoard might represent a similar practice, but on a larger scale."[1] Leahy's point is an important one and when the losses of sword pyramids are considered along with the 'loss' of sword pommels it becomes apparent that there could be something other than just 'carelessness', going on; sword pommels and in some instances – pommel caps, do not just fall off swords, they were made to withstand the rigours of combat and were fixed firmly to the weapon. The pommels, and possibly pyramids and other objects, too, could have been deliberately taken and buried, perhaps sacrificed to a god, possibly a deity of war, Woden, Saxnot , Tiw or perhaps a being whose name has been lost to us, to thank the deity for victory over the fallen foe. It is also possible that the decorative elements being disposed of had belonged to the dead enemy and the weapon was being stripped of its former associations and 'de-personalised'. This may have happened to inherited weapons too, the son, nephew or other receiver of a legacy, may want to make the weapon his own. A possible scenario is that the owner of the item carried out a ritual in which he placed his own possession in the soil as fulfilment of a vow or a plea for aid and, of course, there are other scenarios that can be envisaged. Whatever, the reason the weapon is being distanced from previous holders by the removal of certain alienable parts which were then, probably ritually buried in a place that would have had significance for the individual. It has been noted elsewhere in our book that a votive practice of depositing bladed weapons in watery places was a fairly common practice before our period and was returned to during the 8th century and beyond into the later middle ages. Such deposits are rarely recorded during the 5th to 7th centuries, perhaps that is because alternative arrangements were taking place on dry land? If pyramids are being deliberately disposed of in this way, then it would mean that they had value as a sacrifice and this would increase their significance and meaning.

Incidentally, the PAS now lists at least seventy-three pommels from the 5th to 7th century in its database;[2] there are quite a few from the 8th century onwards, too, but I have not included those. Besides pommels and pommel caps other sword items have been found as isolated finds too. Some of the pommels within the PAS register could have been originally from burials, but most, I think, were not. The example shown in figures 15ii.1a and 15ii.1b is a find now in the Chelmsford Museum, Essex and would appear to be a fragment of a gold and garnet sword pommel.

2. Sword pyramids were always used in pairs

This assumption is a perfectly good one to make, because in graves, at least in England, they are usually found in pairs, and such finds as those from Sutton Hoo Mounds 1 and 17 re-inforce that idea, but it was not a universal practice and sometimes pyramids were used singly (or in other quantities), perhaps even with a backing, like some sword-bosses have been. The Broomfield burial in Essex only yielded one garnet and gold pyramid,[3] Broomfield was excavated in 1888 but not to modern standards and could have been visited by robbers long before the 18th century, so it might not be a good example as the other pyramid may have been taken, although the pyramid from Tuddenham appears to have been the only one in the grave.[4] Another single example seems to be from Kent, Sarre grave 211, where only a single pyramid is recorded.[5] The pyramid from

[1] Leahy and Bland, 2009. 12.

[2] A list of seventy-four pommels and pommel caps from the 5th to 7th centuries, together with their unique identifying numbers can be found in Appendix 10. There are eleven more listed from other sources. Bringing the tally to 85. The PAS database has several more from post 7th century, but these have not been included. The numbers result from searching the PAS website during May 2017. Eight detached sword-rings can be found in Appendix 11 along with a few other sword related parts. I have not included any pommels from the Staffordshire Hoard within the sample from the PAS.

[3] BM Guide to Anglo-Saxon Antiquities 1923.

[4] Kennet, 1977.

[5] http://inventorium.arch.ox.ac.uk/grave.php?site_ID=Sar&grave_ID=Sar211

grave 1, at Welschbillig, in the Rhineland, Germany[1] looks as though it was a single example, but a nearby grave, 585, contained two pyramids, one of metal and another of bone.[2] In northern Italy, single pyramids have been found at Nocera Umbra: graves 106, 111, 115, 137, 138, 156 and 163, although graves from the same cemetery, 98 and 132, both contained a pair. Single pyramids were found, on swords, in graves at the cemetery at Weingarten, Germany, grave 478 had a decorated pyramid while a plain pyramid was included within grave 616.[3] The pyramid found at Uncleby (48) was not only on its own but appears to have been found in a woman's grave who was buried with ear rings and a girdle hanger.[4] Davidson counsels caution regarding the Uncleby grave as the burial was never properly documented and she could see no reason why a woman would have a pyramid at her waist, as they were men's items, however, it could have been given to a deceased loved one by a sweetheart, or be a memento of a dear husband, and of course, again, there are other possibilities. The single pyramid from grave 58 at Finglesham (73) also appears to be from a woman's grave.

Incidentally, Menghin, too, records that not all pyramids were found in burials in pairs.[5]

15ii.2. The three pyramids from grave 7010, The Pilgrim's Way, Wrotham, Kent. Picture taken from Stoodley, 2015.

Recently, at an Anglo-Saxon Cemetery on Pilgrim's Way, near Wrotham, Kent three identical pyramids were found with a sword in grave 7010.[6] (figure 15ii.2). It is not entirely clear exactly how they were being employed but it does seem that they were being used in conjunction with a bandoleer or baldric. Perhaps reinforcing the Wrotham find is Davidson's earlier suggestion that the three pyramids found in the cemetery at Faversham, might have all come from the same grave,[7] and having examined them, I agree with her; the three pyramids are very similar to the Wrotham examples in design (all type 1via). The Faversham pyramids are not the prettiest castings, the maker had intended them all to be identical but two of them were deformed during the process. It is a pity that that almost no documentation survives from Faversham; even the excavation date, possibly 1870, is not confirmed.

Below, I suggest that the sword pyramid tradition developed from the wearing of a single sword-boss or bead. If that is correct it would be perfectly normal for sword pyramids to have been sometimes used in numbers other than in pairs, even if such a practice later became relatively rare. It is also possible that, as the Uncleby example possibly shows, they were not always used exclusively with swords.[8] Incidentally, a pair of pyramids was used to act as decorative rivets for a hand strap on the rear of the Sutton Hoo Mound 1 shield; they appear to be of a different design, in some ways, to any pyramids that are known to have accompanied

---

[1] Menghin, 1983. 262.

[2] Collegno, grave 53, in Italy also had a pair of pyramids made from different materials, one of antler and one of wood. I thank Lars Lüppes for this information.

[3] Roth and Theune, 1995. Tafels 170 and 232. Thanks to Matthew Bunker for bringing this to my attention.

[4] Davidson, 1962 (1994).86.

[5] Menghin, 1983. 363.

[6] Stoodley, 2015. 69, plate 5,

[7] Davidson, 1962 (1994). 86. Number 98 in the collection of images. Davidson reports that they are all f different sizes, and to some extent this is true, but seems to result from casting accidents rather than intention.

[8] While it is possible that pyramids may have been affixed to other weapons, such as seaxes, I am not aware of any examples.

swords[1] but are of a similar size. Whether or not the shield examples are relevant to our study of sword pyramids, it as difficult to determine with our present state of knowledge.

Dating and type 4iii.

The dating of many examples, including those from burials, is frustratingly vague, often with a wide timespan, such as '600 to 800 A.D.' (P3 PAS-41A514) or '500 to 700' (P18 WILT-9AB1A4), 410 to 866 A.D (48) and even '6th to 7th centuries'. Such wide date spreads are not particularly helpful, but we can fairly easily distinguish the earliest examples and the latest and try to work in either temporal direction from them. A few sword pyramids have been given narrower date ranges that would seem to be relatively accurate. We know that those from Sutton Hoo Mound 1 were buried during the very late 6th to the first quarter of the 7th century and, although they could have been in use for a considerable time before deposition, most of the other complex cloisonné and filigree pyramids, types 1iii and 1v respectively, were probably made around the same period as the Mound 1 examples, perhaps with some of them being made a little later during the 7th century? Of course it is possible that the pair from Mound 1 was made long before the burial and the pyramids could have been quite old when buried, so even with very high status examples we should exercise some caution.

15ii.3a. A conical pyramid from Nydam made of copper-alloy and iron, backed with a button of wood. Illustration used with the kind permission of Andreas Rau.

It is certain that pyramids were used during the 6th and 7th centuries, and it seems that their use persisted until the 8th century or even later. When the practice started depends to some extent on what we consider to be a sword pyramid, the example from Nydam, mentioned in the previous chapter, could be a boss or a prototype pyramid, perhaps both;[2] (figure 15ii.3a) it does have a bar underneath, with two slots either side, just like a sword pyramid, and is in fact cone shaped, but it sits on a 'bead' of whitish material, in this case of wood, rather like a sword-boss. There is another of these metal cone 'pyramids' in the deposit, although the second example is not directly associated with a bead.[3] (figure 15ii.3b) Rau illustrates both of the two possible pyramid/bosses together with an elaborate, perforated crystal bead with a silver 'boss' button that has a staple that goes all the way through the bead and extends out the other side[4] (figures 15ii.4 and 15ii.5) and a cylindrical silver, gold and niello 'boss' with a staple.[5] I am not suggesting that the latter two are necessarily to be regarded as 'pyramids', just trying to show that in some ways the bead and boss traditions could have been diverging and taking the form of pyramids at the time of the Nydam deposits – 200 to 400 A.D. I do not agree with Davidson's assertion that pyramids replaced the 'balls' of stone or crystal, as such items continue to be attached to swords after the development of sword-pyramids.[6]

---

[1] The shield pyramids are gilded copper-alloy and all four rising sides are slightly concave.

[2] Rau, 2010b. tafel 48. (See also Rau, 2010b. 111).They are 48, 6. See Appendix 12v for an illustration.

[3] Rau, 2010b, tafel 48. 5. This example's surface detail resembles the wrist clasps sometimes found in burials from Norway and Anglo-Saxon England, however, the wrist claps do not have the bar underneath but were fastened in a quite different manner.

[4] Rau, 2010b. 113 and tafel 48.1.

[5] Rau, 2010b. 112 and tafel 48.2.

[6] Davidson, 1962 (1994). 85. Davidson had been discussing the attachment of beads and stones by a strap to swords.

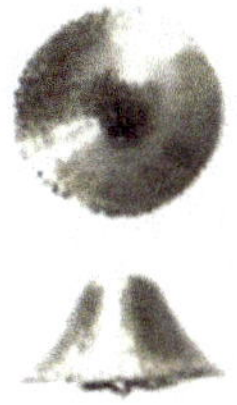

15ii.3b. (top) The second conical pyramid from Nydam, this time made of silver. Illustration used with the kind permission of Andreas Rau.

15ii.5. (above) Cylindrical silver, gold and niello 'boss' with a staple from Nydam. Photograph Paul Mortimer

15ii.4. Crystal bead backing a silver disc and copper-alloy staple from Nydam. Photograph Paul Mortimer.

Another example from Nydam is quite different in character and looks more like a dome shaped button. It is deeply chip-carved which divides the object up into four main sections which are surmounted by a circle; this looks rather like a 'Mercedes' symbol, having three lines radiating from the centre to the circumference.[1] (figure 15ii.6) This, too, has a bar underneath that makes two slots for a tape or cord to secure the object. It also has its 'cone divided into four fields, rather like most of the more conventional pyramids. This example has two cousins that have, ostensibly, been found in England, 8 and 37, in the typology above. The first, 8, came from the collection of M. Braham and Lord McAlpine and was, apparently acquired during the 1980s but was recently sold at auction by TimeLine. Very little more can be said of its provenance, but it is of silver-gilt and is divided into three main sections and decorated with Style I type decoration. (figure 15ii.7). The other 'English' example was sold by Dirk Kennis, an antiquities dealer who lives in Belgium, and is 37 in our sample. Its circumference is divided into four fields all of which are embellished, again, with something resembling Style I decoration. Two of the sections seem to contain a man's face, in one he is smiling and in the other he is not so happy. The remaining sections portray slightly different versions of what could be an 'eye' or a radiating sun symbol, perhaps both. (figure 15ii.8a to 15ii.8f). This object is supposed to have come from Norfolk and what gives the claim of an Anglo-Saxon origin some credence is that the 'eye' symbol from this pyramid appears in a very similar form on the Chessell Down scabbard throat band.[2] (figure 15ii.9)

---

[1] Rau, 2010b. tafel 12 and tafel 48. 3.

[2] Thanks to Stephen Pollington for this observation. The Chessell Down sword is, again 5th century, see Menghin, 1983. 138. Evison, 1967. 114, figure 11. An image of the sword can be seen in Appendix 12v.

15ii.6 (above)
15ii.7 (right)

15ii. 8a 15ii. 8b

15ii. 8c 15ii. 8e

15ii. 8d 15ii. 8f

15ii.6. (top left) 'Pyramid' from Nydam. Now in The Nationalmuseet in Copenhagen. Photograph Paul Mortimer.
15ii.7. (top right) The example from the collection of M. Braham and Lord McAlpine.
Photograph courtesy of TimeLine Originals.

15ii.8a to 15ii.8f. Six views of a pyramid once owned by Dirk Kennis. The four sides can be clearly seen, as can the views from the top and the underside. Photographs courtesy of Dirk Kennis.

15ii.10. The squat pyramid from Sandby Borg, Ölund, Sweden. Photo courtesy of Daniel Lindskog.

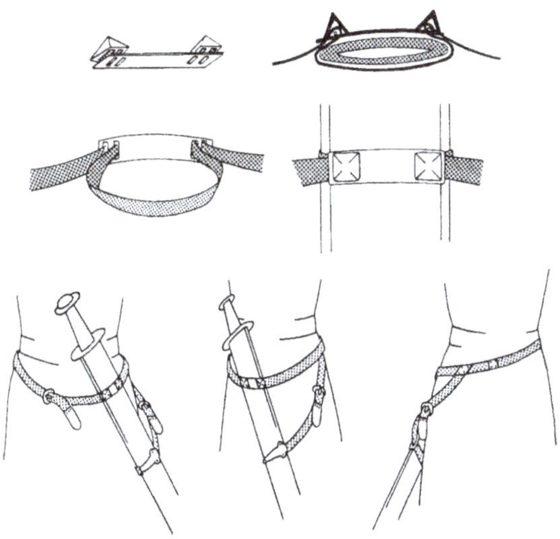

15ii.11. Menghin's suggestion of how pyramids may have been used to connect the sword to the belt based on evidence from some finds.

There are other instances of a similar style of pyramid, none of which are from England. One was recently discovered (2010) at Sandby Borg, on the island of Ölund, Sweden. Sandby Borg was an oval shaped fort and a massacre took place there during the 5th century.[1] (figure 15ii.10) Only a very small part of Sandby Borg has so far been excavated but one of the finds was of a sword-pendant or button, very similar in style to those we have just referred to. It, again like the later more conventional pyramids is divided into four sections by beast's heads, each of which is separated by a triskele. Again, the bar at the bottom gives the impression that it performs an identical function to a sword pyramid. However, at least two more examples of this type of ornament exist in the record and they are both virtually identical to the one from Sandby Borg. One was still mounted on a sword when it was retrieved; the sword was found in grave 5 at Taurapilis in Lithuania, not too far from Sandby Borg. The weapon was elaborately decorated with a pair of metal sliders, a throat band and a chape. The pyramid was being used singly and was mounted on a cylinder of chalcedony[2] in the same way as the sword-bosses discussed in the previous chapter, *Additions, Beads, Bosses and Pyramids*. The cemetery is 5th century. The final example comes, again, from Sweden, this time from Finnestorp.[3]

I suspect that these examples are the origin of the pyramid tradition in that the bead and boss traditions are diverging and evolving and that the sword pyramid is a development of the sword-boss. If this is so, then the examples from Nydam, and the others from type 4iii, are the earliest that we have so far, and would appear to point to an origin during the 5th century. It is interesting that neither the sword-boss, nor the sword pyramid replaced the addition of beads to swords, as the use of sword-beads and bosses continues.

We will return to the issue of dating again below, but first we will ponder the pyramid shape and the fact that most, but not all, pyramids that are found with swords are in pairs.

Utilitarian?

Sword-bosses do not originally appear to have been used in a practical way, unless an amuletic or supernatural purpose can be called 'practical'; that possibility was briefly discussed in the chapter, *Additions,*

[1] Söderström, 2015.
[2] Menghin, 1983. 205.
[3] There are pictures of these in appendix 12v.

Beads, Bosses and Pyramids. Their main purpose seems to have been decorative, although within that possibility they may have signalled achievement, allegiance, an appeal to a deity, or perhaps something else, too. Their function with some swords may just have been as an aid to suspension, especially when used in pairs, but this does not seem to always be the case. Pyramids can be used as 'toggles' or buttons to secure the sword to the sword belt or baldric, Menghin includes several appropriate examples and illustrates clearly how he felt they would have operated.[1] (figure 15ii.11) To use pyramids in this way it is better to have them in pairs. Perhaps that is how the practice of using pairs of pyramids began; that is as a practical method of securing the sword to a belt or baldric. Later, it seems, they were also used in other ways and the Sutton Hoo, Mound 1 set would seem to be an example of a pair being utilised in a non-practical manner, at least as far as attaching to the sword ensemble to the belt or baldric is concerned.[2]

The power of 3 (or 9 or 12 or even 4)?

As for the development of the shape from a pyramidal cone, or dome and perhaps back again, I can only speculate at this stage. Pyramidal plates appear associated with swords and with horse mounts. Menghin provides examples of some with swords at Donzdorf, grave 80,[3] and Wettelsheim, grave 1,[4] both of these examples are from Germany and in both instances the pyramid plates are in pairs. Such plates add another dimension to the subject of pyramids and display and provide another direction to explore, but for reasons of space we will not be doing it here, except to say that pyramid plates are used to connect items together, they possess holes, perhaps for rivets, although exactly how they were used is not always entirely clear and, as far as I am aware, no such finds have been recorded from England. (See figure 15ii.12).

15ii.12. Pyramid plates and a possible way that they were used on a sword.
Reproductions, construction and photograph by Lars Lüppes (Hakun Risti).
Used with the kind permission of Lars Lüppes.

It is a problem that sword pyramids do not seem to be mentioned in the literature,[1] neither are sword-bosses, although ring-swords are, which could provide another clue as to how they were regarded. At this

---

[1] Menghin, 1983. 150 to 151. He also shows the Mound 1, Sutton Hoo pyramids being used in a non-practical way on page 146. See also, Marti, 1995. 88 to 92. Marti reports a find from Altdorf UR, Pfarrkirche St. Martin, Switzerland that has a reasonably intact set of pyramids and suspension system showing how they were used to connect the belt to the sword.

[2] See chapter, Additions, Beads, Bosses and Pyramids and the chapter, Case Studies I, The Sutton Hoo Mound 1 Sword.

[3] Menghin, 1983. 283. Menghin indicates that there are many more of these, all from the Rhine/Danube areas in his distribution map of pyramid plates, Karte 23 Beschläge mit Pyramidenknöpfen Liste C. III (Menghin, 1983).

[4] Menghin, 1983. 286

[1] A problem is that if they are, we have no idea what terms would be used to describe them. There are a number of OE words in the poetry that do not have clear and unambiguous translations. Steve Pollington personal comment.

stage I can only point out that the lack of a reference might be suggestive, and such a reference may have been unnecessary as the purpose of the objects was so well known. As the surviving literature is generally much later than the period when items such as sword pyramids first became familiar, it is also possible that by the time the poems and stories were being recorded, the use of such items and what they conveyed had been forgotten.

The triangular nature of pyramids, I think, is important and may take us a little further towards understanding what the pyramids symbolised. At least seventy-three[1] of our pyramids (37%) have perfect triangles embedded within their design. I have only included actual triangles that occur within the sides or tops of those pyramids that are decorated, I have not counted plain pyramids, nor the sides of plain, pointed pyramids, nor any with stepped cloisonné 'triangles'. Most of the remaining pyramids have sides consisting of truncated triangles, which are not included. Many sides of pyramids are decorated with triangles and they are often shown in groups of three dominant triangles per side (type 1via). This could just be an interesting design feature as it is easy to divide the side of a truncated pyramid into three triangles. Interestingly, the ring-and-dot motifs depicted on the sides of the three pyramids (P53, P62 and 48)in the list are also in groups of three arranged to form a triangle. A further pyramid, 89, has dots arranged to form three rough triangles on each of the four sides. P14 has each of its cloisonné sides divided into three fields and three sided knot work on two sides of P16 and all those pyramids in the M17 series (most of these are type 1iv. Pyramid and one from type 1vii[2]) have sides divided into three areas as does the cloisonné pyramid from Bedfordshire (59) but the M17 series do not contain precisely portrayed triangles.

15ii.13. pommel from Laon. It has a similar design to the Sutton Hoo, Mound 17 pyramids and its relatives. Photograph courtesy of Jean Enser. Thanks to Marie-Thérèse Cariou for putting me in touch with Jean Enser.

It is also interesting that images can occur in threes, sometimes on the swords themselves. For instance, groups of three ring-and-dots occur on several sword pommel caps from Kent, an example is the ring-sword from Dover Buckland (grave C),[3] and another, the sword from Finglesham, grave 204.[4] Three recent PAS finds of sword pommels exhibit the same feature, SF-4EFC81 (49 Appendix 10) (both sides) WILT-5C2531 (20 Appendix 13) which has one ring-and-dot on one side, three on the other and two on each shoulder (eight in all), and IOW-1D68F3 (48 Appendix 10), which has ring-and-dots in each corner of the pyramidal pommel on both sides. All three examples are of copper-alloy. Similar designs also

[1] P1, P2, P3, P6, P7, P9, P10, P12, P25, P27, P28, P33, P40, P43, P48, P51, P56, P57, P58, P59, P60, P63, P64, P66, P75, P76, P77, P80, P81, P85, 12, 13, 17, 18, 20, 21, 22, 23, 24, 27, 28, 29, 31, 33, 34, 39, 42, 46, 51, 52, 54, 55, 56, 67, 68, 72, 77, 79, 80, 81, 82, 83, 84, 85, 92, 9, 94, 98, 101,103, 108 and 109.

[2] The M17 similarity only occurs on the two opposite sides and, like the other designs, is described in niello rather than cloisonné.

[3] Evison, 1967. 107, figure 4 and plate VIII.

[4] Menghin, 1983. 222.

occur on pommels from the Continent.[1] Cloisonné triangles are included on the scabbard throat band on the sword from Selmeston,[2] and the throat band from Högom, Sweden has three similar garnet triangles.[3] Three circles appear on the magnificent cloisonné throat band on the sword from Fèrébrianges, Marne, France,[4] and each circle is further sectioned into three areas by a 'y' shape, rather like those on the M17 type pyramids.[5] (figure 15ii.13) Incidentally, one very interesting sword pommel from the River Bain in Lincolnshire has an enamelled red, solid triangle outlined by a yellow surround only on one side.[6] (figure 15ii.14) This is not an exhaustive list and other instances of images of threes on swords or on swords or their furniture could be added.

15ii.14. River Bain pommel. It has a red enamel triangle inset in one face of the pommel only. Author's photograph.

It is possible that it is just a matter of triangles and the number three being natural partners and perhaps that is their only significance but the ring-and-dot or just dot motifs, especially on sword pommels, are rather suggestive; it would be odd if such symbolism that is so often repeated had no meaning at all and was purely decorative. The motif appears in several different regions too. We know from later literature that importance was attached to the number three and its multiples,[7] and I would, therefore, suggest that the repetition of that number and its connection with triangles had a significance that would have been understood at the time. As for the meaning, we can only speculate. Perhaps the symbolism relates to three significant deities, maybe Woden, Thunor and Tiw or perhaps three or nine worlds? I think it extremely unlikely that the symbolism on the pyramids was used without there being an extramundane reason or reasons for their employment. However, three is a prime number and multiples of three, particularly the number nine, can have special properties, for example, three times three equals nine, and nine is a square number and possibly could be seen to incorporate the properties of a triangle and a square. It could be that the peoples of the 5th to 7th centuries were aware of this and it is its mathematical properties that held their fascination, perhaps they were much more aware of such things than they are often given credit for?

Pyramid P51 is intriguing, the triangles on each side are arranged with three at the bottom and one at the top in such a way that two, relatively large lozenge shapes are formed. The 'lozenges' have a secondary image on them which could be serpents (wyrms) or perhaps lightning strikes; the design might even represent a runic 's'.[8] Pyramid 23 from the Meads has a similar arrangement with two triangles at the bottom and one at

---

[1] A continental example would be the pommel from Langenenslingen, Germany. Evison, 1976. 21. Menghin, 1983. 248.

[2] Evison, 1976. 310.

[3] Evison, 1976. 118, figure 15g.

[4] Evison, 1967. Plate XI. The pommel of this sword has a single ring-and-dot per side. See also, Menghin, 1983. 231 and 313.

[5] See Appendix 9v.

[6] British Museum number 1997,0101.1. See Appendix 10.

[7] For a fuller discussion on the importance of this number in Germanic society, see Mortimer and Pollington, 2011.

[8] From a discussion with Stephen Pollington.

the top of each side, lozenges are marked out and inlaid niello decoration outlines the lozenge but also 'wyrms/'runes'/lightning symbols are marked here too. (Figures 15ii.15 and 15ii.16).

15ii.15. PAS BERK 1DF3D2 (P51). Pyramid found in Harrold, Berkshire.
The relief decoration is niello, and has mercury gilding. The red is either glass or enamel.

When viewed from the top, most of those pyramids where triangles dominate a 'star' shape can be seen, although the peoples of the 5th and 7th centuries could have perceived it as something quite different, nevertheless, I do not think that this an accident and the image might display some sort of understood symbolism, too.

An interesting feature of many swords from the 5th to 7th centuries in England is that, when viewed from the front or rear, they have pyramidal shaped pommels and pommel caps, especially those of the so-called 'cocked hat' type. There is no compelling practical reason for this form on top of a sword grip, so could the shape be linked in some way to the use of the square-based, flat-topped sword pyramid? It is interesting that pyramid 48 from Uncleby, being slightly rectangular echoes the shape of a pommel cap.

15ii.16. One of the two pyramids from the Meads, Sittingbourne, Kent. Note the niello design that has become detached from the main body of the pyramid. Photograph Paul Mortimer.

There are other designs that occur on the sides of the pyramids, and the 'omega' form is the next most frequently observed image to triangles in our sample. It is seen on P15, P22, P23, 40, 50, 53, 75, 100 and possibly 95 (figure 15ii.17) but quite what it means is not possible to say although it could, perhaps be phallic. The two omegas on the sides of 75 are not identical; one has a carefully included dot inside the apex loop but its opposite number is empty. Pyramid 32 has a more developed omega design, which seems to incorporate a possible serpent or interlace extension and might be connected to Style II interlace, while the Style II 'triquetra' on the sides of P12 may have been intended to convey a similar idea.

Some pyramids have stepped designs, with P13 and P82 and possibly others, it is achieved through the use of niello, but with P14, 59 and 35 the steps are the result of garnet cloisonné, while P72 imitates this effect with incised lines in the metal.[1]

Style II interlace occurs on 41, 50, 64, 86, 99, P5 and P12, it could be construed as 'serpents' (wyrms) writhing, but wherever

[1] Other cloisonné pyramids have steps, but they are not a major feature of the design, most of these occur in type 1iii.

it appears, except on P12, the designs on the adjacent sides are different to each other, although usually only slightly. Pyramid 32, with an 'omega' and a 'wyrm' below, might contain a similar image. Style I artwork appears on 8, 37, 69 and P33.[1] It is probably not an accident either, that the pyramids display deliberate choices of colours in their design. I think that some of the designs could show allegiance to a military cult or fraternity or indicate that the wearer came originally from a particular war band, but it could be that they show a form of rank, status or even be a type of campaign badge, or indicate that something has been achieved, perhaps a task completed or even a 'skill's badge. The later very high status pyramids, mostly of gold (quality A), I suspect have a different function from the more 'ordinary' versions and are all about displaying status within the symbolism of the pyramids through the use of high quality components and craftsmanship. Unlike ring-swords they appear in the highest status burials, at least that is what Sutton Hoo Mound 1 seems to indicate. The magnificence of many of the category A pyramids may just be due to the artists who made them showing off their talents in achieving the near impossible and nothing more. Whatever, pyramid designs may show the situation is likely to be even more complex when we factor in the possibility of organic pyramids made for example from bone, ivory or even those of stone.[2]

15ii.17. PAS NMS-1E5834 (P12) This example has a form of the 'omega' shape drawn as a triquetra knot on each of its four sides. Photograph used with the permission of the PAS.

Several of the pyramids in type 1vi have small design elements added to their basic design. Both P51 and 23[3] have similar 'snake', lightning bolt or possible 's' rune situated on their sides either within a lozenge or a diagonal modified to look like one. P51 was found in Bedfordshire and 23 from Kent, but it is, I think, highly likely that we are seeing a shared symbolism. Other type 1vi pyramids with additional embellishment are P63, P76, 17 and 81, all of which have two parallel lines carved into the lozenge shapes. P76, 85 and 51 have chevrons with engraved lines at ninety-four degrees to the direction of the arms of the chevrons. The latter two sets of examples might not signify anything at all apart from an urge to fill a space, but on the other hand they could be subtle signifiers.

37 - an intriguing design

In some ways the most interesting pyramid of all is number 37, briefly mentioned above, because it has four complete and different images plus a group of four studs at the top. The images on the sides of the object are in pairs and the components of each pair are on opposite sides of the pyramid. There are two faces that comprise one pair of images. It is probable that they are wearing helmets as the brows and nasal do resemble those on some helms, but this may not be the case. One of the faces appears to have a broad and exaggerated smile, while the other face is grim. On either side of each face are 'eye' or sun symbols both of which have radiant lines moving down from the eye/sun; it could be significant that one 'eye' has five radiant lines, but the other has six. It is also interesting that a very similar symbol occurs on the scabbard throat band of the

[1] Other designs include scales = P16 - 'Vs' = P39 - 3 sided knots = P16 and some very peculiar shapes on P41.

[2] So far stone examples are known only from the Continent.

[3] The design on 23 was executed in niello.

Chessell Down sword.[1] (figure 15ii.9 and 15ii.10). Are we looking at the eyes of Woden, a sun symbol or both, or something else? Each 'eye' is different, perhaps reflecting the much later story of Óðinn depositing an eye in Mimir's well and that eye being altered forever. The opposition of the, 'inanely' grinning face and the doleful one is intriguing but difficult to interpret. That these images had some reason for being depicted, I think, is certain and I feel, probably refers to the stories and ideas circulating at the time; it is unfortunate, if this is so, that the story seems lost to us. What of the four studs? Perhaps just a reminder that four, too, could be a powerful number, and almost all the pyramids have the number four as an integral part of their design as well, as most have four faces. It is certain that four times three equals twelve and twelve is a multiple of three and many pyramids, as shown above, have four sides with three images on each side.

15ii.18a to 15ii.18f. The Cederlind pyramid (95). Photograph courtesy of Classical Numismatic Group Inc.

[1] See Appendix 9v.

The only other pyramid that has a human face on its sides, again on just two sides, is 95 (figure 15ii.18a to 15ii.18f) where a pyriform head appears between two 'wings'/'hands'/'feet'. There are differences between the two faces but nothing significant or that seems deliberate, just differences that have occurred during the crafting.[1] The other two sides figure a strip made into a loop, which could be an animal or, perhaps, an omega.

Number 100, too, is unique, so far, in that it features a triskele on two sides. Another unusual feature is that the similarly designed sides are adjacent to each other rather than opposites.

There is more work to be done in this area of the significance of numbers and it might require new ways of looking at the information; a few more clues and finds like 37 would assist us.

15ii.19. (left) The Collingham jewel. Note the two birds' heads. Photograph Lindsay Kerr.
15ii.20. (middle) The high status pyramid from Ezinge, Holland.
Again note the two birds' heads incorporated into the design.
15ii.21. (right) Staffordshire Hoard (14 in our list or K1166)

One other design that has echoes elsewhere in the Germanic art of the 5th to 7th centuries is the double birds' beak motif on the Collingham jewel (from Newark) (7); (Figure 15ii.19)it looks like a reference to ravens or predatory birds. The sword pyramid from Ezinge, near Groningen, Holland, (Figure 15ii.20) has what appears to be a very similar device and a garnet and gold pyramid from the Staffordshire Hoard (14 in our list or K1166) (Figure 15ii.21) has two birds entwined, embedded within the sides, which could, possibly, also include a man's face when viewed upside-down, two birds that become a man could have echoes of Óðinn[2] and his two avian companions, Hugin and Munin (Thought and Memory) who flew around the World and reported what they had seen to their master at the end of each day. The design has an echo in a detector find from January 2012 and registered in the PAS database, possibly a pendant consisting of two birds which possibly form a man's face as well (SF-92CD45).[3] (Figure 15ii.22). The piece was located in Suffolk, England and found in 2012; it was included in the PAS database on 26th February 2014. This particular find is similar in shape to some finds of 'horned men', some of which are also logged in the PAS

[1] The form of the face appears elsewhere in the art work of northern Europe at this time, for a full discussion see Mortimer and Pollington, 2013.

[2] Woden and Óðinn, are generally thought to be the same deity, although this has been questioned; see Shaw, 2002, particularly pages 35 to 41.

[3] See Appendix 9v. and see: https://finds.org.uk/database/artefacts/record/id/516316

database.[1] Horned men, appear relatively frequently in the art of the 5th to 7th centuries, including on helmet pressed plates such as the helmet from Sutton Hoo (Suffolk, England) and those from Valsgärde, Sweden.[2] Helmbrecht[3] and Raynor[4] include a survey of such figures and discuss their purpose and possible links to the Other world and to Woden/Odinn.

15ii.22. PAS (SF-92CD45) find of a bird/man face pendant of cast copper-alloy. It was found in Suffolk on 1st January 2012, at Ousden, St. Edmundsbury. Photograph courtesy of the PAS.

Finally, one hundred and seventeen[5] (59%) of the pyramids have, or have had gems, mostly square ones, set in their apex. In most cases the gem is absent having fallen out at some stage of the pyramid's existence, but in forty-one examples garnets are present and in one (Lissington 29) the gem is malachite. One other, (Coddenham 82) has a very small depression once filled with white material.[6] Four of the one hundred and fourteen have round settings and one (P19) has a cabochon garnet at its apex. A further four pyramids (these are not include in the one hundred and five), 14, 30, P3 and P34 have chequers of millefiori in square settings, instead of a gem.[7]

The similarity between some type 1ii pyramids

Pyramids 57, 16 and P38 all appear to be very similar, although we don't have dimensions for 57 and 16. However, there are differences 57 and P38 have their garnets inset while 16 has a proud garnet. 57 has smooth sides whereas 16 and P38 have rougher sides. P38 has an exposed foil when viewed from the underside, the others do not. The garnet in 16 is not quite square with the body; and the one on 57 is slightly offset too, but not as much as 16. The garnet in P38 seems to be set centrally. 57 has casting marks on the inside that do not appear in the same places as those in P38 -- the inside of 16 is rough, but seems to be without casting faults.

A note or two regarding type 1xi

This class of sword pyramids, if that is what they are, do not have a bar under the pyramid and resemble some of the organic pyramids from the Continent in that they have a round hole pierced through the top. It is probable that they would be secured with a cord and knot as has been found on examples from Italy and Germany.

---

[1] See particularly, LEIC-40DB05, NMS559, HAMP2432. Other horned men are recorded in the database but shaped differently: NMS-4C3AD7, HAMP-B292C2 , NMS-F90626, FAHG-8EAAA3, YORYM-024D31, BERK-4F2E17. A very similar piece was previously sold by TimeLine Originals, as noted on the PAS notes..

[2] See Mortimer, 2011, (49 to 83) for a full discussion of the helmet pressbleches.

[3] Helmbrecht, 2007 and Helmbrecht, 2007 – 2008.

[4] Raynor, 2010.

[5] 1, 7, 10, 11g, 12g, 13, 15r, 16g, 17, 21, 22, 23g, 24g, 25, 26, 27g, 28g, 29, 31g, 32g, 33 (Staffs Hoard 4 garnets saltire), 34g, 35, 38, 39, 40, 41g, 46, 47g, 49, 50g, 51g, 52g, 54g, 55g, 57g, 58, 59, 60, 62, 63r, 64, 67, 68, 69, 70, 73g, 74, 75g, 77, 78g, 79g, 81, 82w, 83g, 84, 85g, 87g, 89, 90, 91, 92 (yellow glass), 93, 94g, 97g, 99g, 100g, 101, 105, 108g, 111, P1, P2, P3, P5g, P6, P7, P9g, P10, P14, P15, P16g, P17, P19g, P20, P22, P24g, P23g, P25, P26, P27, P28, P29, P30, P33, P38g, P39, P40, P41g, P43, P48, P49g, P50g, P51, P53, P54g, P55, P56, P58g, P60, P61, P63, P64g, P67g, P72, P75, P80, P81, P82g, P84 and P85 (g = garnet, r = round, w = white material)

[6] West, 1998. 22.

[7] One pyramid, 20, has a single ring and dot inscribed in the top.

There are examples in the PAS database which are similar but seem to come from a later period and I have not recorded them here. The better the gilding quality, the more likely they are to be earlier.[1]

Status and craftsmanship

The quality of workmanship varies considerably along with the materials and it is reasonable to suggest that the level of finish often indicated status, although this could be the 'status' of the achievement or whatever rather than the status of the individual. I think that there are at least four different levels of craftsmanship that we can discern in the pyramid sample that I have been able to assemble and I have categorised them as, A: the finest quality, B: high quality, C: medium quality and D: poor quality. To some extent the separation is subjective, but I have taken into account the materials used and the finish, but the images used to view the pyramids, their state of corrosion and the level of conservation could have affected my judgement and others handling the same objects might arrive at different conclusions. Not quite all the sword pyramids in our sample are included in the analysis of quality and, therefore, the percentages will not add up to a perfect one hundred per cent.

Those of the finest quality in category A, include all in type 1iii -- those decorated with complex cloisonné, and type 1v, Pyramids with filigree, 73 (type 1iii)and the gold, octagonal P20 (type 3), which would have been magnificent with all its missing gems in place. Other top quality pyramids are the gold example, P2 from type 1ii (decorated mainly plain with a gem on top) and 11, the pyramid from Broomfield and 87 (both type 1iv) and 12 one of those from the Staffordshire Hoard (type 4ii). (22/197 = 11.2%)

Perhaps the next level down, category B, those of good quality, should include those from type 1iv. Decorated with simple cloisonné, plus P1, P4, P10, P19, P68, P77, 8, 9, 12, 28, 29, 36, 37, 50, 51, 52, 53, 54, 66, 68, 72, 95, 97, 100, 104, 105 and 107 should be included, too, because of the care taken in achieving the precision of their triangles. (50/197 = 25.3%)

I have put the following into the medium category, C: P6, P7, P8, P9, P12, P13, P15, P16, P22, P23, P25, P27, P28, P32, P33, P34,P39, P40, P43, P46, P48, P51, P54, P56, P58, P59, P60, P63, P66, P69, P72, P75, P76, P78, P80, P82, P85, P86, 3, 15, 16, 17, 20, 22, 23, 24, 34, 39, 40, 46, 51, 57, 60, 63, 67, 71, 75, 77, 79, 80, 81, 82, 83, 85, 92, 93, 94, 103, 106, 108 110 and 111 (70/197 = 35.5%)

The poorer quality pyramids, category D, are: P18, P21, P24, P30, P31, P35, P36, P37, P38, P41, P42, P44, P45, P47, P52, P53, P57, P62, P70, P71, P73, P74, P79, p81, 2, 4, 6, 18, 19, 21, 42, 43 (by far the worst quality of all in this sample), 44, 45, 48, 56, 65, 69, 74, 84, 88, 96 98, 101, 102 and 110. (47/197 = 24%).

Some pyramids are formed from a solid cast, while most of the finest quality pyramids are built up from many components, and others, although less complex, are still made from more than one piece of metal. Some pyramids exist only as the outer casing that once fixed onto a more solid base and it is possible that some of the very plain pyramids may have constituted a base.

Damage

A few pyramids are quite badly damaged and have what look like 'cut' lines, perhaps where they have been struck with a heavy or sharp object (3, 5, 28, 32, 47, 78, P9, P23, P55, P56 and possibly P67). Several others have the bar at the bottom missing; it is possible that they have forcibly removed from whatever held them. Some of the ways in which this small group have been damaged resembles some of the damage to items found within the Staffordshire Hoard. It must be admitted that the damage could have been due to excavation.

---

1 Personal comment Stephen Pollington.

Dating again

It would seem, then, that the pyramids in type 4iii (squat, button-like, with the exception of 97), could be at the start of the tradition and these can be quite securely dated to the 5th century and possibly even be a little earlier. The circular base gives way to the square based pyramids by the beginning of the 6th century, I think. The few pyramids in our sample that have Style I decoration (P33 and 69) were probably made during the 6th century and possibly even in the first half of that century. Most of our sample cannot be securely dated to a narrow date range, and must be given wider date span of the 6th to the first quarter of the 7th century. The burial in Mound 17 at Sutton Hoo of a young horseman with his sword and set of pyramids has been dated to the last quarter of the 6th century by Evans.[1] It is possible that all those pyramids with a similar design came from the same period or a perhaps a little earlier, although their use could have continued well into the early 7th century.

Most of the category A pyramids and those with Style II decoration are probably quite late from the final quarter of the 6th perhaps until the middle of the 7th century when it seems a fashion for the highest quality pyramids may have developed.

The shape of pyramids does evolve, I think, over time although the cone or mound shape is always retained even if it becomes more subtle. Some of those with round or multi-sided bases, I would suggest were made towards the end of 6th or even during the 7th century. It is interesting, too that the sword pyramid might have begun as a circular cone shape at the start of the tradition and that there could have been a tendency to return to it at the end of the pyramid tradition.

The latest sword pyramids would appear to be those with elements of Insular Style art applied to them and there are three examples, 9, 15 and 36. Both 9 and 36 I have assigned to category B, they are generally very well made and are from type 4ii; that is they have flat conical sides, despite their circular bases. Sword pyramid 15 is from category C and the design is not so well executed, it is square based and from type 1vii, but like the other two in this date range has steep sides. These examples are possibly from the mid to late 7th century, perhaps even from the early 8th century.[2] The lead-alloy pyramid with an equal arm cross surmounting the squat cone body (P78) I believe to be quite late too, probably from the end of the 7th century to the beginning of the 8th.

Finally, we come to what arguably could be the most controversial of the 'sword pyramids' in the sample and certainly the latest, P65 and 66, the 'thimble' shaped objects. Both are of silver-gilt and well made, although the finish of P65 is much finer than that of 66. Both are probably post 7th century and if they were employed as sword pyramids then they are evidence that the tradition continued well into the Christian period. P65 was discovered in Melton, Leicestershire and 66 comes from Ketsby, Lincolnshire. Both of these resemble, in both form and decoration, the richly decorated, 8th century mounts found under a slab in the chapel on the Isle of St Ninian, in the Scottish Shetland Isles. The St Ninian's mounts are Pictish and although separated from most sword pyramids from Anglo-Saxon England by a considerable period of time and a fair distance, do have the same type of bar at their base that pyramids usually have. It is possible that P65 and 66 are prototypes for the examples from Shetland. It has to be said that the St Ninians mounts, and indeed P65 and 66 might not have anything to do with swords.

---

[1] Evans, 2005. 243. The burial itself is dated to the first quarter of the 7th century. Carver and Fern 2005. 301.

[2] Webster and Backhouse, 1991. 290.

Conclusions

One interesting fact concerning the finds from the Staffordshire Hoard is that the find represents at least eighty gold or silver hilted swords,[1] yet there are only five pairs of matched pyramids, so it would appear that wearing pyramids was not a universal practice and only certain men wore them on their weapons. Some men could have used pyramids of wood or bone. Although none as yet have been found in England, considerable numbers have been found on the Continent. Such precious hilts as are found in the Hoard are likely, I suggest, to have been paired with precious pyramids if they had them. However, it would seem that not every sword wearer used pyramids, which is something that is also borne out by the evidence from sword graves; only two graves out of one hundred and sixty-nine sword graves in my cemetery sample had metal-based pyramids, Sutton Hoo Mounds 1 and 17.

Although there are tantalising clues, we might never know the true significance of the pyramids and why they were associated with weapons, what is certain is that military men[2] throughout the ages have used visible badges and distinctions to show rank, achievement, skills, experience, membership of an association, unit affiliation, receipt of awards, attendance on a course, etc.; not all insignia is about rank and hierarchy. These signs are often quite subtle and sometimes only of significance to those *in the know*. In modern armies it might not just be a badge but its position, or a slight difference in its shape which is enough to convey the message. It could even be shown by the cut of a particular uniform element; soldiers quickly get used to noticing these subtle indicators and quickly learn what they mean when it is important to them.

I suggest that like their modern equivalents (badges, medals and such like) sword pyramids could have performed a number of different functions and that although the meanings will have changed over time, they appear to have been significant and were certainly not for everyone. Certain motifs appear again and again, for example, three triangles, patterns with three in them, the 'y' shaped M17 series. I suggest that this separation into 'types' would appear to support the idea of meanings recognised through time and space. Such designs can be dispersed over quite a wide area; the Mound 17 pyramids were from a grave in Suffolk, the Coombe Bissett (Salisbury Race Course) examples from Wiltshire, some 180 miles away, while the example from Scunthorpe, Lincolnshire (26) is again some seventy miles from Sutton Hoo. Those pyramids that display three triangles per side are more numerous and spread over an even greater area,[3] from Hampshire (P48, 46 and 67) and Wiltshire (P59 and P75) to Oxfordshire (P77), Warwickshire(P9 and 94), Berkshire (P40, P81), Kent (23 and 79), London (80), Hertfordshire (P27), Cambridgeshire (P28, P43 and 24), Suffolk (P25 and 82), Norfolk (P56 and 39), Yorkshire (P10, P60 and P80), Leicestershire (P76), Buckinghamshire (P6) and Dorchester (P7).[4] It is true that some of the triangles on the faces of some pyramids are crafted in a different style to some others and are of a lower quality, but in many there appears to be a remarkable consistency in the way that they were made and the symbols displayed.

---

[1] Cool, 2015. 87. Cool seems to indicate that there are fourteen sword pyramids from the Hoard, ten of which constitute five pairs. Pieta Greaves reported that there are only ten. (Personal comment). At the Warrior Treasures conference at the Royal Armouries, Leeds on Saturday 11th June 2016, Pieta Greaves who is responsible for conserving the Hoard, revealed that it may represent one hundred and fifty swords.

[2] What follows may well hold true for other elements of society, too.

[3] There are at thirty-nine of these P6, P7, P9, P10, P25, P27, P28, P40, P43, P48, P56, P59, P60, P75, P76, P77, P80, P81, P84, 21, 22, 23, 24, 28, 34, 39, 46, 51, 54, 67, 68, 72, 77, 79, 80, 82, 83, 84 and 94. A few more, P33 (Oxfordshire), P63 and 17 (both Wiltshire), 81 (Somerset) and 85 (no location noted) have three triangles on two sides only; the alternate sides have different arrangements. 89 has three triangles per side marked in dots and 48 a single triangle per side formed by three ring-and-dots.

[4] Quite a few (21, 22, 28, 34, 51, 54, 68, 72, 77, 83 and 84) have no find place noted.

Whether such items show allegiance to a cult, ideology or an achievement or something else is a matter for speculation until better evidence arrives. It is always possible that some designs only had meaning for the individual owner and his friends, rather like many modern tattoos.

Table 15i. Pyramid find locations
PAS list

P1 Sleaford, North Kesteven, Lincolnshire	P21 Ingleby, South Derbyshire	P41 Happisburgh Norfolk	P61 Coddenham, Mid Suffolk
P2 Shouldham, near Swaffham, Norfolk	P22 West Lindsey, Lincolnshire	P42 Near Eye, Mid Suffolk	P62 Manton, North Lincolnshire
P3 Near Dorchester Dorset	P23 Drax, Selby, North Yorkshire	P43 South Cambridgeshire	P63 Witham Friary, Mendip, Somerset
P4 Calbourne Parish, Isle of Wight	P24 Worlington, Forest Heath, Suffolk	P44 Stonegrave, Ryedale, North Yorkshire	P64 Staffordshire (Hoard)
P5 Bury St Edmund, Suffolk	P25 Barham Mid Suffolk	P45 Sleaford, North Kesteven, Lincolnshire	P65 Melton Leicestershire
P6 Princes Risborough Wycombe Buckinghamshire	P26 Near Diss, Mid Suffolk	P46 Aunsby and Dembleby, North Kesteven, Lincolnshire	P66 Feckenham, Forest Heath, Suffolk
P7 Wimbourne St Giles, East Dorset	P27 North Hertfordshire	P47 Near Eye, Mid Suffolk	P67 Folkingham, South Kesteven, Lincolnshire
P8 Maunby, Hambleton, Yorkshire	P28 South Cambridgeshire	P48 Thruxton CP, Test Valley, Hampshire	P68 South Cambridgeshire
P9 Wellesbourne, Stratford-upon-Avon, Warwickshire	P29 Humshaugh, Northumberland	P49 ?	P69 South Ferriby, North Lincolnshire
P10 Folkton near Scarborough, North Yorkshire	P30 Raithby, East Lindsey, Lincolnshire	P50 Hollingbourne, Maidstone, Kent	P70 Chapel St Leonards, East Lindsey, Lincolnshire
P11 The Deverills, Wiltshire	P31 Everleigh, Wiltshire	P51 Harrold, Bedford, Bedfordshire	P71 Aylesbury Vale, Buckinghamshire
P12 Roughton, North Norfolk	P32 Isle of Wight	P52 Everingham, East Riding of Yorkshire	P72 Near Woodbridge, Suffolk Coastal District
P13 Near Hunstanton – Heacham, King's Lynn and West Norfolk	P33 Stanton St John, South Oxfordshire	P53 Wyre Piddle, Wychavon, Worcestershire	P73 Chiseldon, Swindon, Wiltshire
P14 Alderton, South Northamptonshire	P34 Hinton St Mary, North Dorset	P54 West Wratting, South Cambridgeshire	P74 Minterne Magna (Minterne Minerva?), West Dorset
P15 Newbold Pacey, Stratford-on- Avon, Warwickshire	P35 Saxton with Scarthingwell, Selby, North Yorkshire	P55 Caistor St Edmund, South Norfolk	P75 Kingston Deverill, Wiltshire
P16 Near Gainford, County Durham	P36 Woodbridge, Suffolk Coastal	P56 Wretton, King's Lynn and West Norfolk	P76 Scampton, West Lindsey, Lincolnshire
P17 Chearsley, Aylesbury Vale, Buckinghamshire	P37 Bourne, South Kesteven, Lincolnshire	P57 Beeston, Breckland, Norfolk	P77 Crowmarsh, South Oxfordshire

P18 Chiseldon, Swindon, Wiltshire	P38 Kirby Bellars, Melton, Leicestershire	P58 Kingston Deverill, Wiltshire	P78 Herningby, East Lindsey, Lincolnshire
P19 Boughton Aluph, Ashford, Kent	P39 Isle of Wight	P59 Alton, Wiltshire	P79 Lenton Keisby and Osgodby, South Kesteven, Lincolnshire
P20 Bembridge, Isle of Wight	P40 West Berkshire	P60 Kilham, East Riding of Yorkshire	P80 Kilham, East Riding of Yorkshire
P81 Berkshire	P82 Berkshire	P83 Cambridgeshire	P84 Leicestershire
P85 Bainton, Yorkshire	P86 Nottinghamshire		

Non PAS list

1. Mound 6 Sutton Hoo, Suffolk	26. Scunthorpe, Lincolnshire	51. ?	76. ?
2. East Yorkshire	27. Staffordshire, (Hoard)	52. ?	77. ?
3. Leaming Bar, North Yorkshire	28.?	53. ?	78. Tuddenham, Suffolk
4. Maltby by Louth, Lincolnshire	29. Lissington, Lincolnshire	54. ?	79. Pilgrim's Way, near Wrotham, Kent
5. Heckington, North Kesteven, Lincolnshire	30. Sutton Hoo Mound 1, Suffolk	55. Wester Craigie, Dalmeny, West Lothian	80. Forest Gate, Greater London
6. Suffolk	31. Dorchester Dorset	56. ?	81. Witham Friary, Somerset
7. Danethorpe Hill Farm, Collingham, Nottingham, Nottinghamshire	32. ?	57. ?	82. Coddenham, Suffolk
8. ?	33. Staffordshire (Hoard)	58. Coombe Bisset, Salisbury Racecourse, Wiltshire	83. ?
9. Bawtry, Nottinghamshire	34. ?	59. Lidlington, Bedfordshire	84. ?
10. Mound 17 Sutton Hoo, Suffolk	35. Sarre, Kent	60. Lincolnshire	85. ?
11. Broomfield, Essex	36. ?	61. Dorchester, Dorset	86. Selsey, West Sussex
12. Yorkshire	37. Thetford Norfolk	62. Coddenham, Suffolk	87. ?
13. Staffordshire (Hoard)	38. ?	63. Coddenham, Suffolk	88. ?
14. Staffordshire (Hoard)	39. Norfolk	64. ?	89. Kent
15. ?	40. ?	65. Monmow Street, Monmouth, Monmouthshire	90. ?
16. ?	41. Rendlesham, Suffolk	66. Ketsby, Lincolnshire	91. ?
17. Trowbridge, Wiltshire	42. Oxfordshire	67. Winchester, Hampshire	92. Wittenham, Oxfordshire.
18. North Essex	43. ?	68. ?	93. Kent

19. Norfolk	44. ?	69.?	94. Alcester Worcestershire
20. Wiltshire	45. ?	70. ?	95. ?
21. ?	46. Winchester Hampshire	71. ?	96. Kent
22. ?	47. ?	72. ?	97. Lincolnshire
23. The Meads, Sittingbourne, Kent	48. Uncleby, Yorkshire	73. Finglesham, Kent	98. Kent
24. Cambridgeshire	49. Diss, Norfolk	74. ?	99. ?
25. Near Snettisham, Norfolk	50. ?	75. Melton Ross, North Lincolnshire	100. County Durham
	103. ?	102 Brigg, Lincolnshire	101. Oxfordshire
106. Oxfordshire	107. Durham (?)	108. Canterbury, Kent	109. ?
110. Ripon, Yorkshire			

Table 15ii. Pyramids county summary

PAS registered pyramids – total 85

Bedfordshire 2	Derbyshire 1	Kent 2	Northumberland 1	Suffolk 10
Berkshire 2	Dorset 4	Leicestershire 2	Nottinghamshire 1	Warwickshire 2
Buckinghamshire 3	Hampshire 1	Lincolnshire 13	Oxfordshire 2	Wiltshire 7
Cambridgeshire 5	Hertfordshire 1	Norfolk 7	Somerset 1	Worcestershire 1
County Durham 1	Isle of Wight 4	Northamptonshire 1	Staffordshire 1	Yorkshire 10

Non PAS list pyramids – total 111[1]

Bedfordshire 1	Isle of Wight 0	Nottinghamshire 2	West Lothian 1
Cambridgeshire 1	Kent 8	Oxfordshire 3	Wiltshire 3
Dorset 2	Leicestershire 1	Somerset 1	Worcestershire 1
Essex 2	Lincolnshire 10	Staffordshire 4	Yorkshire 5
Greater London 1	Monmouthshire 1	Suffolk 9	County Durham 2
Hampshire 2	Norfolk 5	Sussex 1	Unrecorded 42[2]

Comments

There are at least two factors that will affect any subsequent discussion regarding this sample of pyramids. First of all we should remember that not all the pyramids are metal detector finds but most of them are. Some of the latter could be from disturbed burials but the majority are not. Secondly, the number of finds in any county is likely to be related to the intensity of local metal detector use and there may well be better organised clubs or individuals active in particular areas, so the sample will be spatially skewed. We should

[1] Some of these pyramids have been recorded by the PAS (all those from Staffordshire for instance) but I collected them in the other list first.

[2] 8, 15, 16, 21, 22, 28, 32, 34, 36, 38, 40, 43, 45, 47, 50, 51, 52, 53, 54, 55, 64, 68, 69, 70, 71, 72, 74, 77, 83, 84, 85, 87, 88, 90, 91, 95, 99, 102, 103, 104,105 and 109.

also bear in mind that most of the data from the PAS is likely to be more reliable than information gained through eBay or some other sources.

It is true that some counties cover greater area than others, for instance Lincolnshire is a considerably larger county than Essex or Suffolk. However Suffolk has produced almost as many pyramids as Lincolnshire (nineteen) as Lincolnshire (twenty). Other counties with significant numbers are Norfolk with a total of twelve; Yorkshire has eleven, Wiltshire with ten finds and Kent with seven.

Clusters?

Three of the Suffolk pyramids come from the cemetery at Sutton Hoo and two more (P36 and P72) are from nearby Woodbridge, just across the River Deben. Four pyramids from the same county appear to form a cluster around the small village of Coddenham, Suffolk, not many miles away from Sutton Hoo (P61, 62, 63 and 82). The Isle of Wight, which is one of the very smallest counties and has four recorded pyramids (P4,P20, P32, and P39) is quite rich in other types of finds from the period as well. Two pyramids have been found in the vicinity of Diss, Norfolk (P26 and 49) and there are two found close to each other in Kilham, in the East Riding of Yorkshire.

However, the evidence for the existence of deliberate 'clusters' is slight, although those from Coddenham are intriguing, and it is likely that those that appear to exist are probably due to more active metal detecting in that region than anything else.

List of known pyramids

Notes.

1. As noted above, I have compiled two separate lists of sword pyramids, one from the Portable Antiquities Scheme database (PAS) and a second from various other sources. The number in brackets at the start of each entry indicates the position in each of the lists that sword pyramid occupies; those with a P before the number are from the PAS site, those without are from the second list. Images of most of the items in the sample can be seen in Appendix 9.

2. When pyramids have been located in graves, they are usually found in pairs; if they are of the same design, I have only counted them as one for the purposes of this study.

3. (M17) accompanying a description shows that that example has a similar decorative theme to the pair retrieved from Mound 17 at Sutton Hoo. There are eight of these, P17, 10, 25, 26, 47, 58, 70, and 78 (see below). All of these have mainly white and red, and sometimes blue decoration separated by a distinct 'y' shape.

4. The number at the end of each entry just counts the number of entries in each category.

5. The letter in brackets after most entries A, B, C or D, refers to my subjective assessment of quality. More details follow the typology below.

6. A set of alpha-numeric characters follows each of the bracketed P numbers; that is the entry's PAS unique identifying code. Anyone with access to the database (https://finds.org.uk/) can look up the entry up for themselves.

7. Where a pyramid has different designs on adjacent or opposite sides, in most cases this has been highlighted using text in **bold** type.

8. There are a small number of other pyramids that I am aware of, but as I have been unable to locate an image or a detailed description of them, I have not included them in this collection.

9. Throughout this study I have avoided a discussion on the actual structure and manufacture of sword pyramids. The reason for this is that, in most cases, I have only had access to images and not the pyramids

themselves. However, there is potentially much to be learned from a close analysis of the craftsmanship involved in their making. Some comments do appear in the descriptions.

10. Although I have tried to record the measurements of the pyramids, this has not always been possible. I have not attempted to categories the sizes of the examples in the image collection, but that could provide interesting results should that be done in the future.

11. This typology only contains brief descriptions of each pyramid, in many cases more information can be found in Appendix 9. Almost all the information in the appendix is abbreviated from whatever information was available with the original image.

12. Abbreviations:

UKDFD = United Kingdom Detector Finds Database.

SB = Sue Brunning – a pyramid that was listed by Sue Brunning of the British Museum in her Flickr page.

BM = British Museum.

Type 1. Flat-topped square-based

Type 1i. Plain: iron, cu-alloy, silver, silver-gilt cu-alloy gilt, gold[1] (figure 15ii.23)

15ii.23. Type 1i. Flat-topped, square and plain (96). Photograph Paul Mortimer.

(P11) WILT-4008EB Cast copper-alloy. (D) 1

(P8) NCL-74CD04 An early Anglo-Saxon pyramidal mount from a sword-scabbard, made from cast copper-alloy. (C) 2

(P21) DENO-08BD33 Early-Medieval cast copper alloy pyramidal mount, almost pointed. (D) 3

(P31) WILT-761BB7 An Early-Medieval (500-700) cast copper alloy and silvered pyramidal mount with some edge damage. Everleigh Wiltshire. (D) 4

(P32) IOW-298744 An incomplete cast copper alloy strap mount of Early-Medieval (Anglo-Saxon) date (500-700). Although sides are flat, base is almost circular and has a flange. (C) 5

(P42) SF-970F42 A cast copper-alloy pyramid mount of Early-Medieval/Anglo-Saxon date near Eye Suffolk, rather steep sides. (D) 6

(P44) YORYM-416687 A copper alloy mount of early-medieval date. The mount is pyramidal with a flattened apex. Stonegrave Ryedale Yorkshire. (D) 7

(P47) SF-9769F8 A worn and corroded iron pyramid mount of Early-Medieval/Anglo-Saxon date. It may have had garnets in the four flattened corners at the base (only one now discernible). Near Eye, Suffolk. (D) 8

(P73) WILT-8D2697 A cast copper-alloy pyramid mount. (D) 9

(2) Copper ally, gilt - Near Swine, East Yorkshire SB UKDFD 7354. (D) 10

[1] It is possible that some of the pyramids in this category may not have originally been plain and may have had details that have become obscured.

(4) Copper-alloy mount in the form of a hollow pyramid with flat top and an internal bar for attachment 22mm x 22mm. Maltby by Louth, Lincolnshire UKDFD. (D) 11

(6) From Suffolk, courtesy Alexander Haigh. Type 1i. (D)12

(43) Bronze, 4.6 grams. 16.61 mm. An Anglo-Saxon tinned bronze pyramid mount Worthopedia™ ANGLO-SAXON TINNED BRONZE PYRAMID MOUNT 000551 sold online Mar 07,2012. (D) 13

(71) eBay find. (C) 14

(96) Peter Fischer collection copper-alloy. 14mm x 14mm x 9mm. Kent. (D) 15

(109) Copper-alloy. Measurements 20mm x 20 mm, height 155 mm. Provenance unknown – found unmarked in the collection. Recorded in, *A Summary Catalogue of the Anglo-Saxon Collections: Non-ferrous Metals (British Archaeological Reports)* 1993. (Page 215 – 36.27) Arthur MacGregor and Ellen Bolick. Oxford. (D) 16

(102) Listed as: 'Rare Anglo-Saxon bronze pyramid mount. 6th-9th century. 18mm'. Sold by nottscollectables51 on eBay 28th May 2017. Notts Collectables, 2, Repton Road, West Bridgford, Nottingham, Nottinghamshire, NG2 7 EJ. Found near Brigg, Lincolnshire. Bought by Matt Bunker. (note: it would appear that the securing bar was soldered on separately and has been removed – precisely how, it is not possible to say. This pyramid appears to have four equally spaced indentations in the top. If this is confirmed it is a feature not found on any other pyramid in this corpus. (D) 17

Type 1ii. Decoration consists mainly of single gem on top (figure 15ii.24)

(P2) NMS-A61494 Gold pyramid mount from sword, square base with flattened corners caused by triangular settings at the base. (A) 1

(P34) DOR-A7E806 An incomplete silver pyramidal scabbard mount with worn edges. The mount is hollow and almost square in cross-section millefiori square in top. Hinton St Mary Dorset. (C) 2

(P38) LEIC-4BD061 Early medieval copper-alloy pyramidal mount from a sword-scabbard, 16mm high and 15mm wide. Harborough Leicestershire. (D) 3

15ii.24. Type 1ii. Decoration consists mainly of single gem on top (P38) (LEIC-48D061). Photograph courtesy of the PAS.

(P54) CAM-78D756 A copper alloy pyramidal sword-belt or scabbard fitting dating to the early 7th century AD West Wratting, Cambridgeshire. (C) 4

(16) Copper alloy, garnet - from an 'old collection'1. Sold via TimeLine Originals. SB. (C) 5

(57) Nigel Mills Saxon and Viking Artefacts, 2001. 99. Very similar to LEIC-4BD061 in PAS list. This one from Kent. (C) 6

(65) 24 Monmow Street, Monmouth, silvered-copper-alloy or silver remains of bar still visible and gem or glass at top. 12.7x13 mm height 8 mm bar diameter 2.3 mm. The National Museum of Wales. (D) 7

Type 1iii. Decorated: cloisonné, complex (figure 15ii.25)

15ii.25. Type 1iii. Decorated: cloisonné, complex (13). Photograph courtesy of Andy Pilkington.

(P3) SOMDOR-F381A4 A gold strap fitting from a scabbard or sword-belt, set with cloisonné garnet and opaque pale blue glass over pointillé gold foil. (A) 1

(7) Gold, garnet - Danethorpe Hill Farm (British Museum) Art Fund (Collingham jewel). SB. (A) 2

(13) Staffs Hoard K377 (Birmingham and Stoke Museums). SB. (A) 3

(27) Gold, garnet - Staffordshire Hoard (Birmingham and Stoke Museums). SB. (A) 4

(30) Sutton Hoo Mound 1. (A) 5

(31) Dorchester Dorset gold and garnet SB 4 small triangles at corner of base. (A) 6

(61) The lost pyramid of Dorchester – gold and garnet (18th century unambiguous sketch is all that survives). Exhibited at the Society of Antiquaries of London on Thursday 28th November. (A) 7

(64) Webster and Backhouse, 1991. 57. Gold cloisonné style II, **designs differ**. (A) 8

Type 1iv. Decorated: cloisonné, simple, often slabs of gems/glass on sides and top (figure 15ii.26)

15ii.26. Type 1iv. (M17). Decorated: cloisonné, simple, often slabs of gems/glass on sides and top (78). Photograph Paul Mortimer.

(P17) BUC-90E4A7 A copper-alloy pyramid mount from an Anglo-Saxon scabbard. Aylesbury Vale, Buckinghamshire (M17). (B) 1

(P14) NARC-F627F6 Silver pyramidal mount from a sword-scabbard set with garnets in gold borders. (B) 2

(P49) SWYOR-29F9E5 An Anglo-Saxon copper alloy sword pyramid mount. It measures 24.17mm and weighs 5.03g. It dates to the 7th century AD. (B) 3

(P29) NCL-925284 An early Anglo-Saxon pyramidal mount from a sword-scabbard, made from cast gilded silver with garnet inlay. Humshaugh, Northumberland. (B) 4

(P26) SF-9242E2 Sword pyramid of early Anglo-Saxon date, made from silver and parcel gilt. Each side of the pyramid is decorated with a large garnet-inlaid field that is framed by a fine gold frame. Diss Suffolk. P50, P26 and P55 have a triangular a pronounced 'v' cut from the apex of the garnet triangles on the sides. (B) 5

(P50) KENT-1A2A84 An Early Medieval (Anglo Saxon) silver pyramidal scabbard fitting (late 6th - early 7th. P50, P26 and P55 have a triangular a pronounced 'v' cut from the apex of the garnet triangles on the sides. Hollingbourne, Maidstone Kent. (B) 6

(P55) NMS-FFF757 Early Anglo-Saxon copper-alloy pyramidal mount from a scabbard or sword-belt, each of four faces occupied by trapezoidal setting for missing probable garnet. P50, P26 and P55 have a triangular a pronounced 'v' cut from the apex of the garnet triangles on the sides. Caistor St Edmund, Norfolk. (B) 7

(P61) SF1479 Copper alloy pyramidal mount. Each face of the mount retains the remains of cloisonné cellwork which would have originally held cut garnets (now missing). Coddenham, Suffolk. (B) 8

(P64) STAFFS-3CE936 Staffordshire Hoard Length: 20.5 mm Height: 21.8 mm Width: 18.8 mm Weight: 18.29 g Lichfield – small triangles on corners on the base and prisms along sides. (B) 9

(P67) LIN-105D91 An incomplete Anglo-Saxon sword pyramid. Each of the four sides would have had a triangular garnet with a circular cell in the centre. Only two sides retain their garnet; one of these is missing its setting in the circular cell, the other it is filled with **blue enamel**. Owned by Justin Mercier. (B) 10

(1) Mound 6 Sutton Hoo, UKDFD. SB. (B) 11

(10) Mound 17 Sutton Hoo, the cloisonné design on the side of the pyramid is not only found on other pyramids (marked M17, in this typology) but is also found on both sides of the sword pommel from Laon, Aisne, France.[1] (B) 12

(11) Broomfield princely burial Essex (BM). SB. (B) 13

(25) Copper alloy, gold, garnet - Near Snettisham, Norfolk. Sold via TimeLine Originals. (M17). Recorded in Hammond, 2009. 74. Fig. 1.4.2.2-c. (B) 14

(26) Copper alloy, gold, garnet, glass - Scunthorpe, Lincolnshire. Sold via TimeLine Originals. SB (M17). (B) 15

(35) Grave 211 Sarre. (B) 16

(38) A rare Saxon Bronze Pyramid Mount with rock crystal in the cells. 7th C A.D. and 14mm squared. DRG Coins & Antiquities (del550) eBay sale May 2015. (B) 17

(47) Tuddenham, Bronze pyramidal stud, inlaid with white, possibly shell, paste and red-coloured glass fragments. (M17). (B) 18

(49)sold on eBay. (B) 19

(58) Coombe Bisset, Salisbury Racecourse Evison, 1963. 71 (richly furnished 59) and Swanton, 1973. 176. 7th c. (M17). (B) 20

(59) Pyramid from Bedfordshire. (B) 21

(62) Square pyramid from Coddenham, Suffolk. (B) 22

(70) eBay find(?). (M17). (B) 23

(78) Museum of Archaeology and Anthropology, Cambridge. 1895.136 (Cat ID 14277) Tuddenham, Suffolk 1.6 x1.68 cm – height 1 cm bronze garnet and shell. Little Wilbraham? Linton? Linton Heath? 1.4cmx1.4cm height 1cm. Bronze and garnet. (M17). (B) 24

(87) Saxon gold pyramid mount with garnets, 6th-7th century AD A gold pyramid mount with filigree decoration to the edges. 12 grams, 25mm (1"). (A) 25

---

[1] Evison, 1976. 303, 305 and plate LXIV. Unlike most of the pyramids in this group where the design is multi-coloured, the device on the pommel is picked out with red jewels, probably garnets. **See figure 15ii.13.**

(90) Guy Lown collection. (C) 26

(91) Stephen Pollington Collection. (C) 27

(111) Offered for sale on eBay, August 2017. 14 mm across the base and 15 mm high. Lincolnshire. (c) 28

Type 1v. Decorated: filigree (figure 15ii.27)

15ii.27. Type 1v. Decorated: filigree (33). Photograph courtesy of Andy Pilkington

(P5) Unique ID: SF5196 Gold casing for a pyramidal scabbard mount in the form of a truncated pyramid, garnet in top. **Style II designs opposite sides the same but adjacent sides subtly different.** Near Bury St Edmunds, Suffolk (Moyse's Hall Museum, Bury St Edmunds, Suffolk). P5 is very close in design to 99 and similar to 86 and 41. (A) 1

(5) Gold - Heckington, North Kesteven, Lincolnshire SB. (A) 2

(33) Staffs Hoard tri-lobed garnet sides K302. (A) 3

(41) Rendlesham. **Slightly differing designs**. Similar in design to P5, 86 and 99. (A) 4

(55) Wester Craigie, Dalmeny, West Lothian ASSAH v9 4 (Proudfoot, Edwina and Aliaga-Kelly, Christopher, 1996. 5) and Truckell and Cormack, 1973. Plate II. (A) 5

(76) The side of a pyramid, the rest is missing. MCS5638 British Museum. Cross shaped garnet positioned as a saltire surrounded by filigree. Gold and garnet. (A) 6

(86) The Selsey Pyramid. Just the remains of a carcass decorated with Style II filigree and now in the British Museum. (A) 7

(99) Very few details at present but will be on PAS. Close in design to P5 (SF196) and similar to 41 and 86. (A) 8

Type 1vi. Decorated: chip-carved or cast, triangles dominate sides[1]

Sub-type 1viA. Designs with just three triangles per side[2] (figure 15ii.28)

(P6) BH-40414E A copper-alloy pyramidal mount, dating from the early part of the Early-Medieval period – sides mostly triangles and variations. (C) 1

(P9) WMID-77A22D An incomplete gilded silver pyramid mount of Early Medieval dating (first half of the 7th Century AD . Stratford-upon-Avon, Warwickshire. (C) 2

(P10) PAS-CF049B Silver-gilt pyramidal sword fitting with niello inlay: Folkton near Scarborough, triangles on sides, square garnet inset in top. (B) 3

15ii.28. Sub-type 1viA. Designs with just three triangles per side (P40) (BERK-D29AE3). Photograph courtesy of the PAS

(P25) SF-2563E6 A complete copper-alloy Anglo-Saxon sword pyramid mount. Barham, Mid Suffolk. (C) 4

(P27) BH-460DE6 An Anglo-Saxon silver, parcel gilt, pyramidal mount, presumably from a sword scabbard. Dates from the late 6th to 7th century AD. Hertfordshire. (C) 5

(P28) CAMHER-9583E3 Anglo-Saxon silver, parcel gilt, pyramidal mount, presumably from a sword scabbard South Cambridgeshire. (C) 6

(P40) BERK-D29AE3 A cast copper alloy Anglo-Saxon pyramid mount. West Berkshire. (C) 7

(P43) CAM-6D5C12 A silver Anglo-Saxon pyramidal mount with gold. This item is very similar to Treasure number 2007T188 which has PAS number CAMHER-CAMHER-9583E3. (C) 8

(P48) HAMP-2C1D52 A slightly squashed copper-alloy pyramidal mount of early-medieval (Anglo-Saxon) date. Thruxton CP, Test Valley, Hampshire. (C) 9

(P56) NMS-C19A31 Incomplete Early Saxon copper-alloy pyramid-shaped mount from sword belt. Wretton, Kings Lynn, Norfolk . (C) 10

(P59) PAS-E8E442 Silver-gilt strap mount in the form of a truncated pyramid, the apex set with an unpolished garnet over waffle patterned gold foil. (C) 11

(P60) PAS-E8BEB6 Pyramidal mount, hollow with a square cell at the apex, now empty, and a bar across the open base. Kilham, East Riding, Yorkshire. Sides 3 triangles, square gem missing from top. (C) 12

(P75) An early medieval gilded silver pyramidal mount, probably from a sword scabbard. Between the triangles are two parallel incised grooves forming two chevrons. (C) 13

(P76) LEIC-9E22BD Early medieval (Anglo-Saxon) silver pyramidal mount. 580 to 650 A.D. Length: 16 mm, Height: 8 mm, Width: 15 mm, Weight: 2.94 g. Scampton, Lincolnshire. (C) 14

[1] I have split this category into sub-categories based on the way that the triangles are configured. Whether the different designs have purpose is yet to be determined.

[2] Many of these have significant chevrons separating the triangles.

(P77) BH-B72FCB An Anglo-Saxon copper-alloy pyramidal mount. Dimensions of base: 15.3mm by 14.6mm; height: 8.2mm. Weight: 4.26g. Crowmarsh, Oxfordshire. (C) 15

(P80) YORYM-850552. Pyramidal strap mount, hollow with a square cell at the apex, now empty, and a bar across the open base. H: 9mm; base: 10x9mm. Kilham, East Riding of Yorkshire. (C) 16

(21) Copper alloy. From an 'old collection'(3). Sold via TimeLine Originals. SB. (D) 17

(22) Copper alloy. From and 'old collection'(4). Sold via TimeLine Originals. (C) 18

(23) Silver, gilt, gold, garnet - Grave 174, The Meads, Sittingbourne, Kent. This pyramid has 'snake'/lightning bolt/'s runes' on the modified diagonals (lozenges?) separating the triangles. Similar to P51. SB. (C) 19

(24) Museum of Archaeology and Anthropology, Cambridge. 1948.1655 (Cat ID 5016). (C) 20

(28) Silver, gilt, gold, garnet. From an 'old collection'. Sold via TimeLine Originals. SB. (B) 21

(34) From eBay (Thanks to Lindsay Kerr). (C) 22

(39) Found Norfolk. Sold on eBay, Jan 15, 2012 Worthopedia™ Antiquities. (C) 23

(46) UKDFD : 47962 copper alloy cast silvered --near Winchester Hampshire 15mm x 15mm base; 8mm high, 6mm x 6mm top; 10mm slope. Date recorded Tue April 14, 2015 Validated by Rod Blunt. (C) 24

(51) Nigel Mills Saxon and Viking Artefacts, 2001. 100. (C) 25

(54) Timeline . Timeline. Recorded in Hammond, 2009. 73. Fig. 1.4.2.2-a. (B) 26

(67) Silver scabbard belt fitting Anglo-Saxon, late 6th-early 7th century AD. Unearthed with a metal detector at Headbourne Worthy, Hampshire in 2003. (C) 27

(68) eBay find(?). (B) 28

(72) Smith, R.A.: A Guide to the Anglo-Saxon and Foreign Teutonic Antiquities in the Department of British and Mediaeval Antiquities (British Museum Publications, 1923). Page 44 and 45. The pyramidal mount illustrated at 1:1 measures 12.5mm in height and 15 x 17mm at the base. (B) 29

(77) TimeLine Anglo-Saxon mount 6th-7th century AD. 4.25 grams, 16mm (3/4") Sold Wednesday 9th September 2015 - Saturday 12th September 2015. (C) 30

(79) Wrotham three similar pyramids on one sword, all the same design. Grave 7010. Plate 5. Stoodley, 2015. (C) 31

(80) Forest Gate pyramid. BM. (C) 32

(82) Coddenham, Suffolk. West, 1998. 21 and figure 21.22 -- traces of white inlay in top depression. (C) 33

(83) 2.88 grams, 10mm (1/4"). Very similar to 34 but this one has damage to one of the vertical edges [left front in this picture] and 34 does not). They could be a pair(?). (C) 34

(84) eBay December 2015. Seller: UK - 07734902353 INTERNATIONAL – 447734902353 Ancient-Artefacts-UK. 'bronze' foil insert, 15.8 x 16mm. (C) 35

(85) For sale on eBay 06 Nov, 2015 01:02:34 GMT by aspen-man. (C) 36

(94) Pyramid kept in Alcester Museum and found in Bleachfield Street. Information and photo supplied by Andy Pilkington. (C) 37

(P81) BERK-63AB1F. Silver-gilt (D) 38

(98) Faversham pyramids (three – all very similar design – counted as one here) grave 1247 (?) Discovered 1870 (?) BM no. 1247.d'70, 1247.e'70, 1247.f'70. All copper-alloy. D 15mm x 15mm height 10mm (typical). (D) 39

(105) Timeline auctions March 2017. A parcel-gilt silver sword pyramid mount with chip-carved triangular gilt recesses framing a niello-bordered chevron to each face. Two with additional niello lines on two of the opposite chevrons forming a small lozenge. 2.30 grams, 13mm (1/2"). (B) 40

(P84) LEIC-4BFA72. It's silver with niello inlay and gilded recesses. It is from the Luffenham area of Rutland. Dimensions and image attached. Length: 14 mm, Height: 7 mm, Width: 13 mm, Weight: 2.54 g. (B) 40

(P85) YORYM-AEA546. An Early-Medieval scabbard mount dating to the period c. AD 500-700. The apex of the mount is recessed, possibly to accommodate glass, a stone or enamel, and a white substance is present. Traces of silver gilt are present on the surface. The mount measures 14.64mm in length, 13.75mm in width, 8.25mm in thickness and weighs 3.66g. Bainton, Yorkshire (C) 41

Sub-type 1viB. Three triangles along the base and one at the top, with lozenge shapes between them on each side (figure 15ii.29)

(P7) DOR-E105ED[1] A copper alloy early medieval pyramid shaped mount, possibly from a scabbard, dating – sides mostly triangles and variations. (C) 42

(P51) BERK-1DF3D2 An early Anglo Saxon silver pyramidal scabbard fitting dating to the early 7th century. Harrold, Bedford. This pyramid has 'snake'/lightning bolt/'s rune' designs on the 'lozenges' separating the triangles, as has 23. (C) 43

(P58) WILT-9195E0 Copper alloy and garnet silvered sword pyramid. Kingston Deverill, Wiltshire. (C) 44

(P69) NLM-07F328 Copper alloy Mount. Cast hollow pyramidal mount. (C) 45

15ii.29. Sub-type 1viB. Three triangles along the base and one at the top, with lozenge shapes between them on each side (P7) (DOR-E105ED). Photograph courtesy of the PAS

(18) Copper alloy. North Essex. Sold via TimeLine Originals. Recorded in Hammond, 2009. 74. Fig. 1.4.2.2-d. SB. (D) 46

(20) Copper alloy and tin (?). There is a single ring-and-dot inscribed in the top. Wiltshire. Sold via TimeLine Originals. SB. (C) 47

(52) Nigel Mills Saxon and Viking Artefacts, 2001. 99. (B) 48

(92) Wittenham, near Dorchester on Thames unpublished, found at Long Wittenham, Oxfordshire in a high status hall context. Cupric alloy with yellow glass (crystal?) (C) 49

(101) Announced by Metal Detectives Group on Face Book December 2016. Found Sunday 18th December 2016 at Bampton, Oxfordshire at a Metal Detectives rally. Copper-alloy. (D) 50

---

[1] P7 and P51 have possible 'snake'/'lightning bolts'/'runes' on their sides in a similar way to 23.

(108) Copper-alloy. Pair of pyramids. Three triangles along the base and a fourth at the apex of each side. Foil backed garnet present in recess in apex. Measurements 21 mm x 21 mm – height 85 mm. Recorded in, *A Summary Catalogue of the Anglo-Saxon Collections: Non-ferrous Metals (British Archaeological Reports)* 1993. (Page 215 – 36.26) Arthur MacGregor and Ellen Bolick. Oxford. (C) 51

Sub-type 1viC. Three dominant triangles on two opposite sides and triangles with lozenges on the adjacent sides (See figure 15ii.30)

15ii.30 Sub-type 1viC. Three dominant triangles on two opposite sides and triangles with lozenges on the adjacent sides (P63) (WILT-A0D7F2). Photograph courtesy of the PAS

(P63) WILT-A0D7F2 Early Early-Medieval silver gilt sword pyramid. It is truncated to give a square setting which is now missing. Mendip, Somerset 601 700 -- sides mostly triangles and variations, **but 2 sides one design and 2 another**, square gem missing from top. (C) 52

(17) Silver, gilt. Trowbridge, Wiltshire.SB Website Trowbridge. **Two different designs.** (C) 53 http://www.trowbridgemdc.co.uk/Saxon-Mount-Find.html

(81) Witham Friary. (C) 54

(93) Detector find from Kent in the collection of Peter (Odalar Thorsson) **two different designs**. (C) 55

Sub-type 1viD. Three dominant triangles with apexes meeting in the middle of each face with five subsidiary triangles (figure 15ii.31)

15ii.31. Sub-type 1viD. Three dominant triangles with apexes meeting in the middle of each face with five subsidiary triangles (29). Accompanied in the grave by a sword-ring

(29) Silver, gilt, malachite gen inset in top. Lissington, Lincolnshire (The Collection, Lincoln). SB. (B) 56

Sub-type 1viE. Three triangles on the opposite sides with style I leg/arm on the others (figure 15ii.32)

(P33) SUR-E6B3B4 Pyramidal mount made from silver, partly gilded and inlaid with niello. Stanton St John, Oxfordshire – 2 sides mostly triangles and variations, **the other 2 sides are different**. (C) 57

15ii.32. Sub-type 1viE three triangles on the opposite sides with style I leg/arm on the others (P33) (SUR-E6B3B4). Photograph courtesy of the PAS

15ii.33. Sub-type 1viF. three triangles formed by an outline with an arrow-head shape at the apex (P39) (IOW-7FBDB7). Photograph courtesy of the PAS

15ii.34. Type 1viG. Six triangles per side. (110). Photograph Matt Bunker

15ii.35. Type 1vii. Decorated: chip-carved or cast, other designs on sides (P15) (WAW-7A89A7). Photograph courtesy of the

Sub-type 1viF. Three triangles formed by an outline with an arrow-head shape at the apex (figure 15ii.33)

(P39) IOW-7FBDB7 A complete Early Medieval, seventh-century silver-gilt pyramidal sword strap-mount (AD 600 - AD 700). Curved design an all four sides – row of three at the bottom with a fourth above – all 'v' shaped with rounded outside lines. (C) 58

Sub-type 1viG. Six triangles per side (figure 15ii.34)

(110) Decorated Anglo Saxon bronze pyramid, Six triangles per side, three small ones near the apex and three larger along the base of each side. Ripon, Yorkshire. Measures 14mm X 14mm X 7mm deep. (C) 59

Type 1vii. Decorated: chip-carved or cast, other designs on sides (figure 15ii.35)

(P13) NMS-043037: Early Anglo-Saxon pyramidal mount made from silver and inlaid with niello, in the shape of a truncated four-sided hollow pyramid with a strap bar on the underside. Near Hunstanton. Geometric design on sides, **2 sides one design and the other 2 another. (C) 1**

(P24) SF-DED750 A cast copper-alloy pyramid mount of Early Anglo-Saxon date. Forest Heath, Suffolk. Shallow pyramid – **two sides central small circle with four lines radiating, the design seems to be different on the other two sides**. (D) 2

(P41) NMS-EEF020 Silver hollow pyramid mount, probably from a scabbard. Happisburgh, Norfolk. **Inverted 'v' shape with surround on 2 sides and small 'omega' (?) on others.** (D) **3**

(P15) WAW-7A89A7 A gilded silver early Anglo-Saxon pyramidal scabbard mount. Stratford-upon-Avon. Shallow pyramid – hole in top presumably gem missing – curved design 'omega'design on all four sides. (C) 4

(P16) DUR-BA5975 A cast silver, gilded Anglo-Saxon pyramidal mount from a sword scabbard, inlaid with a single garnet and dating to the late 6th or 7th century AD. County Durham -- fish (?)**scales on two sides and three-sided knot design on the others. (C) 5**

(P22) LIN-05D862 A gilded silver early Anglo-Saxon pyramidal mount from a scabbard. West Lindsey four sides 'omega' design. (C) 6

(P23) SWYOR-F22FB2 A silver gilt Anglo-Saxon pyramidal mount (sword pyramid), Drax, Selby Yorkshire four 'omega' design. (C) 7

(P52) YORYM-AFC732 Identification supplied by Jim Halliday. An enamelled, cast copper-alloy, hollow backed pyramid-shaped mount which is rectangular in plan Everingham east Riding Yorkshire – shallow pyramid – indistinct design but raised line along middle on all four sides. (D) 8

(P53) WMID5233 Copper-alloy pyramid-shaped object. It has a hollow interior, and the upper corner is purposefully missing, therefore forming a sub-square hole. Wyre Piddle, Wychavon, Worcestershire. Three ring-and-dot symbols configured to form a triangle on each of the four sides.[1] (D) 9

(P62) NLM401 Early Anglo-Saxon copper-alloy pyramidal mount from a sword-scabbard, decorated with three ring-and-dot motifs configured to form a triangle on each of the four faces North Lincolnshire. (D) 10

(P72) SF-9F5377 A complete silver Anglo-Saxon mount in the form of a truncated pyramid. **Two alternating decorative schemes, imitating garnet cloisonné, are featured on the faces.** Length: 12.79 mm Width: 12.72 mm Thickness: 6.15 mm Weight: 1.96 g Suffolk Coastal District. (C) 11

(P79) P79. LIN-BD536C AD 500-700. The mount is in the form of a four sided pyramid with a flat square apex. It exhibits ring-and-dot decoration next to the corners on two faces, running in a line of three from the

[1] The sword from Nocera Umbra, grave 74 (Italy) has a pyramid decorated with the ring-and-dot motif. Menghin, 1983. 261.

base towards the apex. The other two faces are quite worn rendering this decoration indiscernible. Lenton Keisby and Osgodby, South Kesteven, Lincolnshire. (D) 12

(P81) Uncleby. York Castle Museum (205.3.47) copper-alloy? 410 to 866 A.D. Excavated in 1868, little information on original excavation. Although slightly squashed, the pyramid seems to be originally rectangular rather than square and the dots, in the ring-and-dots, are pierced. (D) 13

(3) Silver - Leaming Bar, North Yorkshire. SB. UKDFD. (C) 14

(15) Silver, gilt. From an 'old collection'(2). Sold via TimeLine Originals. Indistinct style II on sides. SB. (C) 15

(32) Silver (?) eBay. 'Omega' or serpent or both – on raised plaque. (C) 16

(40)WDRH http://www.wdrh.org.uk/gallery/index.php/Club-Finds 'omega' design on all four sides. (C) 17

(48) Uncleby. York Castle Museum (205.3.47) copper-alloy? 410 to 866 A.D. Excavated in 1868, little information on original excavation. Although somewhat squashed, the pyramid may have been slightly rectangular rather than square and the dots are pierced. (D) 18

(50) Nigel Mills Saxon and Viking Artefacts, 2001. 100. 'omega' design with underlying 'snakes (?)'on two sides and indistinct (in photo) style II on the other two. (B) 19

(53) Nigel Mills Saxon and Viking Artefacts, 2001. 99 'omega' design on four sides. (B) 20

(56) Timeline. Recorded in Hammond, 2009. 74. Fig. 1.4.2.2-b. Design almost scratched in and a little unclear - possibly imitating cloisonné, side different. (D) 21

(60) Lincolnshire – (Lindsey) simple curved and step design – imitation cloisonné. (C) 22

(P66) SF4162 Small and rather crudely decorated silver sword pyramid **but four different straight line designs** on each face– one design – very similar to two of the faces on (56). (C) 23

(69) eBay find. **(Style I?) Designs differ**. (D) 24

(75) Pyramid from Melton Ross.[1] **'Omega' on two sides and a unique design 'stalks' topped by circles on the other sides**. One of the omegas has a dot inside the apex loop. (C) 25

(89) Pair of pyramids from a grave in Kent. Beany Museum, Canterbury. Copper-alloy. Approximately 15mm x 15 mm. Sides decorated with dots that roughly outline three triangles on each side. (C) 26

(100) Offered for sale by der2008_2008 on eBay, December 2016. Two different designs, a triskele on two sides and an 'omega' on the other two. The two similar designs are adjacent rather than opposites, as on most pyramids with different designs. 12mm x 12mm height 8mm. found near Durham. Silver-gilt and niello. (B) 27

(P82) BERK-12CCB3. TREASURE CASE 2016 T1042 Length: 12.2 mm, Height: 6.8 mm, Width: 12.5 mm, Weight: 2.8 g. It was found at Upton Grey, Hampshire. It has niello inlay. Two sides are divided in a similar way to the Mound 17 pyramid, but the other two resemble cloisonné shapes and are divided into three fields. (M17). Type 1vii. (C) 28

[1] Thanks to Kevin Leahy for alerting me to the existence of this pyramid.

Type 1viii. Corrugated sides (figure 15ii.36)[1]

(P30) NCL-7CE767 A cast copper-alloy Anglo-Saxon pyramid mount of the Early-Medieval period, c. 500-600. (D) 1

(P57) NMS-BC07D7 Early Saxon Sword scabbard fitting in the form of a truncated pyramid with integral attachment bar for strap on base, and hollow reverse cu-alloy. Beeston, Breckland, Norfolk. (D) 2

(P71) An incomplete Early-Medieval mount dating to the period c. AD 500-700. The mount is very worn, but there are traces of a white metal coating on the surface. It is undecorated and hollow on the reverse, although there are two worn breaks, each centrally placed on a parallel side, where an attachment strap may have been affixed similarly to DOR-E105ED. It has a green and brown patina. Another similar example on the database is WILT-9195E0. (D) 3

15ii.36. Type 1viii. Corrugated sides (P71). Photograph courtesy of the PAS

Type 1ix. Organic pyramids (figure 15ii.37)

15ii.37. Type 1ix. A pair of organic pyramids from Nocera Umbra (so far none recognised from England). Photograph Matt Bunker

Type 1x. Tauschieren decorated (figure 15ii.18)

(95) From the estate of Thomas Bentley Cederlind. Two faces are decorated with winged busts, two others with a ribbon-like design (omega?). Top has engraved lines that imitate cloisonné. Each face with niello inlay. 13x13x9mm; weight: 3.29g. (B) 1

Type 1xi. Pierced top, round hole, no bar underneath (figure 15ii.39)

(P68) CAM-0909A4 A copper alloy gilt pyramidal mount dating to the sixth or seventh centuries AD South Cambridgeshire – hole in top, fill missing, flange around base. (B) 1

104. Anglo Saxon / Viking bronze pyramid mount / dagger cap Metal detecting detector Anglo Saxon / Viking gilt bronze pyramid mount / dagger pommel, Measures 15mm X 15mm, Ancient Artefacts. (B) 2

(P83) LANCUM-105928. Length: 11 mm, Height: 6 mm, Width: 8 mm, Weight: 0.5 g. 600 – 800 A.D. Swaffham, Cambridgeshire. (C) 3

(106) Copper alloy, gilt - Didcot, Oxfordshire (West Berkshire Heritage Service). SB. (C) 4

(107) Sold on eBay 20 Feb. 2015 23:05:05 H.Esp by styca333. From Durham (?). (B) 5

[1] It may turn out that with most of the pyramids in this category, the 'corrugations' turn out to be poorly formed or damaged triangles. Unfortunately, at the time of writing, I have not been able to handle any of them.

15ii.38. Type 1xi. Pierced top, round hole, no bar underneath. (P68) (CAM-0909A4)

15ii.39. Type 1xii. Other square-based pyramids (74). An example from TimeLine Originals. Photograph courtesy of TimeLine Originals

15ii.40. Type 2.i. Non-tauschieren decoration (P45) (LIN-7AB593). Photograph courtesy of the PAS

15ii.41. Type 2.ii Tauschieren decoration on a pyramid from Giengen an der Brenz, Baden Württemberg, grave 26. Photograph, Matt Bunker

15ii.42. Type 3. More than 4 sides (P19) (KENT-A81B05). Photograph courtesy of the PAS

15ii.43. Type 4i. Plain (P35) (SWYOR-89C624). Photograph courtesy of the PAS

15ii.44. Type 4ii. Conical with flat sides (P70) (LIN-540C0C). Photograph courtesy of the PAS

15ii.45. Type 4iv. Other (P4). Photograph courtesy of the PAS

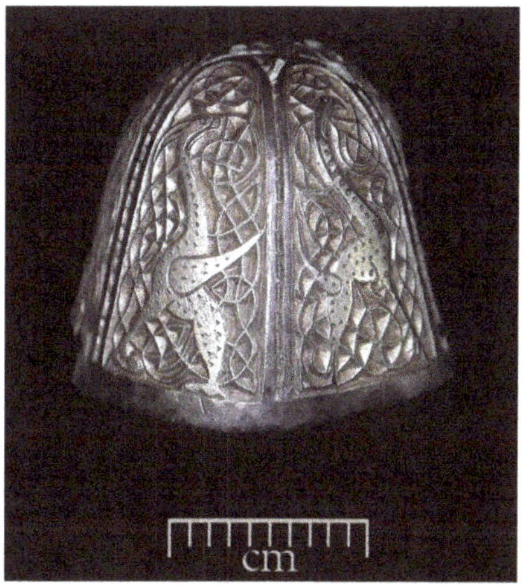

15ii.46. Type 5. Possible sword pyramids. This example (P65) PAS.LEIC-15A132. Photograph courtesy of the PAS

Type 1xii. Other square-based pyramids (figure 15ii.38)

(42) eBay pyramid Mar 26, 2015. Basically a cube, with top corners cut off – top square, therefore offset diagonally compared to base. There are serious doubts about this particular example. (D) 1

(74) Item number: 350040388463 (2) Sold by TimeLine. Steep sided, two adjacent sides with five crystal 'commas' embedded in metal. The other two sides are plain. (D) 2

Type 2. Pointed-top square-based

Type 2.i. Non- tauschieren decoration (figure 15ii.40)

(P12) NMS-1E5834 Saxon copper alloy pyramidal mount, on each of the four sides there is an engraved triquetra, (style II? Perhaps an 'omega'?) within double bordering lines and obliquely notched angles. Base 17.5 x 17.5mm. Height 17 mm. Weight 16.30 g. (C) 1

(P18) WILT-9AB1A4 An Early-Medieval (500-700) cast copper alloy pyramidal mount with some edge damage, flange around base. Chiseldon Swindon. (D) 2

(P45) LIN-7AB593 Incomplete copper alloy pyramidal mount, probably from a sword-scabbard. Sleaford, Kesteven Lincolnshire. (D) 3

(P46) LIN-96C674 An Anglo-Saxon copper alloy pyramidal mount from a sword-scabbard – solid and pointed. Aunsby and Dembleby, Kesteven, Lincolnshire. (C) 4

(P74) A gilded copper alloy pyramidal mount with probable niello inlay. (D) 5

(19) Copper alloy, silver(?). Norfolk. UKDFD. SB. (D) 6

(44) Sold on eBay 05 April 2015. (D) 7

(45) 7th C Saxon pyramidal mount, possibly from a sword scabbard, owned/sold by 'Adam and Lisa'. http://adamndlisasfinds.webs.com/apps/photos/photo?photoid=59191127 (D) 8

(88) Aquired by TimeLine Auctions April 2016. Plain steep sides, cu-alloy. (D) 9

(103) Anglo Saxon Silver Pyramid Mount . 3.5g 10mm H x 13mm W 6th Century. Solid Silver with a fine gold outline on each quarter just still visible in parts. Seller bt07tst. Scarborough (seller). (C) 10

Type 2.ii Tauschieren decoration (figure 15ii.41)

None recorded in England to date although several examples are known from the Continent.

Type 3. More than 4 sides (figure 15ii.42)

(P19) KENT-A81B05 A complete Early Medieval, seventh-century copper alloy-gilt pyramidal, cabochon garnet in top – 6 sides. (B) 1

(P20) IOW-1855E4 Bembridge Pyramidal mount from a sword belt, with an octagonal base and architecturally balanced faces filled with cloisonné cell-work – 8 sides. (A) 2

(P37) LIN-666425 Gilt copper alloy sword pyramid – 8 plain sides. South Kesteven, Lincolnshire. (D) 3

Type 4. Circular-based

Type 4i. Relatively plain (figure 15ii.43)

(P35) SWYOR-89C624 A plain copper alloy probable Early Medieval sword pyramid, pointed. Saxton with Scarthingwell, Selby Yorkshire. (D) 1

(P36) PAS-41A514 An incomplete silver mount, conical in shape and hollow. There are at least seven parallel grooves, which may form a spiral, around the cone. It could be stylistically related to P4, which also has a spiral, although the latter is of gold. Woodbridge Suffolk 600 to 800. (D) 2

(P86) DENO-7DE164. An incomplete copper alloy early medieval 'sword pyramid'. Generally such objects are pyramidal but this example is conical, with a circular hole at the apex, now filled with iron corrosion material and debris. It is hollow, inside the base is the corroded remains of what may be a traverse bar. The outer surface is decorated with a series of panels filled with vertical and horizontal ridges. The edge of the base is irregular, showing damage and chipping. The object has a green patina with dark patches. Probably copper-alloy. The object is 21.5 mm in diameter at the base, tapering to 7 mm at the apex. It is 22.6 mm long. It weighs 18.10 g. Granby, Rushcliffe, Nottinghamshire. Dating: Circa AD 575 to Circa AD 625. (C) 3

Type 4ii. Conical with flat sides (figure 15ii.44)

(P1) LIN-357B4F flat sides, four small triangles at base. (B) 1

(P70) Unique ID: LIN-540C0C An incomplete Anglo-Saxon copper alloy sword pyramid. The mount has a flat circular base with two D-shaped holes extending into the body of the mount. Four panels rise from the circular base and these taper to form a roughly pyramidal mount. Lincolnshire. (D) 2

(9) Silver, gilt - Bawtry, Nottinghamshire (British Museum) – steep pyramid – style II on sides – **two one design and two another. SB. (B) 3**

(12) Copper-alloy, gilt, garnet - Yorkshire steep pyramid – square gem at top – triangles near base. Sold via TimeLine Originals. Recorded in Hammond, 2009. 74. Fig. 1.4.2.2-f. SB. (B) 4

(14) Gold, garnet, glass - Staffordshire Hoard, K1166 (Birmingham and Stoke Museums) complex cloisonné – faces and birds? SB. (A) 5

(36) Anglo-Saxon fitting from a sword scabbard by Anglo-Saxon, 750–799. **Subtly different designs on sides.** Now in BM. (B) 6

(73) Finglesham. Chadwick Hawkes and Grainger, 2006. Four slabs of garnet on sides and gem on top. Squat pyramid. Finglesham, Kent. Now in Ashmolean Museum, Oxford (L11326.32). Attributed to grave 58 and to 500 to 540 AD. Apparently a woman's grave. In many ways this is similar to a pyramid from Lincoln (97) below. Type 4ii. Quality A. (A) 7

(97) Pyramid from Lincoln Museum. Diameter 14mm approx. Silver (?) and garnet. (Photo Matt Bunker). Possibly silver, cloisonné, with a red gem at the top and five garnets (?) arranged around a central sixth to form a circle, but with a vacant 'V' shape at the apex. Very similar in many ways to 73 above. (B) 8

Type 4iii. Squat, button-like, may have Style I decoration (figure 15ii.7)

(8) Gold, silver, squat, flattish - ex Lord McAlpine collection, sold via Artancient (same as TimeLines example) style I designs**, seem to be all slightly different to each other**. The underside has a central bar. SB Sandby Borg type .[1] (B) 1

(37) Dirk Kennis – attributed to Thetford Norfolk . Style I. Four different designs, smiling face, sun/eye 5 rays, man's face unsmiling, sun/eye 6 rays. (B) 2

Type 4iv. Other (figure 15ii.45)

(P4) IOW-B39526 gold beaded whorl, conical. P4 could be stylistically related to P36, which also has a spiral, although the latter is of silver. (B) 1

(63) Almost round, conical pyramid Coddenham - chip carved, a little indistinct – some piriforms, **designs differ**. (C) 2

(P78) SWYOR-AEEA2F A probable Anglo-Saxon, Early Medieval, lead alloy sword pyramid; a 7th century fitting from a scabbard strap. The "sword pyramid" is 21.8mm diameter, 11.6mm high and 16.86g. Herningby, Lincolnshire. 600 to 700 A.D. (D) 3

---

[1] See Rau, 2010. Nydam vol 2 tafel 48,3 and 112 for something very similar (Nydam is in the German part of Jutland), also with Style I and another found in House 40 at Sandby Borg, Ölund, Sweden; see Söderström, 2015 for information concerning the latter.

Type 5. Possible sword pyramids (figure 15ii.46)

(P65) LEIC-15A132 Early-medieval gilded silver object, 26mm high, 29mm wide and weighing 10.39g. The object is in fair condition, but has six small detached fragments accompanying it and some copper alloy corrosion near its apex or crown. (A) 1

(66) UKFD Cast silver-gilt mount in the shape of a hollow conical boss. The field is divided into four principal fields containing deeply carved interlace decoration. (B) 2

Addenda:

Analysis of pyramid finds from graves, single detector finds or other sources

The chart below attempts to give some indication as to the circumstances of the finding of each pyramid in the corpus assembled in Appendices 9i and 9ii and referred to in this chapter. The point of this exercise is to give some indication of the proportions found in graves compared to those found as single finds. It must be borne in mind that in many examples the data is not complete.

The great majority of items in the PAS list (eighty-five of eighty-six) appear to be single detector finds. The exception is a pyramid from the Staffordshire Hoard. This makes almost 99% of pyramids in the PAS list single detector finds.

The non-PAS list is more complex. At least fourteen (14/111= 12.6%) come from graves. Another fourteen (12.6%) are said to have come from 'old collections'. Thirteen (11.7%) of the pyramids are furnished with very little information concerning the circumstances of their finding. Four (3.6%) were found within the Staffordshire Hoard. This makes a total of forty-five. The remaining pyramids (sixty-six, 5.9%) would appear to be single detector finds. For now, I have assumed that all the pyramids sold on eBay, and where I have been able to get at least a rough location, in recent years are detector finds of single pyramids. Whenever I have been able to get information from the seller, this is what they report. It must be borne in mind that the sellers may not have always been completely truthful or accurate in their reporting.

Taking both lists together, a minimum of one hundred and thirt-one (66% overall) appear to be single finds, while only fourteen (14/197 = 7.1%) are definitely from graves. It is possible that some of the detector finds too, are from disturbed graves but it is likely that the majority were not.

Abbreviations:
From a grave – g
From an old collection – oc
Apparent single find – s
Unknown provenance – u
Item was offered for sale on eBay – ebay

PAS	PAS	NON PAS	NON PAS
1 s	44 s	1 g	56 oc
2 s	45 s	2 s	57 s
3 s	46 s	3 s	58 g
4 s	47 s	4 s	59 s
5 s	48 s	5 s	60 s
6 s	49 s	6 s	61 u
7 s	50 s	7 s	62 s
8 s	51 s	8 oc	63 s
9 s	52 s	9 s	64 u
10 s	53 s	10 g	65 s
11 s	54 s	11 g	66 s
12 s	55 s	12 s	67 s
13 s	56 s	13 staffs	68 s ebay
14 s	57 s	14 staffs	69 s ebay
15 s	58 s	15 oc	70 s ebay
16 s	59 s	16 oc	71 s ebay
17 s	60 s	17 s	72 u
18 s	61 s	18 s	73 g
19 s	62 s	19 s	74 u
20 s	63 s	20 s	75 s
21 s	64 staffs	21 oc	76 u
22 s	65 s	22 oc	77 oc
23 s	66 s	23 g	78 oc
24 s	67 s	24 oc	79 g
25 s	68 s	25 s	80 u
26 s	69 s	26 s	81 g
27 s	70 s	27 staffs	82 s
28 s	71 s	28 oc	83 oc
29 s	72 s	29 s	84 s ebay
30 s	73 s	30 g	85 s ebay
31 s	74 s	31 s	86 u
32 s	75 s	32 s ebay	87 u
33 s	76 s	33 staffs	88 u
34 s	77 s	34 s ebay	89 g
35 s	78 s	35 g	90 s
36 s	79 s	36 u	91 s ebay
37 s	80 s	37 oc	92 s
38 s	81 s	38 s ebay	93 s
39 s	82 s	39 s ebay	94 s
40 s	83 s	40 s	95 oc
41 s	84 s	41 s	96 s
42 s	85 s	42 s ebay	97 u
43 s	86 s	43 s ebay	98 g
		44 s ebay	99 s
		45 s	100 s ebay
		46 s	101 s
		47 u	102 s ebay
		48 g	103 s ebay
		49 s ebay	104 s ebay
		50 s	105 oc
		51 s	106 s
		52 s	107 s ebay
		53 s	108 g
		54 s	109 u
		55 s	110 s ebay
			111 s ebay

16. Sword names and vocabulary

Stephen Pollington

Note: runic texts are cited in **bold type**. Reconstructed words not actually found in the sources are prefixed with an asterisk (*).

Our knowledge of the words the Anglo-Saxons used for swords in the 5th-7th centuries is necessarily limited due to the scarcity of written sources from this period. While many Scandinavian sources mention swords with unique names and histories, the references are all later than the period under investigation in this book. While it is not unreasonable to assume that Viking Age (9th-11th century) practices may be a continuation of those of the Vendel and earlier periods, there is again no direct evidence for named swords from this time – the references are mainly Icelandic and date from the mediaeval age.

There are no instances known to me of swords from this period with an inscribed text forming a plausible name. Swords with runic inscriptions are not unknown but these are in every case lightly scratched onto the pommel and generally have the appearance of an amuletic text or apotropaic formula (e.g. **alu**).[1] This is in sharp contrast with e.g. the Iron Age spearheads from the Danish bogs and elsewhere where names such as **ranja** 'attacker' and the like are occasionally found. Later swords with inlaid text on the blade, such as the +VLFBERHT and +INGELRII types, are probably best interpreted as bearing a makers' mark, and the numbers of such swords found support this.

A source of information which may be drawn upon – though it is not without controversy – is the poem *Beowulf*, in which a handful of swords certainly are named. Beowulf exists in a single manuscript dating to circa 980 – 1020 AD. There are grounds for believing that it may have been composed in the mid-8th century, and we cannot know how many copies stand between the first version and the present (copied and revised) text. The world described in the poem is an idealised version of the 5th-7th century in which there are no churches or books or literate priests, and in which men of courage can overcome monsters on behalf of their communities. It is a 'cultural artefact', to use the modern jargon, and reflects the concerns of the societies which cherished it.[2]

The named swords in *Beowulf* are *Hrunting* and *Nægling*, the latter Beowulf's own sword. *Nægling* clearly means '[weapon] associated with a nail or spike' and suggests that it was conceived as a pointed thrusting sword. The name *Hrunting* is of unknown origin but may be a derivative of **hrung* 'pole, staff' (our word 'rung'). In the poem Beowulf forswears the use of his sword when tackling the monster Grendel, either because Grendel uses no weapons (and thus to use weapons against him would not be honourable) or because he knows that iron swords will not bite on the monster's skin. Beowulf borrows Hrunting from its owner when he tackles Grendel's Mother, but again this sword proves ineffectual, and she is finally slain by the *ealdsweord eotenisc* 'ancient sword of the giants' which is magically charged – perhaps it was forged in an earlier age of the world and thus was created to be lethal to the monsters.

Another possible sword-name is *Hunlafing*, which appears in the *Finnesburh* manuscript. Unfortunately the text is unclear as to what the word refers to – it appears in the scene where a warrior places a sword in

[1] Fischer & Soulat, 2009

[2] Girvan, 1971

engest's lap in order to provoke him to action – and it may refer either to the sword or to the warrior. Therefore, *Hunlafing* cannot be confidently added to the woefully short list of Anglo-Saxon sword-names.

Poetic words for swords are relatively abundant, due to the existence of *Beowulf* and a handful of other verses (*Battle of Maldon, Waldere, Finnesburh*). Many poetic names for weapons consist of the simple name with another word prefixed to it, so that *wigbil* is a compound of *wig* 'war' and *bil* 'sword'; but how does 'war-sword' differ from the simple word *bil*?[1] What does the word *wig* add to the meaning of 'sword' and what are its associations? The demands of poetic metre and the alliterative scheme must have determined some of the choices the poets made: a *wigbil* is metrically disyllabic and alliterates on *w-*. The *Beowulf* poet uses it in line 1507 to explain the previous term *hildegicel* 'war-icicle': *wigbil wanian, þæt wæs wundra sum* 'the war-sword [began to] shrink, that was one of the wonders'.

There are distinctions to be made between at least three different types of compound, however: the explicit type, e.g. *wigbil* 'sword for warfare'; the circumlocution, e.g. *beaduleoma* 'battle-light, thing which shines /glistens /flashes in battle = sword'; the metaphoric type, e.g. *wælsteng* 'dead body-pole, staff for carrying dead bodies=spear'. Assignment of words to such rigid classes is not always possible, of course, since we no longer have the opportunity to understand at first-hand what these terms really mean. The sword vocabulary can be summarised thus:

term	meaning	association
sweord	sword	neutral, generic term
heoru	sword	neutral but rare (perhaps only poetic)
bil, bill	'cleaver', sword	long, thin, two-edged weapon (also the name for sickle, hoe, and other cutting tools)
mece	sword	long, two-edged, pointed weapon
hæftmece	'hilt-sword'	as *mece* but with longer hilt (?)
seax	'cutter', knife	shorter, single-edged, thrusting weapon
secg	'blade', sword	(? one-edged) sword
brond	'brand, flame', sword	short sword
wægsweord	'wave-sword'	rippling-patterned (= pattern-welded?) sword
maþþumsweord	'treasure-sword'	sword which is a costly thing
guþsweord	'war-sword'	sword which is used for *guþ*, glorious deeds of arms
guþbil, wigbil, hildebil	'war-sword'	*bil* which is used for glorious deeds of arms, *bil* as weapon rather than as tool
wællseax	'slaughter-knife'	honed and razor-sharp *seax*
hildeleoma	'battle-light'	thing that flashes in *hild*, glorious warfare
hildegicel	'battle-point'	pointed thing for *hild*
guþwine	'battle-friend'	thing that supports and assists the warrior
laf	'leavings, remnant'	sword as legacy, heirloom from glorious ancestors

[1] Brady, 2007

The principal Old English poetic terms for 'sword'

'Cutting' is the basis of most of the words for sword, rather than thrusting or stabbing. The word 'sword' from OE *sweord* is of uncertain origin but appears to be related to a verb **sweran* 'harm, injure, cause pain'. *Bil* means 'biter' or 'cleaver' and is used of a variety of handtools with a cutting edge. *Heoru* is an ancient word, almost lost from the vocabulary of the Anglo-Saxons, also meaning 'cutter'. *Mece* is related to a number of words connected with cutting or chiselling and may have referred specifically to a sacrificial knife. *Seax* is a word for a knife, originally possibly a flint tool.

There are several terms for sword parts which are common to a number of Germanic languages including Old English, Old High German, Old Saxon, Old Norse and Gothic. One such is the Norse term ON *sverð-fetill* 'sword-strap', which appears in Middle High German as *swert-vezzel*; in Old English the two elements appear separately as *sweord* and *fætels*. This suggests that attitudes to swords and the terms needed to talk about them did not change radically in the early mediaeval period.

Brady, C. Weapons in Beowulf: Analysis of the Nominal Compounds and an Evaluation of the Poet's Use of Them, in Anglo-Saxon England, volume 8, Cambridge, reprinted 2007

Girvan, R. Beowulf and the Seventh Century, London, 1971

Holthausen, F. Altenglisches Etymologisches Worterbuch, Heidelberg, 1974

Fischer, S. & Soulat, M. *Runic Swords & Raw Materials – Anglo-Saxon Interaction with Northern Gaul*, Proceedings of the 58th Sachsensymposium, Trondheim, 2009

Orel, V. A Handbook of Germanic Etymology, Leiden, 2003

Runes on Swords and Other Weapons

In this section, runic text is transliterated into Roman script according to the standard methods, presented in **bold type** for clarity.

ON stands for 'Old Norse' and WSax for 'West Saxon'.

Runic texts on weapons are referred to in *Beowulf*, the ON sagas and Eddic verse, but they are not a common archaeological find. The Danish Iron Age bog-deposits contain the greater part of the known examples, where short runic texts sometimes appear on the reverse of the object. This tradition continued for several centuries, with runes most often placed in a position where they could hardly ever or never be seen, such as the back of the flange of a shield boss from Thorsbjerg.

The origins of the custom may lie in the use of imported Roman weapons which equipped many of the Iron Age armies, on the back of which the workshops stamped their factory marks.[1] When texts written in runes came to be used in preference to Roman script, from the mid-2nd century AD onwards, they often appeared in the same position. Perhaps the secrecy and cryptic knowledge appealed more than a formal display of literacy.[2]

Despite the instance in *Beowulf* in which King Hroðgar reads a runic inscription on the hilt of a supernatural sword, clear examples of text – runic or otherwise – on Anglo-Saxon weapons are very rare. The single unequivocal example of text on a blade is the seax from the River Thames with the entire rune-row incised and inlaid in silver, with interlace patterns and the masculine name **beagnoþ** plus some uninterpreted sigils. The weapon was recovered from the river in 1857. It probably represents a 9th c. fuþorc (English rune-row)

[1] Imer, 2010, p.59-60

[2] Losquiño, 2015, p.147

from southern or south-eastern England; it does not feature any of the specifically Northumbrian forms found on the monuments at Ruthwell, Bewcastle and elsewhere. The purpose of the text has aroused some comment since it is not thought to have had any practical purpose, leading to speculation about magical intent. But the runes may have been added for their decorative and prestige value. This is the only known example of runes appearing on the blade of a weapon. Where runes are found, they normally appear on the pommel or the scabbard fittings.

A few examples of runic script appear to have been added at the time of burial rather than before, such as the Chessell Down (Isle of Wight) scabbard mount with its enigmatic **æco:sœri** (I increase pain?) text on a strip of metal added to the reverse of the mount. It therefore did not form part of the sword's furniture while in use. This poses the question: if the runes were added as part of the burial rite, who was supposed to read the text? The implication appears to be that it was a message for the gods or ancestors in the Otherworld rather than for anyone alive.

It may of course be the case that the act of cutting the runes was important and empowering but the text had no communicative value in itself. While this sounds an unlikely reason for adding text to an important object such as a sword, it does account for the very casual nature of many inscriptions. Many runic texts are lightly incised and may plausibly be interpreted as part of the charming of the weapon to increase its efficacy.

The Thorsbjerg scabbard chape of 2nd-3rd c. AD date bears runic texts on both faces: (Side A) **owlþuþewaz** (Side B) **niwajemariz.**[1] The first rune on Side A (**o**) has been interpreted as an ideogram for 'inherited property' with the name of the owner following, but more recent interpretations indicate a transposition of the runes.[2] The name is thus interpreted as **Wolþuþewaz* 'servant of [the god] Wuldor'. The name, meaning 'splendour, glory', appears in later Norse myth as *Ullr*, a god of skiing and hunting. The text on Side B is interpreted as *ni* 'not' and an adjective *wajemariz* 'woe-famed' (cf. Gothic *wajamerein* 'dishonour'). This is an example of litotes: 'not famed for woe' means 'not known for bad deeds' and thus 'of good reputation'.

Some texts were already on the weapon while in use, but still show a lack of planning in their layout. For example, **swart/a** on a shieldboss from Illerup (Denmark) where the writer had to add the final **a** on a second line and perpendicular to the rest of the text.[3] The name is a form of the adjective *swart* 'black, dirty' and may have referred to its maker, a smith with the name meaning 'sooty, grimy'.

The majority of English examples of runic texts on weapons are confined to the south-east – mainly Frankish-dominated Kent – and the swords on which they occur are generally of types found across Europe from the Rhine to the Lower Thames. In the Anglo-Saxon areas of Britain and north-east France, runic inscriptions on weapons remained an important aspect of the aristocratic tradition, alongside other 'ritual' or amuletic symbols, although the great weapon-deposits of the types seen in Jutland are absent from the English record.

Runic inscriptions are associated with swords of the Bifrons-Gilton Type, 6th c. in date and many with a pair of linked rings attached to the pommel.[4] This specific type of sword is found across northern Europe from Snösbäck (Sweden) to Dole in central France, but the greatest concentrations are in Kent and the area north of Paris (France). A Frankish origin for the group has been proposed by Svante Fischer whose meticulous studies revealed the previously overlooked **a**-rune on a sword pommel from King's Field, Faversham (Kent) and a possible inscription **hle[m]** on another from Fréthun 1 (Pas de Calais, France). The word, if it is Old

---

[1] Looijenga, 2003, p.259

[2] Antonsen, 2002, p.146; Losquiño, 2015, p.58

[3] Looijenga, 2003, p.153; Zimmermann, 2010, p.101; Losquiño, 2015, p.62-3.

[4] Fischer, 2008, p.18

English *hlem*, means 'clamour, noise of battle'. Also from Fréthun is a sword pommel with a curious sigil on one face, strongly reminiscent of the *ear* rune but seriffed.[1]

A rare **alu** inscription appears on a pommel from Grenay (Pas de Calais, France), originally discovered in 1913 and largely disregarded ever since.[2] The inscriber appears to have written **lau** in error, but loss due to wear at the top of the pommel cannot be discounted. This word is plausibly linked to 'magical' practices and empowerment. Another **alu** text appears on a pommel from St. Dizier (France), where the extreme wear to the upper edge has left the second upright of **u** separated from the first. Wear to the face has likewise rendered illegible the text on a silver-gilt pommel from Sarre 2 (Kent) where the characters **du** (or retrograde **rm**) have been detected on one broad face, and possibly **c**[...]**ea** on the other.[3] Equally indistinct is the runic group on one face of the pommel in grave 69 at Shorwell (Isle of Wight), where **iec** can be distinguished, but the last rune is unclear.

Another runic pommel, though controversial, is that from Gilton, Ash (Kent), which features two 'arrowhead' motifs on the shorter faces which have been interpreted as a variant **t**-rune.[4] In the same cemetery, in grave 4, was found a pommel with a remarkable runic text of about 19 characters but the wear to the top has obliterated parts of the characters. Interpretations include **eic sigimer nemde** 'Sigimer named [the] sword' (with *eic* for WSax *ecg* 'edge, blade') and **mic sigimer nemde** 'Sigimer named me'.[5]

Weapon-charming was practiced in Scandinavian emigrant communities, in Britain and elsewhere. The magical empowerment of a spear is referred to in *Brennu-Njálssaga* where the warrior, Hallgrímr, has *seiðr* magic worked on his *atgeir* 'thrusting spear' so that no weapon can kill him while he holds it, and it sings when it is about to take a life. The saga does not say whether the magical ceremony left visible marks on the spear, but it seems likely. The empowering of runes was achieved by reddening them with blood, which may have its origins in the ancient practices of hunting rituals.

It is possible that there is a rune on one of the sword pommels in the 7th c. Staffordshire Hoard, but this awaits confirmation at the time of writing. Since the hoard appears to be captured wargear, it is not possible to be certain that it was made in the area where it was found, so the context of this runic find – if such it is – remains uncertain.

Bibliography

Antonsen, E.A. *Runes and Germanic Linguistics*, New York, 2002

Fischer, S. *Les Seigneurs des Anneaux*, 2nd edition, Condé-sur-Noireau, 2008

Hawkes, S.C. & Page, R.I. Swords and Runes in South-East England in Antiquaries Journal, vol.47, 1967

Imer, L.M. Runes and Romans in the North in Futhark, vol.1, 2010

Looijenga, T. Texts and Contexts of the Oldest Runic Inscriptions, Leiden, 2003

Losquiño, I. G. *The Early Runic Inscriptions. Their Western Features*, Berkeley Insights in Linguistics and Semiotics no.92, New York, 2015

Zimmermann, C. How to Do Things with Runes in Futhark vol.1, 2010

---

[1] Fischer, 2008, p.75-8

[2] Fischer, 2008, p.90

[3] Hawkes & Page, 1967, p.1-2, 7; Fischer, 2008, p.116-7

[4] Hawkes & Page, 1967, pl.3

[5] Hawkes & Page, 1967, p.3-4; Fischer, 2008, p.88-90 suggests that an interpretation based on Frankish or OHG might yield results. It is noticeable that the name contains the Frankish element *Sigi* –'victory' as opposed to the WSax *Sige-* .

17. Runes on Swords and Other Weapons

Stephen Pollington

In this section, runic text is transliterated into Roman script according to the standard methods, presented in **bold type** for clarity.

ON stands for 'Old Norse' and WSax for 'West Saxon'.

Runic texts on weapons are referred to in *Beowulf*, the ON sagas and Eddic verse, but they are not a common archaeological find. The Danish Iron Age bog-deposits contain the greater part of the known examples, where short runic texts sometimes appear on the reverse of the object. This tradition continued for several centuries, with runes most often placed in a position where they could hardly ever or never be seen, such as the back of the flange of a shield boss from Thorsbjerg.

The origins of the custom may lie in the use of imported Roman weapons which equipped many of the Iron Age armies, on the back of which the workshops stamped their factory marks.[1] When texts written in runes came to be used in preference to Roman script, from the mid-2nd century AD onwards, they often appeared in the same position. Perhaps the secrecy and cryptic knowledge appealed more than a formal display of literacy.[2]

Despite the instance in *Beowulf* in which King Hroðgar reads a runic inscription on the hilt of a supernatural sword, clear examples of text – runic or otherwise – on Anglo-Saxon weapons are very rare. The single unequivocal example of text on a blade is the seax from the River Thames with the entire rune-row incised and inlaid in silver, with interlace patterns and the masculine name **beagnoþ** plus some uninterpreted sigils. The weapon was recovered from the river in 1857. It probably represents a 9th c. fuþorc (English rune-row) from southern or south-eastern England; it does not feature any of the specifically Northumbrian forms found on the monuments at Ruthwell, Bewcastle and elsewhere. The purpose of the text has aroused some comment since it is not thought to have had any practical purpose, leading to speculation about magical intent. But the runes may have been added for their decorative and prestige value. This is the only known example of runes appearing on the blade of a weapon. Where runes are found, they normally appear on the pommel or the scabbard fittings.

A few examples of runic script appear to have been added at the time of burial rather than before, such as the Chessell Down (Isle of Wight) scabbard mount with its enigmatic **æco:sœri** (I increase pain?) text on a strip of metal added to the reverse of the mount. It therefore did not form part of the sword's furniture while in use. This poses the question: if the runes were added as part of the burial rite, who was supposed to read the text? The implication appears to be that it was a message for the gods or ancestors in the Otherworld rather than for anyone alive.

It may of course be the case that the act of cutting the runes was important and empowering but the text had no communicative value in itself. While this sounds an unlikely reason for adding text to an important object

1 Imer, 2010, p.59-60
2 Losquiño, 2015, p.147

such as a sword, it does account for the very casual nature of many inscriptions. Many runic texts are lightly incised and may plausibly be interpreted as part of the charming of the weapon to increase its efficacy.

The Thorsbjerg scabbard chape of 2nd-3rd c. AD date bears runic texts on both faces: (Side A) **owlþuþewaz** (Side B) **niwajemariz.**[1] The first rune on Side A (**o**) has been interpreted as an ideogram for 'inherited property' with the name of the owner following, but more recent interpretations indicate a transposition of the runes.[2] The name is thus interpreted as **Wolþuþewaz* 'servant of [the god] Wuldor'. The name, meaning 'splendour, glory', appears in later Norse myth as *Ullr*, a god of skiing and hunting. The text on Side B is interpreted as *ni* 'not' and an adjective *wajemariz* 'woe-famed' (cf. Gothic *wajamerein* 'dishonour'). This is an example of litotes: 'not famed for woe' means 'not known for bad deeds' and thus 'of good reputation'.

Some texts were already on the weapon while in use, but still show a lack of planning in their layout. For example, **swart/a** on a shieldboss from Illerup (Denmark) where the writer had to add the final **a** on a second line and perpendicular to the rest of the text.[3] The name is a form of the adjective *swart* 'black, dirty' and may have referred to its maker, a smith with the name meaning 'sooty, grimy'.

The majority of English examples of runic texts on weapons are confined to the south-east – mainly Frankish-dominated Kent – and the swords on which they occur are generally of types found across Europe from the Rhine to the Lower Thames. In the Anglo-Saxon areas of Britain and north-east France, runic inscriptions on weapons remained an important aspect of the aristocratic tradition, alongside other 'ritual' or amuletic symbols, although the great weapon-deposits of the types seen in Jutland are absent from the English record.

Runic inscriptions are associated with swords of the Bifrons-Gilton Type, 6th c. in date and many with a pair of linked rings attached to the pommel.[4] This specific type of sword is found across northern Europe from Snösbäck (Sweden) to Dole in central France, but the greatest concentrations are in Kent and the area north of Paris (France). A Frankish origin for the group has been proposed by Svante Fischer whose meticulous studies revealed the previously overlooked **a**-rune on a sword pommel from King's Field, Faversham (Kent) and a possible inscription **hle[m]** on another from Fréthun 1 (Pas de Calais, France). The word, if it is Old English *hlem,* means 'clamour, noise of battle'. Also from Fréthun is a sword pommel with a curious sigil on one face, strongly reminiscent of the *ear* rune but seriffed.[5]

A rare **alu** inscription appears on a pommel from Grenay (Pas de Calais, France), originally discovered in 1913 and largely disregarded ever since.[6] The inscriber appears to have written **lau** in error, but loss due to wear at the top of the pommel cannot be discounted. This word is plausibly linked to 'magical' practices and empowerment. Another **alu** text appears on a pommel from St. Dizier (France), where the extreme wear to the upper edge has left the second upright of **u** separated from the first. Wear to the face has likewise rendered illegible the text on a silver-gilt pommel from Sarre 2 (Kent) where the characters **du** (or retrograde **rm**) have been detected on one broad face, and possibly **c[...]ea** on the other.[7] Equally indistinct is the runic group on one face of the pommel in grave 69 at Shorwell (Isle of Wight), where **iec** can be distinguished, but the last rune is unclear.

---

[1] Looijenga, 2003, p.259
[2] Antonsen, 2002, p.146; Losquiño, 2015, p.58
[3] Looijenga, 2003, p.153; Zimmermann, 2010, p.101; Losquiño, 2015, p.62-3.
[4] Fischer, 2008, p.18
[5] Fischer, 2008, p.75-8
[6] Fischer, 2008, p.90
[7] Hawkes & Page, 1967, p.1-2, 7; Fischer, 2008, p.116-7

Another runic pommel, though controversial, is that from Gilton, Ash (Kent), which features two 'arrowhead' motifs on the shorter faces which have been interpreted as a variant **t**-rune.[1] In the same cemetery, in grave 4, was found a pommel with a remarkable runic text of about 19 characters but the wear to the top has obliterated parts of the characters. Interpretations include **eic sigimer nemde** 'Sigimer named [the] sword' (with *eic* for WSax *ecg* 'edge, blade') and **mic sigimer nemde** 'Sigimer named me'.[2]

Weapon-charming was practiced in Scandinavian emigrant communities, in Britain and elsewhere. The magical empowerment of a spear is referred to in *Brennu-Njálssaga* where the warrior, Hallgrímr, has *seiðr* magic worked on his *atgeir* 'thrusting spear' so that no weapon can kill him while he holds it, and it sings when it is about to take a life. The saga does not say whether the magical ceremony left visible marks on the spear, but it seems likely. The empowering of runes was achieved by reddening them with blood, which may have its origins in the ancient practices of hunting rituals.

It is possible that there is a rune on one of the sword pommels in the 7th c. Staffordshire Hoard, but this awaits confirmation at the time of writing. Since the hoard appears to be captured wargear, it is not possible to be certain that it was made in the area where it was found, so the context of this runic find – if such it is – remains uncertain.

Bibliography

Antonsen, E.A. *Runes and Germanic Linguistics*, New York, 2002

Fischer, S. *Les Seigneurs des Anneaux*, 2nd edition, Condé-sur-Noireau, 2008

Hawkes, S.C. & Page, R.I. *Swords and Runes in South-East England* in *Antiquaries Journal*, vol.47, 1967

Imer, L.M. *Runes and Romans in the North* in *Futhark*, vol.1, 2010

Looijenga, T. *Texts and Contexts of the Oldest Runic Inscriptions,* Leiden, 2003

Losquiño, I. G. *The Early Runic Inscriptions. Their Western Features*, Berkeley Insights in Linguistics and Semiotics no.92, New York, 2015

Zimmermann, C. *How to Do Things with Runes* in *Futhark* vol.1, 2010

[1] Hawkes & Page, 1967, pl.3

[2] Hawkes & Page, 1967, p.3-4; Fischer, 2008, p.88-90 suggests that an interpretation based on Frankish or OHG might yield results. It is noticeable that the name contains the Frankish element *Sigi* –'victory' as opposed to the WSax *Sige-* .

18. Other marks on sword blades

Paul Mortimer

Sword blades before the 5th century sometimes have inlays or stamps on the blade just below the hilt or on the tang, as evidenced by some of the finds from the Danish bogs. Many of these may be in imitation of the marks and inlays that the Romans seem to have occasionally added to their blades. The marks could be a manufacturer mark, the owner's sign, or perhaps a symbol of allegiance, an appeal to a deity or a combination of some of these or something else. Many of the marks on Roman style swords do seem to be references to gods and to myths. Some examples that Davidson reports will provide examples and she tells of two swords from the bog at Vimose, that have Latin letters stamped into their tangs, one reads 'CAXb' and the second 'TASVIT'.[1] Other swords have symbols such as a star or a scorpion, sometimes below the hilt, but a design might be stamped into the tang instead. Davidson writes that the largest collection comes from Nydam and that some swords there, have what appear to be names marked on them, and again gives examples such as one that has 'RICUS' on the tang, another 'RICCIM', a third with 'COCILLUS' and a fourth with what resembles 'UMORCD' and a stamp with a half-moon with horns.[2] There are examples, too, of both stamped blades and inlays from the bog excavations at Ejsbøl, in Denmark.[3] Davidson could not have known about the swords from Illerup when she wrote her book, but these outnumber the finds from Nydam, and follow similar patterns of what could be names, symbols and stylised figures of what were, probably originally supposed to be Roman gods, many reduced to 'stick men'.[4]

Some swords from the 9th to 12th centuries have blade inlays, the most famous of which are the *Ulfberht*[5] swords and the *Ingelrii* blades, although other names, such as *Leutlrit*, do appear.[6] Many of these names have been cut into the blade and then inlaid with pattern-welded letters; the blades themselves are not usually patterned and in fact, particularly with the Ulfberht blades, tend to be wholly made of carbon steel.[7] The names on such weapons are not thought to be personal names but to be maker's marks or the name of a manufactory.

A number of seaxes have been found that have decorated blades, sometimes quite elaborate, one found in the River Thames bears the twenty-eight characters from the runic futhark, and a further set of runes which may be the owner's or maker's name.[8] There are other examples and at least two from the Continent, less elaborately decorated and thought to be 7th century in date.[9]

---

[1] Davidson, 1962 (1994). 42.

[2] Davidson, 1962 (1994). 42.

[3] Ørsnes, 1988. Tafels 66, 70, 71, 73

[4] Biborski and Ilkjær, 2006. Volume 11, 296 to 302 for discussion of stamps. And volume 12, 207 to 217 for pictures of 65 stamps in tangs and blades and 219 to 221 for pictures of 10 different inlaid figures.

[5] Davidson, 1962 (1994). 46 to 48. Williams, 2012, 117, reports that about 100 Ulfberht swords survive, that includes variant spellings. Williams distinguishes five different groups of decreasing quality among them and suggests that because they were made over a period of about 300 years, the name probably refers to a family of smiths. See Williams, 2012. 116 to 183 for his full analysis of the Ulfberht swords.

[6] Davidson, 1962 (1994). 45.

[7] Davidson, 1962 (1994). 47. This is not necessarily the case for those blades with the name misspelt. See also Williams, 2012. 116 to 183 for a full discussion of this class of swords.

[8] Davidson, 1962 (1994). 43 to 44. This the so called, 'Beagnoth' seax, thought to be 10th century.

[9] Davidson, 1962 (1994). 44.

Spear heads with inlays of bronze, silver and even gold have been found from the 5th to 7th centuries in Anglo-Saxon graves. Some of the marks seem to be runic, probably providing supernatural protection, but they could be maker's marks or have a completely different purpose.

However, sword blades or tangs from the 5th to 7th centuries with inlays, stamps or other marks seem to be extremely rare. Davidson notes an example from a 7th century grave at Nördlingen, Donau-Ries district, in Bavaria, Germany that has 'Oϩ' inscribed on it, although, apparently, the excavator thinks it could be a Roman object.[1] In England Davidson reports on a blade in the British Museum that was found in the River Thames and has had two identical stamps impressed into the blade, probably from a coin die, the impressed marks were subsequently covered with gold foil. The blade is thought to be quite late, probably 8th century.[2]

18.1 The stamps on one side of the River Lark sword, this side has two boars, the other side carries a single example. Photograph by Paul Mortimer taken with the kind permission of the Museum of Anthropology and Archaeology, Cambridge

Another river find, this time from the River Lark, which flows through parts of Suffolk and Cambridgeshire, has three figures of boars punched into the surface of the blade, two on one side and one on the other. The images are detailed and show a, "...plump boar, whose ears, eye, and tail are clearly visible in profile..." and the blade has been attributed to the 7th century.[3] (figure 18.1) Boars are special animals, related to warfare and supernatural beliefs and appear on helmets from Anglo-Saxon England and on pictures of them on the pressbleches of helmets from Sweden.[4] Helmet boars are mentioned as being protective in the poem

---

[1] Davidson, 1962 (1994). 43.

[2] Davidson, 1962 (1994). 49.

[3] Davidson, 1962 (1994). 49 to 50.

[4] Mortimer, 2011. 24, 27, 40 to 42 and for a fuller discussion on the role of boar imagery, 259. Boars appear as crests on the Benty Grange and Pioneer (Wollaston) helmets and are situated above the cheek guards on the helmet from Sutton Hoo. There are a number of stray finds of boars that could come from helmets.

Beowulf.[1] However, the River Lark sword is almost certainly not from the Anglo-Saxon period at all. Stead, who refers to it as the 'West Row sword', claims it for the Iron Age, partly because it seems to have a non-patterned, 'streaky' blade, and because the stamp design has no known parallel from swords from the 5th to 11th centuries in England. On 22nd June 2015 I went to the Museum of Archaeology and Anthropology, (University of Cambridge) to view the sword and have to confirm that it would appear to be from the Iron Age, one of the most convincing details is that the blade is very slightly 'diamond-shaped' in cross-section which is a shape that would be unique if it were to be from our period, but not so for the Iron Age.

A late 6th/early 7th century find of a wooden sword at Arum in West Frisia contains a runic inscription on its blade and it has a burnt blade tip. The runes have been translated as possibly meaning 'return-messenger' but this is still disputed, and the object itself might have been used as a call to war or even a child's toy; in any case, it is not a real sword.[2]

Marks on blades were never a universal practice, just an occasional one, but it is curious that no blades from the 5th to 7th centuries from England appear to carry stamps, inlays or other marks when blades from other times do. It is even odder, when we think about some of the other types of weapons, such as spears, which did carry them. Perhaps it is because almost all swords located from 5th to 7th centuries in England come from burials and were put into the grave inside their scabbards and the resultant corrosive processes could have obscured any trace of blade marking. Perhaps another reason is that until recently few swords had been x-rayed and, even when they were, the examiners tended to be concentrating on the blade structures and such marks may have been missed or might not have shown up using that technology? Ulrich Lehmann, using computer tomography, has found some evidence of marks on a few blades from Germany dated to the 5th to 7th centuries which he has studied and he reports that of the thirty swords[3] from Westphalia, Germany, he has found four examples of simple inlays. Two of the swords possess simple ring-like inlays made of non-ferrous metal (bronze wire and plate): grave 9, Bad Wünnenberg-Fürstenberg (end of 6th century) and grave 6 from Beckum I (also end of 6th century). Another sword possessed a simple ringlike inlay made of welded iron. The blade from grave 11, Dortmund-Asseln (end of 6th century) and that of another sword, a stray find from Leer-Ostendorf (8th century), both show four different inlays made of welded iron: two omegas separated by an 'X'shape and a 'pretzel-like' inlay.[4] La Salvia reports that some of the swords from the 'Lombard' cemetery at Nocera Umbra in Italy had inlays, some of wire, on their blades just beneath tang. They are the swords from graves 6, 15, 122 and 143. The sword from grave 15, "...displays a little circular/spiral shaped motif on one side..." while the example from grave 122 has a, "...a small section *Schleifendamast* visible on it, a wire like ornament, which is very different from other *Winkeldamast*." Graves 143 has an 's' shaped inlay, whilst the sword from grave 6 has two 's' one imposed on the other. The finds are dated to the last third of the 6th century. La Salvia suggests that the double 's' design has much in common with finds from Alemannic graves, further north in Bavaria and reports that it has been suggested that it is the sign of a particular workshop. La Salvia also notes the find of a similar double's' on a Carolingian blade from 8th century from the Netherlands.[5]

---

[1] *Beowulf*, 303 to 304. *Eoforl□c scionon ofer hlēorbergan gehroden golde*..... Above the cheek guards shone the boar images adorned with gold...

[2] Davidson, 1962 (1994). 44 to 45. Davidson recalls Scandinavian practices of sending around a piece of burnt wood as a call-to-arms. She also reports finds of other wooden swords from the bogs at Vimose and Thorsbjerg.

[3] Twenty-eight were from graves while two were found on their own.

[4] Personal communication and Lehmann, 2014. See also Lehmann, 2016. 155 to 158 and 427 to 429 (Bad Wünnenberg-Fürstenberg, Grab 9) and 437 to 439 (Beckum I, Grab 6).

[5] La Salvia, 2007. 55. It is not improbable that such '8th century' swords have blades from an earlier period.

I have not been able to find a complete survey of Continental and Scandinavian swords so I cannot comment on how frequent blade marking with inlays or other methods was outside Anglo-Saxon England, however, the fact that it does occur on sword blades in the Merovingian lands (and on spear heads in England) would suggest that it is likely that the practice of sword blade marking was as prevalent on blades here as it was elsewhere, particularly as it occurs both before and after our period. Perhaps as technologies for examining blades become ever more sophisticated, examples will be discovered?

19.i. The Sword from Mound 1 at Sutton Hoo and the Replica – A Case Study

Paul Mortimer

Mound 1 at Sutton Hoo contained a ship burial dating to around 625 A.D. and probably was the burial of a king of the East Angles. The ship was a large one, some twenty-seven metres long, the biggest ship contained within a ship burial in Northern Europe ever discovered, at least so far. The ship had been laid in a trench in the ground and a wooden burial chamber had been constructed amidships and filled with many items, including the famous helmet and jewellery. Eventually, the ship and chamber had been covered with earth.

In 1938 the owner of the site, Mrs Pretty, asked Ipswich Museum for help in investigating the burial ground and the museum sent along a local archaeologist, Basil Brown. Mound 1 was not the first mound that Brown excavated in the cemetery; during 1938 he excavated Mounds 2, 3 and 4 and had already discovered a ship burial in Mound 2[1] but, like the other mounds explored that year, it had been robbed. The re-opening and robbing probably occurred during the Tudor period and very little of anything that had been deposited was left, although there were tantalising clues that Mound 2 had been a very rich burial and the, somewhat scattered ship rivets indicated that the vessel was about 24 metres in length.

Mound 1 was opened during the summer of 1939, on the eve of the Second World War. Brown unearthed the virtually complete outline of a great ship, but almost all of the wood had been dissolved in the acidic sands of the landscape although, again, the iron rivets were still there. As the dig progressed and the finds increased Brown began to understand the importance of the find and referred back to the museum and more 'eminent' archaeologists were soon sent to take over, although Mrs Pretty insisted that Basil must continue to work on the site.

Once the area of the central chamber was reached, many treasures were revealed; many of them of precious metals and almost all of the finds were unique. One such was the sword.

The sword was discovered, hilt to the west, on the 19th July 1939 but not lifted until it had been photographed in situ on 27th July.[2] Before the chamber roof collapsed, it is likely that the sword, along with many other items had fallen, either from the body of the tomb's occupant or some form of platform. Most, if not all, of the sword's component parts were still present on discovery but the sword was broken in several places as the iron and organic components were badly corroded.

The report on the sword and associated items was not published until 1975 in volume 2 of *The Sutton Hoo Ship-Burial, volume 2 Arms, Armour and Regalia.* This volume, along with the other two, was masterminded by Rupert Bruce-Mitford. It was Bruce-Mitford who very carefully studied the sword and the other components, together with the expert reports and tested various hypotheses concerning the sword and its external decorative elements, including the belt set and put forward his conclusions as to how they would all work together and explained how he had worked that out. The tests were only conducted as drawings on paper and not with reconstructions involving originals or replicas.

My newest replica sword is closely based on Bruce-Mitford's and his team's work plus a re-examination of the evidence and the practical experience of getting all the components to work together.

---

[1] Bruce-Mitford, 1975. 100.

[2] Bruce-Mitford, 1978. 273.

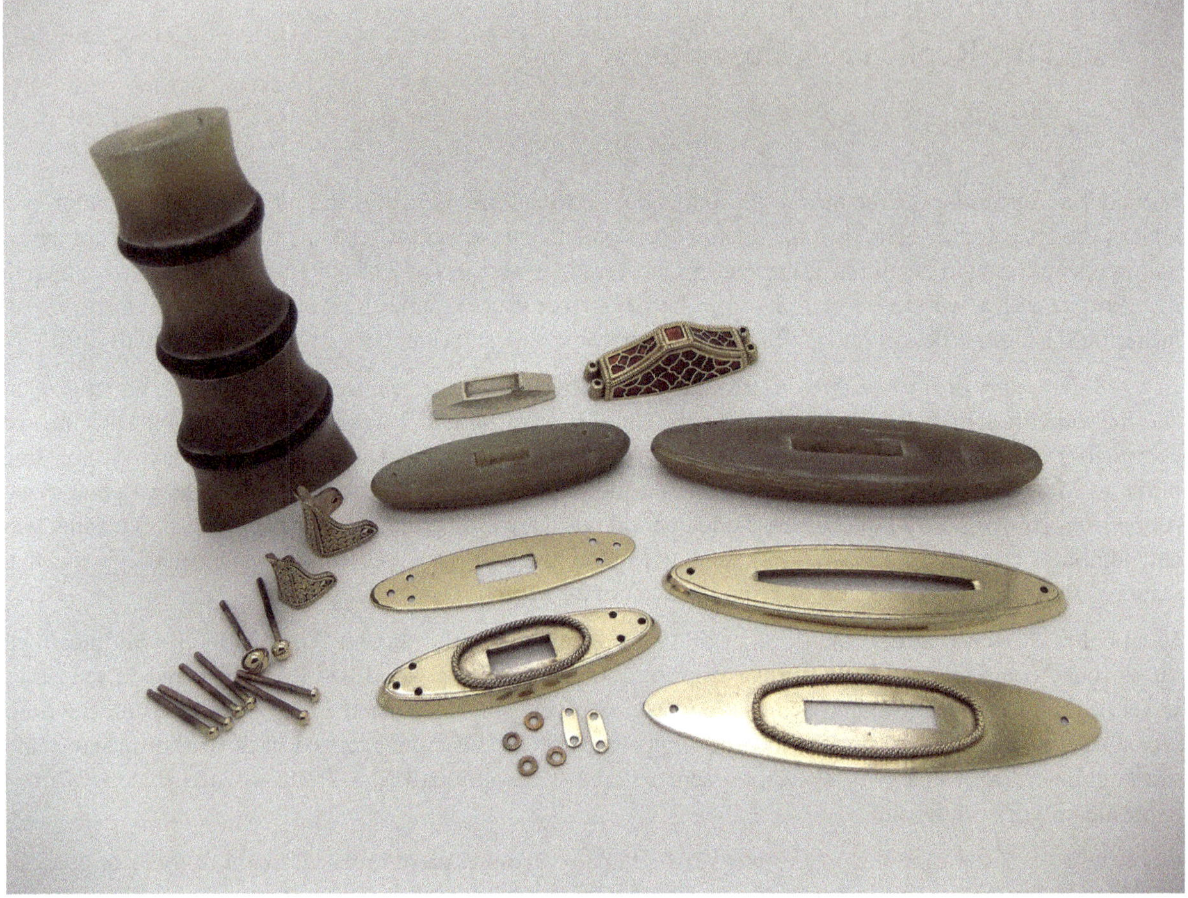

19i.1a. The prepared components for the reconstructed hilt ready to be assembled. Photograph courtesy Dave Roper

19.1b. The hilt assembled

19i.2. The three jewels that Bruce-Mitford felt came from a conjectural cross-strap. I have placed them on the lower suspension strap. One of the jewels can only be seen from the side, but it is identical to that on the right hand side of the illustration.

19i.3a. A closer view of the dummy buckle.

All photos Paul Mortimer

19i.3b. The rear strap distributor provides movement in two different planes.

19i.3c. The buckle from the rear suspension strap (inv 14). Note that there are two separate foil designs behind the garnets.

19i.4. The completed rig showing the sword belt and the lower suspension strap

All photos Paul Mortimer

19i.5. The tablet woven tape woven by Dariusz Wiewiorka from the front. Dariusz thought it would add interest and not deviate from the evidence to pick out a detail in white thread on the upper tape. The colours, in any case, are conjectural

19i.6. The reverse of the sword tapes. The position of the tape has been reinforced with transverse linen thread

19i.7. The complete scabbard by Matt Bunker. The design is loosely based on a detail on a scabbard from Valsgärde 6 in Sweden

All photos Paul Mortimer

19i.8a. (below) The boss strap is quite a complex construction but works very well in practice and although conjectual does not contradict the evidence. Note the different sizes of the two bosses.

19i.8b. (above) The boss strap is actually two straps. The main loop encircles and grips the sword scabbard, but there is a second loop permanently stitched to the rear through which the sword belt is inserted

19i.8c. (right) The second strap can easily be seen in this picture. There is a light brown, folded leather piece to stop the sword belt from getting caught and snagged between the two loops

All photos Paul Mortimer

19i.8d. The boss strap has been lined with split lambskin to provide a buffer between the main loop and the strap and give a little extra friction

19i.8e. The sword strap has been inserted through the second loop and is secured in place with linen tape attached to the sword belt

All photos Paul Mortimer

19i.8f. Another view showing the securing tapes

19i.9. The pyramids on their woollen tablet woven braids, made by Dariusz Wiewiorka. In this picture the pyramids are loose. Photograph Paul Mortimer

19i.10. Peace bands? Photograph Paul Mortimer

19i.11. I tie the pyramids up when I am carrying the sword as if they are allowed to hang loose they could clash and cause damage to each other or to other elements of the sword. I suspect that the owner of the original would have done something similar. Photograph Paul Mortimer

19i.12 (below) The 'wolf' jewel. Bruce-Mitford felt that this may have acted as a counter-plate to the buckle on the main sword belt even though it appears to be of different manufacture. I agree and think that it is possibly an 'heirloom' piece. Photograph Paul Mortimer

19i.13. A view of the rear suspension being worn. Photograph courtesy Lindsay Kerr

The start of the project

I had thought about the sword and belt ensemble for some years and made a previous attempt at recreating it, but experience of using that arrangement made me realise that some of the elements involved in my reconstruction were being used incorrectly and I felt that I needed to go back to the evidence and start again.

I had contacted Vince Evans to see if he would be interested in helping me with the project and I was very pleased when he accepted. Vince, too, believed that we should gather as much information on the original and re-examine all that we knew about the sword. I also spoke with Dave Roper about making the sword hilt as he was familiar with the jewellery techniques employed by the craftsmen who made the original. He had gained experience from replicating most of the garnet and gold jewellery for me already. Matt Bunker had been researching scabbard technologies and, fortunately, was keen to make a scabbard for the blade.

I also wanted to re-consider the way that the belt jewels and sword jewels were employed on the original. As mentioned above I was not, at that time, completely happy with all the conclusions that Rupert Bruce-Mitford had come to in his report on the sword nor was I content with some of the solutions I had used on my first attempt. With Vince making the new blade there was an opportunity to see if ways of mounting the sword could be found that were more in line with the evidence.

The blade

Vince and I were given an opportunity, by the British Museum, to study the sword out of its case[1] and Vince was able, through Janet Lang of the British Museum, to consult the original x-rays of the find. The latter experience was to prove especially fortunate in that Vince did not see some of the things that he was expecting to see, if Bruce-Mitford's and other commentator's (and smith's) ideas about the making of the blade were correct. This led Vince to some fairly radical conclusions. Vince explains those conclusions in his accompanying report on the blade.

On examining the Mound 1 sword, it became apparent that it had not had a fuller, which was another departure from the established 'facts' about the sword; there was no sign of one in the cross-section of the blade at the top of the scabbard/blade mass or on the bottom plate of the hand guard on the hilt. Conveniently, the handle ensemble broke away from the lower part of the sword, long before the grave was excavated and we were able to view the top of the blade. In fact the blade appears to be lenticular on both sides, if perhaps slightly flattened in the central sections.

When compiling the list of different sword patterns and methods of constructing swords blades, it became possible to consider and compare this blade to others from Anglo-Saxon graves and it has to be said that it is not, by any means, the most complex of blades. This is in contrast to many other items within that grave, which are of exceptional quality, and that includes the sword jewels and those belonging to the belt. The blade has four rods per side, as do thirty others in the sample of 214 blades; there are fifteen swords that have five or six rods on each side of their blades. The pattern on the Sutton Hoo is the second most common one, alternate herringbone and straight (AHS).

When considering the blade dimensions it is again not outstanding. The average length of a blade in the sample is 759.5 mms, and there are two modes of 760 mms and 780 mms; the Sutton Hoo Mound 1 sword blade is 720 mms in length.

It is quite probable that the Mound 1 blade had been re-hilted during its period of use, which could have been over a long period. Each re-hilting would have involved shortening the blade a little if the shoulders had been cut to preserve the length of the hilt. It may be that the blade was very famous and old, highly valued because of its age, associations and other glamorous qualities. Unfortunately, such properties cannot be discerned through archaeology.

---

[1] We would like to thank Sonja Marzinzik and Sue Brunning for that opportunity.

The hilt

The original report states that the remnant grip was made from wood;[1] however, there are a number of reasons for thinking that this is unlikely. Firstly the archaeologist Esther Cameron had re-examined the grip and firmly stated that it was made of horn. Also, nearly all such grips on Anglo-Saxon swords of the period, where the material can be identified, are of horn.[2] I therefore asked Dave Roper to make the grip of horn, perhaps two coloured to add interest, which he did. The upper and lower guard were each composed of a sandwich of two gold plates, enclosing an unidentified organic material, for this Dave again used horn. The material has to be very strong, as on the lower guard, where the top of the blade penetrates the lower gold plate of the hand guard sandwich, it is the organic material alone that must take and absorb the stresses from the blade when the sword is in use. (figure 19i.1a)

The two filigree clips were also made of solid gold on the original and are fine and delicate to manufacture. Like almost all 'high status' swords the pommel design on the original sword differs on each side and this needed reproducing, as did the very fine waffle-patterning on the gold foils set behind the garnets. The decorative jewelled pommel cap encases an inner and quite separate under-pommel that forms most of the mechanical connection between the sword tang and the pommel assembly. The jewelled pommel cap is then riveted firmly to the upper guard by some ornamented but strong rivets. (figure 19i.1b)

The belt assembly

Most aspects of the belt were straightforward and Bruce-Mitford's positioning of them can be accepted; it is his placing of the 'extra' jewels (Inv. 12 to 14 together with Inv. 15[3]) that needed particular rethinking. (figure 19i.2) He had put them on a cross-strap that went over the shoulder and then joined the waist belt at both ends.

Initially Bruce-Mitford wondered whether the sword would have been mounted on a baldric or a belt and came to the conclusion that it must have been a belt, once he understood the function of the 'dummy buckle'.[4] This led him to realise that the suspension method would have been the two-point suspension method, with the main support around the waist, and a lower suspension strap attached at the back of the sword belt and that this would then connect with the lower sword scabbard via the dummy buckle at the bottom end and the complex strap-distributor at the top.[5] (figure 19i.3a and 19i.3b) The sword would then rest against the left hip with the hilt projecting forward at a relatively acute angle to the wearer's body. There are several similar surviving belt assemblies in Continental graves.

Bruce-Mitford felt that three jewels, Inv. 12, 13 and 14 (figure 19i.3c) could not have been part of the sword belt assemblage, mainly because of the position where Inv. 12 (the small jewelled buckle) was found in relation to the rest of the sword jewels and equipment.[6] Bruce-Mitford does admit the possibility that Inv. 12 to 14 were mounted on the lower suspension strap but finds the idea 'implausible'; he did not attempt to explain why.[7] There are a number of problems with such a cross strap. First of all, I can find no precedent for such a strap in any other find either on the Continent, Scandinavia or in England; there are shoulder straps, or baldrics, usually carrying a sword, but no straps definitely attached to waist belts. It is interesting that Bruce-Mitford does not provide an example either. It maybe that Bruce-Mitford was influenced by modern military

---

[1] Bruce-Mitford, 1978. 292.

[2] Cameron, 2000. 97.

[3] The inventory numbers are those used in Bruce-Mitford's report.

[4] Bruce-Mitford, 1978. 572. The piece is identified with the inventory number Inv. 10 in the report. Bruce-Mitford had called the suspension method a 'baldric' in his earlier book; see Bruce-Mitford, 1974. 53.

[5] Ruce-Mitford, 1978. 572 to 581.

[6] Bruce-Mitford, 1978. 575. On page 575t

[7] Bruce-Mitford, 1978. 575.

belts for officers in the British Army which do have such an arrangement. As mentioned above, I did try wearing this arrangement with my first replica but found it did nothing to support the weight of the main waist belt and therefore has no purpose; it was also irritating and made my shoulder ache after a while.

Bruce-Mitford's reasoning as to why the pieces could not have been mounted on the lower suspension strap I believe is flawed; he used the positions that the pieces had ended up in the grave, presumably after they had fallen, and tried to work out from that how they were all originally connected. The difficulty here is that so many variables might have been involved in any fall that the resting places may not indicate very much at all. For instance, at the time of the fall, the belt leather or textile could have been so degraded that some of the pieces could have already been freed and fallen, or some other object could have interacted with the belt sets during the fall and possibly moved some of the items nearer or further than they otherwise would have been.

In volume 2 of his report, Bruce-Mitford provides illustrations of fourteen belt assemblies that had the two point suspension system,[1] all of them appear to have had adjustment buckles on the lower suspension straps, some had additional decorative plates as well. It seemed to me that the jewels, Inv. 12 to 14 (and Inv. 15) should have been part of the lower suspension strap and that is where I have placed them on my reconstruction. (figure 19i.4) The decoration on the pieces matches that on the strap distributor, which is surely not a coincidence, and when placed on the strap, the observer's eyes are drawn down towards the dummy buckle and it no longer is an isolated piece at the end of the scabbard. From the experience of wearing the assemblage I can demonstrate, too, I believe, that the jewels Inv. 12 to 15 were not placed on the lower strap just for cosmetic reasons as adjusting the length of the strap is necessary when wearing different garments in order to get the sword to rest at a comfortable angle. The replica Inv. 15 acts as a runner when the lower strap is adjusted at the buckle, as the runner is lower it stops any free part of the strap flapping about and becoming a nuisance.

The top of the scabbard

Many Anglo-Saxon swords (and Continental ones) have binding tapes at the top of the scabbard; these can be of a variety of materials, and usually, where such tapes survive in relatively good condition, such as Broomfield, Essex,[2] it can be seen that they have been applied carefully and accurately. In the case of the Mound 1 sword, the tape consisted of fine, plain weave linen, 0.6 mms wide[3] which covers the top 16 cm of the scabbard. Valerie Fenwick, in Bruce-Mitford's work says that the tape started at the throat of the scabbard and is wound tightly around the scabbard to about 16 centimetres below the top. Esther Cameron believes that Fenwick's description was incorrect and says that the tape wound down from the throat to the scabbard bosses and then began again, winding up towards them from 15cms below the scabbard throat.[4] At each turn of the tape it overlays the layer beneath by about 3mms. On the replica I began winding the tape upwards 16 cm from the scabbard throat and continued all the way almost to the top. Near the scabbard throat the tape was then wound back down about 7 cms to where the bosses would be situated. There are a number of reasons for this possible departure from the evidence. Firstly, when I examined the sword at the British Museum, it was not clear to me whether or not the tape continued behind the bosses so I felt free to do so, also the double layer of tape above the bosses helps to prevent movement of the bosses upwards when the sword is being worn. Continuing the tape beneath the bosses also helps prevent any shifting by the bosses and their mounting strap by providing a little extra friction. (figures 19i.5 and 19i.6) Matt Bunker suggested that to avoid slippage it would be practical to partially coat the linen tapes with beeswax on the underside to help

---

[1] Bruce-Mitford, 1978. 570 to 571 and 573.

[2] Bruce-Mitford, 1975. 464.

[3] Bruce-Mitford, 1975. 449.

[4] Bruce-Mitford, 1975. 449. See also, Bruce-Mitford, 1978. 284. Cameron, 2000. 97. Provides the clearest description of the way that the tapes are disposed.

secure them. I did and it works well but in order to inhibit movement further I also put some securing stitches using linen thread through the tapes on the reverse of the scabbard. Cameron noted some remains of "fine transverse cords, 0.3mm in diameter..."[1] mostly on the reverse and that could have performed a similar function to my stitching.

The replica tapes were hand woven on a tablet loom for me by Dariusz Wiewiorka who had the idea of using two slightly different tapes, one with a very small and subtle design inserted into the tape to add more interest. Although the weave can be detected on the original, the colours cannot; I chose yellow as, together with the red scabbard, that would seem to reflect the predominant colours of the personal items from the grave, which seem to me to be red and gold.

The scabbard

The scabbard upon examination was found to be of 2mm thick willow or poplar with the grain running along the length of the scabbard.[2] The scabbard was lenticular in cross-section and no trace of leather was found on the exterior surface. It was lined with some form of fur, described as "a straight soft type of animal fur",[3] which it seems was not from a sheep, and "...is superficially like that of the densely packed spreads of beaver hairs, similarly interpreted as lengths lying obliquely within the thickness of the skin,.."[4] The animal hairs within the scabbard do lay at an oblique angle to the length of the scabbard. Unfortunately, due to the replacement of much of the animal material within the fur by corrosion products from the sword, it proved to be impossible to determine from which animal the fur actually came.

The scabbard had no metal edge reinforcements or a chape to reinforce the bottom end of the scabbard, although it is possible that something organic was used.[5]

Matt Bunker constructed the scabbard using a carcass of 2mm thick willow that was steamed to the correct shape a technique that has produced an extremely strong scabbard. He lined the scabbard with beaver fur and the hairs are aligned obliquely to the scabbard. Matt points out that the assumption that scabbards makers wanted to protect the blade with the lanolin from the wool or sheep fleece, which many surviving scabbards are lined with, could be just that, an assumption. Perhaps other considerations were more important when linings were included? Beaver fur is practically waterproof and this may have been a consideration.

Although no leather traces were found on the surface we felt that it was likely that the original had been protected with a covering and that it was more than possible that the adverse acidic conditions in the grave had removed all sign of it. Matt chose a very fine vegetable-tanned calf skin to wrap the scabbard carcass in. I wanted a bold design that could be seen from some distance away and he obliged and used foundation moulding techniques to make it stand out. We did consider carving the design directly into the scabbard, but the wood used to make it was so thin that it was unlikely to have worked well. The twin serpent motif design is based on a similar design found on the scabbard of a sword from grave 6 at Valsgärde, in Sweden. The leather, on my replica, was coloured with a dye made from the dried bodies of a female scale insect that lives on the sap of certain trees, especially the *Kermes* oak tree found in the southern and eastern Mediterranean, hence the name of the dye; kermes. A chape of leather was added to the bottom of the scabbard to protect the end from scuffs whenever it is knocked or grounded. (figure 19i.7).

---

[1] Cameron, 2000. 97.

[2] Bruce-Mitford, 1978. 285.

[3] Bruce-Mitford, 1978. 309. Bruce-Mitford, 1975. 441. 'fur-lined scabbard'

[4] Bruce-Mitford, 1978. 285.

[5] Bruce-Mitford, 1978. 285.

The wooden feature

There is one aspect of the sword ensemble that it has not been possible to account for and that is the piece referred to in the report as 'the wooden feature'. This, when found, lay on the front surface and at the top of the scabbard, positioned between the two bosses. It comes to a point some 17.5 to 18.5 cm[1] below the throat of the scabbard and the top surface of the wood is dished, rather like a gutter. It lay over the binding of the scabbard and does not appear to have been part of the suspension arrangements such as the scabbard 'slide' on many swords as it had no '...firm means of attachment to the scabbard...'[2] At the scabbard mouth it would have been 2.3 cm wide[3] but as it passes the bosses the feature narrowed to 1.5 cm wide.[4] Cameron points out that it is incomplete and is 1.5 mm thick.

It is difficult to see what purpose this object had and, as it had no visible means of attachment, it has not been included in the reconstruction. Further work or new finds could help discover its function but it is possible that it never was a permanent part of the sword assemblage.

Mounting the sword on to the belt

Apart from the lower suspension strap, the precise manner of attaching the sword to the belt cannot be discerned from the positions that the sword and belt jewels were found within the tomb. The lack of organic remains further hinders finding a solution. It is possible that the sword had been removed from the belt before deposition although when found the parts were all quite close together. This means that the methods of attachment that I have used are the results of speculation and experiment; the main operating principles I worked with were that the possible solutions should be within the evidence, practical and comfortable.

The boss strap

One essential part of the way that the sword was attached to the belt is the piece that I have called the 'boss strap' (figure 19i.8a to 19i.8f) There is little doubt that there was such a thing, but its precise form and the way that it worked is where there is room to experiment.

The two bosses were found still in situ on the scabbard apparently shifted to the left and 10.5 cm below the scabbard mouth.[5] Each boss consists of a domed garnet and gold cloisonné face and has a staple-like shank on the rear 1.2 cm long. Both were set into some white organic material, possibly bone or ivory but virtually all remains of it had disappeared, except for a residue disc on each boss.[6] The boss on the wearer's left is slightly smaller in diameter than that on the right and a similar difference was found in the diameters of the white material.

Bruce-Mitford felt that the bosses, or studs, as he sometimes calls them, were probably not merely ornamental, but acted as buttons in connection with a strap to secure the sword to the belt.[7] He noticed that the shanks did not pierce the scabbard material itself but had been surface-mounted within the organic white material and so must have been firmly attached to the scabbard.

In my reconstruction I have used the bosses in a similar way to the one that Bruce-Mitford suggests but had real difficulty making the bosses perform as 'buttons'. I had successfully managed that with my previous attempt at reconstruction but could only make it work by inserting ivory pins through the shanks of the bosses and the leather strap and this meant that the backing of the bosses could only be a thin ivory disc. That

---

[1] Cameron says it is 184 mm long. Cameron, 2000. 97.
[2] Bruce-Mitford, 1978. 288 and page 297.
[3] Cameron, 2000. 97.
[4] Bruce-Mitford, 1978. 288.
[5] Bruce-Mitford, 1978. 285.
[6] Bruce-Mitford, 1978. 285 to 288. See also 296.
[7] Bruce-Mitford, 1978. 297.

was a problem because the contemporary evidence for the backing material is that the bosses tend to be supported not with a white disc but with a 'bulb' or near sphere of whitish material.[1] The 'bulb' completely covers the boss shank which makes it hard to get sufficient mechanical force through the 'u' shaped staples to ensure that the strap would be kept tight enough to hold the sword to the belt. The problem went away when I secured the strap by other means and just tied the bosses onto the strap via their shanks. I suspect that the real purpose of the bosses was not as 'buttons' but to be seen and that their significance would be understood by the observer. That idea will be discussed elsewhere within this volume. The bosses on the replica sit on cylinders of polished deer antler.[2]

The strap that I used is of vegetable tanned leather and of sufficient length to go around the sword scabbard and join in the middle. I lined the inside of the strap with split lamb skin to provide a soft surface to interact with the scabbard and the linen tapes and I used plaited hemp cord as a tie to secure the strap around the sword as the hemp is very strong and not likely to fail. Hemp is a material that was available to the early Anglo-Saxons. The bosses are mounted on the strap with plaited leather cords and, as there is very little stress on the cords, this should suffice. The boss strap stays around the sword when it is dismounted from the belt and because of this it is a little more complex than it looks at first glance. At the rear of the strap there is a secondary strap of leather firmly attached; together both straps form a narrow loop through which one end of the sword belt is slid in order to mount it onto the belt. The boss strap also has a length of the same linen tape that is around the scabbard and sewn onto it and this is used to keep the sword ensemble in the correct position once the sword is mounted on the belt. The pyramids are also connected to the boss strap at the rear.

The securing method is of course conjectural, but it is comfortable and practical and accommodates all the evidence and so is possible. It could be that other, similar methods could be used.

The pyramids

The last elements of the Sutton Hoo Mound 1 sword that need to be considered are the pyramids. Unlike the bosses the pyramids are of the same dimensions although the weights are very slightly different. They are hollow and there is a bar that runs across the hole in the base; they are precisely and strongly made. As Bruce-Mitford remarks, "The pyramids can only have been mounted on a cloth or leather strip some 5-6 mm wide or less, depending on the thickness."[3] They were found either side of the sword in line with the scabbard bosses but at a distance of 8 cm on one side and 18 cm on the other from them.[4] As Bruce-Mitford remarks the position may be fortuitous but it does suggest that, "...they were attached to a strap that ran through the bosses transversely."[5] (figure 19i.9)

Bruce-Mitford felt that the pyramids performed the function of 'sword knot' embellishments attached to tapes.[6] I believe that he is correct. On some sword assemblies they do form part of the attachment mechanism securing the sword to the belt, but it is unlikely that this was the case with this sword as the bosses seem to fulfil that purpose here. The supposed casual losses of pyramids, sometimes found by metal detectorists in England, and elsewhere, would seem to show that however they were attached, the straps were not always

---

[1] See for instance Krefeld-Gellep grave 1812 (Menghin, 1983. 227) Krefeld-Gellep grave 1782 (Menghin, 1983. 239 and 240) Niederstotzingen grave 9 (Menghin, 1983. 253) Bulach grave 7 (Menghin, 1983. 258) Ziertheim (Menghin, 1983. 274) and St Dizier (Mortimer, 2011. 113). We will discuss the symbolic purpose of the sword bosses elsewhere in this book.

[2] Thanks to Mark Routledge for the gift of suitable antler.

[3] Bruce-Mitford, 1978.300.

[4] Bruce-Mitford, 1978.301.

[5] Bruce-Mitford, 1978. 301.

[6] Bruce-Mitford, 1978. 581.

that secure.[1] As Bruce-Mitford explains, the pyramids can only have been attached via a 'thin, pliable and narrow' strap as the material has to pass under the bar on the bottom of the pyramid. I feel that it has to have been textile, partly for its strength, but also for decorative effect. On my reconstruction I have used tablet woven woollen braids although passing them through the pyramid bar was not easy. I do not think that such thin leather would have been trustworthy.

Possible reasons for the employment of pyramids will be discussed elsewhere in this book but on this sword I am sure that the owner wanted them to be seen and that it was important for the top surface to be displayed. One suggestion is that the pyramid braids were used as 'peace bands' (figure 19i.10) and such things are mentioned in some of the later literature. It is thought that peace bands would be used to show that the sword carrier had no intention of drawing his sword and was therefore at peace and that this would have been shown visually by tying the straps around the sword hilt; the pyramids would help to make the statement more visible and perhaps more powerful. Whatever, when not being used in that fashion, it seems to me that a sensible owner would tie them carefully to the top of the scabbard in such a way that the pyramids would not be at risk of falling off and would be prominently displayed. (figure 19i.11) There may be other and better interpretations, but I am unaware of any at the time of writing.

[1] I will challenge that idea in the *Riddle of the pyramids*.

19.ii. Sutton Hoo Style Pattern Welded Blades

Vince Evans

Introduction

(See figures 19ii.1 to 19ii.26 for a stage by stage photo essay of the making of this blade)

When I was commissioned to forge a blade based closely on the original sword found in the Sutton Hoo ship burial, I had to wonder where my work would differ from the blades others had forged in the past. It was not my intention to come up with a different hypothesis on how some original Anglo Saxon blades may have been forged, I was merely trying to be as accurate to the original as I could be.

Composite pattern welding in the Anglo Saxon era consisted of forging sword blades from a series of rods of layered steel. The process of twisting the rods created variations of surface patterns that were visible on the finished blade when these rods were welded together.

In my more than 30 years of experience as a bladesmith I have forged a number of composite pattern welded blades but had never been asked to exactly replicate an existing pattern in a blade, and thus, I had not had to conform to the precision of spacing that was required to replicate the Sutton Hoo blade. This project would require a lot of research and experimentation.

Scott Lankton's pioneering work on the replica Sutton Hoo blade that now resides in the British Museum and the paper by Angela Care Evans were obvious starting points.

"The blade and its scabbard had corroded into one inseparable mass, but radiography has revealed that beneath the corrosion products the iron of the blade has survived sufficiently well in restricted areas for an interpretation of its structure to be possible...The radiographs suggest that it was built up of four bundles of seven rods twist-forged in an alternating pattern and lying back to back with four more bundles of seven rods. The bundles of rods twist alternately to right and left, forming a double band of the characteristic herringbone pattern that is one of the most distinctive features of such blades. The twisted bands alternate with straight bands along the length of the blade...In broad terms, and remembering that any interpretation of radiographic evidence may be misleading in one way or another, the proportions of the pattern-welded blade appear to be as follows: the average length of both the twisted and straight areas is 5.3 cm, and the pattern repeats at least eleven times...The cutting edges consist of two long strips of metal each running the length of the blade and forged onto the central pattern-welded core."[1]

As I pored over articles and examined radiographs of original swords, I kept coming back to the thought that there was something that we, collectively as blade smiths, had all been missing as we executed the pattern welding process. While studying the original Sutton Hoo sword radiograph I realised that I was not seeing a complete twist in the twisted portions of the bars. To clarify, when a layered bar of steel is twisted, the layers form a spiral, or a helix, through the centre of the bar. When viewed from the surface, the bar appears as a series of diagonal lines, but if you were to look *through* the bar, you would also see the backside of the twist, as an opposing set of diagonal lines which should form "X"s on a radiograph. The Sutton Hoo radiograph appears to show only one set of diagonal lines, leading me to believe that the back side of the twisted bars had been removed, possibly by splitting the bars in half lengthwise.

---

[1] Rupert Bruce-Mitford, *The Sutton Hoo Ship-Burial*, Vol. 2, pg. 307, British Museum Publications Limited, 1978.

19ii.1. Billets to make the rods. Most modern smiths using current iron alloys stack the metals in layers as pictured here. This is for the Mound 1 sword and hence it has seven layers. Low carbon iron forms the top and bottom layers. Photo Vince Evans

19ii.2. (left) Billet being formed into a rod using heat and forge work. The rod is welded into a whole and gradually shaped until it reaches the correct length and thickness. Photo Vince Evans

19ii.3. Billet rod formed. In this photograph the rod is nearing the required length and thickness. Photo Vince Evans

19ii.4. (left)The required twists are being added to a rod. Photo Vince Evans

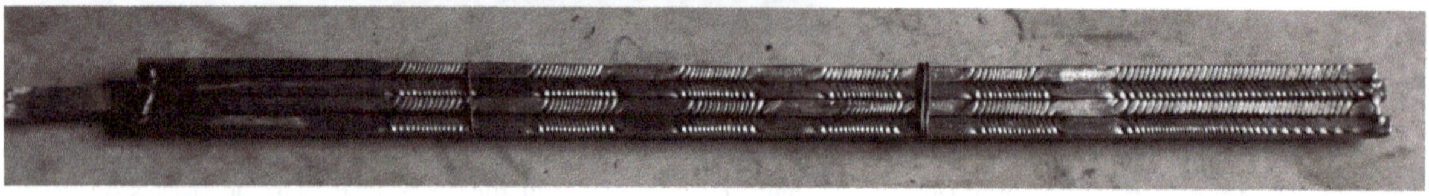

19ii.5. The rods have all received the required twists and counter twists and have been clipped together ready to be forge-welded into the core of the blade. Photo Vince Evans

19ii.6. The clipped together core has been placed in the furnace to get it ready for the hammer work

19ii.7. (below) The heated core is being hit with the hammer and the flux is being ejected under pressure

19ii.8. (bottom) The core is being forged into a whole by hammering while heated to the appropriate temperature

All photos Vince Evans

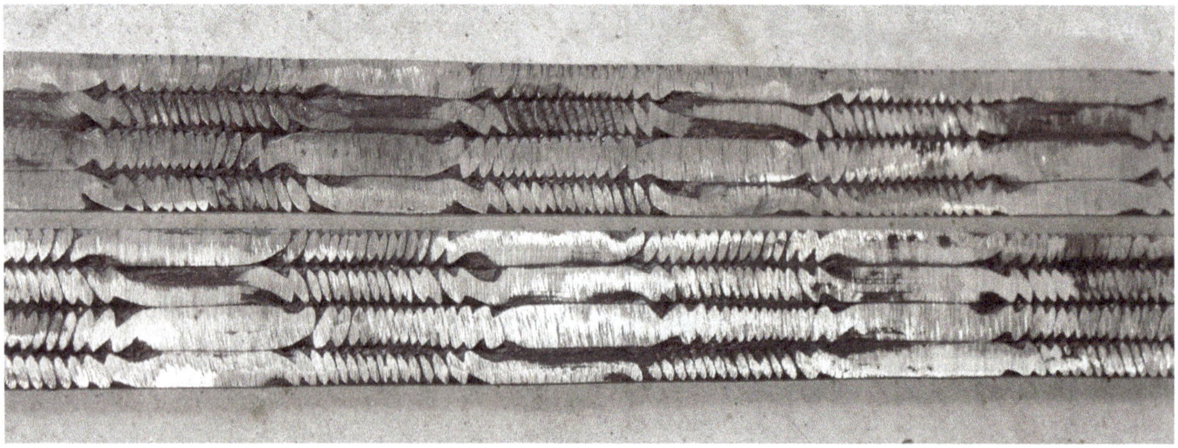

19ii.9. The completed core has been split and will be re-united and forge welded once the two pieces have been off-set. This particular process may not have been a universal practice; some smiths may have preferred to grind the extra metal away in order to achieve the half-rod thickness seen in many swords

All photos Vince Evans

19ii.10. The core is being forged back into one piece

19ii.11. The shape of the tip of the blade is being formed

All photos Vince Evans

19ii.12. The edge being formed and shaped with the hammer

All photos Vince Evans

19ii.13. Vince Evans test-fitting the edge to the core of the blade

19ii.14. The edge is being welded to the core

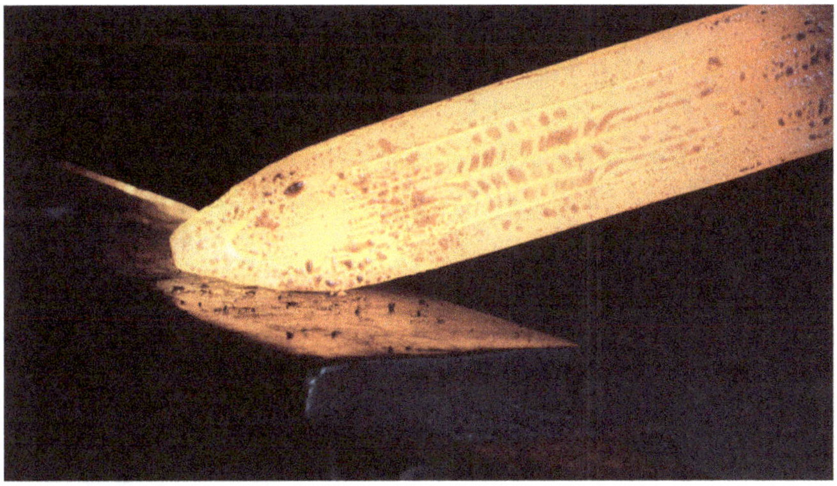

19ii.15. The edge-core welding nearing completion

All photos Vince Evans

19ii.16. (above) The complete blade being compared to the plan made at the beginning of the project

19ii.17. The shape of the blade is being refined in this picture

19ii.18. (top left) The grinding of the blade to the final required dimensions

19ii.19. (top right) Heat treating the blade using quenching

19ii.20. (left) The appearance of the blade after the quench

19ii.21. (below) Polishing the blade

All photos Vince Evans

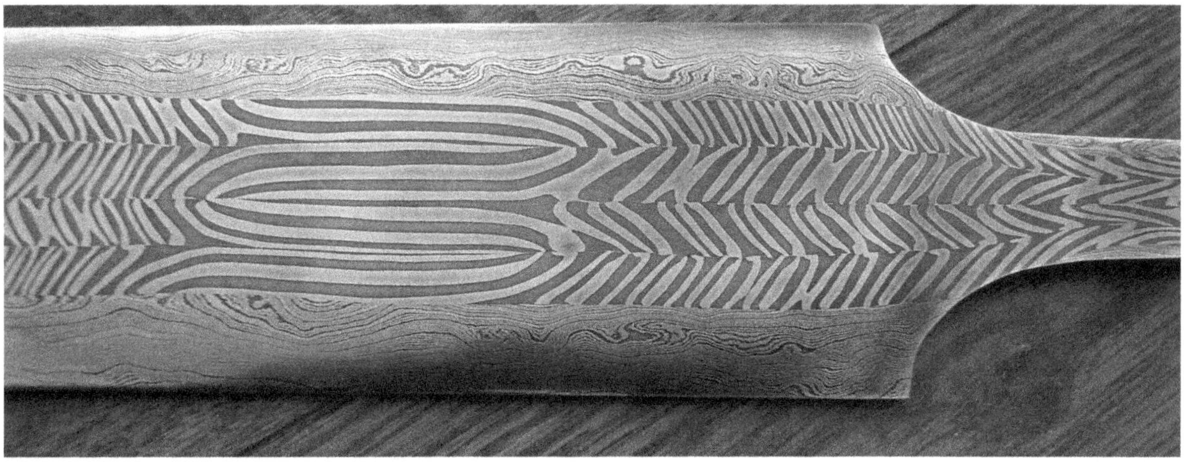

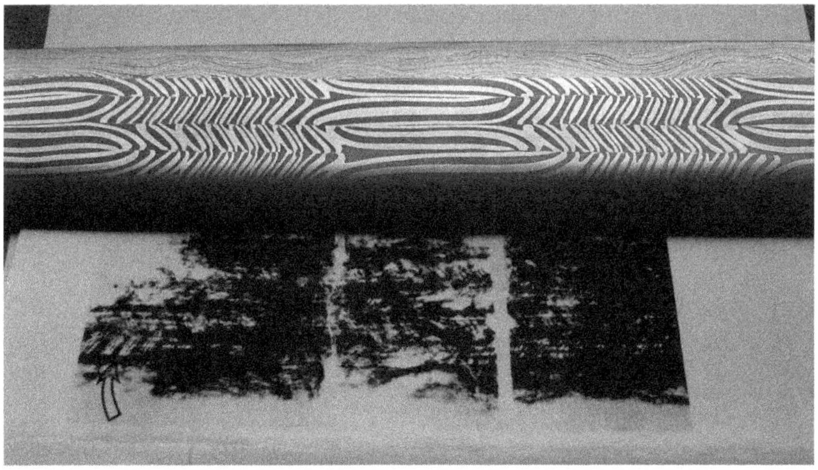

19ii.22. (top) The front of the blade near the tang end

19ii.23. (middle) The other side of the blade, note that twists on the front are backed by straights on this side

19ii.24. (left) Comparing the finished blade to the x-radiograph published in Bruce-Mitford's report

All photos Vince Evans

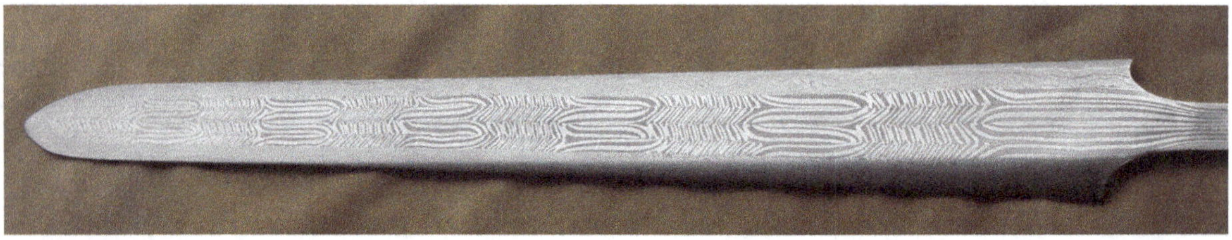

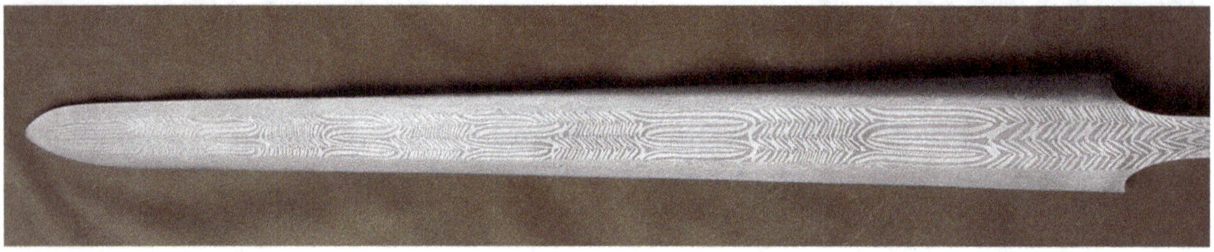

19ii.25. and 19ii.26. The two sides of the blade.

All photos Vince Evans

Scholars have suggested that there are swords which appear to have the centre cores split and inverted.

The process of hot splitting iron and steel is a common practice in blacksmithing. A piece of steel is heated, then taking a hot cutter (a type of chisel with a handle) and a hand sledge hammer, work your way down the length of the steel until a groove is formed. Flip the bar of steel over and repeat the process, making sure your steel is at the correct temperature for splitting. In this way, your bar of steel can be split efficiently. Since I was using modern alloys in my replica blade, I was not certain how the steel would react to hot cutting so I split the bar using a bandsaw. However, I did a test of hot cutting a bar of wrought iron of similar dimensions which produced similar results and was much quicker than using a bandsaw.

When a series of four rods are manipulated in the twist/straight sequence that the Sutton Hoo radiograph shows, you get the appearance of one side of the original blade. On the reverse side of that billet, you would have the herringbone pattern pointing in the opposite direction. It has been generally accepted that you would have to forge two sets of core rods for the pattern to point in the same direction on both faces of the blade.

Having forged composite pattern welded blades that have two full layers of twisted core rods, I have noticed that a problem arises as the blade is forged into the thinner cross sections. All that steel has to go somewhere. What usually happens is that the pattern gets distorted and lengthened as well as forced to the surface during the forging process. With minimal removal of metal from the surface of the blade during the finishing process, you will begin to see what many refer to as a "curving X" or "star pattern." These curving X patterns are not evident on the Sutton Hoo radiograph. What is seen in the pattern of the blade appears to be diagonal lines that form a herringbone.

By splitting one series of four rods lengthwise and inverting one side hilt to point and welding the rods back together, the results are the offset twist and straight series that the radiograph shows and the herringbone pattern points in the same direction on both faces of the blade.

Although, one cannot conclude that this was the method that was used to forge the original Sutton Hoo sword blade, it offers a plausible method that conserves both materials and labour as you are only forging and twisting four core rods instead of eight.

Procedure

The mindset that I chose to work in when I began this project was that I was the 6th century smith being commissioned by the king to create his special sword.

I started with billets that were seven layers of alternating alloys, four high carbon steel (1095), three low carbon steel with 3% nickel content (A203) to enhance the visual pattern of the blade. These billets were forged into long rods, approximately 1 cm square. Our goal was not to replicate the original alloys used in the Sutton Hoo sword but to work in the spirit of the original maker, using steels that I was familiar with on a daily basis, to achieve the vision of the completed blade.

When forged, straight sections of steel will lengthen at a quicker rate than twisted sections so to end up with equal lengths of twist and straight sections, you will need to layout your rods with a shorter straight section and longer sections for the twist. Based on test samples that were forged, I laid out my rods with twisted sections of 4.5 cm and straight sections of 4 cm.

The four rods are marked with dimples to distinguish between twist and straight sections. The dimples remain visible when the steel is red hot.

Four full rotations per section appear to match up visually to what is seen on the radiograph of the original. When twisting the rods, I worked both clockwise twisted rods at the same time. My twist sequence was two full rotations per heat. Limiting the twist sequence to two rotations per heat helps to keep the twist pattern uniform and reduces the possibility of shearing the weld should the steel cool too quickly. It is also easier to keep track of two full rotations rather than working four full rotations per heat. The sequence is repeated for the anti-clockwise set.

When the twisting procedure is completed, the bars are forged back to 1 cm square and annealed. The annealing, or slow cooling process, relieves the stress that builds up during the forging of the steel.

The cooled bars are laid side by side with one clockwise and one anti-clockwise beside each other to create a herringbone appearance. The four rods are bound together with iron wire and heated in the forge to welding heat. The hammer welding is done from one end of the bar to the other, usually point to hilt. All hammering is done on the edges so the rods weld together side by side. When properly welded, you should have a solid bar with the surface pattern visible on the faces of the bar. The bar is heated to critical temperature and left overnight to anneal.

This point in the procedure is where my method differs from what has been done in the past when doing this type of pattern weld. I split the bar of four welded core rods down the centre so that each half of the split had the four rods visible on its face. I then inverted one side of the core bars, hilt to point and bound it back together with iron wire in preparation for welding. The bars are brought back up to welding heat and the two bars are hammered together to form a single bar again. Once it is determined that the weld is sound, the point end of the bar is forged to a slight taper and rounded.

The next step was to forge the billet of pattern weld that would become the cutting edge of the sword. The billet was forged to 512 layers to create a fine layering which would not display much pattern in the edge. After the billet was forged into a bar the same thickness as the core piece, it was split most of the way down the centre, in a sort of "V" shape. While it was still hot, the core was inserted in the "V" and everything was forged to fit together tightly and bound in several places with wire. The unwelded composite now has a twisted core with edge bar running around both sides of the core.

Other methods that have been suggested for the cutting edge are the wrapped tip and the pinched tip. The wrapped tip method uses one long rod that is wrapped around the core bar, but this requires a bar of approximately 1.5 meters long. Working with a red hot bar of steel that long is awkward and may not leave enough material at the point of the blade. The pinched tip method uses two rods for the cutting edge which were pinched and welded together at the point. The Sutton Hoo radiograph did not appear to indicate this pinched tip method.

The final welding step is to heat the bound components of core and edge to welding temperature and to fuse the edge to the core. I strike the point first and work my way back towards the hilt with rapid hammer blows. It usually takes several welding heats as the assembled components are too long to heat everything in one session. Once the edge has welded completely to the core, I then begin at the point again and hammer both edges and face of the billet at welding heat, working back towards the hilt end. The purpose of this procedure is to be sure of complete welding of all of the different components that have been put together during the building of this blade. Now that the billet has been completely welded, the final shaping of the blade begins, forging to complete thickness and width. The tang is forged by narrowing the hilt end of the blade down to narrow proportions that will go through the hilt components. When satisfied with the blade proportions, the blade is heated to critical temperature, checked for straightness and annealed overnight. This concludes the forging process and the blade is ready for grinding, heat treating and finishing.

19.iii. The Bamburgh Blade

Vince Evans

Some thoughts on the forging of a Bamburgh style blade...[1]

I knew going into this project that it was going to be a complex blade to forge. I must admit that it was the most nerve-wracking blade that I have done. Trying to match the pattern on an original blade is more difficult than creating something unique as there are many unknown factors in how the original blade was forged. As the original Bamburgh blade has six core rods visible on each face, with each rod alternating the direction of twists down its length, there are numerous opportunities for error. One mistake would mean that you would have to start all over again which means a lot of material and many hours of labour lost. Only part of the original blade remains so we are speculating on the finished length of the blade and the number of twisted sections.

It was a daunting project with many hours at the forge under extreme heat to accomplish. We chose to work one rod at a time to minimize confusion and we marked each section to be twisted with two dimples as it makes it easier to see when red hot. A chart was made of the twisting sequence and checked off as each twist was completed to help keep track of what was done. Working with so many twists in opposite directions it is easy to get confused and twist the steel in the wrong direction. We found that the job was easier done with two people; one to hold the hot steel in place and keep track of the twisting sequence and the second to do the actual twisting.

Looking at a twisting chart with six rods needing to be twisted 13 times each in a set order and proportion so that they all lined up evenly when done was very daunting. Added to that complexity, the rods are welded to a thin iron core and an edge welded on to that.

I found that for my replica blade, the iron core provided structural stability for the six rods and this may be one reason the ancient smiths used the iron layer between the surface layers.

We allotted the spacing on each rod at 3 cm per twist section and 1 cm per straight section before twisting. One centimeter was the minimum space required for me to fit a tool for twisting. This spacing may still be too long when compared to the original.

The lengths of twisted and straight sections on my finished blade after the final welding and finish forging were as follows: 3 cm sections resulted in approximately 3.5 cm when twisted three full rotations, 1 cm sections resulted in approximately 3.5 cm straight sections. The 3 cm twisted section only lengthened 0.5 cm as that section compresses when twisted and doesn't stretch as much in the final forging process. A straight grain section will lengthen more freely when forged.

The resulting blade is quite impressive and is a testament to the skill and creativity of the smith who created the original Bamburgh blade.

[1] See figures 19iii.1 to 19iii.6 for illustrations of the reconstruction of this blade pattern.

19iii.1 The fragment of the Bamburgh blade

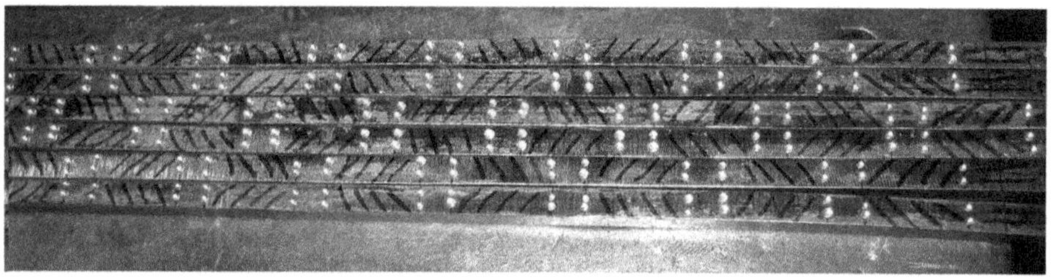

19iii.2 The rods marked for relevant treatments

19iii.3 The centre rods after twisting

All photos Vince Evans

19iii.4 A cross-section of the blade

19iii.5 Obverse of the blade showing the pattern

19iii.6 Both sides of the recreation of a Bamburgh style blade

Photos Vince Evans

19iv. The Vehmaa Sword Recreated

Jesus Hernandez

This is the story of recreating a sword blade

The blade began as a little seed in my mind after reading the book "Swords of the Viking Age" by Ian Pierce. As I turned pages 148 and 149, very much at the end of the book and having already seen many photographs of wonderful pattern-welding, I was stunned by two photographs depicting a sword found in Vehmaa and dating to the 8th century. My first impression on looking at the photographs was that I was looking at two different swords but on reading the brief notes I was shocked to learn it was only one. Ever since I saw that blade I have wanted to try and figure out how the pattern was created. That was many moons ago.

Three or four years before I actually picked up the hammer to make this blade, my interest was again raised after seeing some examples of Viking blades from the 8-10th century at a swordsmith's gathering near Baltimore. But before I set about the task it took me the best part of a year of on-and-off reviewing of my original notes on the pattern and in that time I developed new ideas about how to re-make the blade. At times I was aided by a 3D CAD program that helped me determine the composition, layer count and arrangement of the rods. Although I was able to figure out how the individual patterns for each section were made there were other problems to solve. I suspected that in order to make the original blade, the smith used the same rod albeit with different patterns along the length of the blade. This was simply not possible to do as some patterns required different layer counts and the different steel types needed to be arranged in a different order too. He certainly did not simple butt-weld the different sections as the transition from pattern to pattern is smooth and not abrupt. That had me puzzled for quite some time but one day when I least expected it, I came up with a plausible solution and even if I can't be certain that this how the original was made, I was happy enough with the plans to make some billets and get ready to test my ideas in the fire.

The types of steel that I used were 1095 and 15N20 for the core of the blade and roughly similar amounts of 1095 and W2 for the edge rods. All together I prepared about six kg of steel to begin with and started to run some tests.

The five rods that were to become the core of the blade needed to be twisted and manipulated to very precise dimensions. The day I did the twists I had two pages of carefully detailed directions to show where to start and stop each twist and how many turns to do for each twist. It was nerve wrecking as a single mistake would have thrown the pattern off. Everything needed to line up as close to the layout directions as possible, but it worked.

The next problem was to figure out how much material to remove from either face of the blade in order to expose the pattern as close to the original as I could get. This required quite a bit of planning as the metal needed to be ground to different depths in order to get the correct set of patterns in the right places on the different sections on each side of the blade.

The last weld was the most difficult because I had to attach the wrap-around edge to the core of the blade and this had to be done where the billets were just a quarter inch thick (6.35 mm); there was no room for error as any offset would simply ruin the carefully develop pattern.

Unfortunately, the first blade that I made did not show the correct patterns as a result of me miscalculating the depth. Although the blade was very interesting it was not what I had set out to do and so I started again.

The knowledge that I had gained from my first attempt was very useful and on the second blade I set the centre rods out carefully then welded them together and precisely worked out where the grinding would occur in order to achieve the correct depths.

Once the blade was completed, I took to profiling it on the grinder and then to heat treat the blade. The blade looked really pretty dressed in its tempering colours. (figure 19iv.16) Unfortunately, there was a little bit of a warping towards the point which I tried several times to straighten at tempering temperature over a couple days. It seemed that I had succeeded but when I returned to the shop next morning I could see that the warping had partially returned as the blade had cooled. It was such a little curve that I put it in the vice cold and tried to fix it by bending it to shape. The blade broke in two and all that work seemed in vain! But not all was lost, because this second blade had the patterns at the correct depth. The serpent section was not quite as tight as I wanted but everything else was right and I knew I was on the right track. Although I was a bit depressed for a day or two I went back into the shop the next week to make a new set of billets for a third attempt.

Talk about third time lucky, everything worked very well and the build was a smooth one. My previous experience with the two other blades helped me enormously and I had the sword I wanted with the correct amount of tightness for the serpent and although one of the chevron sections did not have the correct spacing, it was close enough.

I was not planning on hilting the blade or making a scabbard but I needed some sort of sheath for the blade so I made a wooden one and it looked so nice that I was compelled to finish it with a leather wrap and lambskin liner. At that point I just had to make a hilt to complete the project so I made one of those too.

The blade itself weighs 950 grams and the length without the tang was 32 ½ inches (825.5mm). Width of 1 7/8 inches (47.625 mm) at the base and tapering from there. Polishing the w2/1095 edge to show contrast was a little tricky and I ended up using a method similar to the one used by Japanese to polish a sword, which showed contrast between the 1095 and W2 steel.

I was very pleased with myself when the sword was finished!

My respect for the smith who made the original sword is enormous. I cannot be certain that I actually followed the same processes that he used to make his sword while I was making mine, but he knew what he wanted to do and how to do it. I am very grateful to all those who helped me realise this project.

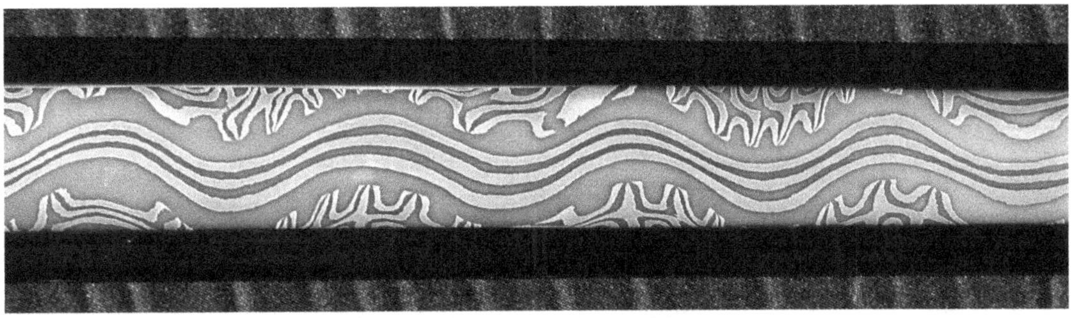

19iv.1. The 'serpent' in the blade

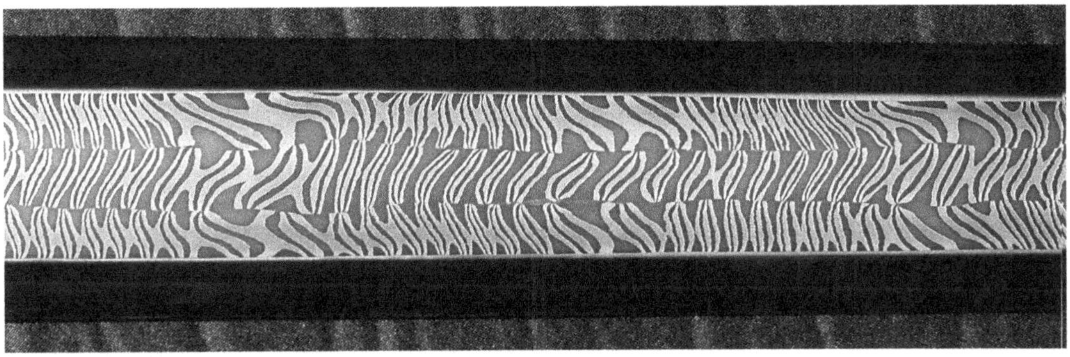

19iv.2. The pattern on the non-serpent side about a third of the way down

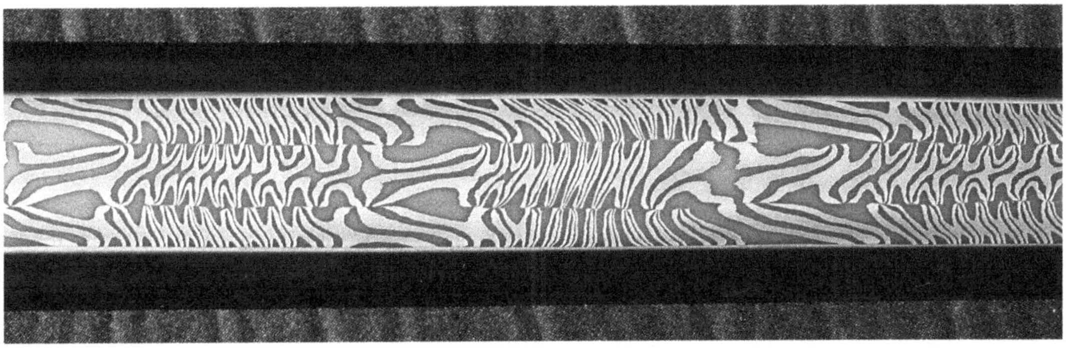

19iv.3. About a third of the way down on the serpent side

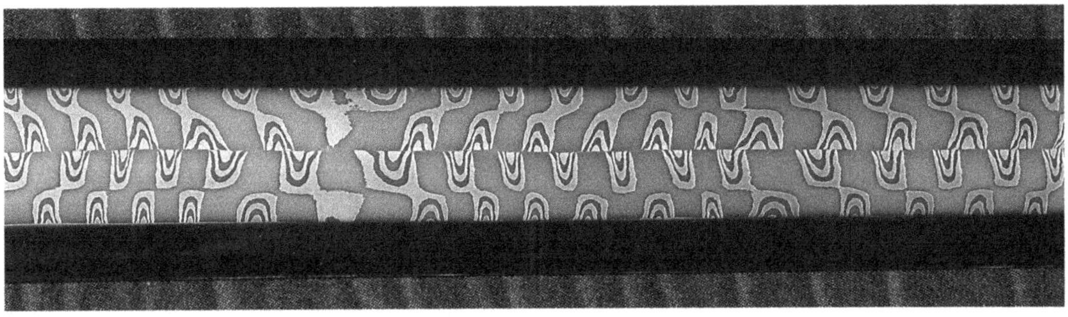

19iv.4. Two thirds of the way down on the non-serpent side

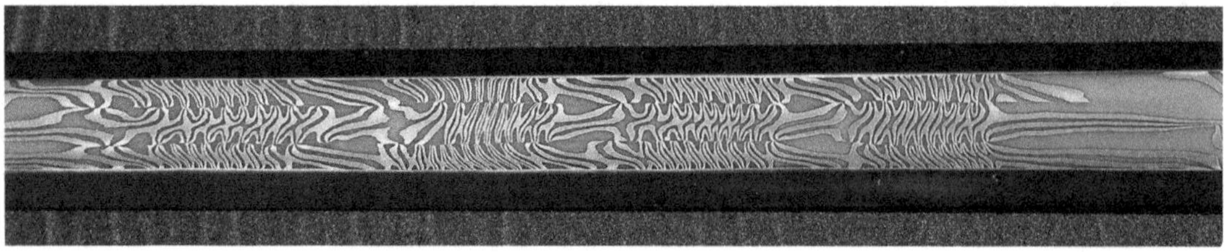

19iv.5. The mid-section of the serpent side

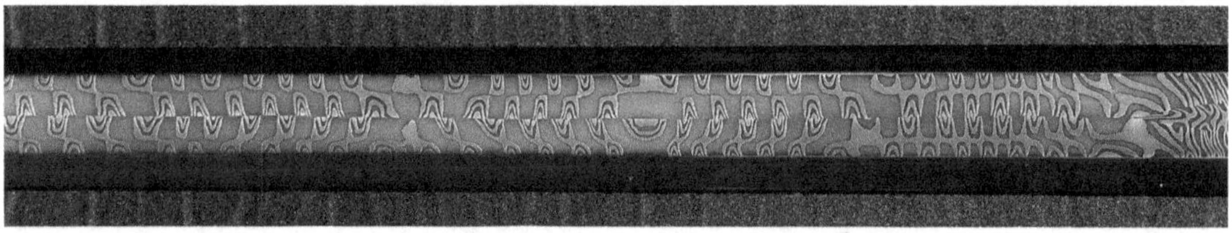

19iv.6. Mid-section non-serpent side

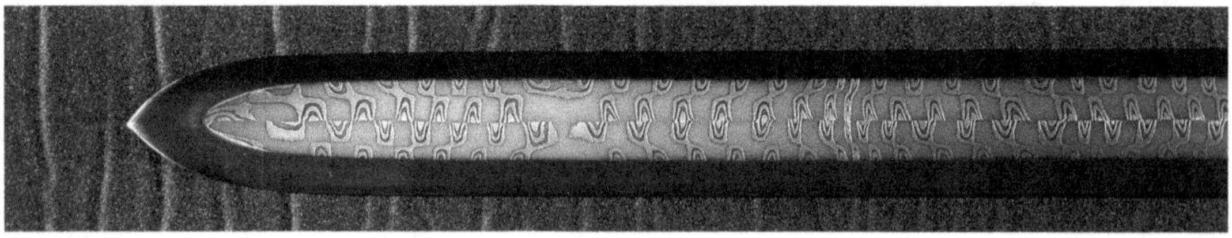

19iv.7. The tip end of the non-serpent side

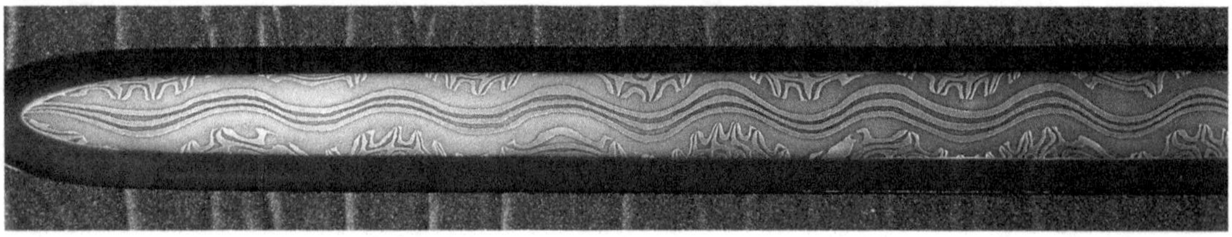

19iv.8. The tip end of the serpent side

19iv.9 The broken second attempt

19iv.10 The plates that will make the rods

19iv.11. Billets made up from the plates. Note that they are of varying thickness

19iv.12. The completed rods

19iv.13. Rods marked up for twisting and other processing

19iv.14. The 'serpent' can be seen in one of the rods

19iv.15. The one piece edge has been wrapped
around the core and the whole is ready for forging

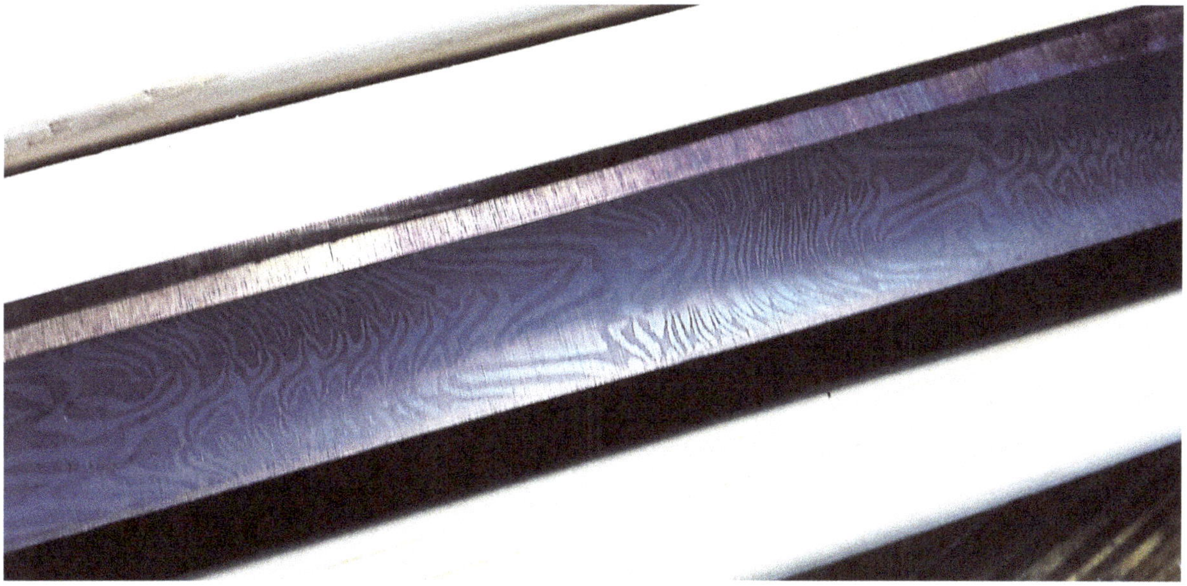

19iv.16. The blade in its tempering colours

19iv.17. The serpent side of the blade

19iv.18. The other side of the blade

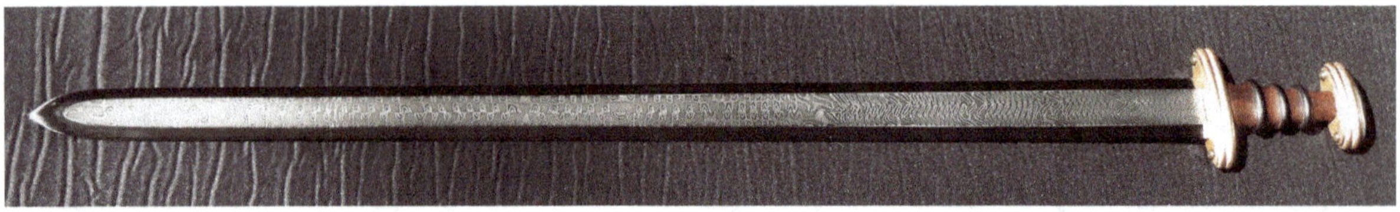

20i. The Sword as a Weapon

Paul Mortimer

"Historically, and to some degree even today in some circles, the popular view of medieval inter-personal and intra-group conflict was considered to be inefficient and brutal. In this light, such affairs were thought to be incapable of being anything but extremely vicious and untutored exchanges. However, in recent years, new research and experimentation has begun to question this hypothesis. Should the nature of inter-personal combat in medieval Europe matter; should this be the subject of study?"[1]

There are a number of major problem when looking at how swords of the 5th to 7th centuries were used in combat. Perhaps the first is that we have all seen far too many films and TV productions that show us how we should view sword fighting, whether it be knights from the later middle ages, gladiators in the Roman world, Vikings, Samurai or fantasy films, we have all been 'educated' and 'know' how sword fighting was conducted. Thrusts and parries, blocks, counter thrusts, cuts, great footwork and speed are all portrayed as the way to fight with a sword, and in fact the portrayal nearly always involves 'fencing'.[2] This view has been compounded by re-enactors and their demonstrations of how combat may have been carried out for real, hashing and bashing at each other, swords blocking other swords using the edge-to-edge technique, but everyone being very careful to 'fight' within the limits of health and safety and their own society rules. As Matzke comments, re-enactment sword-play, "…does little to dispel mistaken beliefs and can easily give an inaccurate view of the past."[3] The second problem is that no-one from the period was thoughtful enough to write a *fechtbuch* or fight manual for us to read, or, if they did, it has not survived. A third problem is the comments of some scholars and other commentators who, presumably because they too think they know how a sword should be used, have suggested that swords from the 5th to 7th centuries were clumsy or unwieldy and may not even have been intended for real combat at all.[4]

I would suggest that these ideas are all based on sets of false premises. Most film sword fighting, if it is based on anything, relies on techniques not necessarily from real fighting but from stage-fighting for the theatre, the object being to create or add to the drama, rather than an accurate show of real techniques. It also tends to use the type of swords from a much later period, the 15th and 16th centuries and usually involves 'fencing', with swords frequently striking each other blade-to-blade. It is questionable how much any of it resembles actual combat of that time or any other but comparing later swords to those of the 5th to 7th centuries is not realistic. It is a bit like (but only a little) expecting a WW1 fighter aircraft to perform like a modern fast jet; the two types of planes both fly but much else about them is very different, as are the way they are/were used in combat. It may be that the gulf separating swords of the 5th to 7th centuries from those of the 15th century

---

[1] Matzke, 2011. 99. Matze is referring to perceptions of combat in the later middle ages, but it applies equally to the perception of fighting in the earlier period.

[2] See also Matzke, 2011, (8.) for similar comments regarding the way that we have been indoctrinated by feature films and the lack of experience of academics. Incidentally, both Matt Easton (Schola gladiatorial) and Roland Warzecha (Dimicator) have carried out valuable work in this area. I am very grateful that Matt Easton has contributed to this volume.

[3] Matzke, 2011. 81.

[4] Gilmour, (2007. 91) quite reasonably asks, "Are we really so confident about what swords meant in Anglo-Saxon society as well as the manner in which they were used,…". Then, (Gilmour, 2007. 107) he says, "Possibly the decorated watery surface and (unnecessary and impractical) great length of these weapons…". Hill and Thompson (2003. 158) state, "The length and clumsy characteristics of these swords may reflect the physique of those using them." Ager too, as recently as 2015 referring to the work of others, suggested that some swords just had a 'ceremonial function'. (Fern, 2015. 123).

and later is of a quite different character to that of the aircraft that I have mentioned, in that the swords of the 5th to 7th centuries were not really primitive, in fact they were highly sophisticated weapons incorporating the highest levels of technology for their time and with a long history of use and development in Europoe and beyond. They were used in a style of fighting which only distantly resembles the way later swords were handled and used and has very little to do with stage fighting.

In real sword-to-sword combat the prime intention was never entertainment but to quickly despatch a deadly opponent, ideally with one blow, before moving on to the next opponent.

Swords and re-enactment

Before we proceed further and because re-enactment events are fairly common in the modern world, we will just spend a while discussing the use of swords in the portrayal of early medieval re-enactment, in order to be clear why such examples of 'living history' can be a problem. First of all I must make it very clear that I am not denigrating the work of anyone in this field, much of that work is of real educational value. I am just pointing out that for many reasons, including the practical and legal, in certain areas of re-enactment limits are placed on the use of weapons. This and the mental outlook of the participants render such events near useless for learning anything about historical methods of fighting. Such displays can be exciting and enjoyable and encourage both those watching and participating to learn more about the period and the use of weapons. However, they are a form of sport and a craft in themselves and hardly reflect the reality of medieval warfare. Some aspects of such displays can be interesting and possibly instructive, particularly when there are large numbers on the field and they are constituted in tactical units and moved around the field in ways to gain advantage. Participants can learn something about the handling of weapons and how they feel and work, but handling a re-enactment 'blunt' is quite different to using a sharp weapon. Most re-enactors are aware of all the shortcomings and are aware of the vast difference between re-enactment and the real thing. They know that they will soon return to their day jobs while real combatants risked their lives in the grisly experience of medieval field combat. Johann Matzke expresses it very well, when he explains that, "Although there is a long tradition of 'living history' and re-enactment groups putting on displays for the entertainment and hopefully the education of the general public, the primary goal of those groups is the promotion of a good show, i.e. entertaining the crowd. There is nothing in itself wrong with that as a goal. However, the general public is ill served by the inaccuracies and fallacies perpetuated by some organisations operating under the guise of 're-enactment'.[1]" Matzke goes further and says, "However, due to balancing the needs of safety and historical accuracy, the re-enactment combat often seen at these events is often grossly inaccurate and doing the public a disservice."[2]

Of course the limitations of re-enactment combat do not just apply to the use of swords, but to all weapons including spears, seaxes, axes and shields. Most societies ban the use of weapons being pointed towards faces, particularly spears, which have to be kept under control and kept low, or very high, even blunt re-enactment spears can cause really serious damage if used incorrectly. Many groups ban head strikes, too, and the only section of the body that can be aimed at tends to be the front trunk; the bit wearing armour.

The reality

As we will see from our brief study on the signs of trauma on the bones of actual battle casualties, the head was frequently the main target and having a spear pointing directly at your eyes would have been intimidating, and was meant to be.

---

[1] Matzke, 2011. 79.

[2] Matzke, 2011. 80.

The blade-to-blade combat shown at many displays was extremely unlikely and swords from the 5th to 7th centuries, and earlier, were not designed for that type of fencing.[1] One feature of the 5th to 7th century swords that makes them particularly unsuitable for fencing is that the lower hand guard was not designed to protect the sword hand from another sword that slides down the blade sword. This would have been important if a combatant tried to block an opponent's blow with a blade-block. The lower guard's main function, apart from helping to provide a secure grip, was to prevent the hand from slipping onto the blade, if the blade of the sword hit an obstruction, say a shoulder blade, while thrusting with the sword.[2] The main defence against someone attacking with a sword was to get out of the way or use a shield to stop the blow. Of course blade-to-blade parrying happened sometimes and there are existing blades that show evidence of it,[3] but it was not the first defence and would have been used rarely and reluctantly as it can damage the sword blade severely. It was not only swords from our period that were not designed for fencing; in fact it applies to just about all swords until the very late medieval period and perhaps even later until full hand protection was developed with the coming of basket type hilts and similar defences.

Many who hold a reasonably accurate reproduction sword from our period will remark on its lack of 'balance' or how 'heavy' it is; this is because there was no attempt to counter the blade's weight with a heavy iron pommel and due to the blade's edges being parallel or tapering only slightly inwards towards the tip. Once again this applies to almost all swords until the late medieval period at least. We have seen that there were different styles of blades,[4] mostly to do with their width but also the length, and these factors would to some extent have affected the feel of the sword but all swords so far found from our period were 'blade heavy', that is their mass feels as though it is concentrated towards the tip. The 'heaviness', I believe, is why several commentators have suggested that swords from this time were 'unwieldy' or 'clumsy'. In fact they are wrong, and as Loades says, "Swords like Raedwald's were never clumsy or awkward, they could be swung fluidly and with speed, but they did require a powerful sword arm to wield them effectively. Once launched at an adversary there is a terrifying weight and power behind the stroke."[5] A skilled swordsman from any period needs to be trained to learn the characteristics of his weapon and how to use it to its best advantage and to build up muscle memory so that he can act without conscious thought. He also needs to be fit to be able to perform at his best.

Sharp weapon combat at any time, including the 5th to 7th centuries, was fast and brutal, individual fights often concluded within seconds of their beginning; they were not intended to be entertaining or pretty. A warrior in battle would want to dispatch each opponent as quickly as possible and move on; weapons, and that includes swords, were designed to maximise blood loss and cause terrible tissue damage to an opponent; once hit with such a weapon a person would be in serious trouble even if they continued to live and fight for a while.

---

[1] The same is true for earlier swords and for a good deal later than the 7th century.

[2] Loades, 2010. 31. Loades is discussing the 'crossguard' (lower guard) on the Greek xiphos sword and Roman gladius, but it applies to the small lower guard on 5th to 7th century swords too. Loades remarks that it is the back of the hand that is really vulnerable and that was not really addressed until the introduction of closed hilts in the 16th century A.D.

[3] See Mäder, (2001. 183, figure 53.) for an example of a blade from, Germany, with a fragment of another sword deeply embedded in the edge.

[4] See chapter on *Blade shapes*.

[5] Loades, 2010. 86. 'Raedwald's' sword that Loades refers to is the Mound 1 sword from Sutton Hoo. Incidentally, commentators have also claimed that swords from the later medieval period, too, were clumsy, heavy and unwieldy, see, Matzke, 2011. 48 to 50. Matzke calls some of these assumptions 'bizarre' – and says that they do show a complete misunderstanding of the ways in which swords were used.

Is there evidence that warriors were trained?

Learning to fight with a sword would not have been taught in isolation, it would have been integrated with the use of other weapons and other skills, such as wrestling and fisticuffs, as well as general fitness and strength. To an extent we have already discussed this issue in the chapter entitled, *The background: Warfare and society, kings and the aristocracy and beliefs*, but we will return to it and extend some of that discussion a little further here.

To many in the modern world, the idea of training for martial arts would appear to be an oriental one, but this is not really so. We have already mentioned the fight books from the later Middle Ages and the renaissance which are clear evidence of such a tradition in Europe and that tradition would have extended way back into history and beyond, as long as a military existed, there was a need for young men who could handle themselves. Matzke believes that it was only the coming of the Industrial Revolution, first to Britain during the 18th century and then elsewhere, and the introduction of firearms on a large scale to the battle field that tended to cause those skills to fade away. Matzke argues that the need for a highly trained soldiery with many developed martial skills was replaced by the requirement to be able to drill, load and fire a musket.[1] Matzke also says of the Industrial Revolution that, "This drew a vast number of people from the rural areas where some martial activities had been kept alive (for example, singlestick, the often brutal competitions with wooden staves had been a staple of country fairs…..and put them to work for many hours of the day in factories, mills and mines. This had a hugely disruptive effect on the social patterns of rural Europe and left only the upper classes with the time to train in any kind of martial activity."[2]

Matzke believes that before then aspects of martial training were widespread throughout most of Europe during the medieval period. He elaborates, "…the documentary evidence suggests that the teaching of the martial arts in medieval Europe was a widespread practice and that it reached most social classes. The physical evidence is also highly suggestive of the idea that these martial skills were not the sole preserve of only the noble combatant. To the contrary, this evidence suggests that a significant percentage of the combatants engaged in intra-group and inter-personal conflict did so as trained professionals."[3] Throughout his study Matzke considers the evidence, mainly from contemporary records, and builds a convincing case that his suggestion that training was provided for youths from many parts of society was indeed the case. He mentions that the apprentices of both London and York were well known to purchase weapons and even to take part in riots and that by the 1540s schools with trained masters were found throughout England.[4] Matzke points to attempts to curtail these activities through the laws, but that they seem to have been largely unsuccessful and says that, "This behaviour demonstrates that, despite repeated legislation against the practice, a quasi-legal tradition amongst those of less than noble status continued for several hundred years"[5]

If Matzke's thesis is true, and I have no reason to believe that it is not, then it beggars the question of how far back this tradition of training young men from different sections of society reaches, and has implications for our period. Indeed, Matzke wonders the same thing, too, and thinks that he has found evidence of this same tradition in *Beowulf* . When the hero, Beowulf is struggling with the monster, Grendel, Beowulf achieves an unbreakable arm lock:

---

[1] Matzke, 2011. 50 to 51.
[2] Matzke, 2011. 51.
[3] Matzke, 2011. 7.
[4] Matzke, 2011. 42.
[5] Matzke, 2011. 42. See also pages 55, 56 and 97.

> Instantly that monster, hardened by crime, realized that never had he met any man in the regions of the earth, in the whole world, with so strong a grip. He was seized with terror. But for all that, he was unable to break away. He was eager to escape to his lair, seek the company of devils, but he was restrained as never before. (Beowulf 750 to 756).

Eventually, after failing to break Beowulf's arm grip, the monster escapes by leaving his arm and shoulder behind and eventually dies because of it. Matzke sees the use of such a common grappling technique by Beowulf as clear evidence that training in these arts would have been widespread in the earlier period.[1] The audience hearing the poem would have recognised the technique. The dating of the writing of Beowulf is still somewhat contentious[2] but perhaps it does not really matter in this case because if Beowulf was reckoned to be using a recognised wrestling move from an earlier period it demonstrates continuity. Matzke provides other literary examples but they tend to be later than the one from Beowulf.[3]

However, as Matzke comments, "Any attempt to reconstruct the martial techniques of Europe prior to the blossoming of evidence in the 1300's remains highly suspect in light of this absence of any written evidence and scarcely any pictorial evidence."[4] So we must exercise some caution but also some common sense, we know that societies were highly militarised during the 5th to 7th centuries and we know that there was a demand for fighting men and that it was a way to advance in society, so it is highly likely that many youths received training and it has always been the way of many young boys and young men to play-fight and seek adventure throughout the ages.

---

[1] Matzke, 2011. 37 to 38.

[2] Various dates have been suggested from the 8th century to 10th. See Newton, 2004 and Neidorf, 2014.

[3] Matzke, 2011. 38 to 40.

[4] Matzke, 2011. 60.

20ia. How early Anglo Saxon swords were used

Matt Easton

Introduction

When approaching the question of how early Anglo-Saxon swords were used in combat, there are three primary underlying contextual factors which must be remembered.[1] First, the sword was usually not a primary weapon in most types of warfare throughout the early medieval period. That place was instead occupied by the spear. The sword is certainly a better weapon in some specific situations, for example, when pressed very close in a melee, when fighting inside buildings or other enclosed spaces, or when pursuing fleeing opponents. It was also convenient to wear as a sidearm – as a secondary weapon when the spear was lost, broken or the environment made the spear inconvenient. However, the spear was usually the primary weapon, even where swords were available and worn as sidearms, throughout various periods and regions. The spear has greater reach than the sword and is able to attack around, above and below the opponent's shield, from a distance at which the swordsman can only defend and try to close the distance with their adversary. The thrust of a spear, whether in hand or thrown, is very powerful and can easily penetrate thick clothing and some types of armour. The advantage of the sword is really in the close fight, and the fact that it can be worn and carried anywhere easily (unlike the spear). It should be remembered that the sword and shield would often have to fight against the spear and shield in actual combat applications, whereas a lot of modern reconstructions have focused heavily on swordsman versus swordsman.

Secondly, in this period, and indeed for a long period before and after this time, the sword was used almost always used in conjunction with the shield. All ideas of single sword use should be set aside in the realisation that the sword or spear was used as part of a cooperative weapon set, together with the shield. Only in some unusual self-defence situations, when the shield was not being carried, or when the shield was broken in combat, might the sword be used alone. The shield was a primary weapon in combat, and must always be accounted for in analysis of combat in the early medieval era. Shields drastically change how the weapon in the other hand is used, making some things possible and other things impractical. The use of the shield fundamentally dictates the combat style, from how attacks are delivered and the stances used to stand in, to the footwork employed by the fighters. The shield was used collaboratively with the sword and spear, offensively and defensively, and whatever fighting systems were employed in this period, it is clear that the weapon set should be viewed as a whole unit.

Lastly, and perhaps most importantly, we should recognise from the outset that we only have the most basic details of how early Anglo-Saxon swords may have been used. Thus we have little on which to base theories about the individual fighting systems of this period. This is true of all personal combat systems up until the first fencing treatises appear in the 14th century. Indeed we know very little about how weapons were precisely used until fencing treatises first appear in the late-medieval era. The written, artistic and archaeological sources for combat in the Anglo Saxon period, such as they are, do not furnish us with anything like the level of detail that we have from the later medieval explicit instructional fencing treatises,

---

[1] In the illustrations Matt is using a later sword from the Viking Period. However, the basic characteristics of swords remained similar from at least the 5th century until, at least, the 10th. This longevity would seem to indicate that the design was a successful one.

which describe minute details of physical movement and martial arts theory. Even the sagas, poems and other sources of the later Anglo Saxon and Viking era only furnishes us with very basic descriptions of heroic combat.

Later medieval combat treatises can certainly be used to help inform our opinions and experimentation with earlier periods, but we should always remember that they are much later in date. And even when those treatises were being written there were diverse methods for the use of those later weapons, with styles varying between individuals and cultures. There are late medieval and early renaissance treatises detailing techniques for using the sword and shield together, but those swords and shields are very different to those used in the early Anglo Saxon era and the context is very different, even if some underlying martial principles always remain constant. Even where we can say some things about early-medieval fighting styles, we should remember that wherever martial arts styles are developed, there are variances between and within different styles. For example, one Mercian Thegn may have used his sword and shield in a slightly different way to another Mercian Thegn. For historical comparison, the sword and buckler was an extremely popular weapon set for late-medieval and early-renaissance civilians and soldiers for over three centuries right across Europe, with a lot of treatises being devoted to that weapon set's method of use. Yet we can see that there were varying systems used across the period by different teachers and treatises, they did not all use the same weapons in the same way. It would be highly dubious therefore to assume that early Anglo-Saxons all used their weapons in exactly the same way.

However, these cautions accepted, we can look at the weapons themselves and ask expert practitioners how they would best use them. We also have archaeological evidence of the wounds caused by weapons and we have contemporary texts. Our knowledge of later martial fighting systems, which used fundamentally similar weapons, enables us come up with some plausible estimations of how early Anglo-Saxons swords might have been used in combat.

Sometimes we can infer details of weapon usage from specific details in weapon design, as well as differences to the weapons of other areas and times. Examples of this, which we can verify across a range of historical primary sources, are the Indian sword called 'tulwar' of the 17th to 20th centuries, where we know that the particular hilt design was employed with a special form of cutting action, or the development of the early-Renaissance Italian sword hilt, where we know that finger-rings developed due to the more common practice of looping the finger over the front guard (quillon). Certainly there are details on Anglo-Saxon swords which may possibly indicate how they were used.

Let us therefore consider an overview of the early Anglo-Saxon sword in greater perspective, as a sword compared to other historical swords. There are some strong characteristics which immediately jump out to someone familiar with using swords. We find a relatively short and confined hilt, a distinctive 'H' hilt shape, a broad and flat, occasionally fullered, blade, and a point of balance often relatively far from the sword hand.

Anglo Saxon sword hilts

Let us first consider the exact form of the upper and lower guards, which more or less form a capital 'H' shape. This is distinctive to European early medieval swords and must again inform the conclusions we make regarding their use in combat. This hilt type is different to most late-Roman swords, where the pommels are usually close to spherical or rounded. While the grips of both types are equally short, the shape of the pommel does not seem to lead to the same ways of gripping and moving the sword as with the Germanic weapons.

The hand guard

Sword guards are there to protect the sword hand in a number of ways. Most people would think of the protection offered from an enemy weapon, either by a direct attack to the hand (a disabled sword hand would usually result in the swordsman being quickly overwhelmed) or by an opponent's weapon sliding down the blade and onto the hand. However the sword guard is also there to keep the hand securely seated in the hilt, preventing it slipping up onto the blade during heavy impacts. Projecting beyond the height of the knuckles also protects the hand from painful or injurious collisions, such as with shield or other hard bits of equipment. A broken finger or knuckle is no small matter when trying to wield a weapon effectively and there is no solid evidence of protective gloves being worn in this era. On the contrary, the weight of evidence suggests bare hands were the norm. Looking at historical accounts from later periods, and indeed from modern conflicts and crimes, it is clear to see that hand injuries are both common during close combat with bladed weapons, often leading to incapacitation.

The shortness of the guards on early-medieval (and earlier) swords is probably largely indicative of the way the sword and shield were almost always used together. Swords to be used with shields have a different set of requirements than swords designed to be used primarily by themselves and equally the types of shields used can change those requirements. Swords generally don't require much of a hand guard when used with a large shield because the sword hand is usually protected by the shield, as the line from the opponent's weapon to the swordsman's sword hand is usually blocked by the presence of the large shield.

Longer hand guards started to develop in the late 10th and 11th centuries, but in the early-medieval period obviously were not seen as necessary. This might indicate that early-medieval boss-held shields were predominantly used with an extended arm (which is supported by much of the available art work), much like later bucklers, which provides a very good level of protection to the extended sword hand and indeed the whole arm, by geometry. As the sword arm is extended, so is the shield arm and the shield covers the vulnerable line to the sword hand and arm. If, in contrast, the sword arm is extended and the shield arm is kept close to the body, the sword hand and arm become more vulnerable as they are exposed in front of the shield with nothing to protect them, unless armour is worn on the arm. At the later period when longer sword guards started to become common we know that the strap-worn tear-drop shield was also coming into vogue. This cannot be held as extended from the body as the boss-gripped shield can, and could explain the requirement for improved hand protection in the same era. There are probably many more factors involved in the change, and the process and motivation for change is probably fairly complex.

Another factor worth mentioning in conjunction with the short hand guards of early-medieval swords is that, as mentioned, the spear was the predominant battlefield weapon of this era, so sword and shield against sword and shield combat might not have occurred with much frequency. Against a spear and shield the hand guard of the sword doesn't really play very much of a role, as the hand is more vulnerable to cuts than thrusts (the spear primarily thrusting when used one-handed). It is also worth noting that even when swordsmen did fight each other, if both were armed similarly then the relative advantages and disadvantages of their weapons would cancel out. At a time when all European swords had minimal hand guards there was little motivation to drive change. In any case, as we have seen, the use of large shields should be considered the principle reason for small hand guards on swords being regarded as adequate in the early Anglo-Saxon era and for many centuries after.

One specific consideration on early Anglo-Saxon hand guards is that many of them were made partially or entirely of organic materials such as wood, horn or bone. This was also true of many late-Roman swords. The use of such materials suggests that the hand guard on these swords was primarily there to prevent hand damage from incidental collisions with shields and equipment. They also kept the hand secure on the hilt. If they had been primarily to protect from the direct or deflected attacks of other swords then they would have been made of iron or other metals at an earlier date. It seems clear that the main thing protecting the swordsman's hand and arm at this date was the shield, not the guard of the sword.

Before leaving the subject of early Anglo-Saxon sword hand guards, one final detail is of relevance. Many early Anglo-Saxon guards feature raised rivet heads which hold together the sandwich-construction. On first sight it might seem that these would rub against the fingers in an uncomfortable way, but it seems from handling accurately-made replicas that if the sword is gripped in the ways described below, this just does not seem to happen.

Handles

The handles of early Anglo-Saxon swords that we have documented, vary in shape from relatively straight-sided, to handles with a narrowing 'waist' in the centre, flaring at top and bottom, to handles with ridges and finger grooves somewhat reminiscent of some Roman swords. In theory, the handles should tell us a lot about how these swords were held, but actually they vary enough and have so little impact on how the sword is actually held and used, that they don't provide us with very much definitive data on sword use at all. Or rather, it's not possible as of yet to draw any such conclusions. What we can say more definitively about the handles of early Anglo-Saxon swords is that they are generally very short. This, in conjunction with the T-junction of the handle and pommel, gives us information to work with.

The grip length is comparable to later Viking-era swords, as well as many late-Roman swords. All of these types of sword hilts share the basic characteristic of being relatively small compared to an adult man's hand, and depending on how exactly we hold these hilts, they pretty well confine the hand and seem to inform and restrict how the sword is exactly used. As an ethnographic comparison, the length of Indian sword hilts on tulwar and khanda is in the same size range as early-medieval sword hilts and we know from 18th and 19th century sources that this was to suit the particular style of use of those swords, both encouraging a specific style of cutting and maximising the effect. The restricted size of early medieval sword hilts is so conspicuous when comparing to other European swords from across history, that it suggests the way they were used must in some way have been connected to the size and shape of the hilts. It is too notable and specific a detail to ignore.

Pommels

Mirroring the hand guard was the pommel. Unlike on later period swords, it seems that early Anglo-Saxon sword pommels were not there to provide a counterbalance to the weight of the blade. Early Anglo-Saxon pommels tend to be relatively light, and in any case with such short handle lengths the weight added at the pommel end has comparatively little balancing effect on the hand.

What is most notable about pommels of this era is that they form a T-junction with the bottom of the grip. As mentioned above, this means that specific ways of gripping the sword must be used. The pommels of early Anglo-Saxon swords probably tell us more about how these swords were specifically used than any other constituent part, though they must of course be considered as part of the whole and not in isolation.

We must consider why these pommels are nearly always created to form this T-junction. It is not conducive to a normal way of gripping and swinging an object in the way most people would apply to a stick, or indeed other kinds of sword.

How to hold early Anglo Saxon swords

Given all of the above and wanting to establish how early Anglo-Saxon swords might have been held, we can be confident about a few things. Firstly, simply holding the handle in a so-called hammer fist and swinging the sword from the pivot of the wrist, elbow and shoulder, as most people would swing a stick, is not comfortable or very effective. Trying to swing this type of sword in that way results in the user getting a pommel bar painfully rammed into their own hand/wrist. In addition to this, the so-called sabre grip with the thumb up the back cannot be used because the hilts are too short. There is also no good evidence for the sabre-grip in any of the illustrations of the period. Some people have argued for the index finger to be looped over the front of the guard, essentially moving the hand further from the pommel and creating more space on the handle. There are a few objections to this in general, the principal one being that the majority of pre-1000AD art shows swords being gripped with all the fingers and thumb around the handle of the sword. However, there are a handful of artistic representations from the period where the index finger does seem to be extended either over the front guard, or at least upwards from the handle. Also, the edges of these swords are usually sharp all the way down to the guard, not having anything like a later 'ricasso', so the index finger would not usually be very secure in this position.

So, the above being the case, and emulating what is most commonly shown in illustrations of the time and in later Anglo-Saxon, Frankish and Italian art, let us wrap all the fingers and thumb around the grip of the sword. Instead of adopting the hammer grip, as used with the disk-pommelled Indian tulwar to facilitate the famous drawing cut, let us angle the grip into what some people have dubbed the 'handshake' grip. In this, the grip of the sword lies diagonally across the inside of the hand, rather than straight across it. If we now slide our hand down to the lowest point on the handle, so that the little finger sits snug in the crook between the handle and the front projecting bar of the T-junction pommel, the rear of the T-junction extends flat against the inside of the heel of the hand, at the bottom of the palm. This now feels comfortable and when moving the sword in a cutting motion the rear extension of the T-junction simply slides back and forth against the palm or heel of the hand. While we cannot say with certainty that this grip was universally used with these swords, or indeed at all, it does work and does make some sense.

The short hilt must have offered some advantages in order for it to be the universal design across much of Europe for several centuries. Indeed when holding the sword in the way described above, it becomes apparent that there are some benefits to it. It helps with edge alignment, which is important when it comes to cutting effectiveness. A sword blow that lands with bad edge alignment just doesn't cut well, if it cuts at all. This grip also seems to improve weapon-retention. That is, it makes it less likely that the weapon can be dropped through fatigue, accidentally, or from a blow, as the little finger is locked into the front of the T-junction. Lastly, it places the hand right at the end of the swung lever, which makes for very powerful cuts. It is not a grip that assists a very nimble or versatile use of the sword but it is a secure grip which produces very effective and powerful cuts. This perhaps fits in the context of when and where it was used; almost always used with a shield which made nearly all the defensive actions while the sword acted almost entirely as a weapon of offence.

Finally, it must be said that it is physically possible to use these swords in a simple hammer grip. (see figures 20ia. 1 & 2) However, study of other ethnographic weapons shows that a peculiar feature of design usually ties to a specific method of preferred (if not universal) use. If we use these swords with a regular hammer grip it hurts the hand. Indeed, our ancestors probably had tougher hands than most of us but that still would not explain why the T-junction pommel was so prevalent for so long; it must have served a purpose beyond hurting soft hands. However, before leaving the subject of the hilt and how to hold it, the caveat should be added that in all forms of swordsmanship it is normal for the grip to constantly shift slightly on the handle to suit the moment. For instance, it is common to change the grip slightly between cutting and thrusting and in the context of sword and shield combat it is quite common to thrust around the edge of an opponent's shield. In doing so, the so-called hammer grip is often necessary to get the required angle. In the case of thrusting there is no discomfort caused by these pommels when using the hammer grip, unlike with the swinging action of the cut.

The blade and balance

Early Anglo-Saxon sword blades vary in length, but are relatively long for the period, being comparable to many other one-handed swords from later centuries. They are generally double-edged, broad, quite flat (of various cross-sectional shapes), with parallel or close-to-parallel edges, ending fairly abruptly in broad points. At first sight and handling this suggests an emphasis on cutting, which is supported by the point of balance usually being far from the hand. Such a balance reduces point-control, but increases striking power.

This is not to detract from the thrusting potential of these swords though and unlike other ethnographic cut-centric sword types, such as the 'khanda' from India or the 'jimpul' from Borneo, the early Anglo-Saxon sword was pointed and this point would be very effective at thrusting through all sorts of material. With the increased focus on cutting power, the ability to use the point was not abandoned as it was in other sword designs. This suggests that the point was utilised in combat at this time. Having said that, we know from historical accounts that despite Afghan 'choora' having sharp tapering points, they were rarely ever used for thrusting, so form does not always inform method of use. We can surmise that the point was used to thrust in early Anglo-Saxon combat and later English art would support that, but it is likely that their fighting style was heavily focused on cutting. The shape of the point, with a broad blade to near the point and then a quite sudden gradient to the point itself, also assists cutting at the extremity of the blade. Broad points like this can cut effectively nearly up to the end of the blade, almost to the point itself. Narrower, more tapering blades, seldom cut well near the tip due to a variety of factors. Therefore a longer tapering blade often has no reach advantage in giving cuts over a shorter, broader blade, as the effective cutting portion of the broader blade can be further from the hand.

Lastly, before considering how the sword and shield were specifically used in combat, armour should be mentioned. In this period it is likely that the sort of person who could afford a sword would often have head and torso protection, though evidence for limb defences is very scant. Swords cannot under normal circumstances cut through metal armour such as a mail byrnie or iron helmet. When such things are described in sagas and poems it is usually in an extremely heroic context. The same is true in later medieval texts and art with Biblical representations. This was the superhero fiction of the time. In normal combat, if an area is well armoured then the person cannot be directly wounded with a sword there (though they may be bruised, bashed and concussed!). The most effective way to use a sword against an armoured opponent is to target the areas that are not armoured. We should also remember that while sword-owners at this time may have had protective armour, many of their poorer opponents would not. However, even where armour is not present, thick clothing and certain types of headwear can provide an effective defence against cutting, thrusting and impact. For instance, during the Battle of Balaclava in 1854, many cavalry troopers escaped with minor injuries because they were protected from sword blows by their winter clothing and uniforms. It is surprisingly difficult to cut through some clothing with swords, regardless of how sharp they are. Turbans in India and even the feathered hats of Highland Scots are documented as turning aside sharp sword blows.

fig. 1 & 2 (below)

Figures 20ia

figs. 3 & 4

figs. 6 & 5

Figures 20ia

figs. 7 & 8

Figures 20ia

figs. 9 & 10

figs. 11 & 12

Figures 20ia

What early Anglo-Saxon sword and shield combat looked like

If we accept the probable features of early Anglo-Saxon sword use described above, which is that the sword was primarily a cutting weapon with capacity to thrust, and held in the way described above, then we can start to look at how the sword and shield might have been used together.

All fencing and martial arts systems are built upon fundamental principles that are common to all combat. Essentially these are timing, distance, geometry and psychology. These are complex topics that are outside the scope of this article. However, we can get an impression of what sword and shield fighting probably looked like at this time by describing the likely movement and positions of the fighters.

In most shield-based systems, whether Greek hoplites, Roman gladiators, Norman knights or Renaissance targeteers, it was normal to stand with the shield-side presented forward most of the time. (See figures 20ia.3 & 4) This means the left foot and shoulder are forward for a right-handed swordsman. This is quite natural to most people who pick up a shield and weapon. When moving around, the fighter can pass feet, as in normal walking, or can advance, retire, turn and traverse with the shield-side foot staying in front, this would be the left foot for right-handers. The problem with bringing the weapon-side foot in front is that it brings your shield further from the opponent, where it is less able to protect, as well as bringing your vulnerable weapon-side forward into an exposed position. Standing with both feet and shoulders at the same distance from the opponent is not generally advisable for many martial reasons. For example, it makes you into a large target and if the opponent charged into you, there is no backwards stability to resist falling on your back. Having one foot in front of the other is therefore advisable and seems to be supported by the historical artistic evidence. It is also generally shown in fencing systems to keep the knees, or at least one of the knees (usually the front one), slightly bent when 'on guard' so that you are ready to move in any direction.

The shield itself should be held in the fist (all early Anglo-Saxon shields were boss-held as far as we know) and in single combat the front edge should be held forwards usually, protecting the 'outside' line. This type of boss-held shield is unstable in the hand; it can pivot in two directions even from fairly weak impacts. If the shield is held face-on, then the opponent can easily strike either side of the shield and the shield will just pivot, creating an opening to the attacker. The defender can use this feature to create deliberate openings and opportunities of their own to counter-attack, however holding the shield face-on is generally a risky way to try and use the shield unless it is curved like a Roman 'scutum' or conical/convex like later Anglo-Saxon and Frankish shields, or indeed Renaissance 'rotella'. As these shields are not flat they don't pivot and leave a big opening. With the front edge of the shield leading and angled at about 45 degrees to the body, the user is fully protected on the 'outside' line and this position is much more able to resist the exploitation of the easily-pivoted boss grip. By now extending the shield arm to varying degrees, you also have a long reach to your defence. The fist being in the centre of the shield, you are able to reach a long way with the leading edge of the shield and this can be used to defend with distance and geometry quite easily, as well as offensively to strike the opponent and their shield. Importantly it also confers far greater protection to the legs than when the shield is held close to the body. When held right against the body the shield can only protect what it covers, but by extending the shield arm the coverage provided by the shield becomes far greater. Legs are a very vulnerable target in unarmoured shield-fighting systems, so using this geometry makes a lot of sense and seems to be supported by historical art sources in most cases. We also see these principles used in later detailed documented shield fencing systems, such as sword and duelling shield as shown in Talhoffer and Kal's treatises, or sword and shield in Manciolino's and other Bolognese treatises.

If we take the above as generally correct, we can now look at plausible positions we'd expect to see our early Anglo-Saxon swordsman in. These positions, or 'guards', are starting and ending positions, as well as transitional points. They are reference points for framing a movement system that makes sense with certain weapons, in a given context. We have our right-handed early Anglo-Saxon warrior standing with his left foot and shoulder forward, his shield in the left fist with the left arm extended to varying degrees. His only exposed parts are his head and his left leg. If either is threatened they can be avoided with distance, or by moving the shield slightly up or down as appropriate, or indeed with the sword if desirable. The opponent now needs to open our swordsman's defence with feints (false attacks), or by physically moving the shield if they want to successfully land a strike.

If we look at other systems of sword and shield, or sword and buckler, use from other periods then we see certain positions to hold the sword were common across time and region. We see some of these positions shown in later Anglo-Saxon, Italian and Frankish art, so it is highly likely we would have found similar positions in the early Anglo-Saxon world. There are simply only a limited number of sensible guard positions to use with a sword and shield or buckler. The use of the shield to a very large degree limits where you can reasonably hold the sword in preparation to defend or attack. Probably the two most common positions being with the sword hand near the right hip and the point towards the opponent, with the tip near the inside (See figures 20ia 7 & 8) and to the right of the shield, and a position with the sword hand raised above the right shoulder, with the blade pointing upwards and sometimes backwards, loaded for a downwards cut. (See figures 20ia 5 & 6). These two positions are shown in a lot of non-technical art and are in many fencing treatises. Other positions that are quite common are (i) with the sword extended down to the rear, like a tail, (ii) with the sword pointing down at the ground in front, (iii) with the sword hand under the shield arm to the left and the sword pointing backwards, in something like the scabbarded position, and (vi) with the sword hand over the shield arm and the sword pointing backwards to the left.

From any of these guard positions, the swordsman would be looking to cut or thrust at any available openings. To create openings in the opponent's guard position through pretend attacks (feints) or manoeuvring, or to physically create openings into which to attack, for example by bashing the edge or boss of their shield into that of the opponent, particularly trying to exploit that pivot point in the boss-grip.

It should also be noted that there are many ways to defend with both shield and sword. For example, a swung sword cut can be received on the flat face of the shield, or the edge (which is more likely to result in the shield breaking), however the edge of the shield can also be used to intercept the sword arm if it is exposed - this both defends and offends at the same time. Equally, a sword at this time could have been used to defend in a number of ways, either by blocking an incoming weapon, or cutting at the incoming sword hand/arm if it were exposed beyond the cover of the attacker's shield. Sword cuts and thrusts can also be diverted or deflected in a number of ways, not only stopped directly. In cases where the incoming attack is dealt with by the sword, the shield can be used as the weapon of attack in that instant.

Finally, we shouldn't forget that in this type of close combat all forms of attack and defence might be called upon. (See figures 20ia 9 & 10, 11 & 12) The sword pommel and guard could be used to strike, particularly at the head, neck and face at very close range and strikes and grappling with the body were normal in armed combat throughout the ancient, medieval and renaissance eras. Particularly in shield fighting the combat can come very close and barging with the shoulder, kicks to the shield and suchlike are all tactics that are occasionally necessary.

Conclusion

On the surface of it, we know almost nothing about how swords were actually used in the early Anglo-Saxon era but when we start to collate the available evidence there is a surprising amount that we can say with a high degree of probability.

The forms of the weapons themselves, placed in perspective and context, tell us about how the weapons probably were and were not held; what sorts of attacks and defences they were geared towards; the preferences of their users. The normal presence of shields sets a lot of the context for how the swordsmen of the time would have stood and moved. Practice with these weapons and study of other comparable systems that are documented shows that there are universal principles to armed combat and even specifically to the use of swords and shields together. There are simply only a certain range of positions, techniques and tactics that are logical with the given weapon set. It was probably a style of fighting that had a lot in common with later medieval and renaissance sword and shield/buckler combat, but with a particular breed of swords that were specialised to strong cutting actions and a type of shield suited to active defence and offence. A caveat to this assertion is that historical people did not always do what was most effective or logical and we should always remind ourselves that we can never say with certainty how they did or didn't use their weapons. Equally, there were probably a variety of 'styles', both between individuals and varying trends between groups. But hopefully this article has demonstrated that there are a reasonable number of things we can surmise about sword use in England at this time.

20ii. Evidence of wear and tear on swords

Paul Mortimer

Many of the commentators referred to in this book have shown that much of the evidence points to most swords remaining in use for a long time, and that old swords that had acquired their own histories were especially valuable. If this was truly the case, and I have no reason to believe otherwise, then a sword with a long life would have had to be cared for, perhaps repaired, sharpened and modified during its period of use. In other words, it would need to be curated.

We will take a brief look at each of the major elements of the sword and attempt to understand some of the effects that long periods of use would have on them.

Scabbards

Scabbards were designed to protect the blade from the environment, including the weather, and from knocks and scrapes; it had to protect the wearer and his garments from the sharp edges of the sword as well. Over time a scabbard would wear out and have to be repaired or replaced. Wearing a sword has demonstrated to me, and others, that the end of the scabbard is particularly vulnerable as it will come into contact with the ground and other obstacles. Some scabbards had metal chapes to strengthen that part of the scabbard, but they seem to be gradually falling out of general use during the 6th century, although it is not possible to say that they disappeared completely by the 7th century. There are only five swords equipped with them in my sample from the cemeteries, which represents around 3%.[1] It is difficult to say whether my sample is typical or not but it is probable that some form of organic reinforcement was used to add strength to the bottom of those scabbards without a metal chape. On my replica of the Mound 1 sword ensemble, an extra layer of leather has been superimposed on the carcass of the scabbard to act as a chape. It has received quite a few scuffs but it works very well and can easily be replaced when it is worn out. Unfortunately, we do not know whether chapeless scabbards received similar treatment because the evidence from the originals is so sparse due to their deterioration in the grave and the prevailing soil conditions in most of England, and for that matter, elsewhere. I have not been able to track down a complete scabbard from the 6th to 7th centuries anywhere. Some of the earlier bog finds in Denmark include virtually intact scabbards, but many of those have metal chapes and in most cases any thin leather coverings that they may have had have disappeared and left few traces. We can only make reasonable suppositions about what the scabbard-makers of the 5th to 7th centuries would have done. It is possible that scabbards without metal chapes were often left without any extra protection, but that would have shortened the life of the scabbard.

Other areas of the scabbard were subject to wear and tear too, and that is why there are edge protectors, often of metal, attached to the section of the sword where the belt or baldric would meet the scabbard. If they were completely successful it would be expected that they would be included on all sword scabbards but, as we have seen in our survey, that was not the case with only 17 (10%) definitely having them.[2] Others could have had organic edge strengtheners which have dissolved in the soil or have not been recorded. Quite significant numbers of the swords in our sample (25 of 168 – 15%) benefitted from being bound with tape, mainly, at the top of the scabbard but frequently extending down to the area where the suspension mated with the scabbard and even beyond. This would appear to be a practice that started during the 6th century

[1] See Appendix 2 and chapter 10.

[2] See Appendix 2 and chapter 10.

and continued into the 7th, although it is difficult to be precise because of the problems associated with dating swords from graves. From practical experiments it does seem to provide reinforcement for that part of the scabbard that can be subject to considerable stress. It also protects the scabbard from the method of suspension. The evidence seems to show that the tape was usually applied very skilfully and, in the case of the Mound 1 sword from Sutton Hoo, was a of very fine weave,[1] which would indicate that it was an important addition. Tape, once it starts to deteriorate or becomes dirty, can easily and swiftly be replaced.

The use of metal throat or mouth bands at the top of scabbard to provide strength to another vulnerable section of the scabbard, too, seem to diminish as the 7th century approaches, but 18 swords in the sample have them (11%).[2] They are often highly decorative as well as functional but why tape tended to be used instead on later swords is a bit of a mystery. Perhaps the tape provided visual clues to the observer concerning the wearer's allegiance, status, or wealth, through colour, the way it had been applied or from what material it was made.

Eventually a scabbard would be beyond repair and have to be replaced, roughly how long one would last is difficult to speculate, even with practical experiments and evidence from the cemeteries it is hard to come to definite conclusions. I have had two scabbards that I have used for a period of around ten years that are still very serviceable and they could well last a lot longer but I do not wear them as often as they may have been worn, nor do I regularly wear a sword while riding a horse, nor do I ever take one onto a real battle field. Hence I have not been able to subject a scabbard to the same wear and tear that an early medieval sword would have experienced.

Other factors might help to hasten a scabbard's deterioration such as a particular environment which could attack the glues holding it together or rot the skin covering, if it had one. In addition, microbes could attack the wood of the carcass. I feel that it is reasonably safe to suggest that a sword worn frequently over a decade or more would need a scabbard replacement every now and then. A sword owner would probably have two or more different scabbards made for his sword, perhaps ones to suit a particular environment or to keep the sword in while travelling.

Blades

Blades are susceptible to wear and tear and need maintaining. Keeping them clean and free of unwanted materials such as dust, liquids, including blood and other organic residues resulting from fighting, would be one of a sword owner's duties towards his weapon. Besides cleaning, especially in a dampish environment such as most of northern Europe can frequently provide, I would suggest that he would want to regularly lubricate the blade with a suitable rust preventative.

The fleece lining recorded in many of the scabbard remains was almost certainly used because the lanolin in the fibres would help protect the blade against corrosion. If as reported likely, the Sutton Hoo Mound 1 sword was lined with beaver fur, then that too would have been included due to beaver fur being largely waterproof and its ability to keep external moisture out.[3]

Blades would need occasional sharpening. This could be a problem for those blades that were only provided with a relatively small amount of steel in the edge and therefore, could not suffer the process too many times. The literature does tend to emphasise the sharpness of the swords of heroes, but even a relatively blunt blade will do enormous damage when used with vigour. Dings and dents in edges could be filed out, but a skilled swordsman would avoid getting them in the first place and all the evidence suggests that sword users did not

---

[1] Bruce-Mitford, 1978. 284.

[2] See Appendix 2 and chapter 10.

[3] Bruce-Mitford, 1978. 285.

'fence'. This may have been easier than a modern mind might think having seen sword fighting in Holywood films or modern battle re-enactment.

A broken sword point can be repaired, but probably only once and the blade would need the hilt to be removed to do it. Mick Maxen explained that a fish tail shape can be cut into the bottom of the blade and then the bottom will need to be heated and re-forged back into the necessary sword tip form.[1] This will shorten the blade a little and it will need heat treatment to bring the blade back to hardness and a useable condition. It may also require polishing and re-etching. How prone swords were to breakage at this point is difficult to say.

Hilts could be broken or be replaced for other reasons. This too will shorten the sword a little as the peened section at the top of the tang that holds the hilt on will need to be filed off. Re-hilting could be the reason why some hand grips seem to be rather small to fit a man's hand, but it is much more likely that a close-fitting grip was preferred because it was better for the style of fighting then employed. Incidentally, it is possible to lengthen the grip again, by filing down the top shoulders of the blade, where it meets the tang, for a few millimetres. I have mentioned elsewhere that the Sutton Hoo Mound 1 sword is shorter by quite a bit than the average blade in the sample that I have collected; perhaps the sword had originally been longer but had had several hilt replacements or suffered a broken tip? Perhaps it has always been the length that we see today.

It is rarely possible to find actual evidence of curation in blades from graves owing to the hostile conditions in the soil of most areas in England and elsewhere and due to the fact that almost all swords were buried within their scabbards and the resulting combination of the remains, corrosion products and other accretions makes them virtually impossible to read. It could be that as more and more blades are subjected to computer tomography scanning and other new techniques, more evidence will be found.

Hilts

Hilt assemblies have received the most attention in recent years, as far as evidence of wear and repair are concerned. In 2014 Sue Brunning devoted part of her thesis to looking at swords from Anglo-Saxon England and Scandinavia from the 5th to 11th centuries.[2] She examined metal fittings in particular and it formed part of her exploration of the concept of the 'living sword'. Long exposure to use and handling would show up as details such as delicate filigree work or chip-carving being worn down and perhaps even flattened or obscured. Solid sections, too, would have wear marks and a higher 'polish' when they have been used a good deal.

Detecting wear is not always such an easy task, as Brunning discovered when examining one of the swords from Valsgärde grave 5 in Sweden. Various commentators have had differing opinions as to whether the sword was worn or new when placed in the grave. Brunning was able to determine by close observation that there are indeed signs of wear on some of the detail on the hilt assembly.[3] Even when found it is not necessarily straight forward to explain what a particular wear pattern means and it is much more difficult to find signs on material that corrodes easily, or on harder metals like iron or bronze. However, with gold, which is usually incorruptible, the purer it is the quicker it will have worn and the easier the recognition and interpretation.[4]

---

[1] Personal comment, Mick Maxen.

[2] Brunning, 2014. 60, 123.

[3] Brunning, 2014. 129 to 130. The Valsgärde 5 grave is still unpublished. That particular sword hilt is one of, if not the most elaborate hilt of all that I have seen. I, too, was able to look at it at close quarters in 2013.

[4] Brunning, 2014. 131.

Sometimes it is obvious that a sword was always kept in the scabbard facing the same way and that its wearer would 'thumb' a particular part of the pommel because it had worn down in that particular place.[1] The ring-sword from grave 204, Northbourne, Finglesham is an interesting case because it has, what looks like, a freshly made pommel but the rings show signs of wear. A conclusion suggested is that the pommel was a replacement but that the old ring set was retained.[2]

The sword[3] from Coombe, Woodnesborough in Kent, the only sword in our sample from a cremation, although completely untouched by the fire[4] and a ring-sword, appears to have a ring-set free of wear with all its details unblemished.[5] When Brunning examined it, she found that the rings were off-centre and had obviously been a late addition to the sword.[6] There are other ring-sets which were added after the pommel had been made. Sometimes the pommel has been altered and parts cut away to accommodate them.[7]

Davidson and Webster suggest that pommels (pommel caps) "…as one might expect, received more wear and tear through handling than any other part of the sword; and evidence for refitted pommels and stray ones….. amply confirms this."[8] I am not so sure that they were right about pommels wearing out more than any other part of the sword, but the possibility that they were replaced on occasion, especially in the case of pommel caps, makes some sense. However, Brunning demonstrates that signs of repair are hard to identify but when seen they are clear signs that work was carried out to extend the sword's life.[9]

One important conclusion that Brunning arrives at is that signs of wear and tear are as much, if not more visible on high status swords than on more modest ones and believes that, "The implication is that the finest swords were not elite trinkets paraded only on rare occasions and statically displayed the rest of the time. On the contrary, they were functional objects in the sense that they were worn, touched and interacted with to such a degree that a permanent mark was left on their fittings."[10]

As Brunning explains, more work needs to be done and finds need to be, "…investigated further from an archaeo-metallurgical point of view, for instance using specialist equipment to examine less well-preserved fittings, and by determining the purity of metals and alloys in order to estimate how long it would have taken for them to become worn. An intensive hunt for battle damage would also be a welcome addition to discussions of sword biography. Finally, the search for curated material could be extended to other weapons because swords were not the only worn, repaired and modified items of war-gear to be deposited in early medieval graves."[11] Experimental archaeology could also help to further our understanding of the conditions that swords and their paraphernalia were subject to in everyday life. It is certain that there is much more to be learnt from this type of study but what has been done clearly shows that swords were worn and used over long periods of time.

---

[1] Brunning, 2014. 140 to 141, 142 and 256. On page 143 Brunning suggests that the sword had a 'public face' and presumably a more private one.

[2] Brunning, 2014. 130.

[3] Davidson and Webster, 1967. 1. It seems that there were actually two swords associated with this cremation, but one disappeared soon after discovery, probably in 1845. An iron spear head, glass and amber beads, and part of a pendant set with garnet or coloured glass. The actual cremains were contained in a bronze bowl and the ring-sword was wrapped in cloth. The objects all lay together besides the bowl. Davidson and Webster, 1967. 10. Davidson and Webster speculate that it was a double cremation of a man and a woman and date it to the last quarter of the 6th century. (page 10 and 11). It was a very unusual cremation indeed.

[4] Davidson and Webster, 1967. 10.

[5] Davidson and Webster, 1967. 23.

[6] Brunning, 2014. 131.

[7] Evison, 1976. 305. Bruce-Mitford, 1978. 133 to 137.

[8] Davidson and Webster, 1967. 23.

[9] Brunning, 2014. 134 and 136.

[10] Brunning, 2014. 133.

[11] Brunning, 2014. 165.

Belt/baldric assembly

Once again, it is difficult to tell from surviving equipment how constant wear and tear would affect belt and baldric assemblies, partly because, as far as I know, such a study has never been carried out, but also because of the fragmentary and delicate nature of many of the finds. All we really have is the experience of re-enactors and experimental archaeologists. Belts and baldrics need maintaining and cleaning, but are subject to wear and tear, and may even have had to be modified or replaced due to weight gain or loss. Leather will eventually perish and textile will wear out. It is unlikely that such items would last as long as a sword and when they changed hands, the new owner would almost certainly update the method of suspension.

21. Swords and trauma

Paul Mortimer

Wesseaxe forð ondlongne dæg eoredcistum on last legdun laþum þeodum, heowan herefliman hindan þearle mecum mylenscearpan. *Brunanburh*, 1.20-4	Forward, the West Saxons the whole day long in mounted troops stayed on the track of the hated peoples, hewed the fugitive foes cruelly from behind with [their] mill-sharpened swords. [1]

At the 7th century Eccles Anglo-Saxon cemetery[2] one of the skeletons reveals an interesting story. It is the skeleton of a young man who had reached the age of about twenty to twenty-five years when he was hacked to death by a swordsman.[3] His bones show at least thirty injuries with the fatal blows being to his head, where his skull indicates some seven blows were landed. Presumably, he was the loser in a fight and his opponent had, it seems, forced him to the ground where, shieldless, he raised both arms in self-defence; his lower arms were made useless with sword cuts and then he was struck on the head a number of times although at some time during the episode he managed to turn his body over to try and alleviate the attack, but his opponent continued to rain blows on his back, presumably until he stopped moving and perhaps beyond that time.[4] There are other possible ways that this man's injuries could be sequenced but we do not know if this was a duel, or if he died in battle, nor can we tell if all his injuries were written on his bones; it might be that some of his wounds were confined to soft tissue and left no trace. Neither do we know why his attacker was so frenzied in his attack. Wenham calls him 'victim II' and says of his head injuries, "Certainly victim II would have died from the first of several blows which severed his brain stem, causing cessation of vital functions."[5]

There are five other examples of men killed after receiving fatal cranial injuries probably inflicted by swords at Eccles, three of these were, it seems, killed with a single cut to the head.[6] The cuts seem to indicate skill and accuracy on the part of the swordsmen inflicting the wounds, and the use of sharp weapons. It is possible that these three, victims I, V and VI suffered other wounds that did not show on their skeletons. The other two slain men, victims III and IV, like victim II suffered multiple injuries; victim III was hit by a projectile that penetrated his spine and had a cut on the left side of his head, while victim IV was struck three times on his head. All the injuries on all six men were inflicted at about the time of death (perimortal).[7] Wenham tells us of the head injuries, "In all cases, the scalp, the meninges which surround and protect the brain, the brain itself and the blood vessels associated with these three structures would have been injured. The immediate effects of the blows to the head would have been profuse bleeding, especially from the scalp and almost certainly a loss of consciousness, due to the force of the blow to the head and its associated shock to the brain. Death may have followed almost

---

[1] In Livingston, 2011. Brunning notes other instances in the literature where sharpening is mentioned. Brunning, 2014.179.

[2] Clay, 2003. 8.

[3] Wenham, 1989. 124, 135, figure 8.17 and 138.

[4] Wenham, 1989. 131.

[5] Wenham, 1989. 128.

[6] Wenham, 1989. 124 and 138. Victims I's and V's ages are estimated at twenty-five to thirty-five years, victim VI's is twenty to twenty-five years.

[7] Both Wenham and Stillwell, (2002. 5 to 10) discuss how it is possible to distinguish between different types of wounds and how they can be recognised from marks on bones, Stillwell's discussion is the more comprehensive.

immediately due to shock to the central nervous system or from loss of blood."[1] Wenham explains that it is not possible to tell whether all the victims were killed in a single incident or over time.[2]

The victims from Eccles are not the only ones that have received such attention in recent years. Stillwell considers some examples from the Sedgeford Anglo-Saxon cemetery which has been carbon dated to 662-881 A.D.,[3] many of the skeletons examined do fall out of our designated period but similar weaponry was still being employed at that time. Stillwell confines his consideration mostly to trauma suffered by the skull and examined one hundred and thirty-four examples from the cemetery.[4] Twelve of the skulls exhibited lesions, eleven of which appeared to be male, one was female.[5]

Two of the victims seem to have had remarkably similar fates to that endured by victim II above, S1016 and S1018 both have multiple cranial injuries and they also seem to have raised their arms against their attackers receiving 'defence blows' on their limbs as a consequence; they too were killed with a sharp blade and the attacks appeared to have continued after the men had fallen to the ground and died.[6] Stillwell reports that 82% of the lesions were caused through the use of sharp implements and that 91.6% of those wounds penetrated the endocranium (the inner lining of the skull) and were certainly fatal.[7]

It is possible that not all the damage was caused by the use of a sword and Stillwell does consider other weapons such as knives, axes, seaxes, javelins and spears[8] and the wounds that they would inflict, but still feels that with most of the cranial injuries at Sedgeford, the weapon used was the sword. He tells us that, "The lesions to S1033, S0086, S0023 (the female), S1016, S1018 and S0067 are deep and penetrating lesions. Some of their lesions would have been caused by a sword evidenced by the length and depth of the cuts."[9] He uses a quote from Beowulf to illustrate the sort of force that was being used,

> ...yrringa sloh
> Þæt hire wið hales heard grapode
> Banhringas bræc bile al ðurwod
> Fægne flæschoman[10]
>
> ...he struck such a savage blow that the blade slashed through her neck,
> smashed the vertebrae; it severed her head from the fated body.
>
> *Beowulf* 1565 – 1568 CH

The blows suffered by the deceased at Sedgeford and at Eccles were indeed 'savage' and he reports that S0023, "...had a cut from the left mastoid process to the right mandibular ramus. S1018 had three cuts to the left lateral side, including one to the left mastoid process and left side of the mandible. S1016 had a deep cut made to the left mastoid process. S0086 and S0067 had two cuts made to the posterior aspect of the cranium."[11] Judging by the accuracy of some of the blows it would seem that the swordsmen involved skilfully went about their business.

---

[1] Wenham, 1989. 128.
[2] Wenham, 1989. 138.
[3] Stillwell, 2005. 32.
[4] Stillwell, 2005. 2.
[5] Stillwell, 2005. 30. The female is listed as S0023.
[6] Stillwell, 2005. 30.
[7] Stillwell, 2005. 30.
[8] Stillwell, 2005. 33.
[9] Stillwell, 2005. 33. Sharp swords cut through bone very cleanly.
[10] Stillwell, 2005. 33. He refers here to Underwood, 2000. 47.
[11] Stillwell, 2005. 33.

The Anglo-Saxon execution cemetery at Walkington Wold[1] may provide evidence of the use of the sword to remove heads, some of the executions were botched, several swipes were taken and mostly hit the rear of the victim, probably with him or her bent over, but Skull 8 (aged eighteen to twenty-five years) and Skeleton 7 (twenty-five to thirty years old) both seemed to have been decapitated from the front using a sharp sword or knife as the blade was too thin for an axe.[2] The male occupant at grave 16 from Butler's Field, Lechlade had a cut to his left parietal lobe on his skull, 83 mm long and penetrated 23 mm deep, and it was caused by a sharp, narrow bladed weapon. The cut was clean on one side but the bone was shattered on the other, suggesting that the weapon had been wrenched out. The man would have been killed by the blow, perhaps immediately or very soon after. However, the man from grave 21 was luckier; it seems that he had received a clean, shallow slice to his scalp (82 mm by 62 mm) with an oval-shaped piece of bone being removed but the interior of his skull was not penetrated and he lived on, as the bone had reformed around the cut. The medical examiner (Dr Jones) inspecting the remains and the writers of the report believe that the injury was most probably caused by a fast moving, sharp sword.[3]

There are other examples in the cemetery record but the above will suffice for our purposes in this book but we will end this discussion by looking at an experiment that was carried out recently.

As swords are not used in modern warfare, and elsewhere sword wounds with long sharp blades are thankfully rare, it is difficult to record the effects of such weapons on living bodies; experiments have been carried out on inanimate objects and on dead animals and there is no doubt that sword cuts could be devastating. However, Johann Matzke, recently, attempted experiments with HEMA[4] combatants to try to match up the weapons used during later medieval combat with the type of injuries found on the remains from a number of known battle casualties found in graves from the period.[5] HEMA fighters do not fight like re-enactors; they do not have to follow the same rules of health and safety, because they are able to wear modern, non-historical protective clothing, such as full face fencing masks in order to avoid injury. The HEMA fighter sacrifices authentic clothing and armour in order to preserve accuracy of technique and will fight with full 'martial intent'.[6] Matzke found that the head was a frequent target, together with the forearms and upper body.[7] Other parts of the body might be struck occasionally, but were not the main targets. The HEMA combatants were asked to fight frequently and record where they were hit.[8] Once adjusted for factors like modern protective headgear, Matzke found a high correlation between the injuries suffered by the battle victims and the HEMA fighters.[9] Obviously the modern fighters were protected and did not suffer in the way that medieval combatants did, but the fact that the same parts of the body were being targeted is revealing, and that seems to have been a constant factor throughout the whole period of sword warfare.[10]

A skilled swordsman could wreak devastation on an opponent.

---

[1] Buckberry and Hadley, 2000. 323 to 324. . "...decapitation can only be securely identified osteologically for four to six individuals, but, nonetheless, we can state that between 30.8 and 84.6 per cent of individuals buried at this site had been decapitated."

[2] Buckberry and Hadley, 2007. 319.

[3] Boyle, Jennings, Miles and Palmer, 1998. 45.

[4] HEMA – Historical European Martial Arts. See discussion on the differences between HEMA style combat in Matzlke, 2011 (81 to 82).

[5] Matzke, 2011. 7, and 11. The battle injuries came from graves resulting from conflict at Towton (England), Fishergate (York, England), Uppsala (Sweden), St Mary's (Oslo, Norway), and Wisby (Gotland, Sweden). Matzke, 2011. 61.

[6] Matzke, 2011. 81 to 82. Matzke fully explains the concept of 'martial intent' in his discussion.

[7] Matzke, 2011. 64 to 71.

[8] Matzke, 2011. 78.

[9] Matzke, 2011. 90 to 94.

[10] For a description of Matzke's methods and results, see Matzke, 2011. 74 to 96. The experiments included other weapons besides swords.

22. The Meaning of the sword; an exploration of the ideology of the sword during the 5th to 7th centuries

Paul Mortimer

Hē þæm bātwearde bunden golde
Swurd gesealde þæt hē syðen wæs
On meodubence māþme þȳ weorþa
Yferlāfe.
Beowulf 1900 – 1903[1]

When a man drew his sword for combat he had no way of knowing what the outcome of the fight would be; even a famous fighter could be brought down by a trip or misjudgement, or by a fourteen year old with a bow and arrow lurking on the edge of the battlefield. Unsheathing a sword for a fight was no light act; it required courage, determination, the ability to take a chance, and a little faith. It was a serious step to take. The uncertainty of combat was a powerful feature of the lives of those who through necessity or choice had to engage in it. The outcome of the conflict had consequences for non-combatants, which in many instances must have meant the whole community. It is easy to see how the sword, as the embodiment of the 'personal weapon', became a symbol of personal and martial prowess. However, it is evident that it signified much more than that. As Theuws and Alkemade remark, "The sword was anything but a neutral object whose function was limited to fighting."[2]

Long before the 'Anglo-Saxons', swords acquired whole sets of meanings in the minds of those whose lives were connected to it. The sword was a weapon without any other practical use, unlike a spear or seax, both of which can be used in the hunt; a sword's only practical function was to kill, harm or threaten human beings.

We will now explore some of the ideas about swords that were current during the 5th to 7th centuries and try to understand how those notions would have affected the societies living in England and northern Europe. In doing this we will sometimes use material from outside that time-scale and geographic region but to a large extent any society engaging in the type of warfare that meant getting very close to your enemy and taking part in hand-to-hand combat, is likely to have held similar views regarding the qualities required of the participants and their weapons. The coming of Christianity possibly altered some of the ways those fighters viewed what they were doing and who or what they were fighting for, but the experience and reality of battle did not change; it was still a matter of overcoming your fears, trusting to your strength, skill and training as well as that of your comrades and having faith in your weapons and your luck. A warrior from the Late Roman period had much in common with one from the 10th century and even from much later.

Before we begin to unravel what a sword might have meant to society during the 5th to 7th centuries, we will ask what a sword consisted of. It may seem obvious but, as pointed out elsewhere, swords are more complex than, probably, any other personal weapon of the period, but 'what is a sword'? Does the word mean the blade, the hilt or both combined? What about the scabbard, or the belt or baldric? Are these inseparable from the sword? Hilts themselves can be built from many parts, then there are the additions, beads,

[1] He gave a sword bound with gold to the boat-guard so that he afterwards was worthier on the mead-bench because of that treasure and heirloom. Baker, 2013. 58.

[2] Theuws and Alkemade, 2000. 419. Brunning, 2014. 243 and 267. Brunning echoes the sentiment.

pyramids and bosses; how necessary are they to the sword? 'What is a sword' is a question that has been rarely discussed by scholars, presumably because it seems so obvious, but how many of the above mentioned parts can be taken away, before a sword can no longer be considered, a 'sword'? As we have plenty of finds of detached sword parts[1] it is an intriguing thought, and we must remember that in the poem, Beowulf brings the hilt of the sword that came from Grendel's mother's lair and presents it to Hrothgar, presumably it still represented the whole sword. The blade had melted away after it killed the monster's mother.

Sue Brunning has asked the question[2] and, after a thorough examination of the literature and iconography, has concluded that the 'sword' was ". . . widely perceived as 'blade-plus-hilt.'"[3] She qualifies this when she writes, mainly after looking at iconography from after our period, "...for some artists it was simply 'blade-plus-hilt'; for others – including, but not confined to, those with warrior patrons – it was blade, hilt, scabbard, sword-belt and more." Brunning's original conclusion would appear to be the most useful one; both blade and hilt need to be together for the weapon to be able to used effectively; without a hilt a blade is not a sword. Everything else, the scabbard and its decorations, belts and other paraphernalia, are just accessories, although probably meaningful ones. However, Halsall points out, while discussing the Staffordshire Hoard, that, "It might be salutary to remind ourselves that for all the intrinsic interest and importance – to us – of the items recovered, to early medieval people the important bit of a sword wasn't the bright, shiny pommel and scabbard fittings but what we see as the rusty length of iron: what they saw as the long, sharp pointy bit that you killed people with."[4] I think he is correct; the sword is the blade. Perhaps that is why so many sword pommels have been stripped away from swords and are found by metal detectorists,[5] not least those in the Hoard? The life of the sword would continue with a new identity; its story would go on but with a new 'face' in the form of different hilt furniture. It is possible that removing a hilt from a looted sword was a deliberate act to distance the weapon from its previous owner and allegiance. Burying the redundant parts was perhaps meant to give something to *the powers that be* but that this did not always happen as is apparent in the story of the Heathobard's sword[6] in *Beowulf* where the pommel was seemingly retained so that it could be recognised. In that case, the hilt, too, is an accessory. One further thought is that removal of the pommel and burying it, rather than recycling or re-using it, could be a way of honouring the previous owner and keeping his part in the sword's history separate by giving it to the other world.[7] Removal of selected parts of the sword could also be a way of creating distance from a hated, defeated or dishonoured foe.

"Heald þū nū, hrūse, nū hæleð ne m[ō]stan,
Eorla ǣhte![8]

This chapter is divided into five main sections, i. Swords, power and the economy of honour , ii. The living sword, iii. Sword burial, iv. Women and swords, v. Concluding thoughts, where a summary is attempted. The sections do not necessarily have to be read in the order they occur here, although, it would probably be better to leave the concluding remarks until last.

---

[1] For instance see *Appendix 10: Sword pommel finds recorded by the PAS.*

[2] Brunning, 2014. 44.

[3] Brunning, 2014. 75 and 256.

[4] Halsall, 2010. 5.

[5] See the discussion in *The Riddle of the pyramids*, and evidence in Appendix 10. The same applies, possibly, to pyramids and other parts from swords.

[6] We will return to this episode below.

[7] Discussion with Matthew Bunker.

[8] *Beowulf*, 2247 to 2249a. "Hold now, earth, now that heroes may not, the treasures of princes! (Chickering Jr. 1977. 182 to 183). Thank you Torun Zachrisson for the insight.

i. Swords, power and the economy of honour

At a basic level, someone with a sword had power over the unarmed, the fearful, the less skilled or those unable to fight. The sword also represented political and social power. The swordsman in the service of a lord could show that he had his lord's backing and his weapon, which may be owned by the lord, was a symbol of all the power that the great man could muster, as well as the sword wielder's own skill and prowess.

Swords were a symbol of rank, status or wealth and achievement

An individual who wore a sword gave out messages about just how far up or down the social ladder he was, and that was probably indicated by the ornamentation on the sword and its scabbard or the fact that it was known that he wore a famous and recognisable weapon.[1] It is highly likely that things like pyramids made of precious metals, elaborate bosses, beads and decorated hilts were all signs of some form of achievement; possibly the equivalent of medals today.[2] An Anglo-Saxon commander (*folctogan*) is unlikely to have carried a sword with a plain hilt because they needed to display their status and position and show that they had their lord's confidence and represented his authority. They would have done this by wearing splendid weapons that had been gifted to them for their service. Today a high-ranking officer has special rank insignia on his collar or his epaulettes and doesn't need to wear special weapons to indicate his status, although many such officers are still issued with a sword for ceremonial purposes. Along with the commanders, famous warriors would also carry a weapon that reflected their capabilities, perhaps a recognisable one that would come to share the credit for that particular hero's exploits. A ring-sword was possibly appropriate for the hero, and/or the general.[3] As we have seen from the introductory quote to this chapter from *Beowulf*, the mere possession of such a weapon, whether he had used it in battle or not, would raise the status of the possessor in the eyes of others.

The descriptions of the adornments of swords in Old English poetry are often carefully written and it seems that this aspect of the sword had real relevance to the audience, even if much of the visible parts of the sword were accessories. No other weapon or artefact regularly receives such detailed descriptions as do swords, the same is true of depictions in the artwork of the 5th to 11th centuries.[4]

The current thinking regarding swords represented by the Staffordshire Hoard is that remnants of around one hundred and fifty swords are contained within it. As previously mentioned, there are eighty gold or silver gilt pommels from the find but it is now believed that the numerous gold strips, often decorated with filigree or other adornments, also come from swords and were probably attached to plain horn hilts in a manner similar to those on the Cumberland hilt.[5] (figures 22.1 and 22.2) It could be that such strips were given out by a lord to a retainer for some service, possibly on the battlefield. Such flashy items would be easily seen and would mark the wearer out and distinguish him from those without adornments. Equally, they could have been within the gift of an officer to his men. Are such strips the equivalent of modern day military medals? Perhaps each of the gold lozenges would have their own story that could be re-told again and again in the Mead-hall.[6] Most of the additions would appear on stylistic grounds to be from the late 6th and early 7th centuries, so they may have been a later development.

---

[1] Of course other aspects of a man's apparel could also provide signals of status.

[2] See chapters, *Additions, Beads, Bosses and Pyramids*, and *The Riddle of the Pyramids*.

[3] Ring-swords and their purpose are discussed elsewhere in this volume and we return to the subject below.

[4] Brunning, 2014. 77.

[5] Personal communication with Chris Fern.

[6] In my cemetery study 82% of the swords found in graves did not have significant metal parts on the pommels or hilts, so it is possible that the Hoard probably represents far more swords if undecorated ones are to be considered than one hundred and fifty. However, it is always dangerous to make assumptions based on burial evidence.

22.1 The Cumberland hilt, now in the British Museum. Besides the gold filigree strips, this side has a garnet and gold triangular decoration.

22.2 shows the other side. Both photographs by Paul Mortimer, taken with the kind permission of the Trustees of the British Museum.

It is unlikely that the lowest men in society, slaves, were allowed to carry swords, therefore the sword carrier displayed his status as both a freeman and a fighter. As a symbol of the fighting man, the sword would also show that he had courage, or was thought to have. According to the poetry, however, not all swordsmen turned out to have the requisite bravery, for instance when Beowulf was being hard-pressed by the dragon, most of the warriors he had taken with him shrank into the woods and did not help him, despite being given their war gear and wealth by Beowulf (*Beowulf* 2635 to 2636 CH). The penalty for their cowardice was drastic and complete, they are called 'oath-breakers' and 'lily-livered', they are shamed, deprived of their property and they, and their male kinsmen, were condemned to become exiles (*Beowulf* 2884 to 2891 CH); in that society, their fate was more onerous than death and they were finished, beneath contempt, their reputations destroyed forever. The poet leaves us in no doubt what was expected of a sword carrier![1]

[1] This idea is present, too, in the much earlier work of Tacitus in the 1st century A.D. Tacitus tells us of the Early Germans that, "In addition, infamy and lifelong scandal await the man who outlives his leader by retreating from the battle-line:" Rives, 1999. 83.

Swords may indicate allegiance to a nation, a tribe, a band, a family or clan, or perhaps showed religious affiliation to a particular sect or god

The idea that there was such a thing as an Anglian, Saxon or Alemannic sword is difficult to support; swords circulated and a particular weapon could pass through many different hands and areas during its life. Theuws and Alkemade have demonstrated that the old ideas that a sword could consistently illustrate ethnic origin are unsound;[1] for instance they found almost as many 'Alemannic' scabbards outside the Alemannic area as those within it.[2] They also show that swords and accoutrements were subject to 'mix and match' and that styles were transferable across space and time. However, it is possible that a sword ensemble displayed more subtle, and perhaps relatively temporary, signs of affiliation to a tribe, clan or military society. Perhaps the colours, weave or method of application of the binding tape applied to many later 6th and 7th century scabbards was fulfilling this function? Unfortunately the colours of such things are rarely, if ever, discernible now but I suspect that opportunities to relay such messages would not normally have been missed and the meaning of particular colours could have varied from region to region and clan to clan. It is likely that ethnic differences were displayed in other ways than by wearing types of swords, perhaps better indicators were how a sword was worn, together with styles of clothing? However, it is not possible to take this idea further with our present state of knowledge, apart from registering it as a possibility.

Divine power?

Much of the symbolism on swords, particularly on pommels and on scabbards, possibly relates to notions of the Otherworld and its deities. The triple ring and dot symbolism seen on many swords from Kent and elsewhere, such as the ring sword from Dover Buckland and that from Faversham, are likely to represent an appeal for supernatural protection or assistance, possibly from three different gods. The fact that such symbolism is repeated again and again would indicate that it had a particular set of connotations.[3] It is possible that the swastika on some pommels could be a reference to the might of the thunder god,[4] or a reference to the sun. Appeals to deities on weapons are a frequent occurrence throughout the history of warfare and certainly did not stop with the conversion to Christianity although some of the symbolism changed.

The pairs of outward facing animals that reside atop some pommel caps could perhaps symbolise requests for assistance from the deities that the animals represented or remind the wielder of some quality that the animal possessed; it is also possible that the animals are biting beasts and that the bite mimics that of the two sword edges. Androshchuk, believes that animal symbolism turns the weapon into an animate object[5] and that is possibly why some swords are given names in the literature, while others may not have had them. We will return to this idea again.

Phallic symbols, male power[6]

Swords could, understandably, be seen as symbols of masculinity because of their phallic shape, and in our post-Freudian world this is a fairly natural assumption, where expensive sports cars have also been seen in this way; whether the peoples of the 5th and 7th centuries saw things in quite the same way is an intriguing thought, but it is possible that we might just be imposing modern ideas on the past. The richest grave in Anglo-Saxon England,

[1] Theuws and Alkemade, 2000. 439, 443, 445 and 447.

[2] Theuws and Alkemade, 2000. 439. See also Birch Iversen, 2010. 155 to 157, for criticism of similar notions supposedly indicating 'ethnic' weapons for an earlier period.

[3] Some examples are discussed in the *Riddle of the Pyramids* and listed in *Appendix 10*.

[4] Davidson, 1962/1994. 67.

[5] Androshchuk, 2014. 202.

[6] Brunning discusses the possibility and how others have suggested swords could be viewed as phallic. Brunning, 2013. 33 to 34.

Mound 1 at Sutton Hoo and almost certainly the grave of a king, has a sword of only moderate length, which possibly tells us something about the fecundity of Anglo-Saxon kings, but probably not.[1]

Swords might have been viewed as phallic, and it is true that swords and other weapons are associated with masculinity, prowess and protection, but that is not the same thing as fertility or sexuality. However, we will revisit the idea again later.

The economy of honour

The 'economy of honour' is a term coined by Peter S. Baker,[2] to describe the ways in which the gift and service system operated during the early middle ages, swords were a useful and powerful component of that system; they can be used to impose obligations or to gain favour through gifting and can play an active part in diplomacy.[3] It should be realised that by gifting, the giver too can gain prestige and enhance his (or her) reputation, depending on the circumstances.[4] We will explore this idea and begin by considering some of the information from Beowulf.

What was important about a sword?

In *Beowulf* swords are referred to frequently and several of the weapons have histories, however, we are only told the names of two swords, Hrunting, the sword that Beowulf borrows from Unferth to fight Grendel's mother and his own sword, Nægling, which fails him during the fight with the dragon. When Nægling fails, the poet suggests that Beowulf was too strong for ordinary swords and that is why they fail him. Hrunting, however, didn't break when he used it; instead it was the Mother's hide, which was invulnerable to blows from ordinary swords, that failed.

The poem is emphatic about the sword's only practical purpose, which is to kill men, and there is an episode from *Beowulf* where that is made very clear. The wife of Hygelac, Hygd, is all that a queen should be and she is contrasted with the very beautiful but "proud and perverse and pernicious" Queen Thryth; any man who looked at her was cruelly put to death and we are told, "...the battle sword was brought into play, the damascened sword (*sceadenmæl*[5]) was allowed to settle the matter." (1942). The poem *Waldere*, too, reminds us of the primary function of the sword when Hildegyð (Waldere's companion) remarks,

'Never shall work of Wayland fail
a master of Mimming, a man who knows
the handing of that blade. Bleeding from its wounds,
lords and athelings are laid on the field.'[6]

Mimming,[7] is a famous sword made by the legendary Wayland.

Some of the descriptions of swords and sword parts in *Beowulf* tell us a great deal about their appearance. We are told, "Not least amongst his mighty aids was Hrunting, the long-hilted sword Unferth lent him in his

---

[1] The Mound 1 sword from Sutton Hoo is discussed elsewhere in this volume and there might be good reasons why it may appear undersized.

[2] Baker, 2013. 37. Baker says that it was a phrase in William Ian Miller's study of the Icelandic sagas and was also used by John M. Hill.

[3] Arrhenius, 1985. 196 to 197. Brunning, 2013. 39 and 246.

[4] Shepherd, 1997. 7.

[5] It is not at all clear that *sceadenmœl* means 'damascened'.

[6] Alexander, 1966. 60. Waldere, is the Old English name for Waltharius, who we have discussed elsewhere in this chapter. 232.

[7] Also 'Mímungr'.

need; it was engraved with deadly, twig-like patterning, (*atertanum fah*) tempered with battle-blood. It had not failed any of those men who had held it in their hands, risked themselves on hazardous exploits, pitted themselves against foes. That was not the first time it had to do a hard day's work." (1455 CH) and then again of the same weapon, "...He swung his sword of battle with a mighty rush, and his hand did not fail in the stroke, for the sword with curving patterns (*hringmæl*) sang a greedy song of battle upon her head. Then the visitor found that the battle-ray had no power to pierce and do no mortal injury..." (1519 K). Which was a problem and on realising that the sword was useless in this fight... "...The enraged warrior flung down the sword with twisted patterns (*wundenmæl*) and bound ornaments (*wrættum gebunden*), hard and steely-edged, so that it lay upon the earth." (1531 K). Hrunting was useless against the Mother.

Fortunately, all was not lost, "Then Beowulf saw among weapons an invincible sword wrought by giants, massive and double-edged, the joy of many warriors; that sword was matchless, well-tempered and adorned, forged in a finer age, only it was so huge that no man but Beowulf could hope to handle it in combat." It had a, "...gold-adorned sword hilt...", it was the, "... age-old work of giants,..." Also on the sword-guards of pure gold, it was recorded in runic letters, as is the custom, for whom that sword, finest of blades, with twisted hilt and serpentine patterning had first been made..." Beowulf goes on to slay the mother, although as a consequence of her corrosive juices, the sword blade dissolves and he is left with just the magnificent hilt, which eventually he will present to Hrothgar, the Danish king whose land had been ravaged by the mother and her son Grendel.

Within those descriptions of swords there is a wealth of detail telling us what the Anglo-Saxons and other folk looked for in a sword; it was important that swords had histories, even if those histories were not known; the 'curving', 'twisted' or 'serpentine' decoration was a significant feature of both swords, whether this also meant the patterns on the blade, is unclear from the language used, but it could do. The second sword was very old and made by giants, it had runic lettering on the hilt, just as some swords found in graves have. The descriptions of ornamented hilts, in the poem, could fit some of the known finds. We also know that old swords were valued,[1] and it seems that Hrunting had been used by several different men, so had probably been around for a long time.

The acquired histories of sword are referred to, again, when Beowulf predicts trouble between the Danes and the Heathobards on his return to Geatland. Hrothgar's and Waltheow's daughter, Freawaru, is promised in marriage to the King of the Haethobard's son, Ingeld, in an attempt to make a lasting peace between the two peoples. However, Beowulf says that at the wedding feast there will be an old, drunken Heathobard warrior who will recognise a sword that one of the Danes will be wearing, apparently a ring-sword,[2]as the weapon of one of his old comrades killed by the Danes. The old one will then provoke the son of the deceased into killing the present holder and the peace will be broken.

An example of a real sword that was very old is one mentioned in the Atheling Æþhelstan's will written in 1014. Æþhelstan was the son of Æþelræd Unræd (Ethelred the Unready) and left most particular instruction about the sword, which had belonged to King Offa, who died in 776. That sword was at least two hundred and thirty-eight years old and could well have been very old when Offa came into possession, so it seems that it was not only in the stories that great value was placed on antique weapons.[3]

---

[1] See below.

[2] Davidson, 1962(1994). 126.

[3] Æþhelstan owned eleven swords, Brooks, 2000. 159. It is possible that this sword was the one of Hunnish workmanship that had been given to Offa. Davidson, 1962. 109, mentions it after Akerman, as a possibility. See also Androshchuk, 2014. 194 to 197. Androshchuk discusses all of Æþhelstan's swords as does Härke, 2000. 385.

Swords as *Madmas* and 'the economy of honour'

Madmas is the Old English word for 'treasure', but it isn't simply about gold and silver, as Baker relates, again referring to the poem, *Beowulf*, "...was not about intrinsic value, its value lies in its history, its 'honour', who it belonged to... ", "Treasure' could store up aspects of its past, including previous owners which explains the extraordinary value placed on some objects and why, at times, a man was prepared to risk all for it."[1] Baker again, 'treasure', "...is never completely separated from its previous owner(s) they just add to its history and value." [2]

It is important to remember here the relationship between a warrior and his lord, again from Baker, "In early medieval societies, a man's value was determined by his prowess as a warrior, and everything worth having – status and the things that came with it: wealth, land, a desirable wife – depended on his lord's recognition of this prowess. A king of a nation at peace could not deliver these things to his people, and the results could be dire."[3] It must have been quite difficult for real kings to manage the giving of suitable gifts; gaining them as booty through warfare or tribute from weaker neighbours would have been valuable and essential sources. In the poetry it is never a problem, great kings can always find a unique and important treasure, perhaps a sword, with which to reward the hero and by doing so, add honour to the receiver and lengthen the history of the item given. Incidentally, the poetry never mentions other, more mundane ways of raising revenues either, taxes, tolls or rents for instance; presumably these would not be 'honourable' enough. One tradition that will have helped the king is that, at least in theory, all loot stripped from fallen enemies should be brought back to the lord for his own treasure chest and eventual redistribution. This is illustrated on two occasions in *Beowulf*; when Eofor strips Ongentheow of his war gear, the loot is presented to his king, Hygelac, (2985 – 2990) and when Weohstan kills Eanmund he must have given the arms and armour to Onela who, a little later, is able to to give them to Weohstan (2612 – 2618).[4] More of this below.

In this way, swords, along with other *madmas*, became part of the 'economy of honour' where gifts and honour became commodities that could be circulated, or not, depending on the will of the giver. "...treasure and honour, indissolubly bonded, are traded up and down the social hierarchy in such a way that the participants gain (and occasionally lose) honour with each transaction."[5] A good example of how the giving of a gift can increase a man's honour is provided when Beowulf gives the Danish coast guard who has cared for his ship a sword. He gave "...a sword bound with gold to the boat-guard so that he afterwards was worthier on the mead-bench because of that treasure and heirloom."[6] Such gifts are much more significant than payment, they make a connection that is moral and emotional and the lord who gives gifts is at the centre of the ethical, economic and political system, 'the economy of honour'. As Androshchuk comments, "In Anglo-Saxon sources swords as well as other weaponry very often play the role of regulators of social connections between different generations, between equals and between chiefs and attendants..."[7]

Incidentally, such 'gifts' are not gifts in the modern sense, they are never free,[8] something has to be done, or as in the case of the boat-guard, has already been done in return. Gifts place an obligation on the receiver. Androshchuk explains that the process of gifting was really an essential part of a 'contract' that involved three obligations: to give, to receive and to reciprocate.[9] Ideally, too, gifts must match the deed and the rank of the

---

[1] Baker, 2013.59.

[2] Baker, 2013. 59. Nicolay, 2014. 11.

[3] Baker, 2013. 10.

[4] Härke, 2000. 391.

[5] Baker, 2013. 37.

[6] *Beowulf*, 1900 to 1903 CH.

[7] Androshchuk, 2014. 194.

[8] They are not 'alienable', they cannot be separated from their previous owner(s). Bazelmans, 1999. 17.

[9] Androshchuk, 2014. 202.

person receiving the gift, otherwise it could be humiliating for one party or the other – sometimes this would be done deliberately; humiliation could be a powerful tool! Gifts can also be used to coerce someone into doing something that they did not intend to do or did not want to do. Gifting could be aggressive!

The receiver of a gift was probably not free to give them away, either, as he is the new custodian and perhaps not more than that. On the death of a follower it seems that, under some circumstances, swords, along with other 'gifts', had to be returned to the lord as a *heriot*.[1]

Diplomatic gifts can be used in a similar way, to reward or insult and to try and gain advantage. Theoderic was very pleased with the gift of patterned swords that he received, possibly, from Thrasamund, king of the Vandals,[2] and therefore Theoderic was well disposed towards the giver and his honour had been enhanced by such a gift, which would have had some public display so that his subjects could see that he was being honoured.

However, as Theuws and Alkemade note, swords as gifts, are only given, according to the texts, under specific circumstances;[3] when a lineage or people were threatened, as was the case with Hrothgar and Beowulf, or when relations of subordination or equality were established, for instance gifts between kings or a lord gifting his follower.

Sacrifice

We have already touched upon the idea that relations with the Otherworld were important, sometimes that relationship would require the sacrifice of a weapon, perhaps a sword? It could be that Theuws and Alkemade provide a reason when they write, "Through the sword the owner/user stands in a direct relation to the reproductive forces of life. It is for this reason that the sword takes on a special position in the interaction between the social and the supernatural world."[4] Of course, a warrior can send someone, very quickly, to the Otherworld, through the use of his sword and is always quite close to making the journey himself whilst fighting.

Bladed weapons, and other objects, are found in watery places all over northern Europe and over an extended time period, sometimes a single weapon is found but there are also large deposits that would appear to represent the arms of a losing force. One example of such a deposit is the find from Illerup, Denmark (200 to 500 A.D.),[5] but there are numerous others. Many of the find areas from Scandinavia seem to have been used on several occasions each and many, but not all of the weapons, were deliberately damaged before being deposited.

It is possible that single swords could have been left in rivers in Anglo-Saxon times, one example may be represented by the gold parts of a hilt found dredged up on a river bank in North Lincolnshire, the Market Rasen hilt. Some organic remains of the grip and part of the tang still survived, which almost certainly means that the whole sword was deposited but that the iron blade had, over time or through the action of dredging become detached.[6] Another possible river find is the pommel from the River Bain in Lincolnshire. This remarkable find is of copper-alloy but has a red enamel triangle on one face of the pommel only.[7] (figure 15ii.37) Watery finds from our period are rare, but not so much before and after. Many of the swords that have been recovered in England after the 7th century are river finds, several being found in the River Thames,

---

[1] Härke, 2000. 382 to 383. Discusses heriots and the possibility that there is a reference to them in *Beowulf*..

[2] Mentioned above, with the caveat that the swords could possibly have come from elsewhere. Theuws and Alkemade, 2000. 420. Say that the sword came from the king of the Warni.

[3] Theuws and Alkemade. 2000. 420.

[4] Theuws and Alkemade, 2000. 421.

[5] Biborski and Ilkjær, 2006.

[6] Mortimer, 2011. 100 and 105.

[7] Thanks to Sue Brunning for bringing it to my attention.

alone. Far too many swords have been discovered in watery places for them all to be casual losses or due to battles taking place at fords or rivers;[1] there was a connection between bladed weapons and watery places and that could be due to many factors. Perhaps the fact that water is involved in creating the blade and is also an essential part of life was important? As we have seen, in the chapter on Pattern-welding, the patterns possibly symbolise water. Certainly the surface of a lake, river or pond can be seen as a liminal place; a border between two worlds; the everyday and the world of the Other.

Leaving a bladed weapon in a watery place seem to be a way of communicating with the Otherworld, it would appear to include the idea of exchange, in other words giving something precious to the powers of the Otherworld will mean that the giver will get protection or added strength in return.[2] This occurs long before the 5th to 7th centuries and again after, well into the later middle ages. Quite why bladed weapons were put into rivers, lakes and bogs is a matter for more thought and debate. It is peculiar that during the 5th to 7th centuries, in England, evidence of this practice is quite rare, although not altogether unknown, as explained above. Whether from our period or not, the weapons deposited are often of a very high quality,[3] just like the Market Rasen sword. As already mentioned, the practice continues after that period and much of our knowledge of 8th to 11th century swords comes from river finds, as they are no longer left in furnished burials which almost completely ceases before the end of the 7th century.[4]

In is intriguing that this practice does not die out until well into the Christian period, which may say something about the solidity of the conversion of many fighters and the superstitions of men who regularly risk their lives; perhaps they had not completely given up on the Old Ways.

It should also be remembered that there is a watery sword in the poem *Beowulf*; when the hero has tried to attack the Mother with the sword, Hrunting, and failed, he finds a giant sword. Now although the Mother's lair seems to be air-filled, it is underwater and it is possible that the sword corresponds to the practice of leaving bladed weapons in lakes, rivers and watery places; it is possible that the Mother represents the sort of being that such things were once sacrificed to.

We have, too, discussed in the chapter, The Riddle of the Pyramids, the phenomenon of the many sword pommels (and other sword parts) that have been discovered by metal detectorists in recent years, and the possibility that these are dry-land sacrifices. Certainly, it is improbable that all of them were lost by accident.[5]

Looting, booty, plunder

There were ways of gaining a sword other than being given one by a lord; it could be inherited through the family[6] or be given by a brother-in-arms, or gained through compensation or via a gambling debt. Or it could be looted.

It is only fairly recently that looting has become anathema for modern soldiers and punishable. Looting the dead, with official sanction, was certainly happening at the very least until the Napoleonic Wars and the Bayeux Tapestry shows it occurring, unashamedly, after the Battle of Hastings. In the past and certainly during the early middle ages, looting was essential and an important part of the economy of honour. Gaining

---

[1] Härke, 2000. 378 and 389. Härke reports that John Hines has suggested that during the later Saxon/Viking period, when the deposition of grave goods has ceased that leaving swords in rivers could have been a substitute. While this is possible, the pagan Vikings had restarted leaving good in graves, as Härke himself points out on the same page. Further work is needed to test this idea. See also discussion in Brunning, 2014. 37.

[2] Theuws and Alkemade, 2000. 426.

[3] Androshchuk mentions this too. Androshchuk, 2014. 244.

[4] Hines and Bayliss, 2013. 517 to 528.

[5] See also Appendix 10.

[6] Davidson, 1960. 99. We will look at the idea of family swords more closely in the section on *Women and swords* below.

your enemies battle gear and wealth was a way of being rewarded for risking your all in battle and a way for a young man to advance in the world, especially if he killed a famous foe and gained gear with a known biography. It should be noted however that in one of the few laws of the early medieval period which specifically refers to looting (number 178 of the Gulatingsloven, a collection of Norwegian laws thought to date to the 10th century) it was a "*nithing* crime if a man robs a (fallen) man on the battlefield by stripping off his clothes and taking his weapons"[1]. Being declared a "*nithing*" (somebody without honour) resulted in outlawry and total exclusion from society, sometimes permanently. Under the Gulatingsloven, a man could become such an outcast as a penalty for murders of a particularly dishonourable nature such as breaking into another man's house to kill him; killing a man to whom pledges of safety had been given; burning a man to death in his house; and looting, which is the only dishonourable crime listed that does not involve murder. This harsh prescription against looting the dead was perhaps to ensure that items recovered from the fallen were handed over to those in charge so that they could be distributed in an appropriate and orderly manner as part of the spoils of victory. It would also discourage individuals from keeping items of value for themselves or fighting over who had what.

It is important to realise that at that time a fighter would wear his finery when he went to a set battle, for many, what they took to the fight could represent their entire personal wealth.[2] In doing this the warrior is sending some important messages. Perhaps the most significant is that he is showing that he is a good combatant as he only has this equipment because he is good at fighting, it also announces to friend and foe the honour that he has achieved and that he has his lord's favour, and faith is placed in his prowess. His war-gear is an essential part of who he is and it demonstrates that he is prepared to risk it all in battle; he is wearing his 'treasure' his madmas; it is his personal heraldry! The warrior wants to be seen and remembered, especially if he falls; he knows, too, that if he dies and his side loses it is likely that his war-gear will be looted, and be recirculated as part of the economy of honour and that his own story and how he died, will become part of his gear's history. He will not be forgotten!

Cherniss tells us exactly that, "In the heroic system of values, the plundering of enemies slain is an integral part of worthy conduct. The treasure which a warrior gains by plundering is the concrete representation of the honour which he has won in battle and is, indeed, the only tangible proof of the honour and esteem to which his deeds entitle him. When a warrior is slain in battle, the honour which he has won by his former deeds passes with his war-implements to his slayer…"[3] Of course the successful man of honour would need to boast about his exploits, as Beowulf does on his return to Geatland. This would not be seen as anything other than honourable, necessary and entertaining! If Beowulf had been killed, then he would have expected the *scops* to tell of his exploits in his absence and to go on doing so, in order for him to become part of the history of his people and for him never to be forgotten. So far, for Beowulf, the system is still working!

Further, taking a slain foe's war-gear meant honour for the victor, passing it to his lord would increases that honour and this would enable the lord or king to bank the war-gear on behalf of the nation. We know that defending the hoard was a vital royal function. The Brunanburh poet says of Æthelstan and Eadmund: *land ealgodon, hord and hāmas* – (they defended the land, its treasury and homes).[4] From the hoard the king makes gifts to those who played a part in winning it and in doing so acknowledges and passes the 'credit' due for the killing, as Cherniss has already pointed out above. Thus Hygelac, becomes, *bonan Ongentheow* - Ongentheow's bane, even though it was Eofor who actually slew him.

---

[1] Larson, 1935. 137

[2] Halsall, 2010. 3.

[3] Cherniss, 1972. 92 to 93.

[4] Livingston, 2011. 41. Brunanburh, ASC Aversion, 937.

There were rules concerning looting it seems, at least for one who saw himself as a high-status warrior as this episode from *Waltharius* may show.[1] Walthari was a hostage at the court of Attila the Hun and he became Attila's champion, winning many battles and taking much loot for Attila. Walthari decides to leave Attila without his permission and takes his girlfriend, Hiltgunt (another hostage) with him and the portion of Attila's treasure that Walthari feels that he is entitled to. On their way back to Walthari's homeland, they go near the city of Worms where Gunthari, king of the Franks is – he wants the treasure to make up for what Attila took from his father. Gunthari doesn't listen to the warnings of Hagano (another former hostage of Attila's – who knows what a great warrior Walthari is) and sends a troop of eleven men plus Hagano to attack Walthari. Battle ensues and Walthari kills the eleven and then, "The young man (Walthari) proceeds to strip the slain of their arms and armour, leaving their tunics and other things: from them he took just armlets and brooches, belts and swords, mail coats, too, and helmets. He loaded four horses with these, called his betrothed and set her on the fifth, and mounted the sixth himself."

According to Baker the poet does this in order to show that Walthari is not greedy and takes only the loot that is honourable for him to remove, which is the war gear.[2] The rest is left for others with less honour to take if they want to.

Later Walthari, Hagano and Gunthari, despite maiming each other, are reconciled, and share the loot.

Ranking of gifts in the economy of honour according to *Beowulf*

As previously explained, gifts from one lord to another or a lord to retainer or vice-versa were not 'gifts' in the modern sense, by receiving a gift the receiver was put under an obligation, which was why the diplomatic gift of a magnificent sword like the one that Theoderic receives and Cassiodorus writes about, was so powerful. Something was expected in return. Nor were all gifts equal as this ranking of gifts from *Beowulf* illustrates.

Beowulf kills the beast Grendel that had been ravaging the Danish king's hall and in return Beowulf is given by Hrothgar a standard, helmet, mail-coat, and sword (*Beowulf* 1020-1024). However, Grendel's mother exacts revenge and Beowulf has to kill her and, again he receives gifts from Hrothgar: eight horses and Hrothgar's own saddle (*Beowulf* 1035-1042). Hrothgar has a problem in that he cannot give a gift to Beowulf that is equal to the value to the kingdom of the Danes, which means that he will always be in deficit to Beowulf. But Beowulf does not take advantage of this, he has too much honour and his main interests lie elsewhere.

Waltheow (Hrothgar's queen and the mother of his two sons) was probably well aware of the possibility that Beowulf could claim the Danish lands when Hrothgar dies, after all, Hrothgar wanted to make Beowulf his son, and perhaps that is why she makes splendid gifts to him and refers to her own two sons.[3] She gives to Beowulf rings and a garment and a fine necklace so wonderful that it is compared to the necklace of the Brosingas (*Beowulf* 1193-1201). Is she attempting to put Beowulf under an obligation to her in order to ensure her son's inheritance? It is ironic that she trusts Hrothulf (Hrolf); the informed members of the audience will be aware of the irony. Once again, Beowulf has no intention of ever taking the Danish kingdom.

On Beowulf's return to the land of the Geats, he gives to his king and kinsman, *Hygelac*, a boar standard, (presumably the one that Hrothgar gave him), a helmet, a mail coat, a sword and four bays (horses), apple brown. He also gives a necklace to Hygd, Hygelac's queen, almost certainly the one that Waltheow gave to him (*Beowulf* 2152 - 2174).

In the economy of honour, Hygelac is now obliged to reward Beowulf's generosity and gives him a sword, possibly Nægling and 7,000 hides of land, a hall and throne (*Beowulf* 2190-1299). But *Eofor*, another Geatish

---

[1] Waltharius, 9th to 10th century. Baker, 2013. 36

[2] Baker, 2013. 65 to 66.

[3] Shepherd, 1997. 7.

prince got far more, Hygelac gave him 100,000 hides of land, rings and his own daughter to marry (*Beowulf* 2995-2998). Eofor, it seems, was not so great a warrior as Beowulf, nor did Eofor become king of the Geats as Beowulf eventually did. So what did Eofor do that merited such valuable gifts?

When the Swedes invaded Geatland , the Geatish king Hæþcyn was killed by the Swedish king Ongenþeow. When he became the new king, Hygelac, sent Eofor, and his brother Wulf, to fight against the Swedes. During the fight Wulf was severely wounded, but Eofor slew Ongenþeow and gave Ongenþeow's arms to Hygelac. So it would seem, judging by the events in *Beowulf*, that artefacts that had been owned by a defeated king and gained in battle, were more valuable than those that were obtained in other ways. Again, it is all about the biography of the items and not just their intrinsic value.

Ring-swords

Ring-swords have been found in graves mostly in Germany, the Netherlands, France, Scandinavia, Northern Italy and England,[1] appear in the iconography and are mentioned in the, often, much later literature. They date from the late 5th century to the 7th and on two occasions have been found on objects other than swords.[2] A few swords pommels, or pommel caps, have had them removed prior to burial, but the marks of the ring assembly are still there and in some cases a part of the pommel, or pommel cap, was carved away to fit the missing rings.[3] Ring assemblies have been found on their own, and sometimes attached to a pommel cap that is no longer mounted on a sword. Some pommels have been adapted to accommodate a ring set and this occasionally means that part of the pommel cap has had to be cut away. The rings themselves are nearly always decorated in some way, often with a punch, engraved lines or both. They were obviously of some significance and indicated something, but what was it? The rings do not appear to have ever had a practical purpose and they hardly added to the sword's ability to kill or wound opponents. Some earlier commentators suggested that the rings could have provided a way of binding the sword to the wrist, but this would not work with the rings that are solid and even with the hollow, free-running rings, the gaps are so small as to make a securing strap close to useless.[4] Another suggestion was that they were used to attach sword beads to the hilt, but the evidence is that these tended to be connected most frequently to the scabbard[5] although the much later Icelandic literature tells us that 'life stones' could be carried in a bag separately from the sword; besides the solid rings could not have accommodated any sort of strap,[6] incidentally, the beads may well not be 'life stones'. One other objection to tying a bead to a sword-ring is that a bead flying around while the sword was being wielded could easily get damaged or cause an irritation for the swordsman.

What were the rings for?

When found in graves they are usually only found in relatively high-status contexts, but not, so far as Anglo-Saxon England is concerned, in the highest level of graves. For example, none were retrieved from Sutton Hoo, Taplow or Prittlewell. Evison[7] and Davidson,[8] posit that the two rings are symbolic of two parties to an agreement; for example, that between a lord and his retainer. We do know that oaths were sworn on rings

[1] Steuer, 1987. 209 and 210. See also Menghin, 1983, Mortimer, 2011. 98 to 99, and Brunning, 2014. 26. Brunning gives a useful summary of the scholarly interpretations of the purpose of the rings.

[2] There is a pair on the front of the shield from Mound 1 at Sutton Hoo (England) and another on a drinking horn from Valsgärde 7 (Sweden).

[3] See Brunning (2014, 137 to 138) for examples; Ash, Gilton, grave 56 and Sarre grave 91. On page 138, she explains how the sword from Vallstenarum was heavily modified to accept the rings. In this case the rings are still present. Also see Evison, 1976. 305and Bruce-Mitford, 1978. 133 to 137.

[4] Behmer, 1939. 135.

[5] Mortimer, 2011. 110 to 112.

[6] Davidson, (1962. 75.) reports that the idea came from Böhner.

[7] Evison, 1967.

[8] Davison, 1962.

and it is possible that they were also sworn on sword rings.[1] If this was the case, were the rings worn by the lord, or the retainer, or both? Kings and lords, in the literature, were often referred to as ring-givers; it could be that the 'rings' were the rings on swords. This idea of the rings as a visible sign of an oath sworn to a lord, a contract if you will, and the retainer then providing service for his master, could fit the evidence. As they are not found in the highest level of graves, it is unlikely that they were worn by kings and overlords, but perhaps by their high-ranking retainers.

Probably the most recent study of swords to consider the ring-sword is Brunning's and to further complicate the matter, she points out that in Sweden some graves (Vendel, grave I and Valsgärde, grave 7) have two swords each and in both one is a ring-sword and the other is not.[2] Brunning is also concerned with the problem of rings occurring in other locations, on a shield and on a drinking horn, which she feels, correctly, has not been fully addressed.[3] However, wherever we find sword-rings, we must remind ourselves that, at the moment, all of the suggestions for their use are still little more than just that – suggestions. Brunning wonders, too, which had the most symbolic force, the ring or the sword?[4] She makes it clear that ring-swords were especially prone to modification and that the rings were often added, perhaps a generation after the sword was made.[5] She has examined wear patterns on them and found that at least one example was placed on a sword not long before it was buried (grave 204 at Northbourne, Finglesham, Kent)[6] and others that showed signs of considerable use (grave 7 at Valsgärde, the sixth-century-sword from Vallstenarum, Gotland[7] and the ring-sword from the boat grave at Ultuna, Bondkyrko).[8]

Ring-swords are rare and Brunning is quite accurate when she writes, "Ownership of such a sword would therefore have transmitted powerful messages about that individual's identity: their loyalty, trustworthiness and martial prowess had earned the bestowal of an exclusive token to adorn what was already a rather exclusive piece of war-gear."[9]

Brunning considers other types of evidence, apart from the swords themselves, including the *pressbleches* designs on helmets found in Sweden and on the dies used for making them. The artists of these rather small plates went to considerable lengths to show the details of ring pommels on some swords[10] and they occur on designs from the Vendel XIV and Valsgärde helmets, one of the Torslunda dies from the island of Ölund, and even on a tiny gold foil plate from, Sorte Muld, on the Danish island of Bornholm.[11] Brunning has carefully examined these images and tried to understand the way that the swords were worn, held, whether they were 'sheathed' or 'unsheathed' or 'reversed' or not,[12] and feels that they are depicted as more or less aggressive depending on how they are shown. The line of warriors from the Vendel XIV helmet carrying their ring-

---

[1] Davidson, 1960. 96.
[2] Brunning, 2014. 26.
[3] Brunning, 2014. 27.
[4] Brunning, 2014. 27.
[5] Brunning, 2014. 137.
[6] Brunning, 2014. 127.
[7] Brunning, 2014. 125.
[8] Brunning, 2014. 127.
[9] Brunning, 2014. 245. See also Davidson and Webster, 1967. 23. I am not so sure that swords were such 'exclusive' items. I have explained my thinking in the chapter, *Pattern-welding*. Halsall, 2010. 2, also feels that although swords were 'expensive' they were not rare.
[10] Brunning, 2014. 99 and 232.
[11] Brunning, 2014. 75.
[12] Brunning, 2014. 89 to 90 and 99.

swords in their hands she reports Steuer as suggesting, "...a ritual of submission or homage made by elite retainers to their lord."[1]

In her examination of the appropriate texts looking for evidence of ring-swords, Brunning says that many references are often ambiguous and could be allusions to ring-swords but could equally be about ring mail or neck-rings,[2] although Hrunting and, possibly, the sword of the Heaðobards in *Beowulf* do appear to have been ring-swords (Beowulf, Hrunting 1521 - Heaðobards 2036 to 2049 CH). What she does wonder is whether the word 'ring' in its different native forms, had not become a type of shorthand for the word 'sword' in the later texts, which, if I understand her correctly, is an interesting possibility; that it was swords (and rings) that were given out rather than just 'rings'.[3] Her conclusion is that, "Ultimately, evidence from at least two sources suggests that ring-swords were associated only with certain individuals, who may well have been the oath-bound retainers of kings or war-leaders, as several scholars have argued."[4]

There are, however, other suggestions that could still fit the evidence. It is possible that the rings indicate that the holder has taken an oath that he must fulfil and that when that happens, the rings are removed, and perhaps pass to another. They could equally show that an oath has been fulfilled. Perhaps, they are an indication of a special link between a warrior and a deity and the binding oath was made to a particular god but was removed when the wearer lost faith or changed his allegiance? Or maybe it is a sign that the sword itself is a 'special' one, one that had acquired the status of being 'antique' or 'old'? In the latter case it would be difficult to see why the rings would ever be discarded. Steuer said that they could just be a reward for a loyal retainer.[5]

It is possible that rings were removed when a warrior migrated from one lord to another or changed his allegiance to a deity.

As for why a ring could appear on a drinking horn, a possible explanation is that the horn was used when a warrior drank to his oath, but that on a shield, and probably a king's shield, too, is more difficult to understand, although the principle of binding' could still operate, presumably. It could be that it was, again, a visible sign that the king had made an oath such as the one allegedly made by Hrolf Kraki and others, never to flee iron or fire?[6] Perhaps warriors swore oaths to the king on his shield ring? That would make for quite a visual ceremony.

As Brunning notes, the addition or removal of the ring does seem to indicate that whatever was going on was not always a permanent arrangement,[7] she also feels that the rings possibly indicate different grades of warriors,[8] and I suspect she is right.

I believe that there is a major difficulty with the idea that the rings represented an oath sworn by a retainer to a lord and that is that all such elite warriors would have been bound to their lord in that way, so why does it appear that there are so few ring-swords; would not all such oath-bound warriors have worn one? In which case there should be many more appearing in the archaeological record. My own inclination is toward seeing the rings as a mark of a commander (*folctogan* OE)and of high military office, or of a great champion and I think there are three main clues; one is that on one of the dies from Torslunda, the leading figure carries a

[1] Brunning, 2014. 90.
[2] Brunning, 2014. 245.
[3] Brunning, 2014. 181 and 245.
[4] Brunning, 2014, 245.
[5] Steuer 1987, 226.
[6] Byock, 1998. 61.
[7] Brunning, 2014. 245.
[8] Brunning, 2014. 245

ring-sword but his follower does not. The second clue is their scarcity - in our sample of one hundred and sixty-nine swords, only eight were ring-swords, which represents 4.7% of swords of the sample, and if this is representative only something like 0.08% (8/10,000) of all graves can be expected to provide such a weapon.[1] The third clue is that, at least with the later larger and solid rings, they appear to have been meant to be easily seen. Nothing in the evidence contradicts this notion, as far as I can tell, and even the incident in *Beowulf* concerning the Heathobard sword does not mean that the man carrying the sword was not either a commander or a champion.

A further complication arises from the recent discovery of a completely new type of ring-sword found in the Staffordshire Hoard. Many of the sword pommels were fragmented before their finding and the conservators have taken great pains to put the pieces back together. There are several examples of conventional forms of ring pommels but three of the eighty-six gold or gilt-silver pommels turned out to be ring-swords that have double rings on each side of the pommel; that is each pommel has a total of four rings, two per side![2] This is completely new, nothing like them has been even hinted at before either in England or elsewhere and yet the Hoard contains three. I believe that this reinforces the idea that ring-swords indicate rank and that the double ring-swords may have been carried by the highest level commanders of an army. However, it is still very much a subject for further discussion and, whatever their significance, ring-swords (double or single) were very much an important part of the 'economy of honour'!

One further comment regarding the removal of rings from swords is worth making; so far we have no satisfactory understanding of why they may have been removed. Scenarios such as a comrade, family member, wife, sweetheart taking a keepsake from the sword at the funeral of a loved one are possibilities, and others could be envisaged.

ii. The 'living sword'

In recent years there has been a shift of emphasis when considering artefacts, away from typologies and classification and towards a consideration of the 'life' of an object, or its 'biography', this has led to new insights and a better consideration of how objects were viewed, used and modified during their lifetime. Swords lend themselves well to this type of study and it is obvious from the literature that sometimes swords could be seen as having a 'life' themselves, as we will see. Fairly recently, Theuws and Alkemade have picked up on this theme and make it clear that swords were not seen as inanimate; they "...were described as living beings, inhabited with personal characteristics: they carried fierce names, their deeds were commemorated, their appearance was praised and they were worn and exchanged on special occasions."[3] The same authors point out that existing typologies are actually just archaeological constructs and that the types of classifications that we use, would not have been understood or have been relevant at the time, and that the ancients would have classified their swords in entirely different ways.[4]

In this section, we will try to understand how a sword was viewed, how it could become more valuable and how it was 'alive'.

---

[1] For further details, see the chapter, *Cemetery/sword ratio; a brief summary with a list of swords recorded*. However, frequencies of objects found in graves should not be taken as a reliable indicator of their use in life.

[2] Staffordshire Hoard Newsletter number 9. 4. http://www.barbicanra.co.uk/newsletters.html. In style they seem to be much closer to the elaborate pommels from Valsgärde, Vallstenarum, Vallentuna and Vendel in Sweden than other pommels from England or the Continent.

[3] Theuws and Alkemade, 2000. 419.

[4] Theuws and Alkemade, 2000. 431, 434, 444, 445, 469 and 470. Nevertheless, when trying to understand developments, typologies can be quite useful, and we have used them in this book. Lucy (2000, 15) too, notes the shift from typologies.

Value

Writing of grave 6 in the Swedish cemetery of Valsgärde, where two swords were deposited with one man, Davidson says of the swords that, "…the weapons given to him at his funeral were ancient ones no longer usable, perhaps collected from a store-room or armour. ...it is possible that the rather mean policy of one family is represented here and that the case is exceptional."[1] Other scholars, besides Davidson have recorded similar impressions concerning old or damaged swords in graves, but it is possible that their assumptions about such swords being inferior in some way are wrong and perhaps the practice of putting old swords into graves was probably not 'exceptional'. There are other instances of swords being buried that at the time of burial were unserviceable. The sword from Mound 17 at Sutton Hoo would appear to be an example of such a weapon; when examined closely it is apparent that the tang was not joined to the pommel cap in any way at the time of burial[2] and, if used in that state for any length of time, the blade would come apart from the grip. Of course, we have no idea how the Mound 17 sword got to be in that condition or how old it was, it is possible that the sword fell apart while the young horse warrior had been wielding it during combat; perhaps like Nægling, the sword that failed Beowulf during the dragon fight, the Mound 17 sword had broken when it was most needed. It may, too, have had a long and illustrious history before coming into the possession of the Mound 17 occupant. However, the sword would appear to have received a 'cosmetic' repair before burial, so that it looked intact as it went into the grave. Neither of our two examples mean that such swords were being buried because they were now deficient. It could be that many of the swords that accompanied a man to his grave were in a similar state and it is possible that is why some swords were selected to be buried; that is a sword had to be broken, or to 'die' to qualify for inclusion within some graves. If this is so, it doesn't detract from other reasons that could have applied to the burial of swords that we will discuss below; in fact it may add to them. As Brunning suggests, "…, it seems likely that the placement of old swords in burials was deliberate and meaningful."[3] And it is possible that **only** old swords were put into graves.

The dating of swords in burials is always difficult and assumptions have to be made about the developments of different aspects of swords, but it is obvious that swords actually consist of a several different parts and that those parts can be changed or altered. Hilts will need repair, or require upgrading during the lifetime of a sword, scabbards would wear and get damaged, the ornamentation, especially on scabbards, could be changed easily.[4] Theuws and Alkemade, too, point out that, "The physical appearance of the swords and scabbards was not static and depended on their use and exchange in aristocratic networks."[5]

We know that old, or ancient swords were valued, and we have already visited this quote by Baker, "Treasure could store up aspects of its past, including previous owners which explains the extraordinary value placed on some objects and why, at times, a man was prepared to risk all for it."[6] We could substitute, the word 'treasure' for 'swords', because they could indeed, store up all their past histories and links to owners and great deeds, and this added to the way that they were seen; their stories were remembered and told and probably grew in the telling. As Brunning has demonstrated, old swords were of especial significance and this was not confined to the 5th to 7th centuries, it continued until, at least until the 11th century.[7] Theuws and Alkemade

---

[1] Davidson, 1962/1994. 13

[2] Carver, 2005. 244, see the drawing (figure 101) 217. My thanks to Matt Bunker for drawing my attention to this.

[3] Brunning, 2014. 38. See also Brunning, 2014. 189. Brunning says, "The evidence suggests that old swords were not considered defunct, but were valuable and worth specific mention. Perhaps an attitude existed that 'age equals experience' and, by extension, reliability – an outlook that prevails to some degree today." See also Brunning, 2014. 117.

[4] Brunning, 2014. 38, 120, 143. See also Theuws and Alkemade 2000, 444. Androshchuk, 2014 (197) makes a similar point concerning Viking Age swords.

[5] Theuws and Alkemade, 2000. 435.

[6] Baker, 2103. 59.

[7] Brunning, 2014. 79.

again, on the appearance of valuable swords, "As such, it is not the precious appearance and execution alone which renders the object valuable; in fact they seem to be the logical consequence of its meaning."[1]

The matter of dating swords is even further complicated by the possibility that new swords were actually made to look old and so appear more valuable.[2]

During later Anglo-Saxon times, old swords are mentioned in wills, the most remarkable of which is that of Æþhelstan who left eleven swords in his (about 1015 A.D). We have already seen that Offa's sword must have been very old indeed, but Æþhelstan shows that he knows each one of his swords well, that they have histories and that they are valued. The 'silver hilted sword that belonged to Ulfketel, 'the sword with the silver hilt which Wulfric made' are examples. Æþhelstan is also very careful to leave them to the right people. Some to former owners, others to members of the family or to new owners and even one to the Church.[3]

Value is a construct that changes from one period, community and place to another; what one society feels is precious may not be seen as valuable by another. Swords in our period were seen as valuable and desirable objects that men would, quite literally, kill or die for

Sword biography, character and personality

One of the aims that Brunning sets herself in her thesis is to explore the possibilities surrounding the notion that a sword had a 'life'; an existence with a beginning, an end and some events in between, a little like a human life. She was also interested in the idea that a sword could be seen as able to act independently of its owner; to have 'life' in that other sense.[4] Her comprehensive study includes finds of swords, references to swords in the literature, and contemporary iconography of them from both England and Scandinavia. In considering a sword's life-history or biography she studied signs of wear, tear, repair and other marks of modification. She reports a great deal of evidence, from the archaeology and other sources, which supports the premise that old, even antique swords were valued.

First we will briefly look at three examples of legendary swords that appear to have been valued because of their age and associations. At least two of these swords appear to have some semblance of being 'alive'.

Skofnung

The first sword is the one called Skofnung, owned by Hrolf Kraki (probably Hrōþulf in *Beowulf*[5]) a legendary king of the Danes. The sword appears to have had a long history, first with Hrolf and then, after being recovered from his mound to have had a second life with other owners.[6] Taking a sword from a mound makes some sense if recovered before too much decomposition had taken place. There could be supernatural hazards to face but it avoided the expense of having one made and it was already a proven weapon. Hrolf's story was written down quite late and, Hrolf, if he existed, must have lived during the late 5th century, perhaps early 6th.[7] The sword' Skofnung would have been a weapon made during our chosen period. In the Saga of King Hrolf Kraki, we are told that the sword was, "...the best sword ever carried in the northern lands..."[8]

---

[1] Theuws and Alkemade, 2000. 423.

[2] Fischer, 2008. 26. Fischer is referring to ring-swords here, but the idea has implications for swords without a ring, too.

[3] Androshchuk, 2014. 195 to 196. Brunning, 2014. 221. Brunning says that there were ten swords.

[4] Brunning, 2014. 40, 44 and 49.

[5] Garmonsway, Simpson and Davidson, 1968. 155 to 156.

[6] Davidson, 1960. 109.

[7] Byock, 1998. Vii. Hrolf appears in a number of different literary sources. The saga is late and was written in 14th century Iceland. Many of the sagas characters also appear in *Beowulf*. Saxo Grammaticus, the Danish historian writes of him about 1200 A.D. (Davidson and Fisher, 1980). Garmonsway, Simpson and Davidson, 1968, (155 to 206) provide a comprehensive list of sources that include references to Hrolf.

[8] Byock, 1998. 68.

just after Hrolf had used it to slice off the buttocks of Swedish king Adils. Hrolf uses the sword in his last battle and we are told that it cuts well and rings when it touches bone.[1] We are also informed that it was buried with him after that battle.

Much of the next stage in the life of Skofnung is recorded in chapters nine and ten of Kormáks saga. Skeggi of Midfjardar, was chosen by lot to break into Hrolf's gravemound and plunder it to recover the sword. Much later, Kormák pesters Skeggi to let him use the sword in a duel with Bersi.[2] Skeggi, although, at first, reluctant to let Kormák have it, eventually gives in but warns him that the sun must not shine on the hilt and that he mustn't wield the sword unless he is ready for combat. Skeggi also tells him that if he blows on the blade a little snake will crawl from under the hilt boss. Kormák is careless about the sunshine and the sword's spell was broken and it came out of its scabbard howling.[3] The duel was inconclusive and nobody died, both swords received some damage though.

Skofnung also appears in the Laxdœla saga, where it has come into the possession of Eid of Ás. Eid is the son of Midfjardar-Skeggi. Eid lends the sword to Thorkel Eyjólfsson to kill the outlaw Grim, who had killed Eid's son. Eid warns Thorkel to be careful with the sword as it, "...can only be used under certain conditions: the sun must not be allowed to shine on its hilt, nor may it be drawn in the presence of women. Any wounds it inflicts will not heal unless rubbed with the healing stone that accompanies it."[4] Thorkel attacked Grim and slightly wounded his wrist but Grim wrestled him to the ground and defeated Thorkel. However, instead of killing Thorkel the two became friends,[5] and Thorkel heals Grim's wrist with the stone. Thorkel has adventures, becomes a wealthy and important man, marries but he never returned the sword to Eid. Skofnung is briefly lost when Thorkel's ship is capsized while sailing around Iceland, and all of those on board, including Thorkel, drown. The sword, though, stuck fast in some of the timbers of the ship, and was washed ashore.[6] The recovered sword came into the possession of Thorkel's son Gellir. Gellir lived a long life, but died in Denmark returning from a pilgrimage to Rome, and is buried at Roskilde and it seems that Skofnung was buried with him,[7] which was probably not far from where it was buried in a mound with Hrolf; Roskilde, means 'Hroar's spring' – Hroar was Hrothgar.

Tyrfing

Another legendary sword that had more than one life and was also taken from a mound is the sword Tyrfing. The story of this sword, too, is spread throughout a number of different pieces of literature, including the *Hervarar saga ok Heiðreks* (The Saga of Hervor and Heidreks) and the Poetic Edda.[8]

Tyrfing was made by the dwarfs Dvalinn and Durin for Svafrlami, king of Gardiriki[9] and Óðinn's grandson. Svafrlami had forced the dwarfs to make it for him. They made it so that it would never rust, would cut through iron and stone and would never miss a stroke. But they cursed it so that each time it was drawn it would kill a man and eventually Svafrlami himself. Although Svafrlami heard them cursing the weapon, the dwarfs escaped before he could take his revenge.

---

[1] Byock, 1998. 74.

[2] Bersi also has a named sword, Hviting; that too has a healing stone to accompany it. Hreinsson, 1997. Vol1, 193.

[3] Hreinsson, 1997. Vol1, 194.

[4] Hreinsson, 1997. Vol 5, 89.

[5] Hreinsson, 1997. Vol 5, 88 to 90.

[6] Hreinsson, 1997. Vol 5 ,116 to 117.

[7] Hrensson, 1997. Vol 5, 119. Roskilde, 'Hróarskelda' or 'Hrolf's spring' – perhaps Hrolf Kraki. In any case it seems that it ended up not far from where it would have been buried with Hrolf.

[8] The Poetic Edda is a collection of Old Norse poems. The poems are thought to have been transmitted orally for generations before being written down in the 1270s. See Larrington, 1996, xi.

[9] Garðaríki -- The kingdom of the *Rus* centred on Kiev.

According to the *Hervarar saga,*[1] Arngrim the berserk attacked Gardiriki and fought with Svafrlami who used Tyrfing to cut Arngrim's shield but the sword stuck in the earth. Arngrim cut off Svafrlami's hand, took the sword and slew its owner with it. The sword later passed to his son Angantyr, and his eleven brothers. All of them were slain in a duel with the *Svear* (Swede) champion, Hjalmar and his Norse companion Orvar-Odd on the island of Samsø. Hjalmar was wounded by Tyrfing and died of his wounds after singing his death-song. He was buried at Uppsala in Sweden. The twelve brothers are buried in a mound.

In the poem, "The waking of Angantyr", [2] and in the saga, Angantyr has one remaining child, a daughter, Hervor, who he had with the woman Tófa.[3] Hervor is brought up as a bond maiden and does not know who her parents are or that she is a real shield-maiden and quite fearless. When she discovers her parentage she sails to the Isle of Samsey (Samsø) to find the mound in which her father and his brothers are buried so that she can recover the sword Tyrfing and avenge his death. When she reaches the mound, Hervor taunts the dead until they awake and she demands Tyrfing. Her now undead father warns her that the sword will destroy her kindred, specifically her sons.[4] Despite ghostly flames and her father's predictions she insists and threatens to curse the dead with annihilation.[5] Her courage is rewarded and she takes the sword. According to the poem, the gleaming blade of the sword has two poisoned edges.[6]

Hervor marries King Gudmund's son, Höfund, and they have two sons of their own, Heidrek and Angantyr. Hervor gives the sword secretly to her son Heidrek, and while the two sons are out walking together Heidrek shows Angantyr the sword but as it was unsheathed the curse on the sword forces Heidrek to kill Angantyr.[7]

Heidrek became the king of the Goths but one night while he is sleeping eight of his thralls take Tyrfing and slaughter him with it and flee. Heidrek's son, another Angantyr, catches up with the thralls and kills them, takes the sword and the curse is at last lifted.

Angantyr is the next king of the Goths but his illegitimate half-brother, the half-Hunnic Hlod, wanted half of the kingdom. Eventually this leads to a battle between the Huns and the Goths in which the Huns greatly outnumber the Goths. However, Angantyr and the Goths win because he uses Tyrfing and kills his brother. The dead choke the rivers, which provokes a flood, killing more men and horses.

Both of the above examples come from fairly late literary examples, but they were being read and told in times when swords were still being used as weapons and as symbols of status. Both swords have personalities and are capable of behaving badly. Tyrfing is particularly malicious but eventually settles down to just being a good sword. Skofnung, on the other hand, has to be treated correctly or it will not perform when needed. And of course both swords are named, just as a person is.

Wiglaf's sword

When Beowulf sets out to kill the dragon he takes eleven companions with him, and although he tells them to just witness the fight from the mounds (*Beowulf* 2529 to 2534 CH), when he became hard-pressed and it looked as though he would be defeated the poet makes it clear that they should have acted and come to the assistance of their lord, but nine of them shrank away (the tenth was the man who stole from the dragon and no warrior) into

---

[1] The saga is about Tyrfing and there are actually different versions of this part of the tale in the three different texts of the saga (H, R and U) and also in the *Gesta Danorum*, by Saxo Grammaticus (Davidson and Fisher 1980) and in the Lay of Hyndla and Orvar-Odd's Saga. In version R of the

[2] Taylor and Auden, 1969. 101 to 105. Other elements of the story appear in the Saga of Hervor and Heidrek.

[3] The woman's name differs in different versions and her status does, too. In one she is the wife.

[4] Klevnäs, 2013. 5.

[5] Davidson, 1960. 109.

[6] Taylor and Auden, 1969. 105.

[7] In the other version of the story, Heidrek kills Angantyr with a stone.

the forest (*Beowulf* 2596 to 2599 CH). Only one man came to the rescue, and he was Wiglaf, son of Weohstan, kinsman of Ælfhere. Both Beowulf and he were the last of the Wægmundings. Wiglaf reproaches his cowardly comrades but to no avail. Together Beowulf and Wiglaf defeat the monster, although Beowulf is mortally wounded. Wiglaf delivered the most decisive blow to the dragon by piercing its belly with his sword soon after Beowulf's own blade, Nægling, broke because Beowulf's strength was too great for it. (*Beowulf* 2698 CH) Wiglaf's sword was a famous one with a history but it is not named. The sword was reputed to have belonged to Eanmund, Ohthere's son. (*Beowulf* 2909 to 2926 CH) Eanmund had been slain by Weohstan and was given the sword as a reward by Onella. Eanmund together with his brother Eadgils had rebelled against their uncle, Onella of Sweden; only Eadmund was killed by Weohstan. As Crossley-Holland explains, there is an irony here as Beowulf later helped Eadgils in his revenge against Onela, yet Wiglaf wields the blade in support of Beowulf.[1] Despite its history the weapon does not warrant a name, at least not in that poem.

Skofnung and Tyrfing were possibly not the only swords that were able to act on their own and to have 'life'. Brunning provides examples of images that use artistic techniques to give an impression of swords being active, such as rows of swords angled with blades forward to give an idea of movement, and of outsize swords, both from the Old English Hexateuch.[2] However, Brunning points out that all of those examples are still under the control of a wielder and that it is only from Scandinavia and areas under Scandinavian influence where a sword is depicted acting violently on its own. The 9th to 10th century grave marker from Kirby Hill, Kirkby-on-the-Moor, North Yorkshire is one such, as is the eleventh-century sandstone slab from Tanberg, Norway and she refers to another example from Russia. Brunning fully understands that such depiction were probably just shorthand, possibly for Sigurðr slaying the dragon Fáfnir with the sword Grámr, but feels that, together with instances of swords acting on their own in the old texts, the images are suggestive[3] and show that it is only swords and not spears or other weapons that are clearly shown acting independently of their wielders.[4] Indeed, surely Tyrfing and Skofnung are particularly good examples?

Mouths?

Another quality of animation that swords are said to possess are the ability to 'bite' (*Beowulf*.ll.1521-1522; Rid.5.l.9), although the sword is never referred to as having a 'mouth' or 'teeth'.[5] They do however, have a 'voice'. Brunning explains that swords in the Old English riddles often speak in the first person, but realises that this is possibly just a convention. In *Beowulf* Hrunting sings 'a greedy war-song' (*agol / graedig guðleoð*: *Beowulf* ll.1521-1522) while being used against Grendel's mother; the sword which Abraham wields to sacrifice Isaac in *Exodus* 'roared' when he drew it (*Exodus*.l.408),[6] we have seen too that Skofnung 'howled' as it was drawn from its scabbard when Kormák mishandled it. There are other examples from the Norse literature, such as the 'singing of swords' (*söngr sverða*) which appears in the *Liðsmannaflokkr* and Einarr Helgason's *Vellekla*.[7] Swords may not have teeth, but it seems they do have tongues and a voice. The comparison to tongues is not surprising as their shape can be construed as tongue-like.[8]

---

[1] Crossley-Holland, 122.
[2] Brunning, 2014. 94 to 95.
[3] Brunning, 2014. 96 and 257. On pages 93 to 96.
[4] Brunning, 2014. 239 and 251.
[5] Brunning, 2014. 190.
[6] Both examples provided by Brunning, 2014. 191.
[7] Brunning, 2104. 190. Brunning says that it is possible that the noun *söngr* could denote 'music' rather than a human voice.
[8] See Brunning, (2014. 190) for more discussion on swords and tongues.

Feelings

Swords are sometimes portrayed as having feelings and Brunning gives an example from the Old English poems *Beowulf*, *The Seafarer* and *Andreas*, where the kenning 'sword hate' (*ecghete* or *billhete* - *Andreas*.l.78; *Beowulf*.ll.84, 1738; *Seafarer*.l.70) is used, although it is not clear who or what is doing the hating,[1] as Brunning comments, "…the kenning could signify hate by, of or inspired by swords." The Beowulf poet also mentions the 'terror of swords'(*brogan billa* - *Beowulf*.l.583) although it is unlikely that it is the swords themselves that are 'terrified' and there is also 'edge pressure' (*ecgþræce* - *Beowulf*. l.596),[2] in this instance it could well be the sword that is experiencing such a pressure. Brunning discusses other kennings relating to weapons in battle, including some Norse examples and concludes that it could be the battle, not the weapon that is perceived as the living entity.[3] She has found a convincing kenning that swords had feelings in the Old English poem *Maxims II,* where the poet writes, 'the sword must experience battle, blade opposing helmet' (*ecg sceal wið hellme / hilde gebidan* – *Maxims ll*.16-17) and explains that this is contrasted to the way that other weapons are dealt with as the, "…spear is said to belong in the hand, and the shield boss on the shield to protect its wielder's fingers (*Maxims II*.ll.21-22, 37-38) – in other words, they are the possessions and / or servants of their human owners, and are not shown capable of feeling or acting in the same way as swords."[4]

Animals

As Brunning shows, swords in Old Norse poetry are sometimes compared to animals and she provides examples such as 'serpent of wounds' (*linnr sára Bersi*.Flokk.§3; Eskál.Laus.2a), and 'serpent of the storms of *Yggr* [battle]' (*linnr éla Yggs*: Ótt.Höf.§8) and such as 'wound-salmon' (*bensíkr*: Gsind.Hák.§7) and 'corpse-sea-trout' (hræbirtingr: Tindr.Hák.§6); or to dogs, as with 'helmet dog' (hjalmgagarr: Darr.§3) and 'dog of shield' (gagarr seilar: Tindr.Hák.§4). She feels that it is easy to see why these animals have been used, serpents weave around, are shiny and are dangerous, fish, too, reflect light, dart around and can bite, whereas dogs can cause nasty wounds and are persistent. There is too an Old English riddle whose solution is ambiguous but could be either a sword or a hawk;[5] it is almost certain that the ambiguity is deliberate and that both meanings are not just meant, but linked together in the listener's mind.

Many decorated hilts have animal ornament on them and on their accessories as well and, Androshchuk, as previously mentioned, feels that it is this animal ornamentation that gives life to the sword. Again, the outward facing animals on the pommels of many swords could possibly be a reference to the cutting or biting powers of the edges of the blade.

The question of whether sword could be viewed as being animated and able to act on their own is difficult to resolve absolutely, as Brunning acknowledges,[6] and adds that, "…the idea that a sword could 'stand in' for a person is also pertinent." This is what the sword Mimming appears to do (see below) and it is not the only example. It would appear that the idea of swords being at least a participant in the action and partly responsible for it, seems to me a very valid one.

---

[1] Brunning, 2014. 197.

[2] Brunning, 2014. 197 to 198.

[3] Brunning, 2014, 198.

[4] Brunning, 2014. 198.

[5] Brunning, 2014. 191 to 192.

[6] Brunning, 2014. 239.

Faces

Brunning raises an interesting question and that is whether swords could have been regarded as having 'faces' and if so, would this contribute to the possibility that swords were,"...thought to have their own characters, personalities, social relationships perhaps other person- or animal-like qualities such as names..." [1] We have discussed the possibility that the gold and garnet sword bosses, which are sometimes found on high status swords, could have been regarded as 'eyes' and if so they certainly could contribute to the idea that there was a face buried within the sword ensemble.[2] But Brunning seems to be asking if the faces of animals, or much more rarely of men, and the hilts themselves could be regarded as contributing to the sword's 'face' and its ability to be perceived as alive.[3] One aspect of sword pommels that we have already noted is that all, but a very few decorated pommels, have different designs on each side. This is quite deliberate and is sometimes extremely subtle, as is the case with the Mound 1 Sutton Hoo sword, but in others the differences are very clear indeed; Brunning, ponders this and wonders whether this was being used to help to construct the sword's 'personality'.[4] Her idea that faces change through time and that the hilts would too is very reasonable and by itself would add to the lifelike qualities of a sword.

Later Brunning raises the possibility that not only did a sword have a 'face' but also a 'body'; the blade is the body and the hilt is the face, and like our faces, it is usually on show, whereas the 'body' was clothed most of the time in its scabbard and only revealed to others under certain circumstances.[5] She adds that the blade was, "...the home of the sword's history and personality – inner elements known only by those with which it has the closest relationships: those, for want of a better analogy, who knew what was 'inside the scabbard' both literally and metaphorically." Brunning goes on to suggest that the 'face' or hilt could be embellished or altered over time to create different identities and demonstrate visually the age of the sword, thus contributing to its perceived value.

Named swords and reputation

According to Androshchuk we know of one hundred and seventy-six named swords in the Norse literature[6] and, although there are far fewer in Old English, there are some. Brunning, rightly, mentions the fact that some scholars seem to have assumed that all swords were named but that most of the sword names we have come from 12th century Icelandic literature.[7] We simply cannot know if only some swords or all swords were given names, but when something gets a name it makes it easier for it to acquire a reputation.

A few swords from the 5th to 7th centuries in England have inscriptions, mainly runic, on the hilt or on a part of the scabbard,[8]although some of these are probably names. As Pollington and Brunning[9] point out, the intentions of the inscriber and his or her object are not always clear. One examples is the sword from Chessell Down, Isle of Wight that has a runic inscription on a scabbard fitting which may say *aeco soeri* in Old English, which could be a personal name like 'Acca' or 'Acco', but 'increaser' or even 'augmenter of pain or wounds' have also been suggested.[10] If it is a name, is it the name of the sword, its owner, the maker or of something supernatural? This

---

[1] Brunning, 2014. 40. See also: 145, 233, 252 and 256.

[2] See chapter on *Additions to the sword, beads, bosses and pyramids*.

[3] See also Androshchuk, 2014. 202.

[4] Brunning, 2014. 43.

[5] Brunning, 2014. 233.

[6] Androshchuk, 2014. 202. These have been collected mainly from Icelandic sagas, Old Norse poetry is more reluctant to name swords outright but uses kennings to refer to them, Brunning, 2014. 256.

[7] Brunning, 2014. 41.

[8] See Pollington in the chapter on *Runes on Swords and Other Weapons* in this volume.

[9] Brunning, 2014. 41 to 42.

[10] Brunning, 2014. 41 to 42. All known runic inscriptions associated with swords are discussed by Pollington in the chapter, *Swords and runes* in this book.

same problem of uncertainty regarding the meaning of most of the inscriptions associated with swords means that it is very difficult to provide an example of an actual sword that has its name on it. Brunning speculates on why, if swords did acquire names, there has been little discussion on why such names are not recorded upon the swords, which is an interesting thought that is currently unanswerable. It is possible that the name of a sword was not usually formally recorded. The sword hilt that Beowulf brought back from Grendel's mother's lair, had an inscription which was said to record the name of the person who had commissioned the sword (*Beowulf* 1696 to 1700) but not the sword's name, and perhaps that is a clue to the purpose of other sword inscriptions? The much later Ulfberht and Ingelrii swords have inlaid letters on the blade but these are maker's marks. Incidentally, there are no known swords from the 5th to 7th centuries from England that have blade inscriptions.

The Old English poem *Waldere* has a named sword in it, Mimming, which brings to a total of three, the swords that are definitely named in Old English poetry, the other two are Nægling, Beowulf's sword and Hrunting, the sword of Unferð. Brunning mentions the possibility that the name Hunlafing could refer to a sword or a man, both, apparently, would work in the context of the poem.[1] Perhaps it is significant that other swords are referred to which are not named, the sword that Beowulf gives to the coastguard, Wiglaf's sword and the sword of the Heaðobards are particularly prominent examples; all of these weapons are significant ones and if they had names the poet does not use them. It could be, as Brunning posits, that the naming of a sword was usually just between a warrior and his sword and concerned no-one else.[2] [3]

There are other ways of referring to swords than using their names and Brunning writes that there have been two hundred and ninety-seven kennings listed in both Old Norse and Old English poetry (two hundred and forty (81%) ON and fifty-seven (19%) OE). Kennings that relate swords to ice and light for example are referring to the bright and cold qualities of the blade[4] but also provide a striking image of action; 'battle icicle' *hildegicel* in *Beowulf* (l.1606): *swurdleoma* – 'sword light' illuminating the hall during the battle described in the OE Finnsburh fragment (l.35)[5] are also able to show the 'active' properties of a sword in a fight and bring them to life. There are many others.

We know from the examples of Skofnung and Tyrfing that swords could develop reputations of their own, and that they were not always beneficial to their owners. This is further emphasised in Beowulf when we are told of the sword of King *Healfdene*, which was a 'renowned treasure-sword' (*maere maðþumsweord*), un-named though, carried across Hroðgar's hall while 'many men watched' it being gifted to Beowulf as one of his rewards for killing Grendel (*Beowulf* 1019-1024).[6] When preparing for the fight against the dragon, the elderly King Beowulf boasts of his own sword, Nægling and tells his comrades that, "I shall fight for as long as I live and this sword lasts, that has served me early and late…" (*Beowulf*, 2498 to 2492 CH). The sword Mimming, from the poem Waldere, also had a reputation. The poet relates, "Never shall the work of Wayland fail a master of Mimming, a man who knows the handling of that blade. Bleeding from its wounds, lords and æthelings are laid on the field." (Waldere I. 2 to 5).[7] It is interesting to note that the action is attributed to the sword, not its wielder, Waldere (Waltharius). The poem also tells us, "I know that Theoderic once thought to send it (Mimming) to Widia himself, and along with the blade a wealth of treasure and a troop in escort all clad in gold - as a gift, because in a narrow place once Nithhad's grandson – the lord Widia, Wayland's son –

---

[1] Brunning, 2014. 184.

[2] Brunning, 2014. 42.

[3] Nick Kyme, 2015. Artefacts – in Laurie Golding (editor) War Without End. Nottingham."

[4] Brunning, 2014. 179.

[5] Brunning, 2014. 193 and 218.

[6] Brunning, 2014. 186.

[7] Alexander, 1966. 60.

had saved Theodoric, and through him the Goth slipped the guard of the giants." (Waldere II. 4 to 11).[1] We are not told how the sword came into Waldere's possession or why Theoderic did not give it to Widia, but it is plain that Mimming was a famous sword and had a reputation, certainly in the hands of Waldere it was deadly, killing eleven opponents in single combat and wounding Guntharius and Hagano. It could be, though, that Waldere had two swords, although, if that is so, the other is not named.[2]

It is obviously impossible to tell from grave finds whether the swords found in them had famous reputations but Brunning believes that, "In general, enough points of correspondence occur between texts and archaeology to infer that swords could acquire reputations, particularly on account of their associations with specific individuals."[3]

Summing up the living sword

Swords, especially old ones, could acquire many of the traits of being 'alive' and some undoubtedly had names, but as we have seen this does not seem to have been the case with all swords, and naming could have been a much more informal process than some have assumed. Swords definitely had public 'faces' that could change over time, just like those of people, and aspects of faces would have contributed to the visual identity of swords, such as the garnet and gold 'eyes' on a very few swords. It is true that some parts of the sword were generally kept covered and 'secret', perhaps a little like a person's body, but this could once again just be a case of transferring modern ideas about modesty into the past. Some swords were trustworthy and reliable and perhaps viewed as old companions. For example, Hrunting is referred to in *Beowulf* as a *guðwine*, (OE) 'friend in war'.[4] There is no doubt that older swords gained in value through their associations and history. Certainly swords could 'bite' in a very real sense and cause tremendous damage to people.[5] There were also links to animal imagery that could contribute to the way that the 'life' of a sword was perceived. It must be added that many other objects, including women's brooches, had similar icons embedded within them, and the link that has recently been made regarding brooches and swords in burials is, I suspect, really significant.[6] There is room for much more research and critical thinking here. The literature tells us that swords could make noises and become animated during use but no one has recently been involved in the sort of combat experienced during the 5th to 7th centuries and beyond, when close combat and the use of sharp weapons in quantity was required. Modern re-enactment is really a type of sport and cannot replicate the actual experience at all. It is possible that there could have been real noises emitted or associated with sharps sword cutting through flesh and bone that we are ignorant of. Swords acquire reputations and not always good ones, at least not for everyone, and this is something that again reflects 'personhood'.

One other factor that possibly contributes to the way that swords were viewed as somehow being 'alive' is the world view of the people at the time. We are only able to get glimpses of the beliefs of the pre-Christians, and it is clear that many of those ideas did not completely die out with the conversion and probably persisted for a very long time. While acknowledging that beliefs would have varied from place to place and over time, it does seem that many things were thought to have 'life' and a 'spirit' (for want of a better word) governing them; these are things like trees, rivers and rocks, which in the modern world we consider to be inanimate.[7]

---

1 Alexander, 1966. 61.

2 Alexander, 1966. 59.

3 Brunning, 2014. 235.

4 Brunning, 2014. 40. *Beowulf*, 1810. The line is cwæð: he þone guðwine godne tealde – Crossley-Holland translates the phrase as 'ally in battle. Crossley-Holland, 1999. 60.

5 See the chapter on *Trauma and the Effect of Swords* in this volume

6 See below.

7 See chapter entitled *Background* in this book where beliefs are briefly discussed and other reference material is noted.

iii. Sword burial[1]

It seems obvious that swords were included with particular dead because they were warriors, or of the warrior class, but they are not deposited with everyone who was a fighter,[2] far from it. There is a possibility that sometimes, but not often, they were left with individuals who were seemingly incapable of wielding them.[3] Some of the questions that need answering are: Why were swords buried with some people and not with others? Did the men buried with swords own them? Perhaps others who carried a sword were not buried with them because their sword was only loaned to them? Were swords really as scarce during the early medieval period as the frequency of burial and many scholars have suggested? Hopefully, we will be able to make some progress with these questions by the end of this chapter.

Swords are actually rare in Anglo-Saxon graves, although some commentators say that they occur in around 5% of graves,[4] this is not dissimilar to what I have found in my survey of two hundred and thirty-one burial places representing at least 9,800 graves where only one hundred and sixty-nine swords were found; the ratio would appear to be around 1.7% of all graves, perhaps slightly less. This would perhaps equate to around 3% to 6% of adult male graves and could lead to an expectation that every one thousand graves of all types would yield between fourteen to twenty swords.[5] It is true that some cemeteries were better provided for and that there were many variations in burial rite over time and in different regions. However, throughout the period of furnished burial in Anglo-Saxon England, sword burials were not frequent. They were certainly not for every dead warrior. This has led many to suppose that swords were used by a very small proportion of warriors during this period,[6] which is a view I disagree with. I think that many fighters carried swords and that burial archaeology does not reflect everyday life during the early medieval period or at any other time.[7]

As swords do not occur in graves very frequently it leaves us with a problem of understanding why they were left in some Anglo-Saxon graves; some men are being marked out as being special in some way. We must not be led astray by the very few graves where the occupant supposedly could not have wielded a sword either; the work on this was carried out by Härke and although he gives this aspect of the sword some emphasis, in his data, I could only find one sword that was left with a 'child' (eight to fourteen years old).[8] This is possibly the same grave that Stoodley discusses at Broadstairs I (grave 71), where a sword bearer was calculated to be aged twelve to fifteen years old; the occupant also possessed a shield and two spears.[9] At fourteen, it seems in the Anglo-Saxon world a boy could be given the arms of a man, as appears to be the case with Saint Wilfrid and

---

[1] Härke, 2014. Contains a survey of ideas from different cultures and time periods as to the reasons why grave goods, generally, could have been deposited. Shepherd, (1997. 7 to 8) sees disturbed burials as signs of stress owing to encroaching conversion. Steuer, 2006. 234, asks, "Are weapons symbols of social status and rank, or indeed signs of actual participation in combat, or even in wars?"

[2] Theuws and Alkemade, 2000. 427, 442, 449 and 462. The authors are referring to deposits in northern Gaul during the 5th and 6th centuries, but the situation is the same in Anglo-Saxon England, as Härke, (1989. 52.) demonstrates.

[3] Härke, 1992. 149 and 156. Shepherd, (1997. 5.) discusses this, as do many others.

[4] Ager, Cameron, Spain and Riddler, 2006. 14. The authors don't actually provide a figure for the whole of England; they are discussing Kent and say the numbers of 5% plus are quite high for England in general. Härke, (1997. 119) says that about 12% of weapon burials contain a sword.

[5] See section on Cemetery/sword ration in this volume for further details.

[6] This issues is discussed in the chapters, *Background* and *Pattern-welding*.

[7] See also Halsall, 2010. 2. See the chapter on *Pattern-welding* for the full discussion on numbers of swords made during the period.

[8] Härke, 1992. 156 and 157.

[9] Stoodley, 1997. 183. I have not been able to check this against Härke's original data, but Broadstairs was one of the cemeteries that he studied.

others like Saint Cuthbert and Saint Guthlac who had their own bands of armed followers at around fifteen.[1] When Härke considered the physiques represented by the skeletal remains, where it could be determined, not one with a weak build was accompanied by a sword.[2] Whatever, most sword graves are quite clearly those of fighters and many of them, where they can be aged, were quite young when they died.[3]

It should be noted that the sword burial rite does not occur in many places in Europe during our period, apart from Anglo-Saxon England swords deposition only occurs in northern Gaul, the Alemannic area, Lombard Italy, certain parts of Scandinavia and parts of the Slav territories.[4]

All the data from Anglo-Saxon cemeteries that I have collected obviously refers only to Anglo-Saxon England and frequencies of swords in graves could well be quite different elsewhere on the Continent. However, recent work, particularly from the Merovingian areas of Europe and from England, may be providing clues as to the role that swords played within the sword burial rite. Perhaps, ironically, the information is being revealed through the practice of grave robbery rather than just grave deposition.

It was recognised, even in the 18th century, that the early archaeologists were not always the first people to open up a grave. The early excavators sometimes found examples where the bones and other contents had been piled up, or scattered, and material had been removed and bits of broken artefacts were left in the grave cut. Modern archaeologists have continued to note such finds but they have not received serious, specialist attention until recently.

There would appear to be several different reasons for re-opening graves and a group of archaeologists from a number of different countries have been focussing much of their efforts on exploring this phenomenon and they have been particularly interested in incidents where the re-opening occurred fairly soon after the original burial; that is within a generation or three. They have called themselves the Grave Reopening and Research Group, hereinafter, the GRRG.[5]

Some graves were disturbed because of revenants, or *draugr*, *wiedergänger* or *shadow walkers*[6] as they are sometimes called in different literatures; modern media often refers to them as 'vampire' burials because they occasionally have stakes placed through their bones and often the skulls are placed anywhere but where they should be. Perhaps, more usually the occupants were buried in a prone position or had heavy stones laid on top of their bodies.[7] As interesting as this type of occurrence is, it does not tell us anything about why swords were put into some graves so we will only refer to it when necessary, such as discussing the ideas of scholars who are exploring re-entry into graves. Some of the archaeology seems to reflect what the literature tells us about how the unquiet dead were dealt with.

A form of grave robbery that will not detain us for long is the type that was conducted purely to obtain wealth, such as that practised during the late Tudor period in England and at other times elsewhere. In this case, everything that had value was removed; there is good archaeological evidence that most of the burial mounds at Sutton Hoo were robbed during the reign of the Tudors, probably while Elizabeth I was on the throne, presumably in order to raise funds for her war effort against Spain.[8] In most of the incidents at Sutton

---

[1] Davidson, 1989. 20.Halsall, 2003. 57 and 58.. All of the above mentioned aristocratic, young men gave up fighting and went into the church.

[2] Härke, (1992.157) Härke says that "…swords seaxes and axes not found with weakly built men…"

[3] See chapter, *Analysis of Some Data from Spreadsheet*.

[4] Theuws and Alkemade, 2000. 427.

[5] The Grave Reopening Research Group can be found here, together with information about the individual member's publications: http://reopenedgraves.eu/

[6] For a discussion of this practice, see Klevnäs, 2013. 4.

[7] Fuller discussions of this phenomenon are included in Reynolds, 2009, and Gardela, 2012.

[8] Bruce-Mitford, 1975. 160 to 161.

Hoo, almost everything was taken from the tomb leaving few clues as to what had been left there by the families of the deceased. This is typical of this type of grave reopening and most of the 'thorough removals' tended to happen hundreds of years after the initial burial as happened at Sutton Hoo and elsewhere and the motives of the robbers were usually just financial gain.

Another practice involving grave reopening occurred when a Christian descendant had some of his ancestor's remains removed from a mound to a church yard or the church itself, as happened with King Harald Bluetooth's father. This would appear to be an attempt to 'Christianise' an ancestor. Once again, this tells us nothing about the inclusion of a sword within a grave. A similar practice was that of entering a grave to obtain Christian relics, the bones of a saint, or some other fragment that was perceived as holy.

Although interested in all forms of grave reopening, much of the work of the members of the GRRG has been focused on accumulating information on a different practice which, as mentioned above, involved entering the grave relatively soon after the burial, that is within a few generations, where only some goods were removed. They are also trying to understand the reopeners' motives.

Before considering the most recent research, we should just mention a couple of the ideas that have been put forward to explain this type of grave robbery. Heinrich Härke,[1] who had noted some of this type of interference during his work, supposed that as the custom of grave good deposition effectively took those artefacts out of circulation, those items would need to be replaced and an easy way of doing that was to remove similar items from other graves, perhaps from neighbouring communities.[2] Heiko Stueur thought similarly and that the grave disturbances were probably committed by families who could not afford continually to bury their dead with valuable grave goods, and therefore began 'recycling' artefacts from older burials in order to conserve resources.[3] There is an assumption here that the kin were relatively poor or of diminished resources, a fairly common view of the peoples of early England. However, there appears to be much more going on as we shall see.

The first member of the GRRG, whose work I would like to bring into the discussion is that of Martine van Haperen, even though she does not specifically mention swords. Haperen has surveyed the work of several archaeologists who have reported finding disturbed graves as well as taking part in archaeological excavations herself to study the phenomenon. One archaeologist that she mentions is H. Roosens, who was working in Belgium.[4] Roosens reported that of the one hundred and twenty graves at the cemetery at Rosmeer, at least thirty-five of them had been robbed. Roosens did not include detailed plans of his finds, but some of what he reported makes it plain that the robbers were treating the graves in particular ways and that they were not interested in taking everything; Haperen, shows that in many examples of grave disturbance, valuable objects that the intruders could hardly have missed were often left behind, large iron objects and bits of jewellery, for example. Sometimes quite useful items, such as pots, belt parts, combs and weapons were, instead, deliberately destroyed by the robbers.[5] Human remains were removed at times, too.[6] So it would appear that Steuer's and Härke's ideas concerning the recycling of artefacts were probably not in the minds of those carrying out the intrusions.[7]

---

[1] Härke, 2000. 392.

[2] Härke, 2000. 390.

[3] Haperen, 2010. 15. Harperen discusses Steuer idea. (Steuer, Heiko., 1998: Grabraub, Archäologisches, *Reallexikon der Germanischen Altertumskunde* 12, 516–523.)

[4] Haperen, 2010. 2. Roosens published in 1978.

[5] Haperen, 2010. 3.

[6] Haperen, 2010. 16.

[7] Klevnäs, 2013. 89.

Haperen also noted in her survey of grave robbing in Merovingian areas that the pits used by the robbers, "…often appear to have been dug with great precision, indicating that the diggers knew where the grave lay, either because they had been present at the funeral or because the location of the grave was indicated by a, possibly perishable, grave marker or, in uniform row-grave cemeteries, by the diggers' knowledge of the way the graves were laid out."[1] When discussing the work of Grünewald, she says that he believed that it would have been difficult for the work of reopening the graves to have been carried out in secret and that therefore the local community was quite likely to have condoned what was going on.[2] Haperen does note some textual materials that possibly support his suggestion.[3]

According to Haperen most of the grave reopenings seem to occur on graves dated from the late 6th and 7th centuries, while the vast majority of earlier burials appear to be unaffected. She feels too, that certain artefacts were so closely connected to the famous dead or events that they were of great value to the living and had to be retrieved. The fact that they had resided with the dead for a period could have altered their symbolic associations and made them more powerful. She suggests, too, that the robbings could be a way of fixing the presence of the ancestors among their descendants.[4] In her conclusions Haperen appears to agree with Grünewald who felt that some of the items in the grave were capable of transforming the dead into dangerous living corpses and were therefore removed as she says, "…the recurrent aspects of these practices indicate that they originated from more complex motives than a straightforward desire to obtain the valuable artefacts buried with the dead." She goes on to say, "The deceased is perceived to change from a potentially dangerous entity into a benign ancestor who has the ability to promote the well-being of her living descendants."[5]

What this does not explain is why the earlier graves were not touched and why some cemeteries were left alone.

Grave reopening was a widespread practice in Merovingian areas and elsewhere and swords were often taken during these episodes, see for example the work of Codreanu-Windauer, excavating in Pliening, Germany, who tells us that even dilapidated swords, of no practical use were targeted.[6] We shall have more to say about this later.

It is to the work of another member of the GRRG to whom we now turn, and it is possible that she may be able to tie together some of the things that were happening on the Continent during the 6th and 7th centuries with similar phenomena in England.

Alison Klevnäs published her thesis in 2013 under the title, *'Whodunnit? Grave-robbery in early medieval northern and western Europe'*. It is a work that can help provide some clues as to why very few graves were provided with swords. Although most of the graves that she studied in detail were in Anglo-Saxon England, and most of those were in Kent, her thesis has implications for most areas during the 6th and 7th centuries and for Scandinavia, perhaps for a much longer period.

Basic statistics for Kent

Klevnäs attempts to bring together all that is known of re-entered burials from two main areas, Anglo-Saxon England and Merovingia, although she does touch on other regions too.[7] She looked at evidence from one hundred and twenty burial sites in Kent, England, and found thirty-two of them contained graves that had

---

[1] Haperen, 2010. 10.
[2] Haperen, 2010. 4. See also Haperen, 2010. 18..
[3] Haperen, 2010. 18.
[4] Haperen, 2010. 24.
[5] Haperen, 2010. 28.
[6] Klevnäs, 2013. 15.
[7] Klevnäs, 2013. iii.

been broken into since the initial closing.[1] Of these, eight cemeteries showed considerable evidence of early disturbance and these were studied in close detail.[2] In the eight sites, at least one hundred and eighty-three graves, (actually containing one hundred and ninety-three burials) had been reopened sometime after the initial funeral. This meant that of the total of one thousand one hundred and ninety-seven graves from within the eight cemeteries, some 15% had been disturbed.[3] She also looked at the records of a further ten cemeteries, all of which had one or two graves that had been visited after the initial burial. She records a minimum number of two hundred and one known disturbed graves in Kent and a ratio of 8% (Finglesham) to 44% (Ozengell) re-entered tombs in the more heavily disrupted sites. She does point out that the variation in the rate of disturbance is at least partly due to incomplete excavation and that the rate of disturbance in Kent is consistent with that across the Channel.

Some of the heavily disturbed sites are close to cemeteries that would appear to have not received any attention from grave robbers at all. This is something that also occurs within the Merovingian lands.[4] Why were some cemeteries being robbed and others not all; indeed why did particular graves receive the attention of grave reopeners while neighbouring ones were left alone?

As with the Merovingian graves, Klevnäs notes that the cuts were often quite precisely dug, so the invaders must have been able to discern where the grave was and its orientation. It is quite likely that there were forms of surface marking or structures that showed the position of the graves.[5]

Elsewhere in Anglo-Saxon England, grave robbery seems to be a much rarer practice and does not reach the level of intensity seen in Kent.[6] However, more cemeteries have been excavated and recorded in Kent and from an earlier time than in any other English county.

How long after the funeral was the grave re-opened?

The main way in which the time difference between the cessation of the funeral rites and the reopening can be worked out is by studying the processes of decay and deterioration that affect the corpse and the other artefacts in the grave, and gaining an understanding of how long those processes take (*taphonomy*). It is then necessary to apply that knowledge to the way that the remains are disposed in relation to each other when the archaeology takes place. Other considerations must be taken into account such as local soil conditions and the proximity of various materials and whether or not a coffin or another container for the body was involved. Air filled coffins speed up the process of decay and also provide a clear space for objects to be moved around in by the intruders.[7] Edeltraud Aspöck, another member of the Grave Reopening Research Group, based on previous studies, worked out a rough set of timeframes to assess the span of time that the deterioration takes.[8] From her work she posits that complete skeletisation would occur between ten to thirty-five years.[9]

In Kent, it seems that most of the reopening happened after the bones had become disarticulated but while they were still in good condition; and in quite a few cases the coffins were still solid.[10] Some of the visited graves seem to have contained an unquiet corpse, a revenant, and these were reopened relatively soon after the

---

[1] Klevnäs, 2013. 34.

[2] 1. Bradstow School, Broadstairs 2. St Peter's Tip, Broadstairs 3. Lyminge 4. Half Mile Ride, Margate 5. Primrose Hill, Monkton 6.Finglesham 7. Ozengell, Ramsgate 8. Sarre

[3] Klevnäs, 2013. 35.

[4] Klevnäs, 2013. 35.

[5] Klevnäs, 2013. 38. Klevnäs contains a discussion on what is known about grave markings and what the possibilities are.

[6] Klevnäs, 2013. 34.

[7] Klevnäs, 2013. 40.

[8] Klevnäs, 2013. 43. Klevnäs, 2013 contains a discussion on taphonomy and how it was applied to the Kentish graves. 42 to 47.

[9] For full details see Klevnäs, 2013. 43.

[10] Klevnäs, 2013. 42.

initial funeral as the corpses were relatively intact and could still be manipulated. Most of the disturbed graves were not of this type.

Gender balance

Similar numbers of female and male graves were attacked and had goods removed and although children's graves were sometimes entered, this appears to happen mostly when they have been mistaken for adult graves, generally they were avoided.[1] In the most disturbed eight sites, of the graves that were reopened, 19% were female, 22% male and 15% unsexed (fifty-four, seventy-four and sixty-three graves respectively).[2] Klevnäs does acknowledge the possibility that at some sites the intruders may have selectively targeted male graves but nonetheless, the overall picture is of a fairly complete gender balance.

It is thought that in both Kent and elsewhere the intruders usually did not know who was in the graves that they chose to enter and that they tended to go for better furnished tombs, probably because of the aforementioned surface markers.[3]

Dating

The attacks on the graves seem to be concentrated to a fairly short time period, probably beginning during the late sixth century and continuing into the seventh stopping before the furnished burial practice ceased; so the attacks span around half a century or more. Most of the targeted burials were from the 7th century, but some sixth century graves were not immune and perhaps a few from the 5th century were opened, too.[4] There was no consistency in the period between burial and the grave being opened again.

What was always taken and how we know?

Although Klevnäs believes that weapon graves were not systematically targeted as many weapon graves were left alone, it does seem that swords and seaxes were virtually always taken if they were in an opened grave. As in Merovingia, this occurs even when the weapon was no longer of any use and parts of it fell back into the grave during removal. One of ways that the excavators could tell that a sword had been taken was because of fragments of the sword or its scabbard left behind, another was rust or other stains in the earth or on the bones.[5] As for seaxes, Klevnäs is referring to large knives used for hunting or war, not small knives. Large seaxes are not found in disturbed graves, although, like swords, remnants or stains might remain.[6]

Brooches appear to have been consistently removed from women's graves. The missing brooches, again, are usually recognised through stains on textiles, bones or earth. They can also be presumed missing when sets of female apparel are found that usually include a brooch. They would also be found with the usual compliment of items that accompanied a woman of status into her grave but which tend to be left behind by grave intruders. Other who have written about brooches have noted the fact that women of the 5th and 6th centuries tended to wear multiple brooches and that there was shift to a single brooch during the late 6th and 7th centuries, and to pendants from the mid-7th century.[7]

As this book is about swords we will not be dwelling on the missing brooches from women's graves but it is remarkable that brooches appear to be somehow the female equivalent of swords and we will briefly return to

---

[1] Klevnäs, 2013. 40 to 41.
[2] Klevnäs, 2013. 42.
[3] Klevnäs, 2013. 15 and 42.
[4] Klevnäs, 2013. 42.
[5] Klevnäs, 2013. 66, 72 and 88.
[6] Klevnäs, 2013. 71. Only one seax has been found in a disturbed grave, "Gr 22 at Monkton, where it is described as 'a large iron knife or perhaps a small seax, blade very incomplete'".
[7] See Klevnäs, 2013. 68 for further details and references.

the topic below.[1] For now it is enough to note that in many ways women and their brooches appear to have been as important in society as men and their swords; both combinations carried sufficient prestige for their symbolism to be important. This does not mean that swords and brooches necessarily carried exactly the same significance.

What was sometimes taken?

Bones quite often appear to have been removed, especially the skull,[2] and this is true for both Kent and parts of Merovingia. Klevnäs notes that when skulls were present it appears they may have sometimes been interfered with in some way. Curiously there are a number of Kentish graves into which extra bones, particularly femurs were added. We will look at this again below.

What was hardly ever taken?

Although they are sometimes taken from some graves, spears, shields, beads and pendants were usually left behind, even though their metal content had considerable value. This was especially so in the occasionally use of precious metals in pendants. Female accessories such as girdle hangers seem never to be taken, nor do small utility knives from either sex.[3] Small buckles from belts were consistently left, too.[4]

What was the main purpose of the robbing?

When attempting to understand why swords were included in some burials, it would be useful to know why they became an important and consistent target for removal. They obviously held a very special significance for those that were prepared to take them, along with seaxes and brooches. It is possible, too, that Haperen's suggestions have provided the answers; the artefacts were too valuable to leave as they had acquired added value by being with the dead. Their removal fixed the dead person in place and transformed them into a 'safe' and benign ancestor. The trouble with this set of ideas is that surely they would apply to all furnished burials? In that case, why would only some be selected for reopening? There does not seem to be much difference in the frequencies and actual practice of grave robbing in Merovingia and in Anglo-Saxon England.

We have already looked at the idea of recycling artefacts and shown that the evidence does not really support it. So why, for a period of more than fifty years, were swords and other objects regularly removed from graves?

Klevnäs points out that there is a puzzle; robbing a grave would appear to be transgressive and a terrible act of disrespect but it does appear that the intruders were following certain rules about what should and should not be taken,[5] yet they could be quite careless about the way they left the stuff behind and often did not refill the graves.

Klevnäs remarks that Codreanu-Windauer, writing in 1997 of the Pliening cemetery, felt that the disruptive way that the graves were entered and the way that swords and seaxes were targeted pointed to a motive in the realm of the spiritual and that it was all about an attack on the power and prestige of the dead ancestors and an attempt to ensure that they could not support the living, and it was an indirect attack on their living descendants. The cause, probably involved local arguments between different family groups vying for power.[6] Certainly the reopenings are happening at a time of social change, when there was a shift to greater centralisation of wealth and power away from 'free' independent men and, perhaps, towards a more 'manorial' system.[7]

---

[1] It is interesting, too, that *Androshchuk* seems to make a similar connection in that sword distribution coincides with that of oval brooches in parts of Scandinavia. Androshchuk, 2014. 203.

[2] Klevnäs, 2013. 17, 52 and 57.

[3] Klevnäs, 2013. 89.

[4] Klevnäs, 2013. 71.

[5] Klevnäs, 2013. 72. The bones were often scattered and the graves left open. See Klevnäs, 2013. 17 regarding the way graves were often left.

[6] Klevnäs, 2013. 15.

[7] Klevnäs, 2013. 19.

Building on those ideas, Klevnäs's conclusion is that the intention of the robberies was not to obtain objects for the living, "…but to remove them from the realm of the dead."[1] She believes that this process involved, "…highly specific symbolism. Only certain types of artefact, primarily swords and brooches, are consistently missing from disturbed burials."[2] The robberies were an attack on the prestige of the burying group rather than the specific dead people that were in the graves that had been targeted. She writes, "Grave robbery was a tactic to undermine the use of competitive burial rites and monuments, and in particular the deposition of grave-goods. The remains of past displays were destroyed, and subverted into memories of weakness and defeat."[3]

In addition, Klevnäs, along with others, feels that the evidence supports the notion that the reopenings were committed, "…by people from the same immediate or regional community as the burying group, who shared the same understandings of different grave goods."[4]

The reasons why particular objects like swords and brooches were taken must be due to the ways in which their inclusions within the tombs were viewed, so what was special about them?

Swords were objects whose ownership could be transferred from one person to another; that is they were 'alienable'.[5] Brooches, it seems, were likewise alienable and could be passed down the female line. Utility knives and girdle hangers were too personal to be removed; they belonged with and must stay with the deceased. It has been suggested that swords are seen as symbols of strength and that arming the ancestor protected the living and that brooches are in some ways a female analogy.[6] This model may explain why so few burials needed swords, perhaps only one ancestor per generation needed to be armed? Perhaps, too, depositing a sword was a sacrifice for a community? Whether a sword was included or not might be dependent on the manner of the occupant's death; perhaps something to do with dying in a particular way during combat…or maybe the sword had failed and it too had died?

Thewus and Alkemade, in an earlier paper, seem to support much of what Klevnäs suggested for northern Gaul during the 5th and 6th centuries. They say that swords were included within special graves, possibly that of a 'founding ancestor', in order to provide protection for the community/family; removing the weapon would take that spiritual protection away.[7] They too feel that the sword deposits relate to changing ideas about the lifestyle of the elite and the exercise of power and that, "Depositing weapons in the grave of a man who was turned into an ancestor at the funerary ritual, was thus an act that gave form to newly developing martial aristocratic values relating to new practices in the exercise of power."[8] They go on to write, "What was essential was that the performance of these rituals contributed to the increased sharing of new values by aristocrats whose origins and cultural backgrounds were very different.""

The ideas of Klevnäs and Theuws and Alkemade appear to fit much more closely to the evidence than some of the other suggestions, and it could be that we are now beginning to understand why swords were deposited in so few graves, and why they were targets for removal. It is easy to imagine a scene where the robbing family would put on a display of the items taken from a grave in order to show how they had weakened and shamed their rivals in both this and the Otherworld. To be able to enter and desecrate a sacred place with impunity must have sent a very powerful message. I wonder what they did with the bones they took?

---

1 Klevnäs, 2013. 88.
2 Klevnäs, 2013. 89.
3 Klevnäs, 2013. 90.
4 Klevnäs, 2013. 73.
5 Klevnäs, 2013. 73. Interestingly, others have had the opposite idea; that is the sword was 'inalienable' – see Härke, 2000. 395 and Härke, 2014. 45. The author is discussing the ideas of Reinecke in the latter paper.
6 This possibility comes from Stoodley (1999, 2000) in Klevnäs, 2013. 74.
7 Theuws and Alkemade, 2000. 455, 461, 466
8 Theuws and Alkemade, 2000. 454.

To conclude this section, it is probably better to leave Klevnäs to comment on her own research: "This research demonstrates that, despite the apparently transgressive nature of the robbing, the people who carried it out followed clear rules about what should and should not be taken. It is argued here that these rules were closely related to the reasons that particular artefacts found their way into burials in the first place. This was malefaction within firm bounds: the perpetrators expressed contempt for the individual graves, but not for the furnished burial rite itself."[1]

vi. Women and swords

There is no doubt that women wore and were buried with swords but most, if not all, of the examples known are miniature sword-shaped amulets, and they have been found in England and Scandinavia.[2] Often the miniature sword is not alone and can be part of an ensemble of other miniatures that could represent hammers,[3] various tools, spears and knives, which probably represent deities in some way. The inclusion of such miniatures in women's graves occurs in England from the 5th to 7th centuries and in Scandinavia for a much longer period.[4]

Weaving battens

Weaving battens are another sword-like item that sometimes occur in women's graves and it is possible that they have occasionally been mistaken for swords due to their shape and the fact that they can be pattern-welded. It has been assumed by some that they were old, worn-out swords being reused in the making of textiles. I feel that this is unlikely, mainly due to the special status that swords held for their societies. To turn one into a tool for weaving seems to go against that outlook but there are examples of swords having been converted into weaving batterns. Weaving battens are shorter than swords,[5] at around 50cms in length, they also tend to have a radically different method of attaching to a handle or grip than a sword with its tang and hilt. There is also the problem of their weight making them impractical for beating the threads of material. However, it is possible that swords taken from a rival family's grave site could be dishonoured in this way but most battens appear to have been carefully crafted for that purpose and seem to have had a significance and prestige of their own, but we will not be pursuing that any further here for reasons of space.[6]

There are burials that contain both women and real swords, such as graves 96a and b at Dover Buckland, Kent. This is a double burial, 96a, a man, has a sword, shield and spear, while 96b, the woman (if it was a woman, despite female characteristics) in that grave was obviously not alone.[7] Brunning notes that the sword for 96b had a spear lying between it and the female's body, thus separating them, which could be significant. There are other unusual features about the way the bodies are resting.[8] There are other possible female sword burials at the same cemetery (graves 33 and 93) which have skeletal remains that exhibit both male and female characteristics and are therefore difficult to interpret. Brunning has considered these and the small number of other graves where it is felt that a woman may have been included in association with a sword, and has shown

---

[1] Klevnäs, 2013. 89.

[2] Meaney, 1981. 148 to 159.

[3] Probably Thunor's or Thor's hammers.

[4] Brunning, 2014. 101 and 247. There is a discussion on females buried with swords from my sample of sword graves collected from the survey of cemeteries in the chapter, *Analysis of selected data from cemetery survey*. This subject is also discussed below.

[5] Stoodley, 1997. 71.

[6] Brunning mentions that Harrington, 2008 (*Aspects of Gender Identity and Craft Production in the European Migration Period: Iron Weaving Beaters and Associated Textile Making Tools from England, Norway and Alamannia*. Oxford) has discussed the possible symbolic functions of weaving battens. Brunning, 2014. 35.

[7] Stoodley, 1997.

[8] Brunning, 2014. 152. See Evison, 1987. 349 for a diagram of the grave. See also Stoodley, 1997. 69 to 70, and 168 for his observations.

that this evidence is not always so straightforward.[1] She summarises the archaeological evidence in both England and Scandinavia by pointing out that women and 'unadapted swords'[2] only appear together in double burials with a male, and there is little evidence from the graves to indicate that real women ever wielded actual swords.[3]

It seems that examples of actual women using swords is absent from Old English and Old Norse poetry,[4] as almost all of the women who handle such weapons in the poetry are not 'normal' women;[5] they are exceptional in some way, tending to be Valkyries or some other representative of the Otherworld. We will return to this idea again later. Brunning's survey of the iconography rendered similar results, in that the images demonstrate that the relationship between men and swords is fundamentally different to that of swords and women,[6] and again, sword-bearing women in the images are all extraordinary females.[7]

The status of women

It is important to understand that woman during the Early Medieval period were not necessarily as subservient to men as they appear to have become during the Later Middle Ages. Women could hold land and property,[8] they controlled the home, they could leave their husbands and divorce and, as one story recalled by Procopius indicates (*The History of the Wars* book viii chapter 20), they could at times even lead armies.

Procopius was a high-ranking official at the court of the Byzantine emperor, Justinian and received his information from a member of the Frankish embassy. The incident happened sometime between 533 and 548 A.D. and the high-ranking girl (princess) in the story was a contemporary of Procopius; unfortunately, he neglects to name her. The girl's people were the Angles (*Angiloi*) and they were settled in eastern Britain by the time of the story. Hermegisclus was the king of the Varni (*Wærne* OE, *Varini* L) who lived near the mouth of the Rhine on the eastern side and they were under pressure from the Franks. Hermegisclus agreed to marry the younger half-sister of the Frankish King Theudebert I; her name was Theudechild. Hermegisclus had also sent to the Angles[9] asking for the king's sister as a bride for Radigis, Hermegisclus's young son. The Anglian princess accepted the betrothal gift sent by Hermegisclus, which meant that a binding contract had been made.

While out riding one day, Hermegisclus heard a raven on a tree croaking and he told his companions that it had given him a message that he would die in forty days. He then had to make arrangements for his death, one of which was to make as sure as he could that he could keep the Varni safe from the Franks, so he made his young son, Radigis, promise that he would marry his Frankish step-mother when he died; forty days later the king died and Radigis did as his father had ordered. A message reached the Angles that the princess's marriage was off, but that she could keep the gold that Hermegisclus had sent as compensation.

---

[1] Brunning, 2014. 152.

[2] Brunning allows the possibility that weaving battens were possibly made from altered swords. Brunning, 2014. 247.

[3] Brunning, 2014. 248.

[4] Brunning, 2014. 210 and 211.. There is the exception of Hervor, (Hervarar saga ok Heiðreks) see above.

[5] Brunning, 2014. 219.

[6] Brunning, 2014. 247.

[7] Brunning refers to a, then forthcoming, paper by Hadley, where it is implied that, the presence of swords in the graves of women and those deemed too young, infirm or physically impaired (in the modern sense) to fight, while importantly challenging established ideas, does not subvert the reality that most swords were buried with adult men of fighting age and ability. Brunning, 2014. 45. The sensational announcement that a female, Viking 'warrior' buried with a full set of weapons including a sword (see http://nypost.com/2017/09/08/viking-skeletons-dna-test-proves-historians-wrong/ or http://www.newsweek.com/gender-reveal-viking-warrior-was-woman-dna-analysis-show-662972) arrived too late to be fully assimilated into this chapter. However there are serious doubts regarding the veracity of this find, which was excavated during the late 18th century and not terribly well recorded. For a full discussion see Androshchuk, 2017.

[8] Herbert, 1997. 13. The evidence comes from wills and charters

[9] Procopius does not tell exactly which Anglian (or Saxon, etc.) kingdom the princess came from.

The princess was not interested in keeping the gold for its own sake and would not accept the snub; she had been humiliated and dishonoured and demanded an explanation. No such explanation was sent back to her. She was made of stern stuff and had been spurned so, with the backing of her tribe and kin, she gathered a force to take against the Varni. Procopius says that she took four hundred ships and 100,000 warriors, but these numbers are surely a great exaggeration. However, Procopius also makes it clear that she was the war leader, even though she took her younger brother with her. When her troops landed near the mouth of the Rhine, she instructed her men to build a strong fort to ensure that they had a secure base for any operations that become necessary. Once secure, she sent her brother to find the Varni force, he located them not too far away, encamped close to where the Rhine meets the sea. When she heard the news, she, straightaway ordered an attack which was very successful, defeating the Varni, killing many of their soldiers. Radigis escaped but was pursued for a time by the Angles who eventually gave up and returned to the princess. On their return they were scolded for not capturing Radigis, and sent off again with the instruction that they were not to come back without him!

The Angles eventually found him hiding in the woods and brought him back to the jilted princess. He expected to be killed but she questioned him. Apparently he behaved less than well, and blamed his father, pleaded for mercy and promised to marry her if he was allowed to live. At that she let him free and ordered her troops to treat him kindly and then married him. The Frankish woman, Theudechild, was sent back to Francia.[1]

Incidentally, at no time does Procopius indicate that the princess wielded arms herself but she showed herself to be a formidable woman, so much so that it was the Franks, who to some extent were humiliated by her, who told Procopius the story.

Another formidable woman, who appears in Bede's *A History of the English Church and People*, is the wife of King Raedwald of East Anglia. Raedwald is thought by many to be the royal who was buried in Mound 1 at Sutton Hoo. Again, we do not know her name and she is only referred to twice in Bede's history, but on both occasions it is plain that she exercised great influence over her royal husband. The first instance concerns the future King Edwin of Northumbria when he was being pursued by the then King of Northumbria, Æthelfrith, who wanted to kill him.[2] Raedwald granted Edwin exile in East Anglia. However, Æthelfrith sent messages to Raedwald that he should give up Edwin and offered bribes or war in order to try and force his hand. Edwin is portrayed as a noble soul and even though he becomes aware that Raedwald is going to give him up he refuses to break faith with the king. Edwin later receives a visit from what appears to be a supernatural entity but it is actually Raedwald's wife who dissuades her husband from betraying their guest and she does it on the principle that it would be, "...unworthy in a great king to sell his best friend in the hour of need for gold, and worse still to sacrifice his royal honour, the most valuable of all possessions, for love of money."[3] Raedwald did as she advised, and not only kept Edwin safe but as soon as the Northumbrian envoys had returned to Æthelfrith, he raised an army to assist Edwin in his efforts to gain the throne. The East Anglians met Æthelfrith's forces at the River Idle, around 616 A.D,. and defeated them, Raedwald's own son, Ragenhere, died in the battle but Edwin was put on the throne and became the new King of Northumbria.

On another occasion Bede tells us that Raedwald had received baptism and become a Christian while visiting the King of Kent, but on his return to East Anglia had been persuaded by his wife and 'perverse advisers' to apostatize from the 'True Faith'. We are told that, "...his last state was worse than the first: for like the ancient Samaritans, he tried to serve both Christ and the ancient gods, and he had in the same temple an altar for the holy Sacrifice of Christ side by side with an altar on which victims were offered to devils."[4]

---

[1] Herbert, 1997. 8 to 13. The original story was written in Greek.

[2] Sherley-Price, 1955. 122 to 126.

[3] Sherley-Price, 1955. 125.

[4] Sherley-Price, 1955. 130.

There is no doubt that the two women we have just considered were powerful and respected and had influence over the actions of men. Were they isolated examples? Not according to the literature of the North where such women often play major roles and, even though some of them were legendary and not historical, the position of women is being emphasised and celebrated. We will mention more of them below. Many scholars have noted that women had a special relationship with the divine and were regarded as closer to the supernatural than men in the immediate pre-Christian period, and were respected for that.[1]

The sword at the wedding[2]

In 1960 Hilda Ellis Davidson published an essay with the above title which demonstrated that women and swords may have had a direct and important relationship that endured over time and in many regions of northern Europe. Davidson looked mainly at the literature and remnants of folklore that had survived in the texts and amassed a considerable set of evidence to show that the sword and its inheritance, to some extent, was controlled by the woman of the house. In important ways Davidson's suggestions seem to support the conclusions that Brunning came to regarding women and swords, i.e. that women associated with swords were 'extraordinary'. Davidson's ideas, also harmonise with some of the findings of the GRRG.

Davidson notes a curious custom of a wedding ring being handed to a bride on the hilt of a sword[3] and the poem implies that this is a threat warning the bride of the penalty for infidelity. Davidson does not feel that this was all that was meant and believes that it was probably a threat to both participants, emphasising the sacredness of the wedding and the binding nature of the oaths being sworn. She finds confirmation of this in later texts and earlier practices. She points out that swords and rings have figured in the swearing of oaths in pre-Christian times and refers to the practice of taking an oath of loyalty while laying a hand on the hilt of the overlord's sword, and suggests that could account for the rings on the hilts of swords (discussed above).[4] In the poem *Ruodlieb* (XV, 63) the bridegroom draws the sword and rubs it ('whets'?) on something described as a pyramid (*piramide tersit*) before the ring is placed on the hilt.[5] The suggestion is that it might have been placed on a family tomb or a stone step in order to connect with an ancestor.

Apparently, Jacob Grimm, in his collection of ancient laws, also tells of a sword at a wedding. In this case it takes place in Frisia, where the sword was placed against the threshold of the doorway into the husband's house in order to block the bride's progress. The sword was called the *æftswird* (Old Frisian) 'the sword of marriage' and was felt by Grimm to be warning the bride against being unfaithful.[6] He records another example of a sword at a wedding, also from Frisia, this time a young man with a naked sword walks before the bride. Davidson goes on to report that there are records of ceremonies and games taking place in country districts of Sweden involving a sword and a 'stock' at weddings as late as the 17th century, and that in Norway a custom survived until recent times where the bridegroom would stab his sword into the roof beam to test the 'luck' of the marriage by looking at how deeply the sword had entered the wood.[7]

---

[1] See for example Price, 2002. 111 to 124.

[2] Davidson, 1960. 95 to 112.

[3] Davidson, 1960. 95. The incident is recorded in a Latin poem *Ruodlieb* written by a German monk in 1023 A.D.

[4] Davidson, 1960, 96. Davidson, 1962. 76.

[5] Davidson, 1960. 96. This sounds like a symbolic sharpening of the blade – but could it be a sword-pyramid?

[6] Davidson, 1960. 96.

[7] Davidson, 1960. 97.

The Volsungs

The *Vǫlsung Saga*[1] has an incident where something similar took place, and the 'stock' was a real tree that grew in the hall of the King Volsung and it was called the *barnstokkr*. Incidentally, it seems that the family were protected by a *fylgja* (a female spirit).[2]

King Rerrir had sat on a mound and prayed to Frigg for a son and in reply received an apple, which he and his wife ate; the result was Volsung, the future king. Davidson notes the link between the mound, the tree, the fruit and the birth of children and believes that the connections are significant.[3] When Volsung's daughter, Signy, was to marry King Siggeir of Gautland, the marriage feast took place in the hall and during the feast, Óðínn entered as a huge old one-eyed man wearing a low-hanging hood. He went up to the tree and thrust a sword into it up to its hilt, and said, "He who draws this sword out of the trunk (stock) shall receive it from me as a gift, and he himself shall prove that he has never carried a better sword than this one."[4]

Many tried to retrieve the sword that came to be called, Gram, and all failed, but when Volsung's son, Sigmund tried, it came out easily. The bridegroom, Siggeir wanted to buy it but Sigmund refused the offer. Siggeir was very angry but appeared to let his anger pass. However, when King Volsung and his sons visited Siggeir in Gautland three months later, at his invitation, Siggeir took his revenge, killing Volsung in a battle and taking all ten of the sons prisoner. All were killed except Sigmund who escaped with Signy's help.[5]

The gift of the sword was, as Davidson reminds us,[6] made at the wedding feast and was bound up with the tree; she feels that the tree was a 'guardian tree' closely connected to the fortunes of the family, perhaps an embodiment of its 'luck'. She finds and lists evidence of other such trees associated with families from the British Isles and even the *Odyssey*.[7] Bearing in mind the other traditions that Davidson records, she feels that she understands why Siggeir was so angry about not receiving the sword that Óðínn left; it should have been the bridegroom who collected the sword, not the bride's brother; Siggeir had been grossly insulted![8] The insult must have been what the god intended in order to set off a specific chain of events. These events included a son, Sinfjotli (who dies early), born out of an incestuous union between Signy and Sigmund, Sigmund being killed in battle and another son, Sigurd, born of a different woman, receiving his father's sword in pieces and having it re-forged by the dwarf, Reginn, uses it to slay the dragon Fafnir. Sigurd, was of course, the great hero.

Davidson ponders whether the phallic symbolism of the sword is an important part of its role at weddings and suggests that besides a fertility aspect the sword was there to show the continuation of the family and that the handing down of a 'family' sword was an integral part of this.[9] Davidson, too, mentions the value of old swords and provides a number of examples in the stories where swords are passed down in families and gives both historical and legendary examples;[10] in either case the audience would have understood the relevance and importance of such heirlooms. She also suggests that it could be that a young man did not become a full man until he had received a 'family' sword and recalls the story of Alboin, a Lombard prince, written down by Paul the Deacon. She relates that the tale definitely says that a king's son could not sit and eat with his father until he had received arms from a foreign king, and that Alboin had travelled to the King of the Gepidæ to demand

---

[1] Byock,1999. 37 to 39.
[2] Davidson, 1960. 104.
[3] Davidson, 1999. 97.
[4] Byock, 1999. 38.
[5] Byock, 1999. 40 to 42.
[6] Davidson, 1960. 98.
[7] Davidson, 1989. 98 to 99.
[8] Davidson, 1989. 99.
[9] Davidson, 1989. 99 to 100.
[10] Davidson, 1989. 100.

from him, the arms of the King's son who Alboin had slain in battle. The Gepidæ king recognised Alboin's claim, protected him from his own vengeful warriors and gave him the war gear. Alboin, on his return was then accepted as a full table companion.

Davidson says that there are two principles operating here; first the winning of a sword and other arms through valour, and secondly the recognition by the family of the youth as a full man. She believes, too, that this is what happened to Beowulf, when Hygelac, his lord, gives him an heirloom sword; that is a family sword.[1] The sword of Wiglaf would seem to also fit a similar profile, as would Tyrfing. She goes on to discuss the case of Grettir the Strong, who lived in Iceland during the 10th century. Iceland had no kings, but the same basic principles seem to operate. Grettir was given a sword by his mother. The sword had been procured by her great –great grandfather, Ingimund and the sword was called, Ættartangi, which means – 'tang of generations'.[2] Ingimund, who, like the Volsungs, claimed the protection of a *fylgja*[3] had obtained the sword via a trick and its history after Ggrettir receives it, is given in the *Saga of Grettir the Strong*.

Davidson provides examples of other swords,[4] from the Norse literature, that were handed down through families and then proceeds to discuss the phenomenon of guardian spirits associated with weapons and families. She relates an example from *Hallfreðar Saga*, in which a supernatural being personifies the family's 'luck'; the being is a huge woman and she is associated with the family sword. When Hallfred, the poet, now a Christian, lay dying on board a ship the woman was seen travelling over the waves wearing a mail coat and she is referred to as a *fylgjukona*. Hallfred tells her that he is dying and their association is at end and she asked his brother, Thorvald, if he would receive her, but he refuses, but the poet's son, another Hallfred accepts her and she vanishes straightaway.[5] The dying Hallfred then leaves his sword that he gained from King Olaf, to his son. Davidson says that only after making his pledge to the *fylgjukona*, can the sword and the 'luck' of the family be bestowed on the young man.

Davidson picks up on the theme of a young man being granted a sword by a Valkyrie and tells us that in the Edda poem, *Helgakviða*, Hjorvard's young son sits on a mound and sees nine Valkyries ride past him. One gives him his name, Helgi and presents him with a sword as naming gift, apparently the best of forty-six swords, she says, too, "There is a ring on the hilt, there's courage in the middle, and terror in its point, for him who manages to own it; a blood-dyed snake lies along the edge and on the boss a serpent chases its tail."[6]

Saxo Grammaticus tells us of the hero Regnar who receives a sword from a maiden called Svanhita who rides with her sisters on horseback. "Svanhita wondered at the young man's determination and, banishing from herself the misty clouds of darkness, scattered the shadows obscuring her face to give the air a clear transparency. Then, having promised to present him with a sword to meet all kinds of conflicts, she revealed her wondrous maiden beauty and the fresh radiance of her limbs. He grew so warm to her that she agreed to marry him…."[7] She then gives him the sword and a speech about how he will learn to use it.

Other female beings besides valkyries and fylgja are mentioned in the Norse literature that could be part of the same tradition. Davidson tells of a series of stories where giantesses foster young heroes, become betrothed to them, sometimes have a child with them, but protect the hero, even after they have acquired normal human

---

[1] Davidson, 1989. 100.

[2] Davidson, 1989. 101. The name of the sword is not given in Grettir's Saga (but referred to as *J□kulsnautr* (the gift or heirloom of Jokull)) but is the same sword written about in *Vatnsdæla Saga* where a fairly involved history of the sword is provided.

[3] Davidson, 1960. 104.

[4] Davidson, 1960. 102 to 103.

[5] Davidson, 1960. 104.

[6] Larrington, 1996. 125.

[7] Davidson and Fisher, 1979. 45.

wives. These female members of the Otherworld often give their charges a sword and armour.[1] She notes parallels to the accounts involving the young Hallfred and Helgi where they were free to reject or accept the offers made, but once they accepted they were bound into a contract, which often benefitted the men. She says, "There seems little doubt that one aspect of the Valkyrie, seen as a supernatural woman in armour who protects the hero and gives him a weapon, is based on the idea of a protective spirit of the family, the personified 'luck' of his ancestors."[2]

Another example is found In *Flateyjarbók*, where the ancestor of St Olaf,[3] another Olaf, appeared in a dream to his father's foster-brother and told him to break into the burial mound and take out treasures that were buried with Olaf. Most of the treasures were to be given to the son of Jarl Hakon in return for his help, but the ring, the belt and the sword belonging to the dead king were to be taken to his foster-brother, King Harald. The foster-brother, Hrani, carried out the instructions and fought and overcame the body of Olaf, who despite the dream was reluctant to let his possessions go. Hrani cut off Olaf's head and took his treasures to the hall of King Harald, where everyone was waiting for Harald's wife to give birth. Hrani put the belt around the expectant mother's waist and said that the child was to be called Olaf and that the sword and ring were to be given to the child. Davidson thought that this demonstrates a clear link between the ancestors, the living and their swords, and in this case, the name giving.[4] The mention of mounds makes Davidson remind us of Hervor and the sword Tyrfing which Hervor took from her father's mound and Davidson wonders whether the persistence of such mound robbing stories demonstrates that they were not just wanton acts of tomb robbery.

Offa (Uffa) of Angeln

One final story that Davidson thought was relevant and concerns a sword and a mound is the story of Offa of Angeln, the 4th century king of the Angles.[5] This is a story that is referred to in the Old English poem *Widsið*, the Anglo-Saxon Chronicle (Mss A, B, C for 755 A.D.[6]) and *Beowulf*, but the fullest accounts come later from Denmark and there are two versions that are still extant, one was written by Sven Aageson and the other by Saxo Grammaticus. Both tales attempt to convert the Angle king Offa into a Dane and to absorb his story, and that of the Angles, into Danish history. The Angles had by that time long migrated to Britain and the Danes had moved into their vacated territory and made it part of Denmark.

> Offa ruled the Angles,   Alewih the Danes;
> he was among all men;   the bravest,
> but not more heroic than Offa,   who before he was a man,
> won in combat   most of his kingdom.
> None of his age   showed greater feat of arms.
> With his single sword   he spread his borders.
> Against the Myrgings   he fixed the border
> at Fifeldor.   Henceforth it was held
> by Angle and Swæfe   as Offa won it.
>
> *Widsið*

[1] Davidson, 1960. 105 to 106.
[2] Davidson, 1960. 107.
[3] Olaf II Haraldsson, King of Norway from 995 to 1030, and later known as Olaf the Holy or Saint Olaf.
[4] Davidson, 1960. 107 to 108.
[5] Alexander, 1966. 156.
[6] Garmondsway, Simpson and Davidson, 1968. 222.

In Aageson's account, Uffo was born to Wermundus (son of Froði the Bold) the king of the Danes. Wermundus was a famous hero when younger. Uffo didn't speak for his first thirteen years and it turned out that was because he was ashamed that two of his relatives had killed their father's slayer together and not in single combat. In old age Wermundus became blind – the 'Teuton' king wanted to take the Danes over and gave them the choice between surrender or to allow a battle of champions to decide the issue. Unfortunately the Danes had no such champion and everyone thought that Uffo showed no promise; it is worth pointing out that since childhood his main pre-occupation seems to have been gluttony. Uffo speaks for the first time and says that he will take on the Teuton's champion. Uffo then tries many swords but they all shatter in his hand, he is so strong, so Wermundus retrieves a sword from a mound where had hidden it. He gives the sword to Uffo and tells him that he had much success with the sword when younger. The combat takes place on an island in the River Eider. The Teutons send two men – one a son of the Emperor. Uffo carries two swords and uses an ordinary sword first, which he breaks and then draws the other sword, with which he kills both of his opponents.[1] Saxo's account differs a little but provides more details regarding the killing that Uffo was so ashamed of and the enemies in Saxo's version are the Saxons and the sword is named, 'Skrep' and that is the only sword that he uses during the combat in this version. Incidentally, the blind Wermundus recognises the sound that it makes. At the end of Saxo's version Uffo becomes king of the Danes and Saxons.[2]

As Davidson reports there is archaeological evidence of robbing mounds and provides examples such as Gokstad and Oseberg; it is a pity that she did not have access to the work of the GRRG, with her insight and knowledge of the folklore of many northern lands, she could have brought much to the discussion. In her conclusion, Davidson draws all the threads that she has spun together and explains that many of the stories involve a mother that kept and gave the sword to the son, for instance Olaf's mother kept the sword in a chest until he was of an age, Hervor passes the sword to Heidrek, and Grettir's mother passed the sword to him. Davidson believes that these examples are very significant. She goes on to say that the presence of a sword at the wedding is in keeping with the part played by the mother. Tacitus, writing in the 1st century A.D., says of the Germans wedding practices, "Parents and kinsmen take part, and approve gifts, not gifts intended for feminine luxury or used to adorn a new bride, but cattle, a horse with its bridle, and a shield, *framea*, and a sword. Against gifts such as these a wife is acquired, and in turn she herself brings the man a weapon." Tacitus says that this was the most important part of the ceremony and Davidson notes the inclusion of a sword and the bride's duty to bring the groom a weapon and the bride receiving her gift as a treasure to be stored and passed down to the next generation, which the bride will give birth to. [3]

Continuity?

Davidson sees the tradition recorded by Tacitus as being carried on to later times and believes that the examples of mothers storing weapons and giving them to their sons in the sagas and other literature are echoes of that tradition and linked to that was the practice that a father's arms would be entrusted to his daughter, possibly as part of a marriage dowry. She feels, too, that the young man had to prove himself to be worthy of the family's weapons.[4] As she says, "But what stands out clearly is the belief in the significance of the family as a continuing force, and of the tremendous importance of the family store of 'luck' and valour, embodied in its greatest treasures to each new generation in turn. The idea of a mighty woman in armour, riding through the air or striding over the hills, is one way of picturing this 'luck', and the man who received

[1] Sven Aageson, Historia brevis regum Dacie 'A Brief History of the Danes' -- 1187

[2] Saxo Grammaticus , Gesta Danorum (Danish History) 1200

[3] Rives, 1999. 84.(Germania, XVIII). Davidson, 1960. 110 to 111.

[4] Davidson, 1960. 111.

such a being as his protector was held to concentrate in himself during his lifetime the power and greatness of the family to which he belonged."[1]

Davidson concludes by explaining that the sword should not be seen as just a masculine treasure; it was wielded by a man but the woman was its guardian and it was her task to ensure that it reached succeeding generations. Davidson argues that it was entirely appropriate that a woman's hand should be placed on the sword hilt at the wedding and that it was far from being a sign of subjection of the female to the male, or a threat to either party; the woman was receiving, perhaps the greatest honour that a heroic society could pay her, the future of the next generation.

Conclusions

This consideration of sword ideology during the 5th to 7th centuries has been a long one, and it can be safely said that swords had multiple meanings for the societies of Anglo-Saxon England and elsewhere, and there will, no doubt, be other meanings that we have not yet noticed.

A number of themes keep occurring in our consideration of the ideology of the sword, one is the very real connection of women to swords, whether as 'guardians' and dispensers of family weapons and the family's 'luck', or the finding that women's graves and ornaments were also subject to attack during the period when men's graves were being robbed of swords. This aspect of the ideology of the sword was a surprise to me and needs further thought and discussion. Some of the implications could well be profound and might alter the way in which we see much of what was happening to the societies of the 5th to 7th centuries.

Another theme is the value placed on 'old' swords and how they could focus within themselves much of what was valued by almost everyone within the community. This aspect of the living' sword is an important one. Whether swords were seen by everyone as actually being sentient is difficult to ascertain, especially from the specialised poetic language that has come down to us, but the fact that they could store up the history of the people, the clan, the family and the individual and represent the qualities that were most admired in a heroic society, must have made them extremely powerful and potent symbols in a heroic age.

Another strand that links both of the above themes is the importance of heirlooms. The responsibility of passing on treasured artefacts that contained the history of the greatest members of the family or clan was a huge one, and it seems often borne by a woman. Women were not bystanders, or also-rans during the 5th to 7th centuries, they were essential partners in guaranteeing the future of their progeny and people.

Swords were also a medium of exchange and regulators in the power-relationships in the societies of the north, whether as gifts, loot, or as a sacrifice for victory to the gods. We have seen how 'honour' and obligations were transferred with swords and other gifts from one individual to another, and that humiliation or prestige could be manipulated up and down the social strata. Where this was done effectively the people prospered but such manoeuvres would not always have succeeded and then the folk were vulnerable and they could face tribulations and a possible takeover and absorption and subjugation into another clan-group.

Throughout this consideration of the sword, the idea of the Otherworld has been persistent and never far away, whether it involved stories of a valkyrie, ornamentation on a sword or pattern-welding. We still know far too little about the beliefs and worldview of the people of pre-Christian Europe, particularly in the north. Although such ideas would have varied from place to place and over time, we can be certain that most people believed enough in the Other to deposit their most valued possessions in graves with their loved and respected ones; and we can be sure that that process had real meaning for the observer and the grave robber!

---

[1] Davidson, 1960. 112.

We have seen from our discussion on *Beliefs*[1] that they might not always have been about gods and major deities, and that there was at least one level in the world of the Other that involved a close relationship with aspects of nature and local 'spirits'. It is likely that many of these notions persisted, some as superstitions, long after the advent of Christianity, and it is more than probable that the Church absorbed much of this symbolism into its own practice, not least the wholesale takeover of previous places of worship and the festivals associated with them. Some temples were, on the instruction of Pope Gregory, re-consecrated while others were demolished and churches built where they had stood. Disposal of weapons in watery places is another example of how such old ideas continued for a considerable time into the so-called Christian period.

Animal and human symbolism embedded in the ornament of weapons and other valuable items, such as brooches, were a manifestation of the belief in the Other and possibly incorporated appeals for supernatural protection, strength or other attribute. Although the symbols could have changed after the conversion, the same principle operated, but the Deity was, of course, different. Then there are the patterns on the blades, which appear to have had meaning for the people of the time, we have seen, in the chapter on *Pattern-welding*, that it is possible that those patterns could have been used to symbolise running water (that link with water and swords again) or the serpent, another recurrent theme that occurs again and again in the art.

We have seen that bones went missing from graves but why were they taken? What was done with them? Could Terje Gansum be correct about the use of human bones in the making of the iron and steel for a sword; did the addition of the bones of a sword-wielder, or a famed ancestor, or a fierce animal make the sword more powerful? Were such bones used in combination to give even more prestige to the sword? Does the animal ornament reflect what was in the sword? We can never know for sure but the possibility is intriguing.

Swords were an important symbol to everyone who lived in northern European communities during the 5th to 7th centuries. Every male and female, whether or not they owned or used one, had a direct link to the sword.

[1] See chapter, *Background*.

Appendices

Over 200 pages of additional information, text and images are available at -

www.independent.academia.edu/PaulMortimer

Bibliography

Abels, Richard, 1998. *Lordship and Military Obligation in Anglo-Saxon England*. Berkeley.

Åberg, Nils Fritjof, 1926. *The Anglo-Saxons in England*. Uppsala.

Ackerman, J., 1860. *Second Report of Researches in a Cemetery of the Anglo-Saxon period at Brighthampton*. In Archaeologia v.38 Issue 1. Oxford.

Ager, Barry; Cameron, Esther; Spain, Stephanie and Riddler, Ian, 2006. *Early Anglo-Saxon Weaponry from Saltwood Tunnel, Kent*. CTRL.

Ager, Barry, 2012. *Swords.* In Parfitt and Anderson, 2012.

Ager, Barry and Gilmour, Brian, 1988. *A Pattern-Welded Anglo-Saxon Sword from Acklam Wold, North Yorkshire*. In Yorkshire Archaeological Journal, volume 60, 1988. Bradford.

Ahumada Silva, Isabel; Lopreato, Paola and Tagliare, Amelio, 1990. *La necropoli di Santo Stefano in pertica, campagne di scavo 1987-1988*. Città di Castello.

Alborne, James and Leahy, Kevin, 2000. *The Anglo-Saxon Cemetery at Tallington, Lincolnshire.* In Griffiths, 2000.

Albrethsen, 1997. *Logistical Problems in Iron Age Warfare*. In Jørgensen, Anne Nørgård and Clausen, Birthe (eds); 1997.

Alexander, Michael (translator), 1966. *The Earliest English Poems*. Harmondsworth.

Alkemade, Monika, 1991. *A History of Vendel Period Archaeology*. In Theuws, Franz; (ed).1991.

Anderson, Poul, 1973. *Hrolf Kraki's Saga*. New York.

Androshchuck, Fedir, 2014. *Viking Swords; Swords and Social Aspects ofWeaponry in Viking Age Societies.* Stockholm.

Anlezark, Daniel (editor), 2011. *Myths, Legends, and Heroes; Essays on Old Norse and Old English Literature.* Toronto.

Annable, F. K., and Eagles, B. J., 2010. *The Anglo-Saxon Cemetery at Blacknall Field.*Wiltshire.

Arrhenius, Birgit, 1985. *Merovingian Garnet Jewellery*. Stockholm.

Arwidsson, Greta, 1942. *Valsgärde 6*. Stockholm.

Arwidsson, Greta, 1954. *Valsgärde 8.* Stockholm.

Arwidsson, Greta, 1977. *Valsgärde 7.* Stockholm.

Arwidsson, Greta and Berg, Gösta, 1999. *The Mästermyr Find; a Viking Age Tool Chest from Gotland*. Lompoc.

Atkins, Rob and Connor, Aileen, 2010. *Farmers and Ironsmiths: Prehistoric, Roman and Anglo-Saxon Settlement beside Brandon Road, Thetford, Norfolk.* Oxford.

Bachrach, Bernard S., 1972. *Merovingian Military Organisation*. Minnesota.

Baker, Peter S., 2013. *Honour, Exchange and Violence in Beowulf.* Woodbridge.

Bateman, T., 1861.*Ten Years Digging*. London.

Bazelmans, Jos, 1999. *By Weapons Made Worthy; Lords, Retainers and their Relationship in Beowulf.* Amsterdam.

Behmer, Elis, 1939. *Das zweischneidige Schwert der germanischen Völkerwanderungszeit.* Stockholm.

Bellows, H. A., 2007. *The Poetic Edda: The Heroic Poems*. Chatham.

Biborski, Marcin and Ilkjær, Jørgen, 2006. *Illerup Adal: Die Schwerter: Textband / Katalog, Tafeln Und Fundlisten: volumes. 11 & 12*. Aarhus.

Birbeck, V., 2005. *The Origins of Mid-Saxon Southampton: Excavations at the Friends Provident St. Mary's Stadium 1998-2000*. Trust for Wessex Archaeology Ltd. Salisbury.

Birch, T., 2013, *Does pattern-welding make Anglo-Saxon swords stronger?*. In Dungworth, D and Doonan, R C P (eds) *Accidental and Experimental Archaeometallurgy*. London.

Birch Iversen, Rasmus, 2010. *Kragehul Mose; Ein Kriegsbeuteopfer auf Südwestfünen*. Aarhus.

Blakelock, Eleanor; Mongiatti, S Aude and La Niece, Susan, 2013. *Scientific investigation of an Anglo-Saxon sword hilt from Cumberland (1876,0717.1)*. British Museum. London.

Bliujienė, Audronė (editor), 2007. *ARCHAEOAEOAEOLOGIA BALTICA 8; Weapons, Weaponry and Man Weapons, Weaponry and Man.* Klaipėda.

Bone, Peter, 1989. *The Development of Anglo-Saxon Swords from the Fifth to Eleventh Century.* In Chadwick Hawkes, 1989.

Boyle, Angel; Dodd, Anne; Miles, David and Mudd, Andrew, 1995. *Two Oxfordshire Anglo-Saxon Cemeteries: Berinsfield and Didcot.* Oxford.

Boyle, Angela; Jennings, David; Miles, David and Palmer, Simon, 1998. *The Anglo-Saxon Cemetery at Butler's Field, Lechlade, Gloucestershire; Volume 1: Prehistoric and Roman Activity and Anglo-Saxon Grave Catalogue.* Oxford.

Boyle, Angela; Jennings, David; Miles, David and Palmer, Simon, 2011. *The Anglo-Saxon Cemetery at Butler's Field, Lechlade, Gloucestershire; Volume 2: The Anglo-Saxon Grave Goods Specialist Reports, Phasing and Discussion.* Oxford.

Brady, C., 2007. *Weapons in Beowulf: Analysis of the Nominal Compounds and an Evaluation of the Poet's Use of Them*. In Anglo-Saxon England, volume 8. Cambridge.

Bradley, J. L., 1987. *Legendary Metal Smiths and Early English Literature*. Unpublished PHD theisis, Leeds University.

Breay, C. and Meehan, B., 2015. *The St Cuthbert Gospel: Studies on the Insular Manuscript of the Gospel of John.* London.

Broadberry, Stephen; Campbell. Bruce M. S. and van Leeuwen, Bas, 2010. *English Medieval Population: Reconciling the Time Series and Cross Sectional Evidence*. Warwick University.

Brooks, Nicholas, 2000. *Communities and Warfare 700—1400.* London.

Brooks, Stuart; Harrington, Sue and Reynolds, Andrew (editors), 2011. *Studies in Early Anglo-Saxon Art and Archaeology: Papers in Honour of Martin G. Welch.* Oxford.

Brown, G. Baldwin, 1915. *The Arts in Early England; Saxon Art and Industry in the Pagan Period.* London.

Bruce-Mitford, Rupert, 1974. *Aspects of Anglo-Saxon Archaeology; Sutton Hoo and Other Discoveries*. London.

Bruce-Mitford, Rupert, 1975. *The Sutton Hoo Ship-Burial, volume 1 Excavations, Background, the Ship, Dating and Inventory.* London.

Bruce-Mitford, Rupert, 1978. *The Sutton Hoo Ship-Burial, volume 2 Arms, Armour and Regalia.* London.

Bruce-Mitford, Rupert, 1983. *The Sutton Hoo Ship-Burial, volume 3 (2 volumes) Late Roman and Byzantine Silver, Hanging Bowls, Drinking Vessels, Cauldrons and other Containers, Textiles, the Lyre, Pottery Bottle and other Items.* London.

Brunning, Susan Elaine, 2013. *The 'Living' Sword in Early Medieval Northern Europe: An Interdisciplinary Study*. PhD thesis. University College, London.

Buchwald, Vagn Fabritius, 2005. *Iron and Steel in Ancient Times*. Copenhagen.

Buckberry, J.L. and Hadley, D.M., 2007. *An Anglo-Saxon Execution Cemetery at Walkington Wold, Yorkshire.* In, Oxford Journal of Archaeology 26 (3). Oxford.

Byock, Jesse L. (translator), 2005. *Snorri Sturluson – The Prose Edda*. London.

Bycock, Jesse L. (translator), 1999. *The Saga of the Volsungs*. London.

Byock, Jesse L. (translator), 1998. *The Saga of King Hrolf Kraki*. London.

Cameron, Esther A., 2000. *Sheaths and Scabbards in England AD 400-1100.* Oxford.

Cameron, Esther A., 2006. *Scabbards of Swords in Early Anglo-Saxon Weaponry from Saltwood Tunnel, Kent.* CTLR.

Carnap-Bornheim, Claus von and Ilkjær, Jørgen, 1996. *Illerup Ådal, volume 7: Die Prachtausrüstungen.* Aarhus.

Carver, Martin (editor), 2005. *A Seventh-Century Princely Burial Ground and its Context.* London.

Carver, Martin and Fern, Christopher, 2005. *The Seventh-century Burial Rites and their Sequence.* In Carver (editor) 2005.

Carver, Martin (editor), 1992. *The Age of Sutton Hoo : the seventh century in north-western Europe.* Woodbridge.

Carver, Martin; Hills, Catherine and Scheschkewitz, Jonathan, 2009. *Wasperton a Roman, British and Anglo-Saxon Community in Central England.* Woodbridge.

Carver, Martin; Sanmark, Alex and Semple, Sarah (eds), 2010. *Signals of Belief in Early England.* Oxford.

Cassau, A., 1935. *Ein Feuersteindolch mitl Holzgriff und Lederscheide aus Wiepenkathen.* Kreis Stade.

Cathers, Kerry, 2002. *An examination of the horse in Anglo-Saxon England.* PhD, thesis Reading University.

Chadwick, Sonja E., 1958. *The Anglo-Saxon Cemetery at Finglesham, Kent; a Reconsideration.* In the Journal of Medieval Archaeology, 1958.

Chadwick Hawkes, Sonja; Campbell, James and Brown, David (editors), 1985. *Anglo-Saxon Studies in Archaeology and History volume 4.* Oxford.

Chadwick Hawkes, Sonia (editor), 1989. *Weapons and Warfare in Anglo-Saxon England.* Oxford.

Chadwick Hawkes, Sonia and Grainger, Guy, 2003. *The Anglo-Saxon Cemetery at Worthy Park, Kingsworthy near Winchester, Hampshire.* Oxford.

Chadwick Hawkes, Sonia and Grainger, Guy, 2006. *The Anglo-Saxon Cemetery at Finglesham, Kent.* Oxford.

Cherniss, Michael D., 1975. *Ingeld and Christ: Heroic Concepts and Values in Old EnglishChristian Poetry.* In, *Speculum Vol. 50, No. 3 (Jul., 1975), pp. 482-485.* The Hague.

Chikering Jr. and Howell D., 1977. *Beowulf a Dual Language Edition.* New York.

Clark, Ann, 1993. *Excavations at Mucking volume 1: the Site Atlas.* London.

Clay, John, 2003. *Continuity of Christian practices in Kent, c.410-597:a historical and archaeological review.* In York Medieval Yearbook, ISSUE No. 2, (2003). York.

Cleere, Henry F. 1981. *The Iron Industry of Roman Britain.* Thesis available via Wealden Iron Research Group http://www.wealdeniron.org.uk/research.htm

Coatsworth, Elizabeth and Pinder, Michael, 2002. *The Art of the Anglo-Saxon Goldsmith.* Woodbridge.

Collis, John, 1983. *Wigber Low Derbyshire: a Bronze Age and Anglian Burial Site in the White Peak.* Sheffield.

Conti, Mark-Anthony, Bowlus, Charles R and Braithwaite, Wilfred J, 1997. *Questioning the Accepted Techniques for Sword Forging in Anglo-Saxon England and Frankish Europe.* In Journal of Arkansas Academy of Science Volume 51. Little Rock. http://home.comcast.net/~wjbraith/sword1stDraft.pdf

Cool, H.E.M., 2015. *Contextualising Metal-Detected Discoveries: Staffordshire Anglo-Saxon Hoard (Project 5892) Stage 2 Project Design Version 4.* Submitted 9th January 2015. Barbican Research Associates.

Cowgill, Jane, 2009. *Metalworking.* In Lucy, Tipper and Dickens, 2009.

Cox, Rev. J. Charles, 1891. *On an Anglo-Saxon Cemetery Recently Uncovered near Saxby.* Leicestershire Architectural Society, volume VIII.

Crawford, Sally and Hamerow, Helena; (editors), 2008. *Anglo-Saxon Studies in Archaeology and History volume 15.* Oxford.

Crossley-Holland, Kevin (translator), 1999. *Beowulf.* Oxford.

Danielsson, Ing-Marie Back, 2007. *Masking Moments: The Transitions of Bodies and Beings in Late Iron Age Scandinavia*. Stockholm.

Davidson, Hilda Ellis, 1960. *Patterns of Folklore*. Ipswich.

Davidson, Hilda Ellis, 1964. *Gods and Myths of Northern Europe*. Harmondsworth.

Davidson, Hilda Ellis, (1962) revised 1994. *The Sword in Anglo-Saxon England: Its Archaeology and Literature*. Woodbridge.

Davidson, Hilda Ellis, 1989. *The Training of Warriors*. In Hawkes (ed). 1989.

Davidson, Hilda Ellis, 1993. *The Lost Beliefs of Northern Europe*. London.

Davidson, Hilda Ellis, 1969. *Scandinavian Mythology*. London.

Davidson, Hilda Ellis (edit) and Fisher, Peter (translator), 1980. *Saxo Grammaticus; The History of the Danes* Books *I-IX*. Woodbridge.

Davidson, Hilda Ellis and Webster, Lesley, 1967. *The Anglo-Saxon Burial at Coombe (Woodnesborough), Kent.* Journal of Medieval Archeology, volume 11, 1967.

Davies, S. M., 1985. *The excavation of an Anglo-Saxon cemetery (and some prehistoric pits) at Charlton Plantation, near Downton*. In Wiltshire Archaeology and Natural History Magazine, 79.

Deutscher, Lisa; Kaiser, Mirjam and Wetzler (Hrsg.), Sixt, 2014. Das Schwert – Symbol und Waffe. Beiträge zur geisteswissenschaftlichen Nachwuchstagung, from 19 - 20 October 2012 in Freiburg / Breisgau.

Dickinson, T. and Härke, H., 1992. *Early Anglo-Saxon Shields*. London.

Dickinson, Tania, 2004. *An Early Anglo-Saxon Cemetery at Quarrington, near Sleaford, Lincolnshire: Reports on Excavations, 2000-2001.* In Lincolnshire History and Archaeology volume 39, 2004.

Dickinson, Tania, 1973. *Excavations at Standlake Down in 1954: the Anglo-Saxon Graves.* Oxoniensia XXXVIII 1973

Dinnetz, Mattias Karlsson, 2003. *Notes on the FerrousTechnologyof the Hispano Islamic City of Vascos.* In Nørbach, 2003.

Dinwiddy, K. E. and Stoodley, N., 2016. *An Anglo Saxon Cemetery at Collingbourne Ducis, Wiltshire*. Wessex Archaeology.

Down, Alec and Welch, Martin, 1990. *Chichester Excavations 7; Apple Down and the Mardens.* Chichester.

Drinkall, Gail and Foreman, Martin, 1998. *The Anglo-Saxon Cemetery at Castledyke South, Barton-on-*Humber. Sheffield.

Dronke, Ursula, 1997. *The Poetic Edda volume II Mythological Poems*. Oxford.

Elliott, Paul, 2014. *Legions in Crisis; Transformation of the Roman Soldier A.D. 192 – 284.* Stroud.

Englehardt, Conrad, 1863. *Søderjske og Fynske Mosefund bind I-III.* Slesvig (reprinted 1969.)

Englehardt, Conrad, 1866. *Denmark in the Early Iron Age, Illustrated by Recent Discoveries in the Peat Mosses of Slesvig*. London.

Engstrom, Robert; Lankton, Scott Michael and Lesher-Engstrom, Audrey, 1990. *A Modern Replication Based on the Pattern-Welded Sword of Sutton Hoo.* Michigan.

Ennis, Trevor, 2008. *An Early Saxon Cemetery at Rayleigh, Essex: Excavations at the Former Park School.* Chelmsford.

Espelund, Arne, 2003. *The Four Phases Carbon – Gas – Slag – Metal and their interactions in Bloomery Ironmaking*. In Nørbach, 2003.

Evans, Angela, 2005. *Seventh-century Assemblages*. In Carver (editor), 2005.

Evison, Vera, 1956. *An Anglo Saxon Cemetery at Holborough, Kent.* In Archaeologia Cantiana Vol.70.

Evison, Vera, 1976. *Sword Rings and Beads*. In Archaeologia, volume CV, 1976, Society of Antiquaries of London.

Evison, Vera, 1963. *Sugar-Loaf Shield Bosses*. In The Atiquaries Journal, volume XLIII, part I. 1963. Oxford.

Evison, Vera, 1967. *The Dover Ring-sword and Other Sword-rings and Beads.* Archaeologia 101, 1967.

Evison, Vera, 1987. *Dover: The Buckland Anglo-Saxon Cemetery*. London.

Evison, Vera, 1988. *An Anglo-Saxon Cemetery at Alton, Hampshire*. Gloucester.

Evison, Vera, 1994. *An Anglo-Saxon Cemetery at Great Chesterford, Essex.* York.

Evison, Vera and Hill, Prue, 1996. *Two Anglo-Saxon Cemeteries at Beckford, Hereford and Worcester.* York.

Fern, Christopher J. R., 2015 *Before Sutton Hoo: the Prehistoric Remains and Early Anglo-Saxon Cemetery at Tranmer House, Bromeswell, Suffolk.* Bury St. Edmunds.

Fern, Christopher J. R., 2005. *The archaeological evidence for equestrianism in early Anglo-Saxon England, c.450-700.* BAR International Series 1410. Oxford.

Fern, Christopher and Speake, George, 2014. *Beasts, Birds and Gods; Interpreting the Staffordshire Hoard.* Birmingham.

Feveile, Claus, 2014. *At the geestland edge southwest of Ribe: On the track of a centre of wealth during the first millennium A.D.* In Stidsing, Høilund Nielsen and Fiedel, 2014.

Filmer-Sankey, William and Pestell, Tim, 2001. *Snape Anglo-Saxon Cemetery: Excavations and Surveys 1824-1992.* Ipswich.

Fischer, Svante, 2008. *"Les Seigneurs des Anneaux".* Saint-Germain-en-Laye.

Fischer, Svante; Soulat, Jean and Fischer, Teodora Linton, 2013. *Sword parts and their depositional contexts; Symbols in Migration and Merovingian Periodmartial society*. In Fornvännen 2013, s. 109-122. Stockholm.

Fischer, Svante; Graf, Martin and Soulat, Jean, 2013 *"Runes and Rivers" - Typology and Distribution Patterns of Merovingian Period Runic Inscriptions along the North Sea, the Rhine, and the Danube.* Strasbourg.

Fischer, Svante and Soulat, Jean, 2007. *Runic Swords and Raw Materials – Anglo-Saxon Interaction with northern Gaul*. Paper given at 2007 58th Sachsensymposium.

Fischer, Svante and Soulat, Jean , 2011. *The Typochronology of Sword Pommels from the Staffordshire Hoard*. http://finds.org.uk/staffshoardsymposium/papers/svantefischerandjeansoulat

Frere, S.S.; Bennett, P.; Rady, J. and Stow, S., 1987. *The Archaeology of Canterbury, volume VIII: Canterbury Excavations intra- and extra-Mural Sites 1949-55 and 1980-84.* Maidstone.

Gansum, Terje, 2004. *Roll the Bones – From Iron to Steel.* In Norwegian Archaeological Review, volume 37, no. 1, 2004.

Gardeła, Leszek, 2013. *'Warrior-women' in Viking Age Scandinavia? A preliminary archaeological study*. In, Analecta Archaeologicar Ressoviensia. Rzeszów.

Gardeła, Leszek, 2012. *Entangled Worlds; Archaeologies of Ambivalence in the Viking Age*. Unpublished phd thesis, Aberdeen University.

Gardeła, Leszek, 2009. *A Biography of the Seiðr-staffs. Towards an Archeaology of Emotions*. In Słupecki, and Morawiec, (eds), 2009

Garmonsway, G. N.; Simpson, Jacqueline and Davidson, Hilda Ellis, 1968. *Beowulf and its Analogues*. London.

Geake, Helen, 1995a. *The Use of Grave-Goods in Conversion-Period England c. 600 - c. 850 A.D. volume 1.* PhD thesis, York University.

Geake, Helen, 1995b. *The Use of Grave-Goods in Conversion-Period England c. 600 - c. 850 A.D. volume 2.* PhD thesis, York University

Geake, Helen, 2011. *Accidental losses, Plough-damaged Cemeteries and the Occasional Hoard: the Portable Antiquities Scheme and Early Anglo-Saxon Archaeology*. In, Brookes, Harrington and Reynolds (eds), 2011.

Gerhards, Guntis, 2007. *Traumatic Lesions in Human Osteological Remains from the Daugava Area (Seventh to 17th Centuries AD)*. In Bliujienė, 2007.

Gibson, Catriona, 2007. *Minerva: an Early Anglo-Saxon Mixed-Rite Cemetery in Alwalton, Cambridgeshire*. In Semple, 2007.

Gilmour, Brian, 1990. *Developments in Iron Smithing and Decorative Welding Techniques Found in Anglo-Saxon Swords and Related Edged Weapons*. PhD Thesis, University College London Institute of Archaeology.

Gilmour, Brian, 1991. *A Snake Patterned Sword Blade from West Heslerton, Yorkshire*. Ancient Monuments Laboratory Report 129/91. Royal Armouries, Tower of London.

Gilmour, Brian, 2007. *Swords, Seaxes and Saxons: Pattern-Welding and edged Weapon Technology from Late Roman Britain to Anglo-Saxon England*. In Henig and Smith, 2007.

Gilmour, Brian and Edwards, Glynis, 2009. *Construction of Blades*. In Hirst and Clark part ii, 2009.

Gilmour, Brian, 2010. *Ethnic Identity and the Origins, Purpose and Occurrence of Pattern-Wlded Swords in Sixth Century Kent: The Case of the Saltwood Cemetery*. In Henig and Ramsay, 2010.

Gilmour, Brian, 2012. *Kindi's 'On swords and their kinds'. Commentary*; in Hoyland and Gilmour, 2012.

Gilmour, Brian and Hoyland, Robert, 2006. *Bīrūnī on Iron*, in Hoyland and Gilmour, 2006.

Gingell, C. J., 1978. *The excavation of an early Anglo-Saxon cemetery at Collingbourne Ducis, Wilt*. In Wiltshire Archaeology and Natural History Magazine.

Girvan, R., 1971. *Beowulf and the Seventh Century*. London.

Glover, R, 1996. *English Warfare*. in Morillo (ed), 1996.

Godfrey, E. and Van Nie, M., 2004. *A Germanic ultrahigh carbon steel punch of the late Roman -Iron Age*. In Journal of archaeological science 31, 1117-1125, 2004

Godfrey, E. G.; Vizcaio, A. and McDonnel, J.G., 2003. *The Role of Phosphorous in Early Iron Working*. In Nørbach (ed), 2003.

Gordon, R.B., 1996. *American iron 1607-1900*. Baltimore.

Grandin, Lena and Hjärthner-Holdar, Eva, 2003. *Early Iron Production in the Red Earth Area, South Central Sweden*. In Nørbach, 2003.

Green, Barbara; Rogerson, Andrew and White, Susan G., 1987. *Morning Thorpe Anglo-Saxon Cemetery, Norfolk, Volume I: Catalogue*. Dereham.

Green, Barbara; Rogerson, Andrew and White, Susan G., 1987. *Morning Thorpe Anglo-Saxon Cemetery, Norfolk, Volume II: Illustrated Grave Goods*. Dereham.

Green, Barbara and Rpgerson, Andrew, 1978. *East Ang;ian Archaeology Report No. 7: Norfolk, Bergh Apton*. Dereham.

Griffith, A. 1921. *An Anglo-Saxon Cemetery at Alfriston, Sussex. Supplemental Paper*. In Sussex Archaeological Collections Volume 62, 1921. Lewes.

Griffiths, David, (ed), 1996. *Anglo-Saxon Studies in Archaeology and History volume 9*. Oxford.

Griffiths, David, (ed), 2000. *Anglo-Saxon Studies in Archaeology and History volume 11*. Oxford.

Gurney, D., 1986. *Settlement, Religion and Industry on the Fen-Edge; three Romano-British sites in Norfolk. EAA 31*. Norwich.

Hall, Alaric, 2007. *Elves in Anglo-Saxon England: Matters of Belief, Health, Gender and Identity*. Woodbridge.

Halsall, Guy, 2010. *The Staffordshire Hoard: Warfare, Aggression and the Use of Trophies*. York University. Symposium paper, available: http://600transformer.blogspot.co.uk/2010/05/the-staffordshire-hoard-warfare.html

Halsall, Guy, 2003. *Warfare and Society in the Barbarian West 450 – 900*. London.

Hamerow, Helena, 1993. *Excavations at Mucking, volume 2: the Anglo-Saxon Settlement*. London.

Hamerow, Helena, 2002. *Early Settlements; The Archaeology of Rural Communities in North-Western Europe 400-900*. Oxford.

Hammond, Brett, 2009. *British Artefacts; volume 1 – Early Anglo-Saxon*. Witham.

Hammond, Carolyn (translator), 2008. *The Gallic War: Seven Commentaries on The Gallic War with an Eighth Commentary by Aulus Hirtius*. Oxford.

Haperen, Martine C. van, 2010. *Rest in pieces: an interpretive model of early medieval 'grave robbery*. In, Medieval and Modern Matters, vol. 1 (2010) Turnhout.

Haslam, Jeremy, with Biek, L. and Tylecote, R. F., 1980. *A Middle Saxon Iron Smelting Site at Ramsbury, Wiltshire*. In, Medieval Archaeology 24, 1980.

Härke, Heinrich, 2014. *Grave goods in early medieval burials: messages and meanings*. In Mortality, volume 19. No. 1. London.

Härke, Heinrich, 2000. *The Circulation of weapons in Anglo-Saxon society*. In, Theuws and Nelson, 2000. Leiden.

Härke, Heinrich, 1997. *Material Culture as Myth: Weapons in Anglo-Saxon Graves*. In, Jensen and Nielsen (eds), 1997.

Härke, Heinrich, 1997a. *Final Comments: Ritual, Symbolism and Social Inference*. In, Jensen and Nielsen (eds), 1997.

Härke, Heinrich, 1997b. *Early Anglo-Saxon military organisation: an archaeological perspective*. In Jørgensen, and Clausen, (eds), 1997.

Härke, Heinrich, 1992. *Changing symbols in a changing society: the Anglo-Saxon weapon burial rite*. In, Carver (ed) 1992.

Harke, Heinrich, 1989. *Knives in Early Saxon burials: blade length and age at death*. Medieval Archaeology 33.

Härke, Heinrich, 1989. *Early Saxon Weapon Burials: Frequencies, Distributions and Weapon Combinations*. In Chadwick Hawkes, 1989.

Harrington, Sue and Welch, Martin, 2014. *The Early Anglo-Saxon Kingdoms of Southern Britain AD 450-650; Beneath the Tribal Hidage*. Oxford.

Haughton, Christine and Powlesland, Dominic, 1999. *West Heslerton the Anglian Cemetery; volume 1 The Excavation and Discussion of the Evidence*. Yedingham.

Haughton, Christine and Powlesland, Dominic, 1999. *West Heslerton the Anglian Cemetery; volume 2 Catalogue of the Anglian Graves and Associated Assemblages*. Yedingham.

Hedeager, Lotte, 2010. *Split Bodies in the Late Iron Age/Viking Age of Scandinavia*. In Rebay-Salisbury, Sørensen and Hughes (eds) 2010.

Hedeager, Lotte, 2011. *Iron Age Myth and Materiality; An Archaeology of Scandinavia A.D. 400-1000*. Abingdon.

Hedenstierna-Jonson, Charlotte; 2006. *The Birka Warrior. The Material Culture of a Martial Society*. Doctoral thesis, University of Stockholm.

Hedges, J.D. and Buckley, D.G., 1985. *Anglo-Saxon Burials and Later Features Excavated at Orsett, Essex, 1975*. Journal of Medieval Archaeology volume 29.

Helmbrecht, Michaela, 2007 – 2008. *Figures with Horned Heads; a Case Study of Context Analysis and Social Significance of Pictures in Vendel and Viking Age Scandinavia*. In Lund Archaeology Review 13 – 14 (2007 - 2008). Lund.

Helmbrecht, Michaela, 2007. *Innere Strukturen vonSiedlungen und Gräberfeldern als Spiegel gesellschaftlicher Wirklichkeit?* Aschendorff Münster.

Henig, Martin and Smith, Jo Tyler, (eds) 2007. *Collectanea Antiqua: Essays in Memory of Sonia Chadwick Hawkes.* Oxford.

Henig, Martin and Ramsay, Nigel (eds) 2010. *Intersections: The Archaeology and History of Christianity in England, 400-1200; Papers in Honour of Martin Biddle and Birthe Kjølbye-Biddle.* Oxford.

Herbert, Kathleen, 1997. *Peace-Weavers and Shield-Maidens; Women in Early English Society*. Hockwold-cum-Wilton.

Hill, Paul and Thompson, Logan, 2003. *The Swords of the Saxon Cemetery at Mitcham.* Surrey Archaeological Collections, 90, 147-161, 2003.

Himes, Jonathan B. (editor and translator), 2009. *The Old English Epic of Waldere.* Cambridge.

Hines, John; Nielsen, Karen Høilund and Siegemund, Frank, (eds), 1999. *The Pace of Change; Studies in Early Medieval Chronology.* Oxford.

Hines, John; Bayliss, Alex (editors), 2013. *Anglo-Saxon Graves and Grave Goods of the 6th and 7th centuries AD: a Chronological Framework*. London.

Hinton, David A, 1998. *Anglo-Saxon Smiths and Myths*. Manchester Centre for Anglo-Saxon Studies. Manchester.

Hinton, David A (ed), 2000. *A Smith in Lindsey; the Anglo-Saxon Grave at Tattershall Thorpe, Lincolnshire*. London.

Hirst, Susan M., 1985. *An Anglo-Saxon Inhumation Cemetery at Sewerby East Yorkshire.* York.

Hirst, Sue and Clark, Dido, 2009. *Excavations at Mucking; volume 3, the Anglo-Saxon Cemeteries, part i Introduction, Catalogues and Specialist Reports.* London.

Hirst, Sue and Clark, Dido, 2009. *Excavations at Mucking; volume 3, the Anglo-Saxon Cemeteries, part ii Analysis and Discussion.* London.

Holbrook, Neil, 2000. *The Anglo-Saxon Cemetery at Lower Farm, Bishop's Cleve: Excavations Directed by Kevin Brown 1969.* Bristol and Gloucestershire Archaeological Society volume 118.

Holthausen, F., 1974. *Altenglisches Etymologisches Worterbuch*. Heidelberg.

Hoyland, Robert G and Gilmour, Brian, 2012. *Medieval Islamic Swords and Swordmaking; Kindi's treatise 'On swords and their kinds'*. Oxford.

Hreinsson, Viþar (editor), 1997. *The Complete Sagas of the Icelanders including 49 Tales*. 5 volumes. Reykjavík.

Ijäs, Antti, ????. A Comment Regarding the Handedness of Anglo-Saxon and Viking Swords. https://www.academia.edu/10275283/A_Comment_Regarding_the_Handedness_of_Anglo-Saxon_and_Viking_Swords

Ilkjær, Jørgen, 1991. *Illerup Adal: Die Lanzen Und Speere: Textband / Tafleband, volumes 1 & 2*. Aarhus.

Ilkjær, Jørgen, 2002. *Illerup Adal: Die Schilde: Textband / Tafeln volumes 9 and 10*. Aarhus.

Ilkjær, Jørgen, 2001. *Illerup Ådal*. Aarhus.

Inker, P. A. 2006. *The Saxon Relief Style*. Oxford.

James, Edward; 1997. *The Militarization of Roman Society, 400-700*. In Jørgensen, Nørgård and Clausen, (eds), 1997.

James, Simon, 2011. *Rome and the Sword*. London.

James, Simon, 2010. *Excavations at Dura Europos 1928-1937. Final Report VII.* Oxford.

Jensen, Claus Kjeld and Nielsen, Karen Høilund (eds), 1997. *Burial and Society; The Chronologigical and Social Analysis of Burial Data.* Aarhus.

Johnson, Thomas M., 1995. *Collecting the Edged Weapons of the Third Reich volume 1.* Fredericksburg.

Jones, Lee A., 1997. *The Serpent in the Sword: Pattern-welding in Early Medieval Swords.* In the catalogue of The Fourteenth Park Lane Arms Fair. Morpeth.

Jørgensen, Storgaard and Thompson (eds); 2003. *Spoils of Victory*. Aarhus

Jørgensen, E. and Vang Peterson, P., 2003. *Nydam Bog – New finds and observations.* In Jørgensen, Storgaard and Thompson, 2003.

Jørgensen, Anne Nørgård and Clausen, Birthe (eds), 1997. *Military Aspects of Scandinavian Society in a European Perspective, AD 1 – 1300*. Copenhagen.

Jorgensen, Lars, 2001. *The "Warriors, Soldiers and Conscripts" of the Anthropology in the Late Roman and Migration Period Archaeology*. In Storgaard, (ed) 2001.

Kennet, D. 1977. *The Anglo-Saxon Cemetery at Tuddenham, Suffolk*. Cambridge Antiquarian Society. Cambridge.

Kershaw, Kris; 2000. *The One Eyed God*. Washington D. C.

Kirk, Joan and Marshall, Kenneth, 1956. *A Saxon Cemetery near the Village of Harwell, Berkshire.* Oxoniensia.

Klevnäs, Alison, 2015. *Abandon Ship! Digging out the Dead from the Vendel Boat-Graves* . In Norwegian Archaeological Review, 2015

Klevnäs, Alison, 2013. *Whodunnit? Grave robbery in Anglo-Saxon England and the Merovingian kingdoms.* Oxford.

Klevnäs, Alison, 2007. *Robbing the dead at Gamla Uppasla*. Uppsala.

Kmetič, Dimitrij; Horvat, Jana and Vodopivec, Franc, 2004. *Metallographic Examinations of the Roman Republican Weapons from the Hoard from Grad near Šmihel.* In Arheološki vestnik (Arh. vest.) 55, 2004.

Lang, Janet and Ager, Barry, 1989. *Swords of the Anglo-Saxon and Viking Periods in the British Museum: a Radiographic Study.* Chadwick Hawkes, 1989.

Lang, Janet, 2006. *The Technology of Some Swords.* In Stead, 2006.

Lang, Janet, 2007. *The Rise and Fall of Pattern Welding: an investigation into the construction of pre-medieval sword blades.* Unpublished PhD thesis, University of Reading.

Lang, Janet, 2012. *Metallographic Study: Five Swords.* In Parfitt and Anderson, 2012.

Langton, Scott, 1989. *A Replica of the Sutton Hoo Sword*. In The Anvil's Ring volume 17, number 4. Nashville.

Larrington, Carolyne (translator) 1996. *The Poetic Edda*. Oxford.

Leahy, Kevin, 2003. *Anglo-Saxon Crafts*. Stroud.

Leahy, Kevin, 2007. *'Interrupting the Pots' The Excavation of Cleatham Anglo-Saxon Cemetery.* York.

Leahy, Kevin and Bland, Roger, 2009. *The Staffordshire Hoard*. London.

Leeds, E.T. and Harden, D.B., 1936. *The Anglo-Saxon Cemetery at Abingdon, Berkshire*. Oxford.

Leeds, E.T., 1940. *Two Saxon Cemetries in North Oxfordshire.* In Oxoniensia volume V 1940.

Lehmann, Ulrich, 2009. *Eine merowingerzeitliche Spatha mit Scheide aus Bad Wünnenberg im CT-Scan*, in Archäologie in Westfalen-Lippe 2009. Westfalen.

Lehmann, Ulrich, 2015. *Wurmbunte Klingen – Studies of pattern-welded swords in early medieval Westphalia using computerised x-ray tomography.* In Ruhmann and Brieske (editors), 2015.

vom 19. – 20. Oktober 2012 in Freiburg/Breisgau

Lehmann, Ulrich, 2013. *Wurmbunte Klingen – Studies on pattern-welded swords in early medieval Westphalia based on digital computed x-ray tomography*. Unpublished research paper.

Lehmann, Ulrich, 2014. Die Klinge der frühmittelalterlichen Spatha. In Deutscher, Kaiser, and Wetzler (Hrsg.), 2014.

Lethbridge, T. C., 1951. *A Cemetery at Lackford Suffolk: Report of a Cemetery of the Pagan Anglo-Saxon Period in 1947.* Cambridge.

Lethbridge, T. C., 1936. *A Cemetery at Shudy Camps Cambridgeshire; Report of the Excavation of a Cemetery of the Christian Anglo-Saxon Period.* Cambridge.

Liddle, Peter, 1986. *An Anglo-Saxon Inhumation Burial from Lutterworth, Leicerstershire.* Leicestershire Archaeological and Historical Transactions, 60.

Liddle, Peter and Middleton, Samantha, 1994. *An Anglo-Saxon Cemetery at Wigston Magna, Leicestershire.* Leicestershire Archaeological and Historical Transactions, 68.

Liddle, Peter, 1981. *An Anglo-Saxon Cemetery at Wanlip, Leicestershire.* Leicestershire Archaeological and Historical Transactions, 55.

Liestøl, Aslak, 1951. *Blodrefill og Mál*, in Viking Tidsskrift for Norrøn Arkeologi, Bind XV 1951. Oslo.

Lindow, John, 2001. *Norse Mythology; a Guide to the Gods, Heroes, Rituals, and Beliefs.* Oxford.

Lindqvist, Sune, 1936. *Uppsala Högar och Ottarshögen*. Stockholm.

Livingston, Michael (ed), 2011. *The Battle of Brunanburh; a Casebook.* Exeter.

Loades, Mike, 2010. *Swords and Swordsmen*. Barnsley.

Lucy, Sam, 2000. *The Anglo-Saxon Way of Death*. Stroud.

Lucy, Sam; Tipper, Jess and Dickens, Alison, 2009. *The Anglo-Saxon Settlement and Cemetery at Bloodmoor Hill, Carlton Colville, Suffolk.* Cambridge.

Lüppes, L. H., 2010. *Gedanken zur spätmerowingerzeitlichen Spathaaufhängung - eine zu belegende und tragbare Rekonstruktion*. Mainz.

Lyngstrøm, Henriette, 2003. *Farmer, Smelters and Smiths.* In Nørbach, 2003.

Mackinder, Tony, 2006. *Iron Age Settlement and an Anglo-Saxon Cemetery at Cuxto, Kent; volume 1: The Site Report.* CTRL.

Mackenzie, E., 1825. An Historical, Topographical, and Descriptive view of the county of the Northumberland. Newcastle Upon Tyne.

Mäder, Stefan, 2001. *Stähle, Steine und Schlange. Zur Kultur- und Technikgeschichte von Schwertklingen des frühen Mittelalters*. PhD thesis, Humboldt Universität zu Berlin. Berlin. http://www.schwertbruecken.de/pdf/staehle.pdf

Malim, Tim and Hines, john, 1998. *The Anglo-Saxon Cemetery at Edix Hill (Barrington A), Cambridgeshire.* York.

Mäntylä-Asplund, Sari and Storå, Jan, 2010. *On the Archaeology and Osteology of the Rikala Cremation Cemetery in Salo, SW Finland*. In, Fennoscandia archaeologica XXVII (2010).

Marti, Reto, 1995. *Das Grab eines wohlhabenden Alamannen in Altdorf UR, Pfarrkirche St. Martin*. In, Jahrbuch der Schweizerischen Gesellschaft für Ur- und Frühgeschichte, Band(Jahr): 78(1995). Zurich.

Marzinzik, Sonja, 2003. *Early Anglo-Saxon Belt Buckles (late 5th to early 8th centuries A.D.) Their classification and context.* Oxford.

Matthews, C. L. And Chadwick Hawkes, 1985. *Early Saxon Settlements and Burials on Puddlehill near Dunstable, Bedfordshire.* In Chadwick Hawkes, Campbell and Brown, 1985.

Matzke, Johann Keller Wheelock, 2011. *Armed* and *Educated: Determining the Identity of the Medieval Combatant*. Masters degree thesis, University of Exeter.

Mayer, Joseph, 1856. *Inventorium Sepulchrale: an Account of Some Antiquities Dug Up at Gilton, Sibbertswold, Barfriston, Beakesbourne, Chartham, and Crundale in the County of Kent, from A.D. 1757 to A.D. 1773, by the Rev. Bryan Faussett of Heppington.* Oxford.

McCormick, E., 2016. *Mineral Preserved Organics and Compositional Analysis of Metalwork*. in Dinwiddy and Stoodly, 2016.

McDonnell, Gerry, 1993. *Slags and Ironworking Residues.* In Hammerow, 1993.

McDonnell, Gerry, 1989. *Iron and its Alloys in the Fifth to Eleventh Centuries A.D. in England.* In World Archaeology, volume 20, no. 3 Archaeometallurgy. February 1989.

Meaney, Audrey, 1981. *Anglo-Saxon Amulets and Curing Stones*. Oxford.

Meaney, Audrey, 1964. *A Gazetteer of Early Anglo-Saxon Burial Sites.* London.

Melzer, W., 1991. *Das frühmittelalterliche Gräberfeld von Wünnenberg - Fürstenberg, Kreis Paderborn*. Aschendorffsche.

Menghin, Wilfried, 1983. *Das Schwert im Frühen Mittelalter.* Stuttgart.

Miket, Roger and Pocock, Michael, 1976. *An Anglo-Saxon Cemetery at Greenbank, Darlington.* In the Journal of Medieval Archaeology volume 20, 1976.

Mikkelsen, Peter Hambro, 2003. *Slag – With an Impression of Agricultural Practices*. In Nørbach (ed), 2003.

Miks, Christian, 2007. *Studien zur Römischen Scwertbewaffnung in der Kaiserzeit*. (2 volumes – volume 1 text, volume 2 Katalog und Tafeln). Köln.

Mills, Nigel, 2001. *Saxon and Viking Artefacts*. Witham.

Molloy, Barry (editor)2007. *The Cutting Edge; Studies in Ancient and Medieval Combat*. Stroud.

Mortimer, Paul, 2011. *Woden's Warriors; Warfare, Beliefs, Arms and Armour in Northern Europe during the 6th and 7th Centuries.* Ely.

Mortimer. Paul and Pollington, Stephen, (eds), 2013. *Remaking the Sutton Hoo Stone; the Ansell-Roper Replica and its Context*. Ely.

Morillo, John, 1996 (ed). *The Battle of Hastings*. Woodbridge.

Motz, Lotte, 1983. *The Wise One of the Mountain, Form, Functionand Significance of the Subterranean Smith: A Study in Folklore*. Lorch.

Murray, Jon and McDonald, Tom, 2006. *Excavations at Station Road, Gamlingay, Cambridgeshire.* In Semple, 2006.

Musty, John, 1969. *The Excavation of two Barrows, One of Saxon Date, at Ford, Laverstock, near Salisbury Wilthsire.* In The Antiquaries Journal volume XLIX part 1.

Mühlhäsler, Mark and Hoyland, Robert, 2006. *Swords in Arabic Poetry,* in Hoyland and Gilmour, 2006.

Myres, J.N.L. and Green, Barbara, 1973. *The Anglo-Saxon Cemeteries of Caistor-by-Norwich and Markshall, Norfolk.* Oxford.

Navasaitas, Jonas; Sveikaustkaite, Aušra; Selskis, Algirdas and Matulionis, Eimutis, 2003. *Ironmaking Techniques during the Roman Period in Lithuania.* In Nørbach, 2003.

Neidorf, Leonard (editor), 2014. *The Dating of Beowulf; A Reassessment*. Woodbridge.

Neuffer, Eduard M., 1972. *Der Reihengräberfriedhof von Donzdorf.* Stuttgart.

Newton, Sam, 2004. *The Origins of Beowulf: and the Pre-Viking Kingdom of East Anglia*. Woodbridge.

Nicolay, Johan, 2007. *Armed Batavians Use and Significance of Weaponry and Horse Gear from Non-military Contexts in the Rhine Delta (50 BC to AD 450).* Amsterdam.

Nicolay, Johan A. W., 2014. *The Splendour of Power; Early Medieval Kingship and the use of Gold and Silver in the Southern North Sea Area*. Groningen.

Nikulka, Frank, 2003. *The Development of Iron Production in Northwestern Germany.* In Nørbach, 2003.

Nørbach, Lars Christian (ed). 2003. *Prehistoric and Medieval Direct Iron Smelting in Scandinavia and Europe*. Aarhus.

Nørgård Jørgensen, Anne and Andersen, Hans Christian H., 2014. *Ejsbøl Mose Die Kriegsbeuteopfer im Moor von Ejsbøl aus dem späten 1.Jh.v.Chr. bis zum frühen 5 Jh.n.Chr*. Aarhus.

Nørgård Jørgensen, Anne, 2008. *Porskjær Mosefund; Jernalderen i Nordeuropa*. Aarhus.

Nørgård Jørgensen, Anne, 1999. *Waffen und Gräber: Typologische und Chronologische Studien zu Skandinavischen Waffengräbebern 520/30 bis 900 n. Chr. Copenhagen.*

Nørgård Jørgensen, Anne, 1997. *Scandinavian Military Equipment and the Weapon-Burial Rite, A.D. 530 – 800.* In, Jensen and Nielsen, 1997.

Novum Inventorium Sepulchrale. http://inventorium.arch.ox.ac.uk/home.php

Nuemann, Bernhard, 1927. *Römischer Damaststahl*, in Archiv Für Das Eisenhüttenwesen Fachberichte, 1927. Dusseldorf.

Nyström, I. 1993. *"Vapnö-fyndet", Stiftelsen Västsvensk Anskrift.* Oslo.

Oakeshott, Ewart, 1991. *Records of the Medieval Sword.* Woodbridge.

Oakeshott, Ewart, 1974. *Dark Age Warrior*. London.

Oakeshott, Ewart, 1964 (1997 reprint with appendix). *The Sword in the Age of Chivalry.* Woodbridge.

Orel, V., 2003. *A Handbook of Germanic Etymology*. Leiden.

Ørsnes, Mogens, 1988. *Ejsbøl I; Waffenopferfunde des 4. – 5. Jahrh. Nach Chr.* Copenhagen.

Ottaway, Patrick, J., 2012. *Anglo-Saxon Ironwork.* http://www.pjoarchaeology.co.uk/academic-consultancy/anglosaxon-ironwork.html

Otto, Ton; Thrane, Henrik and Kilde, Hellevand (editors), 2006. Warfare and Society; Archaeological and Social Anthropological Perspectives. Aarhus.

Patrick, Philippa; French, Charles and Osborne, Christine, 2007. *Rescue Excavation of an early Anglo-Saxon Cemetery at Gunthorpe, Peterborough.* In Semple, 2007.

Parfitt, Keith and Anderson, Trevor, (eds) 2012. *Buckland Anglo-Saxon Cemetery, Dover; Excavations 1994.* Canterbury.

Parfitt, Keith and Dickinson, Tania, 2007. *The Anglo-Saxon Cemetery at Old Park, near Dover, Revisited.* In Henig and Smith, 2007.

Paulsen, Peter, 1967. *Alamanische Adesgräber von Niederstotzingen*. Stuttgart.

Peets, Jüri; Allmäe, Raili and Maldre, Liina 2010. *Archaeological Investigations of Pre-Viking Age Burial Boat in Salme Village at Saaremaa*. Tallin.

Peirce, Ian G. and Oakeshott, Ewart, 2004. *Swords of the Viking Age*. Woodbridge.

Pelsmaeker, Sebastiaan B M, 2010. *Weapons of Princes, Weapons of War? An experimental analysis of pattern-welded swords from northwestern Europe, 400-1100 AD*. Unpublished MA Thesis. University of Groningen.

Petersen, Jan, 1919. *De Norske Vikingsverd*_ Oslo.

Petersen, Leif Ing Ree, 2013. *Siege Warfare and Military Organisation in the Successor States (400 – 800 AD)*. Leiden.

Penn, Kenneth, 2011. *The Anglo-Saxon Cemetery at Shrubland Hall Quarry, Coddenham, Suffolk.* Bury St Edmunds.

Penn, Kenneth, 2000. *Norwich Southern Bypass, Part II: Anglo-Saxon Cemetery at Harford Farm, Caistor St Edmund.* Dereham.

Penn, Kenneth, 1998. *An Anglo-Saxon Cemetery at Oxborough, West Norfolk: Excavations in 1990.* Dereham.

Petersen, Leif Inge Ree , 2013. *Siege Warfare and Military Organization in the Successor States (400 – 800 AD).* Leiden.

Pickwoad, N. 2015. *"Binding" in The St Cuthbert Gospel: Studies on the Insular Manuscript of the Gospel of John*. British Library

Plunkett, Steven, 2005. *Suffolk in Anglo-Saxon Times*. Stroud.

Pollington, Stephen, 2016. *Runes: the adoption of Literacy in the Germanic Iron Age*. Ely.

Pollington, Stephen, 2011. *The Elder Gods; the Otherworld of Early England*. Ely.

Pollington, Stephen; Kerr, Lindsay and Hammond, Brett, 2010. *Wayland's Work:Anglo-Saxon Art, Myth and Material Culture from the 4th to 7th Century.* Ely.

Pollington, Stephen, 2008. *Rudiments of Runelore*. Ely.

Pollington, Stephen, 1996. *The English Warrior from Earliest Times to 1066*. Hockwold-cum-Wilton.

Pollington, Stephen, 2011. *The Elder Gods; The Otherworld of Early England*. Ely.

Price, Neil, 2002. *The Viking Way; Religion and War in Late Iron Age Scandinavia*. Uppsala.

Price, T. Douglas; Peets, Jüri ; Allmäea, Raili; Maldrea, Lina and Orasa, Ester, 2016. *Isotopic provenancing of the Salme ship burials in Pre-Viking Age Estonia*. In Antiquity volume 90. Cambridge.

Proudfoot, Edwina and Aliaga-Kelly, Christopher, 1996. *Anomalous Finds and Place-Names of Anglo-Saxon Origin in Scotland*. In Anglo-Saxon Studies. In Archaeology and History volume 9. Oxford.

Rau, Andreas, 2010. *Nydam Mose: die personengebundenen Gegenstände. Grabungen 1989-1999*. Volume 1. Aarhus.

Rau, Andreas, 2010b. *Nydam Mose: die personengebundenen Gegenstände. Grabungen 1989-1999*. Volume 2. Aarhus.

Raynor, Keith, 2010. *The Rempstone Mount*. East Leake.

Rebay-Salisbury, Katharina; Sørensen, Marie Louise Stig and Hughes, Jessica, 2010. *Body Parts and Bodies Whole. Changing Relations and Meanings*. Oxford.

Reynolds, Andrew and Semple, Sarah, 2011. *Anglo-Saxon non-Funerary Weapon Depositions.* In Brookes, Harrington and Reynolds, 2011.

Reynolds, Andrew, 2009. *Anglo-Saxon Deviant Burial Customs*. Oxford.

Richardson, Andrew Frank, 2000. *The Anglo-Saxon Cemeteries of Kent.* Cardiff University, Phd thesis.

Riddler, Ian; McKinley, Jacqueline and Skittrell, Simon, 2006. *The prehistoric, Roman and Anglo-Saxon funerary landscape at Saltwood Tunnel, Kent: The grave catalogue*. CTRL

Riley, Dennis, 2011. *Anglo-Saxon Iron and Steelmaking*. Unpublished.

Rives, J. B. 1999. *Tacitus Germania*. Oxford.

Roach Smith, Charles, 1863(?). *Collectenea Antiqua volume VI.* London.

Rolfsen, P. and Stylegar, F-A. 2003. *Snartemofunnene I Nytt Lys*. Oslo

Roth, Helmuth and Theune, Claudia, 1995. *Das frühmittelalterliche Gräberfeld von Weingarten, Band 1.* Stuttgart.

Ruhmann, Christiane und Brieske, Vera, (editors) 2015. *Dying Gods – Religious beliefs in northern and eastern Europe in the time of Christianisation. Neue Studien zur Sachsenforschung Band 5.* Hannover.

Rupp, Cornelia, 2005. *Das Langobardische Graberfeld Von Nocera Umbra*. Firenze.

Russell, James C., 1994. *The Germanization of Early Medieval Christianity; a Sociological Approach to Religious Transformation*. Oxford.

Salvia, Vasco La, 1998. *The Archaeometallurgy of Lombard Swords; From Artifacts to a History of Craftsmanship*. Florence.

Salvia, Vasco La, 2007. *Iron MakingDuring the Migration Period the Case of the Lombards.* Oxford.

Scull, Christopher, 2009. *Early Medieval (Late 5th – Early 8th Centuries AD) Cemeteries at Boss Hall and Buttermarket, Ipswich, Suffolk.* Leeds.

Scull, Christopher, 1992. *Excavations and Survey at Watchfield, Oxfordshire, 1983-92. (mainly dealing with the Anglo-Saxon Cemetery)*. In Archaeological Journal, 149, 124-281

Semple, Sarah; (ed), 2006. *Anglo-Saxon Studies in Archaeology and History volume 13.* Oxford.

Semple, Sarah; (ed) 2007. *Anglo-Saxon Studies in Archaeology and History volume 14.* Oxford.

Shaw, Philip Andrew, 2002. *Uses of Wodan; The Development of His Cult and of Medieval Literary Responses to it.* PhD Thesis, University of Leeds.

Shearman, Fleur; Bullock, Hayley; Nordgren, Eric and Leseur, 2012. *Sword Scabbards and Hilts*. In Parfitt and Anderson, 2012.

Shepherd, Deborah J., 1998. *The Elusive Warrior Maiden Tradition, Archaeology, and the Social Meaning of Bearing Weapons in North European Society before the Christian Conversion*. Paper delivered at Society for American Archaeology Meeting, Seattle, Washington, 1998.

Shepherd, Deborah J., 1997. *Evidence of Violence from Anglo-Saxon Cemeteries*. Conference paper at 1997 International Medieval Congress, Kalamazoo, MI

Sherley-Price, Leo (translator), 1955. *Bede; A History of the English Church and People*. Harmondsworth.

Sherlock, Stephen J. And Welch, Martin G., 1992. *An Anglo-Saxon Cemetery at Norton, Cleveland*. London.

Sim, D. and Kaminski, J., 2012. *Roman Imperial Armour; the Production of Early Imperial Military Armour*. Oxford.

Słupecki, Leszek P., and Morawiec, Jakob (eds), 2009. *Between Paganism and Christianity in the North*. Rzeszów.

Smith, R. A., 1923. *A Guide to the Anglo-Saxon and Foreign Teutonic Antiquities in the Department of British and Mediaeval Antiquities*. British Museum Publications. London.

Southern, P. and Dixon, K.R., 1996. *The Late Roman Army*. London.

Spazier, Ines, 2003. *The Germanic Iron-Smelting Complex at Wolkenberg in Lower Lausacia, Southern Brandenburg*. In Nørbach, 2003.

Staffordshire Hoard Newsletters. http://www.barbicanra.co.uk/newsletters.html

Starley, David, 2006. *Sword fragment from Bamburgh Castle: Job Code: AM 1991*. Royal Armouries Analysis Record . York.

Stead, I. M., 2006. *British Iron Age Swords and Scabbards*. London.

Stead, I. M. And Rigby, V., 1989. *Verulanium: the King Harry Lane Site*. London.

Stenvik, Lars F., 2003. *Recent Results from Investigations of Iron Production in Northern Europe*. In Nørbach, 2003.

Steuer, Heiko, 2006. Warrior Bands, War Lords, and the Birth of Tribes and States in the First Millennium AD in Middle Europe. In, Otto, Thrane and Vandkilde (eds), 2006.

Steuer, Heiko, 1987. *Helms und Ringschwert Prunkbewaffnung und Rangagzeichen germanischer Krieger. Eine Übersicht*. In Studien zur Sachsenforschung 6 (1987). Hanover.

Stidsing, Ernst; Høilund Nielsen, Karen and Fiedel, Reno (editors), 2014. *Wealth and Complexity; Economically Specialised Sites in Late Iron Age Denmark*. Aarhus.

Stillwell, Ben, 2002. *An investigation into cranial trauma from the Anglo Saxon cemetery at Sedgeford, Norfolk, UK*. Undergraduate dissertation, Bournemouth University.

Stoodley, Nick, 1997. *The Spindle and the Spear: a Critical Enquiry into the Construction and Meaning of Gender in the early Anglo-Saxon Inhumation Burial Rite*. PhD thesis, university of Reading.

Storgaard, Birger; (ed) 2001. *Military Aspects of the Aristocracy in Barbaricum in the Roman and Early Migration Periods*. Copenhagen.

Söderström, Ulrika, 2015. *Sandby Borg; Unveiling the Massacre*. Kalmar.

Soulat, Jean, 2009. *Le Matériel Archéologique de Type Saxon et Anglo-Saxon en Gaule Mérovingienne*. Saint-Germain-en-Laye.

Stolpe, Hjalmar, 1912. *Graffältet vid Vendel*. Stockholm

Sundkvist, Olof, 2002. *Freyr's Offspring; Rulers and religion in ancient Svea society*. Uppsala

Swanton, Michael (translator and editor) 1996. *The Anglo-Saxon Chronicles*. London.

Swanton, M. J., 1973. *The Spearheads of the Anglo-Saxon Settlements*. Leeds.

Symons, David, 2014. *The Staffordshire Hoard*. Birmingham.

Taylor, Paul B., and Auden, W. H. (translators), 1969. *The Elder Edda*. London.

Thålin Bergman, Lena and Arrhenius, Birgit, 2005. *Excavations at Helgö XV*. Stockholm.

Theuws, Franz, (ed, 1991. *Images of the Past*. Lieden.

Theuws, Frans and Nelson, Janet L., 2000. *Rituals of Power from Late Antiquity to the Early Middle Ages*. Leiden.

Theuws, Frans and Alkemade, Monica, 2000. *A Kind of Mirror for Men: Sword Deposition in Late Antique Northern Gaul*. In, Theuws and Nelson, 2000. Leiden.

Timby, Jane R., 1996. *The Anglo-Saxon Cemetery at Empingham II, Rutland*. Oxford.

Tizzoni, Constanza Cucini and Tizzoni, Marco, 2003. *The Late Roman Ironworking Site at Ponte di Val Gabbia III, Bienno (Brescia, Italy)*. In Nørbach, 2003.

Trousdale, W., 1975. *The Long Sword and Scabbard Slide in Asia*. Washington.

Truckell, A.E. and Cormack, W.F., FF.S.A.Scot (editors), 1973. *Transactions of the Dumfriesshire and Galloway Natural History and Antiquarian Society THIRD SERIES, VOLUME L*. Dumfries.

Tylecote, R. F. and Gilmour, B. J. J., 1986. *The Metallography of Early Ferrous Edge Tools and Edged Weapons*. Oxford.

Tyler, Susan and Major, Hilary, 2005. *The Early Anglo-Saxon Cemetery and Later Saxon Settlement at Springfield Lyons, Essex*. Chelmsford.

Tyler, Susan, 1988. *The Anglo-Saxon Cemetery at Prittlewell, Essex: an analysis of the grave goods*. In Essex Archaeology and History volume 19 (1988). Essex.

Underwood, Richard, 1999. *Anglo-Saxon Weapons and Warfare*. Stroud.

Wallace Hadrill, 1971. *Early Germanic Kingship: In England and on the Continent (Ford Lectures)*. Oxford.

Walton Rogers, P., 2005. *The Textiles from Mounds 5, 7, 14 & 17*. In Carver, 2005.

Wang, Quanyu, 2007. *Effects of Relative Humidity on the Corrosion of Iron: An Experimental View*. In The British Museum Technical Research Bulletin, volume 1, 2007. London.

Webster, Leslie and Backhouse, Janet (editors), 1991. *The Making of England; Anglo-Saxon Art and Culture AD 600 – 900*. London.

Welch, Martin, 2000. *Croydon Saxon Cemetery: The Re-discovery of the early Anglo-Saxon Cemetery at Croydon in 1992 and its Partial Excavation in 1999: Trials and Tribulations*. Croydon.

Welch, Martin, 2008. *Report on the Excavations of the Anglo-Saxon Cemetery at Updown, Eastry, Kent*. In Crawford and Hamerow, 2008.

Wells, Peter S., 2012. *How Ancient Europeans Saw the World: Vision, Patterns, and the Shaping of the Mind in Prehistoric Times*. Princeton

Wenham, S.J., 1989. *Anatomical Interpretations of Anglo-Saxon Weapon Injuries*. In Chadwick Hawkes, 1989.

Welton, Andrew J., 2016. Encounters with Iron: An Archaeometallurgical Reassessment of Early Anglo-Saxon Spearheads and Knives. In, Archeological Journal, DOI: 0.1080/00665983.2016.1175891.

West, Stanley E., 1988. *Westgarth Gardens Anglo-Saxon Cemetery, Suffolk: Catalogue*. Bury St Edmunds.

West, Stanley (editor), 1998. *A Corpus of Anglo-Saxon Material from Suffolk*. East Anglian Archaeology (EAA 84). Ipswich.

White, S., 1998. *The Patching Hoard*. Medieval Archaeology 42.

Wickham, Chris, 2009. *The Inheritance of Rome; a History of Europe from 400 to 1000*. London.

Williams, Alan, 2012. *The Sword and the Crucible: A History of the Metallurgy of European Swords up to the16th Century*. Leiden.

Williams, Alan; 2003. *The Knight and the Blast Furnace: a History of Mettallurgy of Amour in the Middle Ages and Early Modern Period*. Leiden.

Williams, Phillip and Newman, Richard, 2006. *Market Lavington, Wiltshire: An Anglo-saxon Cemetery and Settlement.* Salisbury.

Winterbourne, Anthony; 2004. *When the Norns Have Spoken*. New Jersey.

Wright, Duncan, 2010. *Tasting Misery Among Snakes: The Situation of Smiths in Anglo-Saxon Settlements*. In Papers from the Institute of Archeology University College London. (PIA) volume 20, 2010. London.

Index

Some of our other titles

Please see www.asbooks.co.uk for latest availability and prices

First Steps in Old English

An easy to follow language course for the beginner

Stephen Pollington

A complete and easy to use Old English language course that contains all the exercises and texts needed to learn Old English. This course has been designed to be of help to a wide range of students, from those who are teaching themselves at home, to undergraduates who are learning Old English as part of their English degree course. The author has adopted a step-by-step approach that enables students of differing abilities to advance at their own pace. The course includes practice and translation exercises, a glossary of the words used in the course, and many Old English texts, including the *Battle of Brunanburh* and *Battle of Maldon.*

£16-95 272 pages

Old English Poems, Prose & Lessons 2 CDs

read by Stephen Pollington

These CDs contain lessons and texts from *First Steps in Old English.*

Tracks include: 1. Deor. 2. Beowulf – The Funeral of Scyld Scefing. 3. Engla Tocyme (The Arrival of the English). 4. Ines Domas. Two Extracts from the Laws of King Ine. 5. Deniga Hergung (The Danes' Harrying) Anglo-Saxon Chronicle Entry AD997. 6. Durham 7. The Ordeal (Be ðon ðe ordales weddigaþ) 8. Wið Dweorh (Against a Dwarf) 9. Wið Wennum (Against Wens) 10. Wið Wæterælfadle (Against Waterelf Sickness) 11. The Nine Herbs Charm 12. Læcedomas (Leechdoms) 13. Beowulf's Greeting 14. The Battle of Brunanburh 15. A Guide to Pronunciation.

And more than 30 other lessons and extracts of Old English verse and prose.

£15 2 CDs - Free Old English transcript from www.asbooks.co.uk.

Learn Old English with Leofwin

Matt Love

This is a new approach to learning old English – as a *living language*. Leofwin and his family are your guides through six lively, entertaining, topic-based units. New vocabulary and grammar are presented in context, step by step, so that younger readers and non-language specialists can feel engaged rather than intimidated. The author has complemented the text with a wealth of illustrations. There are listening, speaking, reading and writing exercises throughout. Free soundtracks available on the Anglo-Saxon Books website.

£16.95 160 pages

Wordcraft: Concise English/Old English Dictionary and Thesaurus

Stephen Pollington

This book provides Old English equivalents to the commoner modern words in both dictionary and thesaurus formats. The Thesaurus presents vocabulary relevant to a wide range of individual topics in alphabetical lists, thus making it easily accessible to those with specific areas of interest. Each thematic listing is encoded for cross-reference from the Dictionary. The two sections will be of invaluable assistance to students of the language, as well as to those with either a general or a specific interest in the Anglo-Saxon period.

£9.95 256 pages

An Introduction to the Old English Language and its Literature

Stephen Pollington

The purpose of this general introduction to Old English is not to deal with the teaching of Old English but to dispel some misconceptions about the language and to give an outline of its structure and its literature. Some basic knowledge of these is essential to an understanding of the early period of English history and the present form of the language.

£5.95 48 pages

The Origins of the Anglo-Saxons

Donald Henson

This book has three great strengths.

> First, it pulls together and summarises the whole range of evidence bearing on the subject, offering an up-to-date assessment: the book is, in other words, a highly efficient introduction to the subject. Second – perhaps reflecting Henson's position as a leading practitioner of public archaeology (he is currently Education and Outreach Co-ordinator for the Council for British Archaeology) – the book is refreshingly jargon free and accessible. Third, Henson is not afraid to offer strong, controversial interpretations. The Origins of the Anglo-Saxons can therefore be strongly recommended to those who want a detailed road-map of the evidence and debates for the migration period.
>
> *Current Archaeology*

£18.95 296 pages

The Elder Gods – The Otherworld of Early England

Stephen Pollington

The purpose of the work is to bring together a range of evidence for pre-Christian beliefs and attitudes to the Otherworld drawn from archaeology, linguistics, literary studies and comparative mythology. The rich and varied English tradition influenced the worldview of the later mediaeval and Norse societies. Aspects of this tradition are with us still in the 21st century.

£35 70 illustrations 528 pages

Leechcraft: Early English Charms, Plantlore and Healing

Stephen Pollington

An unequalled examination of every aspect of early English healing, including the use of plants, amulets, charms, and prayer. Other topics covered include Anglo-Saxon witchcraft; tree-lore; gods, elves and dwarves.

The author has brought together a wide range of evidence for the English healing tradition, and presented it in a clear and readable manner. The extensive 2,000-entry index makes it possible for the reader to quickly find specific information.

The three key Old English texts are reproduced in full, accompanied by new translations.

Bald's Third Leechbook; *Lacnunga*; *Old English Herbarium.*

£25 28 illustrations 536 pages

Remaking the Sutton Hoo Stone

The Ansell-Roper Replica and its Context

Paul Mortimer and Stephen Pollington

The 7th century ship burial at Sutton Hoo contains many enigmatic objects, none more so than a beautifully-worked stone with metal fittings. It is often referred to as a 'sceptre' or 'whetstone' but it may be neither.

The techniques used in making the stone and fitments display exceptional craftsmanship. So why were considerable resources devoted to creating it?

The making of a museum quality replica stone has provided new information and fresh insights which may help us answer many of the questions that have been asked about this beautiful and puzzling object.

The techniques used in making the stone are explained as is the geometry embedded within the overall design. There is also a critical review of the existing literature on the subject and a series of essays on aspects of Anglo-Saxon society that may be related to the making of the original stone.

£19.95 33 colour illustrations, 106 black & white, 196 pages

Peace-Weavers and Shield-Maidens: Women in Early English Society

Kathleen Herbert

The recorded history of the English people did not start in 1066 as popularly believed but one-thousand years earlier. The Roman historian Cornelius Tacitus noted in *Germania*, published in the year 98, that the English (Latin *Anglii*), who lived in the southern part of the Jutland peninsula, were members of an alliance of Goddess-worshippers. The author has taken that as an appropriate opening to an account of the earliest Englishwomen, the part they played in the making of England, what they did in peace and war, the impressions they left in Britain and on the continent, how they were recorded in the chronicles, how they come alive in heroic verse and riddles.

£5.95 64 pages

Anglo-Saxon Runes

John. M. Kemble

Kemble's essay *On Anglo-Saxon Runes* first appeared in the journal *Archaeologia* for 1840; it draws on the work of Wilhelm Grimm, but breaks new ground for Anglo-Saxon studies in his survey of the Ruthwell Cross and the Cynewulf poems. It is an expression both of his own indomitable spirit and of the fascination and mystery of the Runes themselves, making one of the most attractive introductions to the topic. For this edition new notes have been supplied, which include translations of Latin and Old English material quoted in the text, to make this key work in the study of runes more accessible to the general reader.

£5.95 80 pages

Looking for the Lost Gods of England

Kathleen Herbert

Kathleen Herbert sifts through the royal genealogies, charms, verse and other sources to find clues to the names and attributes of the Gods and Goddesses of the early English. The earliest account of English heathen practices reveals that they worshipped the Earth Mother and called her Nerthus. The tales, beliefs and traditions of that time are still with us in, for example, Sand able to stir our minds and imaginations.

£5.95 64 pages

Rudiments of Runelore

Stephen Pollington

This book provides both a comprehensive introduction for those coming to the subject for the first time, and a handy and inexpensive reference work for those with some knowledge of the subject. The *Abecedarium Nordmannicum* and the English, Norwegian and Icelandic rune poems are included in their original and translated form. Also included is work on the three Brandon runic inscriptions and the Norfolk 'Tiw' runes.

£5.95 88 pages

Anglo-Saxon FAQs

Stephen Pollington

125 questions and answers on a wide range of topics.

Are there any Anglo-Saxon jokes? Who was the Venerable Bede? Did the women wear make-up? What musical instruments did they have? How was food preserved? Did they have shops? Did their ships have sails? Why was Ethelred called 'Unready'? Did they have clocks? Did they celebrate Christmas? What are runes? What weapons and tactics did they use? Were there female warriors? What was the Synod of Whitby?

£9.95 128pages

Plain English – A Wealth of Words

Bryan Evans

The message here is not that we should strive in a narrow-minded way to scorn and drive out all borrowed words – it is that, if we stop and think, we can probably find a plain English word for what we want to say. We will thus keep plain English alive and help it thrive. We have been given a great gift, let us wherever possible use it and keep it safe.

This wordbook outlines the story of English then it offers 'A hundred words to start you off' (*shorten* rather than *abbreviate*, *speed up* instead of *accelerate*, *drive home* rather than *emphasize*, and so on). In the main part of the book will be found over 10,000 English words that are still alive and well, then a list of some 3,600 borrowed words, with suggestions about English words we might use instead. It is hoped that this book will help readers think about the words they use, and in doing so speak and write more clearly.

£12.95 328 pages

The Origins of the Anglo-Saxons

Donald Henson

This book has three great strengths.

> First, it pulls together and summarises the whole range of evidence bearing on the subject, offering an up-to-date assessment: the book is, in other words, a highly efficient introduction to the subject. Second – perhaps reflecting Henson's position as a leading practitioner of public archaeology (he is currently Education and Outreach Co-ordinator for the Council for British Archaeology) – the book is refreshingly jargon free and accessible. Third, Henson is not afraid to offer strong, controversial interpretations. The Origins of the Anglo-Saxons can therefore be strongly recommended to those who want a detailed road-map of the evidence and debates for the migration period.
>
> *Current Archaeology*

£18.95 296 pages

The Elder Gods – The Otherworld of Early England

Stephen Pollington

The purpose of the work is to bring together a range of evidence for pre-Christian beliefs and attitudes to the Otherworld drawn from archaeology, linguistics, literary studies and comparative mythology. The rich and varied English tradition influenced the worldview of the later mediaeval and Norse societies. Aspects of this tradition are with us still in the 21st century.

£35 70 illustrations 528 pages

Runes: Literacy in the Germanic Iron Age

Stephen Pollington

The fuþark or runic alphabet has been studied for several centuries. Scholarly opinion about the origins and purposes of the writing ranges from the purely pragmatic to the highly symbolic, with a great deal of speculative fringe literature.

This new survey takes into account recent finds from Britain, Scandinavia and the Continent together with new interpretations of old finds. It investigates the reasons for the creation of the script. The culture which devised the runes must have been familiar with one or more existing alphabet. What combination of factors impelled the script's creation? How was it transmitted from generation to generation? Who used it, when and how? The author divides the thousand-year history from inception to widespread adoption into phases, each with its own characteristic usages. The study traces the runes' transition from 'hieratic' to 'demotic', from the secret of a closed social class to the common property of entire societies.

£35 428 pages

Organisations

Þa Engliscan Gesiðas (The English Companions)

Þa Engliscan Gesiðas is a historical and cultural society exclusively devoted to Anglo-Saxon history. The Fellowship publishes a quarterly journal, *Wiðowinde,* and has a website with regularly updated information and discussions. Local groups arrange their own meetings and attend lectures, exhibitions and events. Members are able to share their interest with like-minded people and learn more about the origins and growth of English culture, including language, literature, archaeology, anthropology, architecture, art, religion, mythology, folklore and material culture. For further details see www.tha-engliscan-gesithas.org.uk

The Sutton Hoo Society

Our aims and objectives focus on promoting research and education relating to the Anglo Saxon Royal cemetery at Sutton Hoo, Suffolk in the UK. The Society publishes a newsletter SAXON twice a year, which keeps members up to date with society activities, carries resumes of lectures and visits, and reports progress on research and publication associated with the site. If you would like to join the Society please see www.suttonhoo.org

Wuffing Education

Wuffing Education provides those interested in the history, archaeology, literature and culture of the Anglo-Saxons with the chance to meet experts and fellow enthusiasts for a whole day of in-depth seminars and discussions. Day Schools take place at the historic Tranmer House overlooking the burial mounds of Sutton Hoo in Suffolk.

For details of programme of events contact www.wuffings.co.uk

Places to visit

Sutton Hoo near Woodbridge, Suffolk, IP12 3DJ

Sutton Hoo is a group of low burial mounds overlooking the River Deben in south-east Suffolk. Excavations in 1939 brought to light the richest burial ever discovered in Britain – an Anglo-Saxon ship containing a magnificent treasure which has become one of the principal attractions of the British Museum. The mound from which the treasure was dug is thought to be the grave of Rædwald, an early English king who died in 624/5 AD.

This National Trust site has an excellent visitor centre, which includes a reconstruction of the burial chamber and its grave goods. Some original objects as well as replicas of the treasure are on display.

2 miles east of Woodbridge on B1083 www.nationaltrust.org.uk/sutton-hoo

West Stow Anglo-Saxon Village

An early Anglo-Saxon Settlement reconstructed on the site where it was excavated consisting of timber and thatch hall, houses and workshop. There is also a museum containing objects found during the excavation of the site. Open all year 10am (except Christmas) Last entrance summer 4pm; winter 3-30pm. Special provision for school parties. A teachers' resource pack is available. Costumed events are held on some weekends, especially Easter Sunday and August Bank Holiday Monday. Craft courses are organised.

For further details see www.weststow.org

The Visitor Centre, West Stow Country Park, Icklingham Road, West Stow,
Bury St Edmunds, Suffolk IP28 6HG Tel. 01284 728718

www.ingramcontent.com/pod-product-compliance
Lightning Source LLC
LaVergne TN
LVHW060017210926
848481LV00013B/3732